Richard van Basshuysen

Gasoline Engine with Direct Injection

Richard van Basshuysen

# Gasoline Engine with Direct Injection

Processes, Systems, Development, Potential

With 399 Figures

Ulrich Spicher
and 21 co-authors

VIEWEG+
TEUBNER

Bibliographic information published by the Deutsche Nationalbibliothek
The Deutsche Nationalbibliothek lists this publication in the Deutsche Nationalbibliografie; detailed bibliographic data are available in the Internet at http://dnb.d-nb.de.

This book is based on the 1st edition of the German book "Ottomotoren mit Direkteinspritzung" edited by Richard van Basshuysen and now offered to the English speaking public by the authors.

1st Edition 2009



Editorial Office: Ewald Schmitt | Gabriele McLemore

Vieweg+Teubner is part of the specialist publishing group Springer Science+Business Media.
www.viewegteubner.de



Cover design: KünkelLopka Medienentwicklung, Heidelberg
Typesetting: KLEMENTZ publishing services, Gundelfingen
Printing company: MercedesDruck, Berlin
Printed on acid-free paper

ISBN 978-3-8348-0670-3

# Preface

Mobility connects.

Mobility – the opportunity to travel to more distant destinations, to explore other continents, even to conquer the universe – this has long been the dream of mankind. First horses were used to overcome the constraints of time and space. Boats and ships conquered the world's oceans and the most remote countries. Then the wheel was invented and, eventually, motorized transport.

Motorized transport is not free of controversy, as it has changed and revolutionized every aspect of life on our planet as never before. The consequences are immense. Important ecosystems have been thrown out of balance and environmental compatibility has not been restored in many places to this day. In addition to the uncontrolled exploitation of nature that has depleted the earth of natural resources, environmental destruction – due, for example, to extensive occupation of land and contamination of ground, water and air – also endangers the planet for all living beings.

We engineers, who are responsible for the blessings and curses of this development and are very much responsible for the current situation, also have the duty to restore environmental balance wherever it has been lost. The aim of this book is to make a small contribution to this purpose. How can we drastically reduce resource consumption and pollutant emissions of our automobiles? What noise reduction measures can be adopted in our all too noisy world?

This book covers these and other questions and points out the most promising developments for gasoline engines. Towards the end, a comparison with alternative drives is drawn.

The book has been written especially for product developers and those in charge of manufacturing in the automobile and supplier industries, and for their service providers. It is intended to serve as an important guidebook for science, research and teaching.

The editor would like to thank the co-authors and, representing them, Prof. Ulrich Spicher and his associates, who have made a major contribution to the contents of this book. All authors and their companies or institutions are listed in the credits. They have not only reproduced the latest knowledge with impeccable professional competence and commitment but have, time and again, also made predictions for the future. Nearly 400 pictures illustrate the contents and 427 up to date reference works are listed to encourage further reading.

Thanks also to Robert Bosch GmbH, without whose professional and material support this work would not have been possible.

Finally, the publishing house of Vieweg, especially the editorial team, have done an excellent job in the shortest possible time, making the book extremely up to date, for which I am also very grateful.

And now I wish this book the attention it deserves.

Bad Wimpfen
April 2007

Richard van Basshuysen,
Member of VDI, the Association of German Engineers

# About the Author and Editor in Charge

**Dr. Richard van Basshuysen**, Member of VDI, the Association of German Engineers, was born in Bingen on the Rhine in 1932. After completing an apprenticeship as an automotive mechanic, he studied at the Fachhochschule Braunschweig in Wolfenbüttel, a University of Applied Sciences, from 1953 to 1955, graduating as a mechanical engineer. A university degree in engineering, the Diplom, was conferred on him in 1982. From 1955 to 1965, he was a research assistant with Aral AG in Bochum.
In 1965, he joined NSU AG, where he took over as manager of the testing department for engine and transmission development including the development of the Wankel engine, and was appointed as the assistant manager for vehicle testing. In this role, he was jointly responsible for the development of the vehicles Prinz 4, NSU 1000 and 1200, RO 80 and K70. In 1969, NSU AG was taken over by today's Audi AG. With Audi, he was development manager for comfort class vehicles and manager of engine and transmission development. At the same time, he was a member of the supervisory board of Audi AG as the elected representative of executive staff. His most important development was the first exhaust-emission controlled passenger car diesel engine with direct injection and turbocharging, which he implemented against significant opposition, even from within VW. Since the engine requires 20% less fuel than its predecessor, which was a prechamber engine, and produces higher power and torque, it has established itself worldwide. In Europe, its market share grew from approximately 12% in 1989 to more than 50% in 2007. Following his active career in the automobile industry, Dr. van Basshuysen founded an engineering consultancy in 1992, which he still runs. He currently works as the editor of the internationally recognized technical-scientific journals ATZ (magazine on automobile technology) and MTZ (magazine on engine technology), as a consultant to international automobile manufacturers and engineering service providers and as an author and editor of technical-scientific books, many of which have also been translated into English. Since 2006, he has also been the editor and co-author of the internet portal www.motorlexikon.de, together with Prof. Fred Schaefer. Furthermore, he is a member of various advisory councils, boards and committees, such as the Association of German Engineers (VDI) and the Austrian Society for Automotive Engineering (ÖVK). He has authored or co-authored more than 60 technical-scientific publications. His editorials have been published in ATZ and MTZ since 1991. In 2001, he was awarded the highly remunerated Ernst Blickle Prize 2000 for the development of the innovative diesel engine with direct injection and the VDI's BENZ-DAIMLER-MAYBACH medal of honor for his "outstanding engineering performance in the development of the passenger car diesel engine with direct injection and his longstanding commitment as editor of ATZ/MTZ and as advisory board member of the VDI's Society for Automotive and Traffic Systems Technology". In 2004, an honorary doctorate was conferred on him by the University of Magdeburg for his lifetime achievements.

# Index of Authors


| | |
|---|---|
| Baretzky, Ulrich, Diplom-Ingenieur | AUDI AG, Neckarsulm<br>www.audi.de |
| Baron, Ulrich | Deutsche BP AG, Bochum<br>www.bp.com |
| Brodesser, Kay, Diplom-Ingenieur | Mann + Hummel GmbH, Ludwigsburg<br>www.man-hummel.com |
| Clade, Michael | CATEM, GmbH & Co. KG, Herxheim<br>www.catem.com |
| Eichlseder, Helmut, Univ.-Prof. Dr. | Technical University of Graz<br>www.portal.tugraz.at |
| Elsaesser, Alfred, Dr. of Engineering | MAHLE International GmbH, Stuttgart<br>www.mahle.com |
| Häntsche, Jan Patrick, Diplom-Ingenieur | University of Karlsruhe (TH)<br>www.uni-karlsruhe.de |
| Hatz, Wolfgang, Diplom-Ingenieur | AUDI AG, Ingolstadt<br>www.audi.de |
| Heidenreich, Thomas, Diplom-Ingenieur | University of Karlsruhe (TH)<br>www.uni-karlsruhe.de |
| Issler, Wolfgang, Dr. of Engineering | MAHLE International GmbH, Stuttgart<br>www.mahle.com |
| Kubach, Heiko, Dr. of Engineering | University of Karlsruhe (TH)<br>www.uni-karlsruhe.de |
| Meinig, Uwe, Dr. of Engineering | SHW Automotive, Bad Schussenried<br>www.shw.de |
| Menne, Rudolf, Prof. Dr. of Engineering Dr. h.c. | Ford Forschungszentrum Aachen GmbH,<br>[Ford Research Center], Aachen<br>www.ford.de |

| | |
|---|---|
| Muenz, Stefan, Dr. | BorgWarner Turbo Systems Engineering GmbH, Kirchheimbolanden<br>www.turbos.bwauto.com |
| Puck, Alexander, Diplom-Ingenieur | MAHLE, Ventiltrieb GmbH, Stuttgart<br>www.mahle.com |
| Scharnhorst, Christian, Dipl.-Betriebsw. | Robert Bosch, GmbH, Stuttgart<br>www.bosch.de |
| Schmalzl, Hans Peter, Dr. | BorgWarner Turbo Systems Engineering GmbH, Kirchheimbolanden<br>www.turbos.bwauto.com |
| Seiffert, Ulrich, Prof. Dr. of Engineering | Technical University of Braunschweig<br>www.tu-braunschweig.de |
| Spicher, Ulrich, Prof. Dr. of Engineering | University of Karlsruhe (TH)<br>www.uni-karlsruhe.de |
| Stoffels, Harald, Diplom-Ingenieur | Ford-Werke GmbH, Cologne<br>www.ford.de |
| Walther, Dieter, Dr. | Deutsche BP AG, Bochum<br>www.bp.com |
| Wild, Stephan, Dr. of Engineering | Mann + Hummel GmbH, Ludwigsburg<br>www.mann-hummel.com |
| Xander, Benedikt, Dr. of Engineering | University of Karlsruhe (TH)<br>www.uni-karlsruhe.de |

# Index of Companies and Universities

## Companies

| | |
|---|---|
| AUDI AG, Neckarsulm | Ulrich Baretzky, Diplom-Ingenieur<br>Wolfgang Hatz, Diplom-Ingenieur |
| BorgWarner Turbo Systems Engineering GmbH, Kirchheimbolanden | Stefan Münz, Dr.<br>Hans Peter Schmalzl, Dr. |
| CATEM GmbH & Co. KG, Herxheim | Michael Clade |
| Deutsche BP AG, Bochum | Dieter Walther, Dr.<br>Ulrich Baron |
| Ford-Werke GmbH, Cologne | Harald Stoffels, Diplom-Ingenieur<br>Rudolf Menne, Prof. Dr. of Engineering Dr. h.c. |
| MAHLE International GmbH, Stuttgart | Wolfgang Ißler, Dr. of Engineering<br>Alfred Elsäßer, Dr. of Engineering |
| MAHLE Ventiltrieb GmbH, Stuttgart | Alexander Puck, Diplom-Ingenieur |
| Mann + Hummel GmbH, Ludwigsburg | Stephan Wild, Dr. of Engineering<br>Kay Brodesser, Diplom-Ingenieur |
| Robert Bosch GmbH, Stuttgart | Christian Scharnhorst, Dipl.-Betriebsw. |
| SHW Automotive, Bad Schussenried | Uwe Meinig, Dr. of Engineering |

## Universities

| | |
|---|---|
| Technical University of Braunschweig | Ulrich Seiffert, Dipl.-Betriebsw. |
| Technical University of Graz | Helmut Eichlseder, Univ.-Prof. Dr. |

University of Karlsruhe (TH)

Ulrich Spicher, Univ.-Prof. Dr.
Jan Patrick Häntsche, Diplom-Ingenieur
Thomas Heidenreich, Diplom-Ingenieur
Heiko Kubach, Dr. of Engineering
Benedikt Xander, Dr. of Engineering

# Chapters and Contributions

Lower consumption, more power?
Yes
Bosch gasoline direct injection.
4
3
5
1/min x 1000
2
6
7
5.2 l/100 km
Innovation from Bosch: Gasoline direct injection cuts $CO_2$ emissions and fuel consumption by up to 15%. Together with turbo-charging, it also generates greater power over the entire speed range compared with conventional injection systems. At low rpm, that adds up to 50% more torque, for a lot more driving fun. www.bosch.de/ja
BOSCH
Invented for li

# Contents

# 1 Introduction

Diesel engines with direct injection systems have more or less replaced indirect injection diesel engines in only 15 years, primarily for reasons of consumption, torque and power output. Gasoline engines are about to follow suit. Although engineers have been working hard to refine direct injection for almost a century now in order to improve power output, it is only in the last ten years or so, since Mitsubishi introduced gasoline direct injection in Europe, that the technologies have been available to launch this concept successfully. The difficulties to be overcome are much more complex than those for diesel engines with direct injection, however. Gasoline engines with direct injection will therefore take longer to penetrate the market, but the gains to be made in power output and torque, and the high potential for reducing fuel consumption and pollutant emissions, make this move virtually compulsory.

On closer examination, three development approaches are being pursued, with varying results:

1. Direct injection with homogeneous mixtures
2. Direct injection with stratified mixtures
3. Direct injection with homogeneous mixtures and controlled auto-ignition, charge compression ignition.

1: The development of direct injection with more or less homogeneous mixtures is comparatively easy. It does not require special fuels, since the existing exhaust-gas treatment concept can be retained. Thus, this process is suitable worldwide. It focuses on optimizing power output, torque and dynamic behavior. Influenced by the considerable initial success it has enjoyed in motor sports, this concept especially suits sports cars and other powerful cars.

2: Direct injection with stratified mixture has been developed primarily to reduce fuel consumption and $CO_2$ emissions. In addition to Mitsubishi's wall-guided concept, air-guided processes have also been introduced in series production in Europe. The expected fuel consumption advantages of these processes failed to materialize, however, especially in the higher load and engine speed ranges. Therefore, there are justified doubts with regard to these concepts, as they are too expensive and complex to be implemented worldwide, because of the increased cost for exhaust-gas treatment in the hyperstoichiometric range and because they require sulfur-free fuels. Taking these aspects into consideration, attention has been turning back to the spray-guided process for some time now. Though it imposes even higher demands on mixture formation, reductions in fuel consumption of up to 20% can be expected. This is the same remarkable magnitude of change that was achieved in diesel engines with the shift to direct injection and would leave the diesel engine with a volumetric consumption advantage of only 10 to 15%.

3: In the field of research, processes with direct injection using homogeneous mixture and controlled auto-ignition with charge compression ignition have been studied for a long time. These processes are also known as "HCCI" (*Homogeneous Charge Compression Ignition*) and "CAI" (*Controlled Auto-Ignition*) and are being developed as new combustion processes for gasoline and diesel powered engines. Due to the great potential of concepts with minimum emissions and good combustion efficiency, high hopes are placed on them. In combination with fuels developed specifically for these combustion processes, a long-standing dream could come true: an optimal new combustion process that combines the advantages of the two types of engine could emerge from the existing combustion processes for gasoline and diesel engines.

# 2 History of Gasoline Direct Injection

In terms of technological evolution, the steam engine can be considered the predecessor of the combustion engine. In a steam engine, which is a heat engine with external combustion, the thermal energy of the combustible gas is supplied to the working medium through a heat exchanger, while in a combustion engine, it is characteristic of the engine to largely convert the chemical energy of the fuel into mechanical work in the engine. During the development of the combustion engine, the real challenge – besides the ignition of the working gas – was mixture formation, i.e. the metering, preparation and mixing of fuel and air.

An illuminating gas engine designed by Jean Joseph Etienne Lenoir in 1860, without mixture compression but equipped with electric spark ignition, can be seen as the first successful invention of an internal combustion piston engine. In 1867, Nikolaus August Otto patented the first atmospheric gas engine, which bears his name. Based on this engine, he invented a four-stroke engine at Deutz in 1876, although its characteristics (four-stroke process, precompression of the air/gas mixture) had already been described by Beau de Rochas in 1862. The four-stroke engine marks an important development milestone for the creation of the first automobiles with gasoline engines in 1886, which were developed by Carl Benz and Gottlieb Daimler as well as by Wilhelm Maybach. In the first vehicle engines designed by Benz and Daimler, the mixture formation took place in float carburetors, which were developed independently. The intake air was conducted through a gas column held at a constant level by means of a valve ball. In both carburetor concepts, fuel evaporation was enhanced by heat supply.

A fundamental step for the solution of the mixture formation problem in gasoline engines was the invention of the Maybach spray carburetor in 1893. In this mixture formation concept, considered as groundbreaking for decades, a fuel jet was directed through an inlet tube with a reduced cross-section. The reduced cross-section accelerated the airflow, resulting in the intake of the fuel as well as the atomization of the fuel released from the nozzle. This carburetor also had a constant fuel level maintained at the nozzle by means of a valve ball.

The history of gasoline injection development is almost as long as the history of the carburetor. As early as 1877, Nikolaus August Otto patented the idea of injecting fuel directly into the combustion chamber [1].

In 1884, the Hallesche Maschinenfabrik produced stationary engines with fuel injection for the first time [2]. **Figure 2.1** shows the cross-section of a cylinder and the injection system of this engine as designed by Johannes Spiel.

Fuel flows from gasoline tank *a* through the suction valve *b* into the cylinder of the injection pump. The plunger *c* of the injection pump is actuated along with the inlet valve *d*

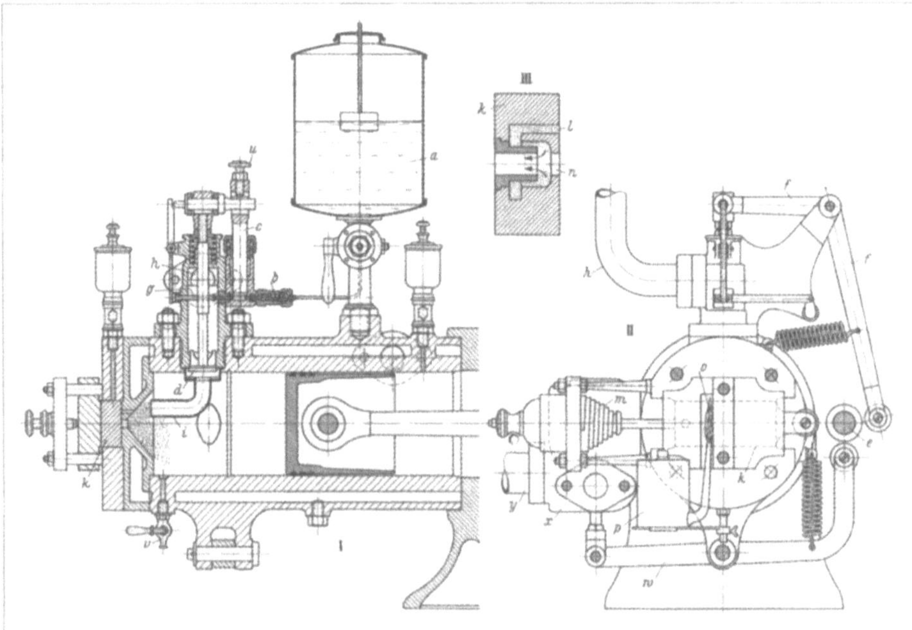


**Figure 2.1:** Sectional drawing of Johannes Spiel's 1884 gasoline engine with fuel injection in the intake manifold [2]
*a* gasoline tank, *b* suction valve, *c* plunger of the injection pump, *d* mechanically operated mixture intake valve, *e* camshaft, *f* bell-crank lever, *g* pressure valve of the injection pump, *h* air supply, *i* manifold for mixture supply, *k* ignition slide valve, *l* hollow space for the flame, *m* spring for ignition slide valve movement, *n* ignition opening in slide valve, *o* ignited flame, *p* spirit container, *u* manually operated screw for increasing gasoline requirement during start-up, *v* drain cock, *w* lever for exhaust valve control, *x* exhaust valve, *y* exhaust pipe

using a valve lifter lever connected to the camshaft *e*. During the compression stroke of the pump plunger *c,* the gasoline previously drawn in is injected in the inlet port by the inlet valve through pressure valve *g*. The pressure valve of the injection pump, which functions like an injection nozzle opening outwards, enables a sealing with a conical sealing face, through which a conical fuel spray is generated. The injection quantity can be varied by changing the stroke of the injection pump plunger using a manually operated bolt. The systematic generation of charge stratification in this engine is particularly noteworthy. To ensure that the air/fuel mixture reaches the vicinity of the igniter directly, the inlet port opens into the combustion chamber. The port mouth is directed towards the cylinder head. This leads to the displacement of the residual gases present in this area, which are difficult to inflame. As in the engines developed by Nikolaus August Otto, inflammation of the air/fuel mixture takes place by means of a flame igniter operated by the camshaft.

As stated in [2] and in the context of the development of vehicle engines, the year 1884 also saw Wilhelm Maybach's observation about using a "carburetor pump" instead of an "evaporation device" for the metering and atomization of fuel. The gasoline pump, preserved in the form of a sketch, shows a pump plunger diameter of 2 mm and a 7 mm stroke, and the suction and pressure pipes being operated by a flat slider.

Other milestones in the development of manifold injection systems, according to [3], are the series production of stationary engines with manifold gasoline injection by Deutz in 1898/1901, the start of the successful use of manifold injection systems in Grade two-stroke aircraft engines and Antoinette and Wright four-stroke aircraft engines in 1906, and the start of tests on manifold injection systems at Bosch in 1912 and at Pallas in 1914.

Initial tests on direct fuel injection for gasoline engines had already begun in 1898. The process developed by Haselwander [4] was based on an open combustion chamber concept and the engine ran on petroleum; **figure 2.2.** The fuel was injected shortly before the piston reached the top dead center, as a result of which a vaporized rich mixture reached the spark plug. The required injection pressure was generated using a stepped piston during compression. However, the lack of technological possibilities made these initial tests unsuccessful.

**Figure 2.2:**
Haselwander engine (1898)

To the best of the author's knowledge, gasoline direct injection was implemented in practice for the first time at Junkers in 1916 [5, 6]. The starting point for these developments at Junkers in 1914 were the efforts, in the context of the growing aviation and engine industry, to develop a diesel engine for aircraft propulsion that was economical and less prone to catching fire. As an engine concept, a horizontally positioned two-stroke opposed-piston engine with two mechanically coupled crankshafts was selected. This concept later proved itself as aircraft diesel engines in Junkers' Jumo 205 and 207 series. Although Junkers were convinced of their competence with the diesel process in these engines, tests with the first four-cylinder test engine (type: M 0.3, $V_h = 7.238\ l$) in early 1915 were largely unsuccessful. Due to technical problems with the diesel injection system and the demands of the war ministry regarding the use of gasoline or benzene in aircraft engines, among other factors, all subsequent experiments with aircraft engines were conducted with gasoline as fuel at Junkers until the end of the first world war in 1918. The basis for tests on gasoline engines, besides a corresponding single-cylinder unit, was a newly designed six-cylinder engine (type: M 0.8, $V_h = 14.137\ l$) with which test bench trials were started in 1915. The initially favored mixture scavenging proved to be uncontrollable due to fatal engine damage caused by backfiring in the crankcase. Therefore, Junkers decided to inject the fuel directly into the engine cylinder. This development was lead-managed by Otto Marder. Due to various problems with the power transmission and the adverse weight-to-power ratio of the M 0.8 engine, development continued in 1916 with newly designed six-cylinder engines (type: Fo.2, $V_h = 17.105\ l$) with larger displacement and a corresponding single-cylinder test engine. **Figure 2.3** shows a photograph of this engine in horizontal position. The sectional drawing of this engine in **figure 2.4** shows the position of the injection nozzle and the two spark plugs. The tangential arrangement of the scavenging and exhaust ports leads to the combustion air forming a swirl flow that supports the mixture formation.

In addition to the piston, already known as a weak point, the engine load control proved to be particularly problematic. Under the guidance of the indefatigable Otto Marder, the injection pump was optimized in tedious design and test work. As early as 1917, standard Junkers' Fo.2 engines with gasoline direct injection [6] were delivered to the navy for speedboats. Serious problems occurred in Fo.2 engines with a compression ratio $\varepsilon = 10$ due to the extreme thermal load on spark plugs, which caused the electrodes and ceramic insulators to overheat. A solution, though technically very complex, was pursued in the form of a water cooling system for spark plugs. This part of the development was undertaken by Bosch. Even reducing the compression ratio to $\varepsilon = 6$ did not satisfactorily solve the problem of spark plug overheating of the aircraft engine until the end of 1918.

Besides the tests on the Fo.2 engine, a 260 PS Mercedes aircraft engine was converted to a gasoline direct injection engine following the proposal of the Directorate of Aircraft Production. A valve pump with a maximum delivery stroke of 0.37 ccm per cylinder and pump stroke was used as injection pump which was provisionally mounted on the exhaust side and operated using a chain. The results of the tests performed on this engine in the summer of 1918 can be summarized as follows [6]: the engine with the injection system starts more easily than the corresponding carburetor engine after manual rotation of the injection pump. A low idle speed of 300 $min^{-1}$ can be maintained without problems. The control of

**Figure 2.3:** Junkers Fo.2 engine in horizontal arrangement [Source: Deutsches Museum, Munich]

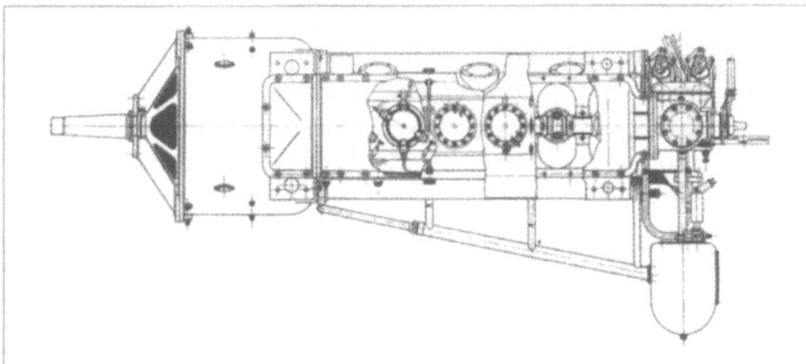

**Figure 2.4:** Sectional view of Junkers Fo.2 engine [6]

the metering of injection quantity at all load points is good, i.e. mixture formation takes place without misfires or retarded ignitions. Appropriate modifications to the intake system could improve the full load power output of the injection engine by 15% as compared to the carburetor engine.

In October 1918, shortly before the development work was suspended due to the end of World War I, this engine was presented to the representatives of the Inspectorate of Aviation Troops. A visit report written by Otto Marder on this occasion [5] shows that Junkers had a good command of gasoline direct injection. In addition to the advantages already mentioned, the report also emphasized the reduced risk of fire, improved adjustment of the mixture formation to the particular flying altitude and reduced demand for installation space.

It is apparent that the advantages of gasoline direct injection proven by Junkers more than a decade earlier formed the background for the fact that in the early 1930's, major aircraft engine manufacturers in Germany concentrated on the development of gasoline direct injection, in agreement with the responsible ministries. It proved beneficial that Bosch, Junkers and l'Orange had already gained development experience and product engineering expertise with regard to the exact metering and preparation of smaller fuel quantities in the field of diesel injection without a compressor. The starting point of this development in the early 1930's was the assignment given by the then Ministry of Transport to the German Laboratory for Aviation (DVL) in Adlershof for a systematic analysis of fuel injection in igniter aircraft engines. The tests were carried out under the direction of Kurt Schnauffer, first on BMW IV and Siemens Sh 13 single-cylinder engines and a DKW two-stroke engine. Conventional diesel injection pumps with eccentric shafts and pintle nozzles, as well as some with hole-type nozzles, were used during these tests. A DVL report by K. Schnauffer dated September 1, 1932 [8] and other DVL reports [3] emphasize the basic usability of gasoline direct injection in four-stroke gasoline engines, with better results being achieved during injection in the intake stroke compared to operating the engine with a carburetor. In case of fuel injection during the compression stroke, useful results were achieved only with separate combustion chambers. In contrast, the results of corresponding tests on two-stroke engines were negative. Further tests on a six-cylinder aircraft engine – BMW Va at DLR – showed a power increase of up to 17% and an initial reduction in fuel consumption of 3% by 1934. In the same year, these successes prompted the Ministry of Aviation to encourage the aircraft engine industry (Argus, BMW, Daimler-Benz, Junkers, and Siemens) to pursue further development and to introduce gasoline direct injection in series. This created a broad basis for the development of gasoline direct injection in Germany in the following years, with important contributions also being made by the suppliers of the actual injection systems (Bosch, Deckel, Junkers and l'Orange) and the manufacturers of fuel filters and supply pumps. As an example, the development of gasoline direct injection in Daimler-Benz aircraft engines is explained in detail.

In March 1934, development work was started for gasoline direct injection on a single-cylinder unit of the DB 600 engine at Daimler-Benz, based on a diesel injection pump and using pintle nozzles initially. During the tests, however, it became evident that it was advisable to inject fuel into the zones with the strongest airflow during the intake stroke. In this context, first l'Orange and later also Bosch developed multi-hole nozzles with which a considerable reduction in fuel consumption and improvement of the knocking limit could be achieved as compared to the pintle nozzles used previously.

**Figure 2.5** shows a cross-section of the hole-type nozzle developed by l'Orange. Due to the sensitivity of the nozzle to fuel contamination, a small disc-type filter was integrated in the injection nozzle, at Daimler-Benz' suggestion.

One of the results of these tests was that there were no problems with regard to wear of the pump elements during gasoline or benzene operation when the fuel was finely filtered. It therefore turned out that there was no need to add oil lubricant to the fuel; this had initially been regarded with apprehension and considered as undesirable with regard to reducing knocking. In order to prevent fuel leakage by forming an effective barrier between the pump elements and the cam gear on the one hand, and to increase the lubrication of the pump elements on the other hand, Bosch developed lubricating oil pumps with oil barriers based on the "Eugen Ludwig Müller" system, which proved themselves in the following years. **Figure 2.6** shows a pump element with oil blocking developed by Bosch and **figure 2.7** shows a cross-section of the complete pump element of the Bosch injection pump for DB 603 and DB 605 engines. The design of these injection pumps is similar to the diesel injection pumps. The

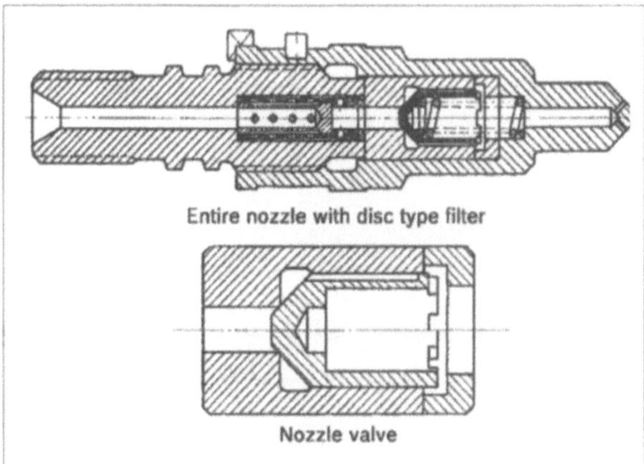


**Figure 2.5:**
Cross-section of the l'Orange 9-2261 injection nozzle used in Daimler-Benz aircraft engines [3]

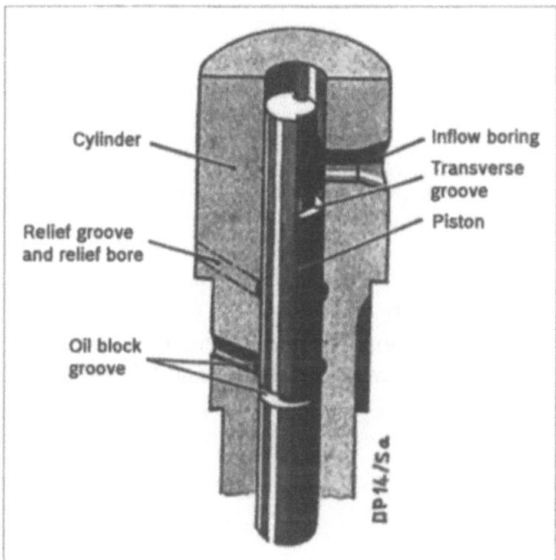


**Figure 2.6:**
Cross-section of a pump element with oil blocking developed by Bosch [3]

quantity of fuel injected is varied by the rotation of the pump plunger through a toothed rack (control rod). The plunger helix is designed so that load-dependent delivery is the result.

In 1935, Hans Scherenberg, who was then working under the guidance of Fritz Nallinger, started to develop gasoline direct injection systems for aircraft engines for series production at Daimler-Benz in close cooperation with Bosch. Their development work focused on careful tuning of the gas exchange cycles, the compression ratio, the charge pressure and the valve timings. **Figure 2.8** shows the position of the injection pump, the injection nozzle and the spark plugs, with a cross-section of the Daimler-Benz aircraft engine DB 601 as an example; **figure 2.9** illustrates the mixture formation conditions in the Daimler-Benz aircraft engines. The result of the extensive tests on the 12-cylinder DB 601 engine

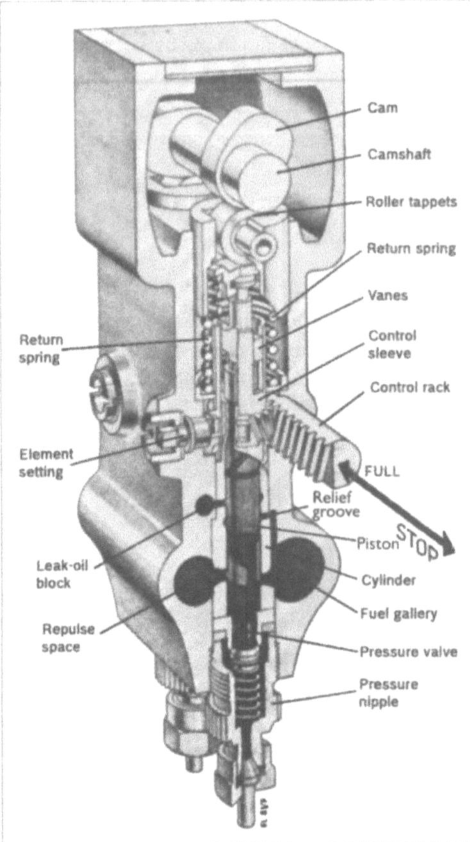


**Figure 2.7:**
Cross-section of a pump element of the Bosch injection pump for DB 603 and DB 605 engines [6]

($V_h$ = 33.8 *l*) was, as shown in **figure 2.10,** a considerable power increase with simultaneous reduction in fuel consumption as compared to a similar carburetor engine.

In 1937, the first aircraft engine with gasoline direct injection, the DB 601 A by Daimler-Benz, went into series production almost at the same time as the aircraft engines Jumo 210 G and 132 F developed by Junkers and BMW respectively, which were also equipped with gasoline direct injection. Different world records set in the following period, e.g. the record for the absolute speed of 746.606 km/h set by Hans Dieterle on a Heinkel He 100 aircraft on March 30, 1939 and by Fritz Wendel on a Messerschmitt Me 209 at 755.11 km/h on April 27, 1939, underlined the large potential for the power increase of aircraft engines with the use of gasoline direct injection. Both record-making aircrafts were operated with a DB 601 fuel injection engine in which the power was increased to 1544 kW (2100 hp) with high pressure supercharging and the injection of methanol upstream of the supercharger.

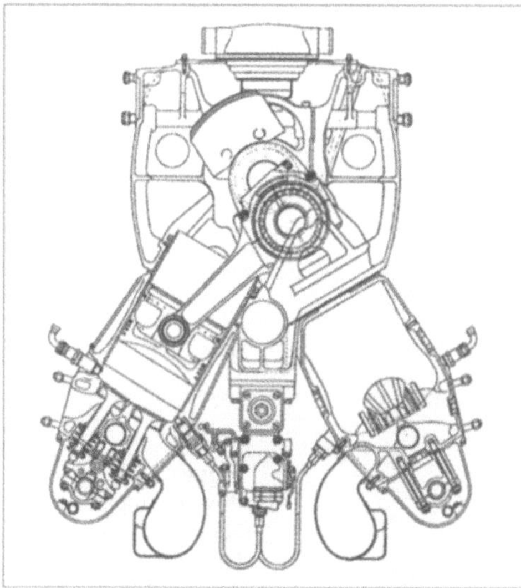

**Figure 2.8:**
Cross-section of a DB 601 E 12-cylinder aircraft engine developed by Daimler-Benz with a displacement of $V_h = 33.8\ l$ [3]

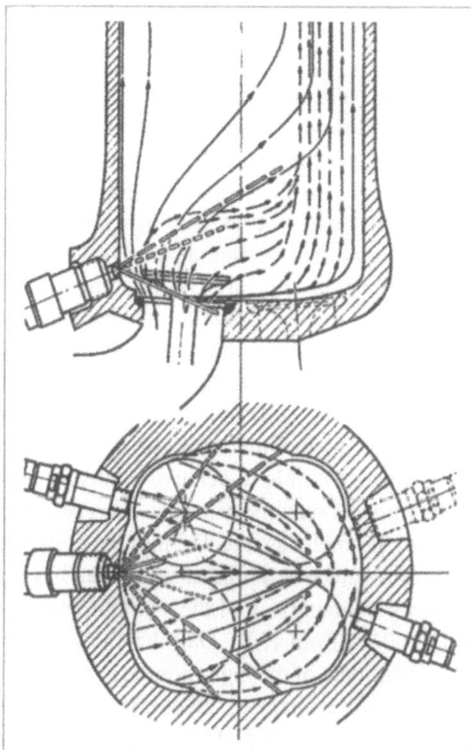

**Figure 2.9:**
Sectional view of a cylinder of the 12-cylinder aircraft engine DB 601 and DB 603 illustrating the air movement and fuel spray [3]

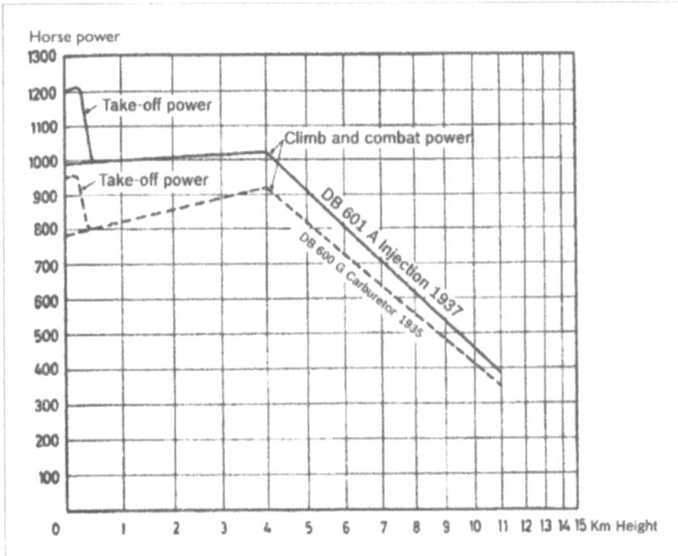


**Figure 2.10:** Power comparison between Daimler-Benz DB 600 and DB 601 A aircraft engines (in 1935/1937) [3]

From the start of series production in 1937 to the end of development in 1945, the power of the fuel injection aircraft engines at take-off and at different altitudes was drastically improved step by step. An important prerequisite for this was the possibility of realizing large valve overlaps, created by the direct injection of fuel into the cylinder, without having to accept large losses of fresh air, as is inevitably the case with external mixture formation. Complete residual gas scavenging and effective internal cooling of the cylinder were achieved in connection with superchargers gradually developed for higher and higher flying altitudes by means of a valve overlap of more than 100° crankshaft angle between the start of injection and the closing of the exhaust valve. However, the large valve overlaps required careful design of the intake manifolds, as the limitation of the residual gas content in the cylinder is a basic prerequisite for acceptable idling behavior of the engines. Due to the short fuel and air prereaction times and the precise fuel metering for each cylinder of the engine brought about by direct injection, it was possible to achieve relatively high compression ratios and a general reduction in the engine's sensitivity to fuel. For good mixture formation without substantial loss of fresh air, the fuel was injected into the cylinders with a pressure of approximately 50 bar with short injection timings, immediately after the closing of the exhaust valve. **Figure 2.11** shows a comparison of the valve timings and injection timings of the DB 601 A (1937) and DB 601E (from 1939) Daimler-Benz aircraft engines. **Figure 2.12** gives an overview of the development of the high-altitude performance of Daimler-Benz aircraft engines from 1935 to 1945. As explained, direct injection was an important prerequisite for the dramatic increase in high-altitude performance within just a few years.

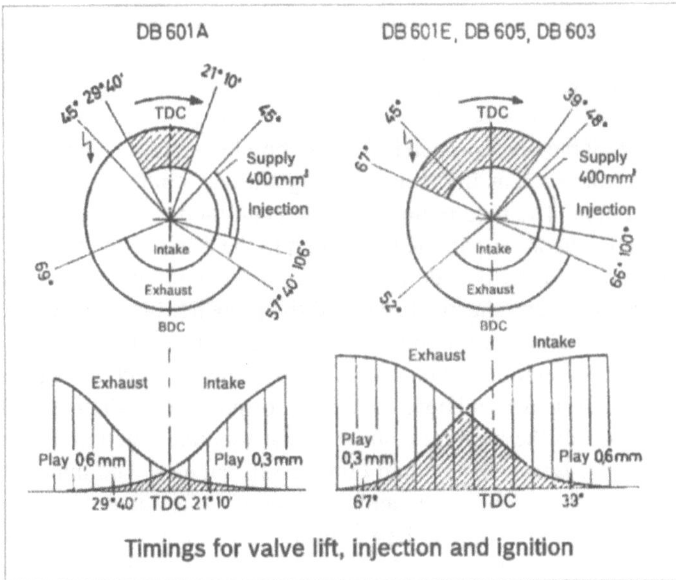


**Figure 2.11:** Valve timings and injection timings of Daimler-Benz DB 601 A (1937) and DB 601E aircraft engines (from 1939) [3]

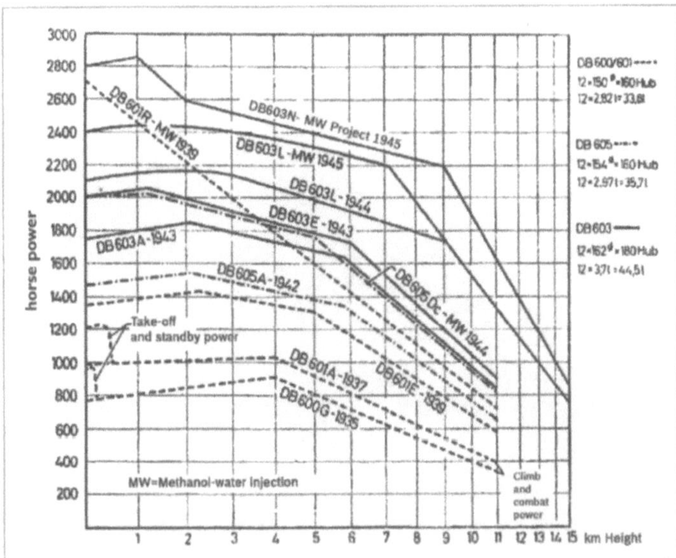


**Figure 2.12:** Increase in high-altitude performance of Daimler-Benz DB 600 to DB 603 N aircraft engines from 1935 to 1945 [3]

Other manufacturers of aircraft engines also successfully developed aircraft engines with direct injection [6]. The 12-cylinder aircraft engines Jumo 210, 211 and 213 from Junkers were equipped with injection pumps developed internally by Junkers, whose basic design was similar to the Bosch injection pump developed for Daimler-Benz aircraft engines. The injection pump for the two-row 14-cylinder radial BMW 801 aircraft engine was supplied by Deckel. In this pump, 14 pump elements were arranged in the pump housing in a drumlike formation. The metering edges of the pump plungers were tapered from both sides so that the start and end of delivery could be changed by rotational movement of the pump plungers (variation of injection quantity).

The aircraft engine and the variable pitch propeller form a complex system which is affected by several external factors such as flight state (flying altitude, climb flight, dive, etc.) in the actual flight operation. The requirements for the open-loop control or close-loop control and for the engine operation parameters related to the flight state including the charge pressure, the air/fuel ratio, the throttle valve position, the point of ignition, the switch position of the supercharger and the positioning of the propeller blade. Since the regulation of these parameters called for an extremely high degree of attention from the pilots and thus involved the risk of operating errors, some German aircraft injection engines were equipped with complex mechanical mixture controllers or predictors. **Figure 2.13** illustrates the functioning mode of the mixture controller that was integrated in Bosch injection pumps for the Daimler-Benz DB 603 and 605 aircraft engines. The injection quantity for this controller was dependent on the ambient pressure, the charge pressure and the temperature of the charge air. **Figure 2.14** shows a corresponding schematic diagram of the command master of the two-row 14-cylinder radial aircraft engine BMW 801. The functions of this device, i.e. charge pressure control, engine speed control,

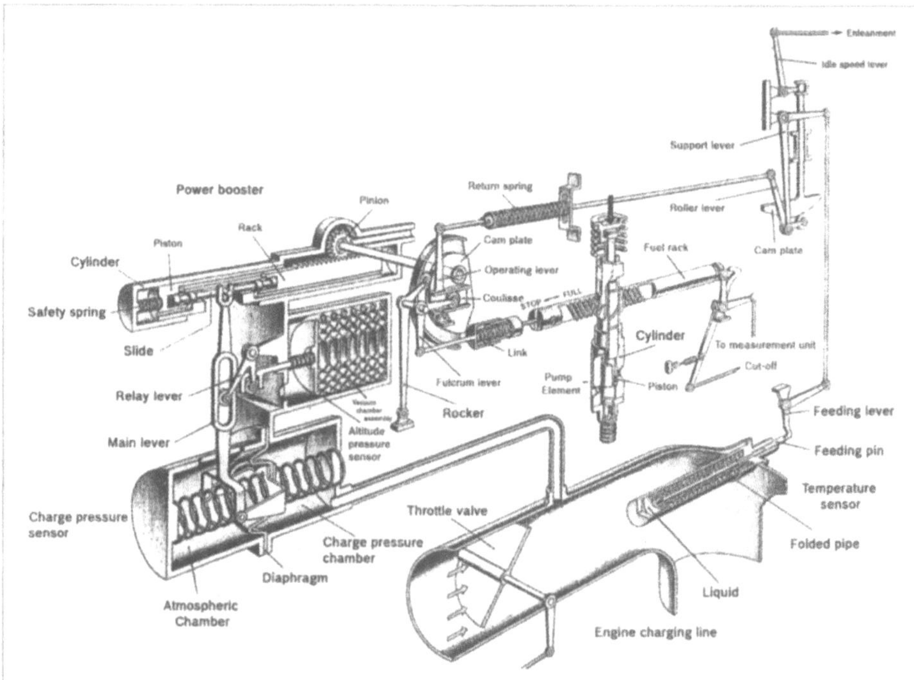


**Figure 2.13:** Illustration of the function of the mixture controller integrated in the Bosch injection pump of Daimler-Benz DB 603 and 605 aircraft engines [3]

mixture control, automatic throttle valve mechanisms, ignition control, idling control, supercharger switching and pressure oil supply of the controllers with power booster, are combined in a unit that is easy to replace. In the BMW engine 801TJ with exhaust-gas turbocharger, built under close-to-series conditions until the end of the war, the predictor was also used for turbine control.

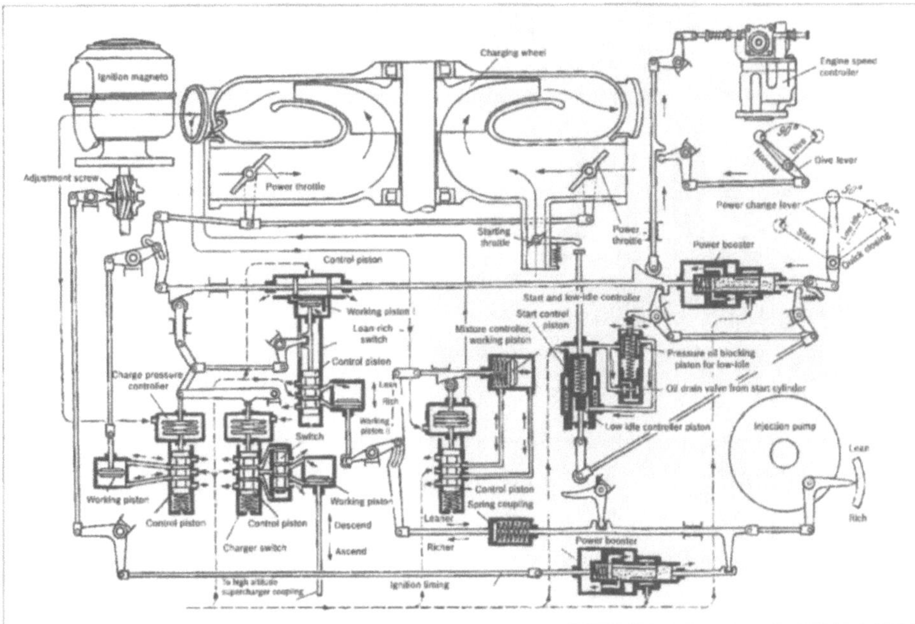


**Figure 2.14** Schematic diagram of the command master of the two-row 14-cylinder BMW 801 radial aircraft engine (1939) [9]

As early as the 1930's, the good results achieved in aircraft engines with gasoline direct injection provided impetus for tests on fuel direct injection in automotive engines. The gasoline direct injection process with open combustion chamber and charge stratification was first realized by Hesselman [10, 11] around 1930. In the field of automotive engines, this Swedish engine was widely in use as a result of various companies obtaining licenses [10, 12]. The aim of the Hesselman engine shown in **figure 2.15** was to combine the advantages of diesel engines and gasoline engines. It is a combustion process with air compression, non-homogeneous mixture formation, spark ignition, and mixture quality control at full load, part load and in low idle mode. The distinctive feature of the process was the delayed fuel injection in the compression stroke, the fuel being formed as a swirl flow. Swirl generation was realized by baffle valves. The mixture preparation took place during the transportation from the injector to the spark plug situated opposite, where the end of the incoming mixture cloud was inflamed. The injection timing could be varied for load control, with the end point of fuel injection remaining constant. At low load, throttling was required due to otherwise unstable combustion. The engine featured good power and fuel consumption data as well as the ability to burn a great variety of fuels. The combustion process can also be classified as a hybrid or multi-fuel process but it is mentioned here because of its technical proximity to gasoline direct injection.

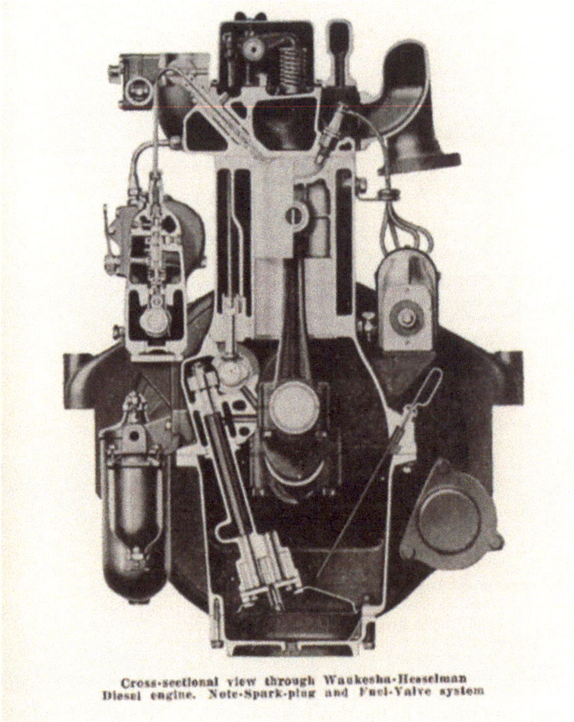


**Figure 2.15:**
Hesselman engine (1930) [10]

Daimler-Benz was one of several companies that tested single-cylinder engines for racing cars in the 1930's. When World War II broke out, all development activities were discontinued [13] and were resumed only after the end of the war, mainly for two-stroke engines for use in vehicles. In two-stroke engines, fuel direct injection offers considerable potential for reducing fuel consumption, which is especially due to the scavenging losses which result from the operating principle. For this reason, Bosch worked on gasoline direct injection for a DKW "Meisterklasse" engine and was able to achieve good results in tests with a vehicle equipped with this engine. When World War II started, these development activities also came to a halt. In 1949, based on experience gained with aircraft engines, the development of gasoline direct injection for two-stroke engines was resumed in close cooperation between Bosch and Gutbrod and between Bosch and Goliath. The development started optimistically but turned out to be more difficult than initially thought. Difficulties cropped up, mainly relating to the high pump speeds required for two-stroke engines, injection pump control over a large engine speed range and the noise emissions of the injection pump and nozzles. As described in [3] and [10], various development steps had to be implemented in order to be ready for series production. These steps included the transition from injection cams to eccentric cams, the selection of the injection nozzle position in the cylinder head so that the injection spray was directed opposite the rising scavenging flow, the development of a new throttle valve housing including overrun fuel cut-off, the increase of the engine compression, the determination of the most suitable position for the spark plugs and the adjustment of the inlet and exhaust system. Nozzles with conical valve seat opening outwards and without a leakage line were used as injection nozzles.

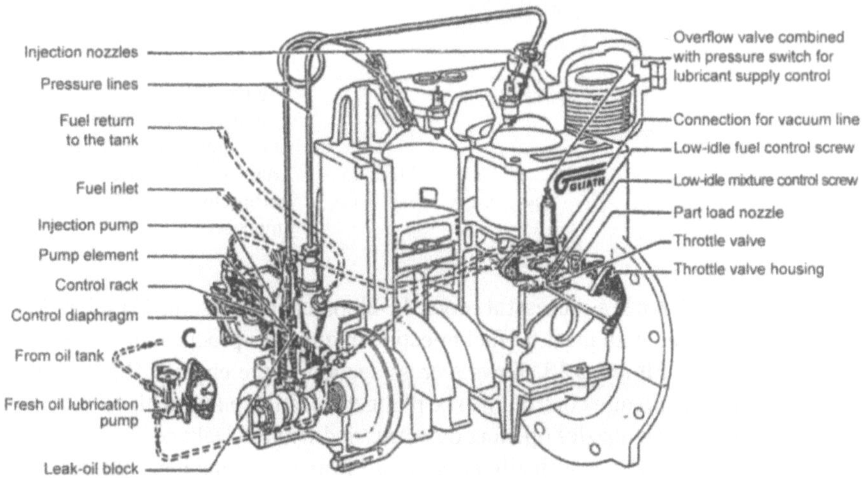


**Figure 2.16** Structure of the Goliath GP 700 engine with gasoline direct injection [3]

**Figure 2.16** shows the structure of the Goliath GP 700 engine with gasoline direct injection, which was presented along with the 600 ccm Gutbrod Superior Engine at the International Motor Show (IAA) in 1951 and went into series production a few months later. **Figure 2.17** shows the engine power of the 600 ccm Gutbrod Superior engine at different engine speeds and corresponding fuel consumption in comparison with the fuel consumption of the corresponding carburetor engine. In spite of the increase of the power output per liter from 29 kW to 33 kW, it was possible to reduce the specific fuel consumption significantly, especially at low engine speeds. By increasing the displacement to 900 ccm, Goliath developed a 900 ccm fuel injection engine with a power of 29 kW, which was included in the Goliath Express utility vehicle from August 1955 on. The high costs, especially for the injection system, as well as the limited importance of low fuel consumption at that time, were the reasons why gasoline direct injection did not become established for two-stroke engines.

Against the background of the positive results achieved with gasoline direct injection in two-stroke engines, Daimler-Benz started the development of gasoline direct injection for a 3.0 *l* four-stroke passenger car engine at the beginning of 1952. By the end of the year, there were already so many positive results available that vehicle tests of this engine con-

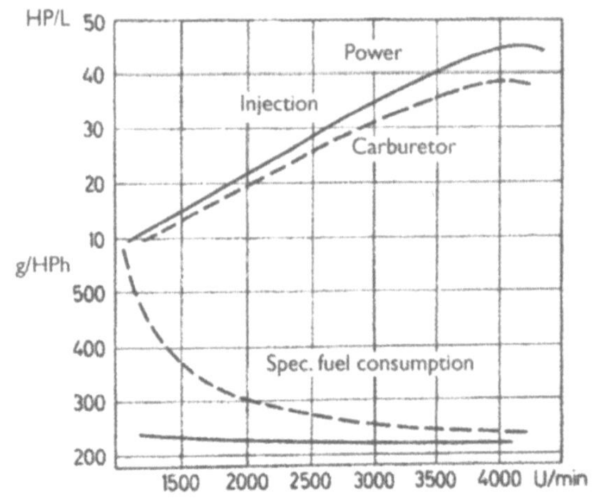


**Figure 2.17:**
Comparison of power and fuel consumption between the 600 ccm Gutbrod Superior engine with carburetor and with gasoline direct injection

cept could be started. After a successful series development process, the Daimler-Benz 300 SL sports car equipped with the 3.0 *l* fuel injection engine was produced in series from the summer of 1954 on. **Figure 2.18** shows a cross-section of the engine. The throttle valve housing used for the mixture control, the six-plunger injection pump with oil block in-line design, and the injection nozzles opening outwards and without leakage line were supplied by Bosch and differed only marginally or not at all from the components used in the two-stroke engines mentioned above in their basic design. For the adjustment of the air/fuel mixture at high altitudes (mountain passes), a barometric cell was integrated in the controller of the injection pump. The injection pressure of the fuel injection nozzles was approximately 45 bar. **Figure 2.19** shows a vertical cross-section of the injection nozzle used.

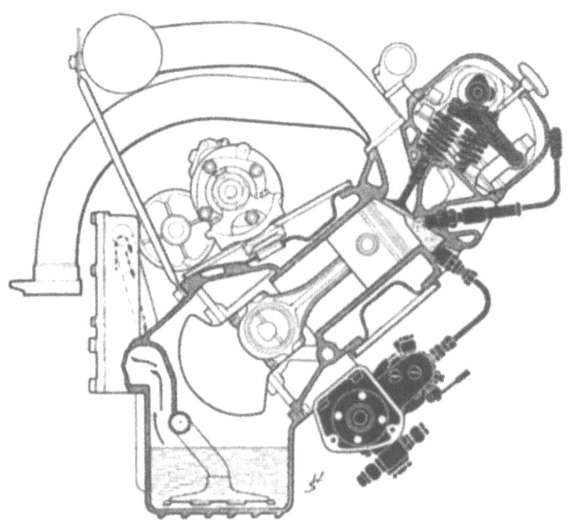

**Figure 2.18:**
Cross-section of the Daimler-Benz 300 SL engine with direct injection [3]

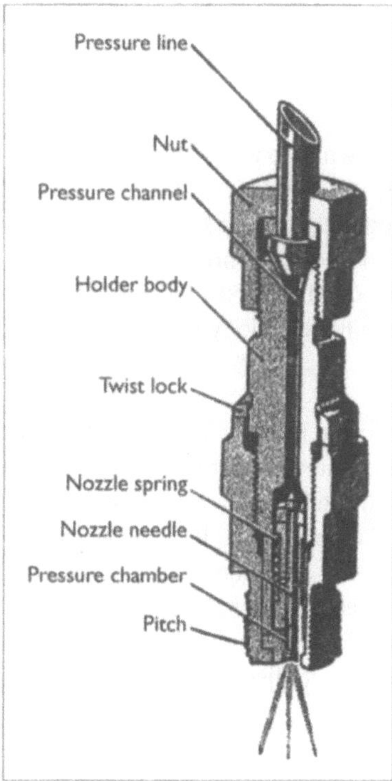


**Figure 2.19:**
Cross-section of the outward opening injection nozzle developed by Bosch [3]

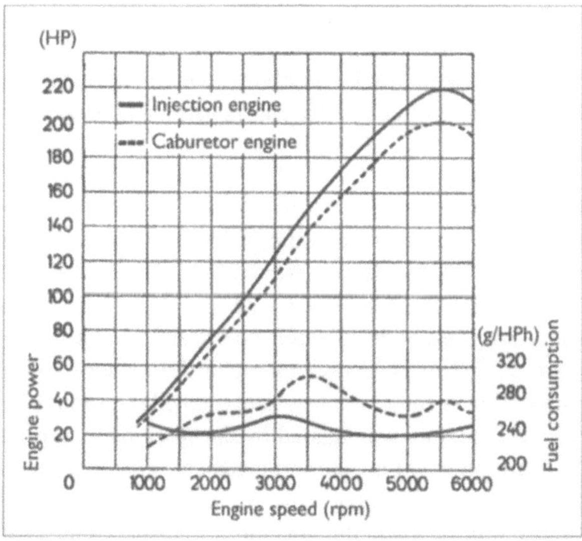


**Figure 2.20:**
Comparison of power and fuel consumption between Daimler-Benz 300 SL engines with carburetor and with direct injection system (as in 1953) [3]

**Figure 2.20** shows a comparison of power at full load and fuel consumption for DB 300 SL engines with carburetor and gasoline direct injection. In spite of a considerable reduction in fuel consumption by means of direct injection, the maximum engine power was increased by about 10%. The points to be highlighted with regard to this engine concept are high elasticity at low engine speeds and the direct, instantaneous response of the engine to load changes.

In the 1950's, the potential of gasoline direct injection shown here was successfully implemented by Daimler-Benz and Borgward in car racing [3, 14]. The 1950's also saw different approaches and tests on direct injection, such as corresponding tests at NSU with different injection systems [15, 16, 17] which, however, did not go into series production. Development activities in gasoline direct injection were abandoned in the 1960's and 1970's due to relatively low fuel prices and the development of injection technology in the direction of manifold injection in particular.

Research on gasoline direct injection continued solely in the research departments of the petroleum and automobile industries and in research institutes. A part of the combustion process shown below conforms to the characteristics of the combustion process in gasoline engines only in the broadest sense. This is due to the relatively strong local non-homogenous air/fuel mixture caused especially by retarded injection after combustion has started (considerable components of droplet combustion). Such engines are generally called hybrid engines or multi-fuel engines. For a historical chronological approach and because of their technical similarity to classic gasoline engines with direct injection, these engines are also described here.

The first stratified charge concept with fuel direct injection was introduced by Texaco as early as 1949, and called the TCP process (*Texaco Combustion Process*). The aim of this process was not to increase the specific power but to reduce fuel consumption [18, 19]. The process was later further developed under the name TCCS (*Texaco Controlled Combustion System*), **figure 2.21,** until 1986 and was characterized by the ability to burn a great variety of fuels and by a reduction in fuel consumption of up to 30% during part load operation as compared to a reference engine [20, 21].

Moreover, from the end of the 1960's to the mid-1980's, the MAN FM process (spark ignition with spherical combustion chamber) represented a stratified charge process which was derived from the diesel engine. It combines the characteristics of the classic diesel and gasoline engine processes and is thus a hybrid process. **Figure 2.22** shows the shape of the combustion recess in the piston and the positions of the injection nozzle and the spark plug. Here the fuel was injected in the form of an intense spray in a rotation-symmetric piston recess. A swirl, supported by the combustion chamber temperature, enabled re-evaporation of the fuel on the piston recess wall. The mixture was ignited by the spark plug. Various tests dealt with the optimization of spark plug and injector arrangement. The best results were achieved with configurations where one spark plug with two parallel pin electrodes was placed at the combustion chamber wall opposite the injector or one spark plug with three ground electrodes was mounted at the circumference of the spray. Different variants were implemented in vehicles.

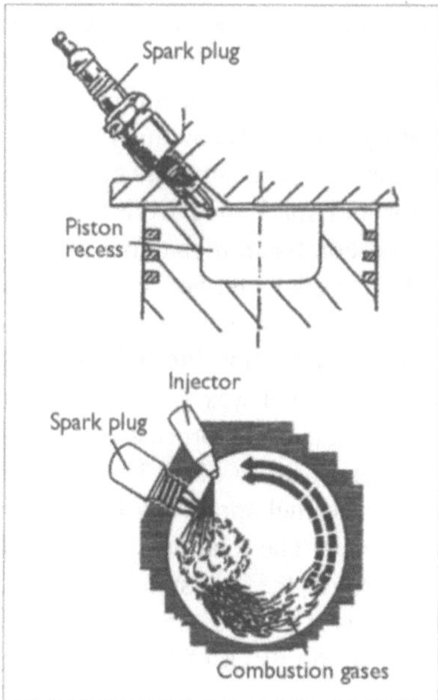


**Figure 2.21:**
Texaco Controlled Combustion System [20]

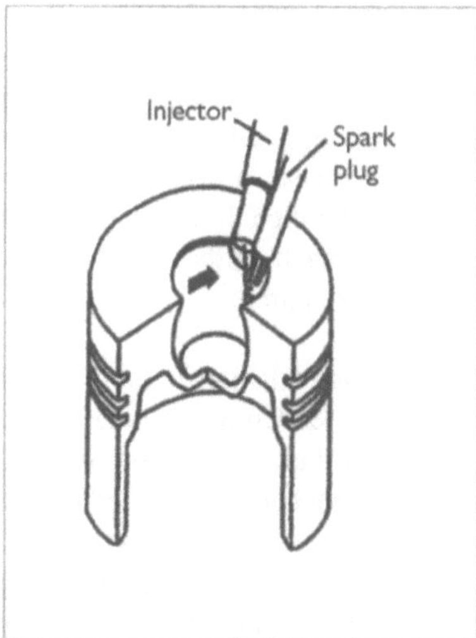


**Figure 2.22:** MAN FM process [22]

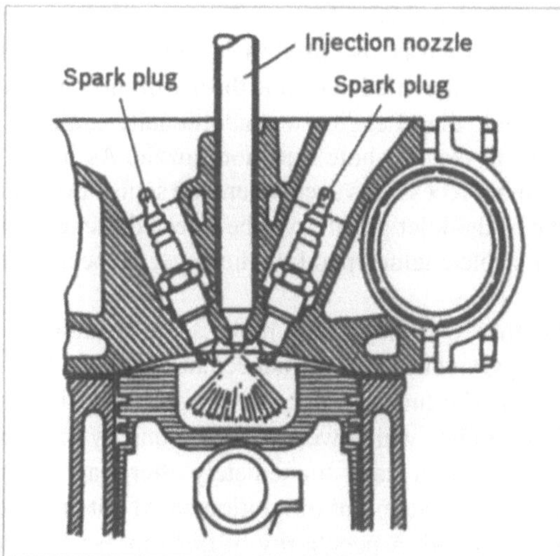


**Figure 2.23:** Ford-PROCO [25]

Ford also worked on gasoline direct injection from the mid-1950's onwards and introduced the FCP process (*Ford Combustion Process*) [24] in 1968. The aim of this process was to reduce fuel consumption. By shielding the intake valve in the cylinder head, a swirl flow was formed which was further intensified by the squish flow starting before the top dead center. The fuel was injected from an injection nozzle with a large spray dispersal angle (approximately 100°) in the combustion chamber, the start of injection for full load operation being at the bottom dead center and for part load operation for the stratified charge, shortly before the top dead center. An injection valve with an oscillating nozzle needle was used to improve the fuel distribution so that the mixture cloud was broken multiple times during the injection. The injector and spark plug were arranged so that the injected fuel was as much in the center of the combustion chamber as possible. The process took place without any throttling. It was characterized by optimum fuel consumption, low pollutant emissions and high specific power. A further developed variant shown in **figure 2.23** was introduced in 1972 under the name PROCO process (*Programmed Combustion*) [25, 26]. The goal of the development was the reduction of exhaust-gas emissions by means of measures aimed at the engine. The process was characterized by the use of high exhaust-gas recirculation rates of up to 50%. The injector was positioned at the center of the combustion chamber. The fuel spray was ignited by two spark plugs placed on the sides of the spray circumference. The rotation-symmetrical piston recess was at the center of the piston. In order to achieve the lowest possible $NO_X$ emissions, exhaust gas was supplied even at high engine loads. The process was examined in several engine and vehicle tests. Further developments of the PROCO process can be documented with the help of publications up to the 1980's [27].

The AD process (*multi-fuel direct injection*) presented in 1972 [28, 29] by Klöckner-Humbold-Deutz was also used in several vehicles. This process, which was derived from the diesel engine, tried to enable the use of fuels with a low cetane number. For this purpose, an available diesel engine was additionally equipped with a spark plug, which was mounted close to the two-hole injection nozzle. As in the MAN-FM process, a deep combustion chamber recess was used where the squish flow additionally intensified the swirl generated due to the inlet geometry. The injected fuel evaporated before reaching the piston recess. For stable engine operation, however, a special spark plug was required.

Another swirl-supported process is Mitsubishi's MCP process (*Mitsubishi Combustion Process*), which was introduced in 1976 [30, 31]. The process operated with a large piston recess. The fuel was injected against the swirl flow. This led to fast evaporation of small fuel droplets which were carried along by the swirl flow and reached the spark plug in a well-prepared state. Immediately after reaching the spark plug, the mixture was ignited. Since the exact point of ignition could not be exactly predetermined, long ignition timing was required. A peculiarity of this process was the possibility of changing the geometry of the injection spray for full load operation and part load operation. For this purpose, a pintle nozzle was used, with which different pintle strokes could be realized. At part load, a hollow conical spray with large spray dispersal angle and small fuel droplets was used. This led to effective mixture preparation and mixture transport to the spark plug within a short period of time. At full load, a fuel spray in solid conical pattern with a small spray dispersal angle and great penetration depth was used. This prevents the formation of a vaporized mixture that is too rich. The interval between start of injection and ignition remained constant in terms of crankshaft angle. Engines with this combustion process were used in harvesting machines in Japan.

In 1987, the swirl-free combustion process OSKA [32] was introduced. The fuel was injected into the combustion chamber with an intense spray. As shown in **figure 2.24,** the fuel atomization took place via a baffle plate in the combustion chamber, with a mixture cloud being formed in the center of the combustion chamber as a result, similar to the PROCO process. Ignition took place by means of a spark plug with exposed spark position installed close to the center of the combustion chamber.

Volkswagen also developed processes with deep piston recesses and close arrangements of injector and spark plugs, similar to AD and TCCP. In several variants, different spark plug and injector positions were tested. Here, too, the mixture first reaching the spark plug was ignited. The ignition point was coupled to the start of injection. Advanced variants of the engine were introduced in the research vehicle IRVW-FUTURA (*Integrated Research Volkswagen*) in 1989, among others [33]; **figure 2.25.**

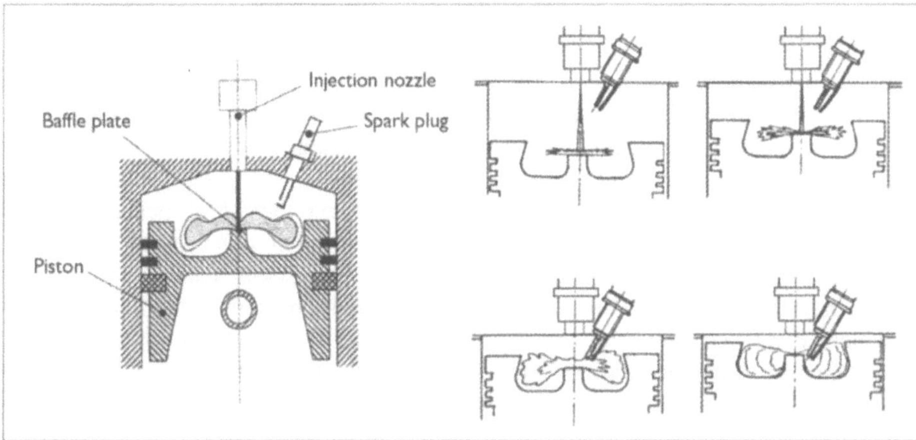


**Figure 2.24:** OSKA combustion process [32].

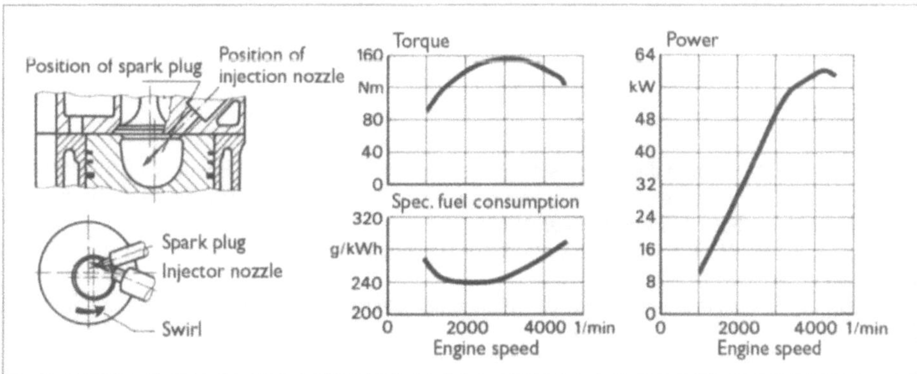


**Figure 2.25:** IRVW-FUTURA [33].

All these gasoline direct injection concepts tested in research departments so far presented with the problem that coordination between injection, mixture formation and combustion was possible only at individual steady-state operating points. It was not possible to satisfactorily represent dynamic engine operation with rapid load changes for all the stratified charge concepts, due to the mechanically operating injection systems. Only at the end of the 1980's did technical papers and publications on two-stroke engine concepts with internal mixture formation, especially those written by the Australian company Orbital [34], Japanese automobile manufacturers (Subaru [35], Toyota [36]), and AVL [37, 38], bring the technical potential of gasoline direct injection back into the focus of engine developers. Although the two-stroke engine concepts presented at that time did not go into series production, mainly due to reasons related to exhaust-gas emissions and the continued use of powertrains to drive passenger cars, they formed the background for the successful series introduction of direct injection in two-stroke outboard engines (Evinrude, Mercury, Yamaha) and two-wheelers (Aprilia). In particular, the possibility provided by the use of direct injection of stratifying the charge in the combustion chamber, with the option of operating vehicle gasoline engines with excess air at part load with reduced pumping loss, led to the introduction and launch of passenger car four-stroke engines with gasoline direct injection by Japanese manufacturers (Mitsubishi [39], Toyota [40]) from the mid-1990's and later also by European manufacturers (VW [41], Audi [42]). Bosch also introduced a test vehicle with spray-guided direct injection and stratified charge mode [43]. It was due to the limited fuel consumption reduction during vehicle operation, the cost-intensive exhaust-gas treatment for lean operation ($NO_X$ storage catalyst) and the risk of increased particle emissions in the exhaust gas for retarded injection (short mixture preparation time), that these initial concepts did not become established on a broad basis. Instead, various automobile manufacturers now favor concepts with continuous operation with a homogeneous, stoichiometrically controlled mixture ($\lambda = 1$), the trend ranging from wall-guided and air-guided mixture formation processes to spray-guided direct injection processes with injection pressures of 200 bar and greater. Current publications [44, 45] show that gasoline direct injection in combination with turbocharging offers great potential for reducing the power-to-weight ratio and fuel consumption, and will be increasingly used in both performance and downsizing concepts in the coming years.

## Literature

[1] *Hucho, W. H.*: Schläft die europäische Autoindustrie? In: Automobil Revue, Heft 33, 1997
[2] *Sass, F.*: Geschichte des deutschen Verbrennungsmotorenbaus von 1860 bis 1918. Berlin, Göttingen Heidelberg: Springer-Verlag, 1962
[3] *Scherenberg, H.*: Rückblick über 25 Jahre Benzineinspritzung in Deutschland. In: MTZ 16 (1955), Nr. 9, S. 245–254
[4] *Düll, R.*: Übersicht über den heutigen Stand niedrig verdichtender Einspritzmotoren mit Fremdzündung. In: ATZ 38 (1935), Nr. 6, 7, 8
[5] *Ittner, S.*: Dieselmotoren für die Luftfahrt. 1. Auflage. Oberhachingen: Aviatic-Verlag 1996
[6] *von Gersdorff, K.; Grasmann, K.*: Flugmotoren und Strahltriebwerke. 3. Auflage. Koblenz: Bernhard & Graefe-Verlag, 1995

[7] *Wagner, W.*: Hugo Junkers – Pionier der Luftfahrt – Seine Flugzeuge. Koblenz: Bernhard & Graefe-Verlag, 1996
[8] *Schnaufer, K.*: Bericht über Versuche mit Brennstoffeinspritzung in Zündermotoren. DVL-Bericht vom 01.09.1932
[9] *Teku, K.*: Technische Kurzberichte, Triebwerke. Doppelsternmotor BMW 801. Deutsche Luftwacht, Ausgabe Luftwissen 10, S. 156–157 (1943)
[10] *N.N.*: Der neue Hesselman-Motor. In: ATZ 33 (1930), Nr. 21, 22
[11] *N.N.*: Motoren mit geschichteter Ladung, 1. Teilabschlussbericht: Einteilungsschema und Auswertung des veröffentlichten Materials zu Schichtladeverfahren. Forschungsberichte Verbrennungskraftmaschinen, Heft 179 (1975)
[12] *Schweitzer, P.*: Hybrid-Motoren. In: MTZ 24 (1963), Nr. 5
[13] *N.N.*: Zweitakt-Benzineinspritzung bei Gutbrod und Goliath. In: MTZ 13 (1952), Nr. 10, S. 213–221
[14] *Hütten, H.*: Schnelle Motoren seziert und frisiert. 10. Auflage, Stuttgart: Motorbuchverlag, 1994
[15] *Froede, W.*: Benzineinspritzung für kleine Zylinder von Ottomotoren. In: ATZ 57 (1955), Nr. 8, S. 213–227
[16] *Bönsch, H. W.*: Der schnellaufende Zweitaktmotor. 2. Auflage, Stuttgart: Motorbuch Verlag
[17] *Dröschel, H.*: Die Unterbrechereinspritzanlage. In: ATZ 58 (1956), S. 335–341
[18] *N.N.*: Die Beseitigung des Kraftstoffklopfens durch das Texaco-Verfahren. In: ATZ 52 (1950), Nr. 5
[19] *Barber, E.; Reynolds, B.; Tierney, W.*: Elimination of combustion knock – Texaco Combustion Process. SAE Quarterly Transactions, Vol. 5. No. 1, 1951
[20] *Mitchell, E.; Alperstein, M.*: Texaco controlled combustion system – multifuel, efficient, clean and practical. SAE Technical Paper Series No. 740563, 1974
[21] *Lewis, J.*: UPS multifuel stratified charge engine development program – field test. SAE Technical Paper Series No. 860067, 1986
[22] *Rixmann, W.*: Der MAN-FM-Motor. In: ATZ 67 (1965), Nr. 10, S. 359–360
[23] *Norris-Jones, S.; Russel, J.*: 'FM' – A high efficiency combustion system for the future light duty engine? SAE Technical Paper Series No. 820760, 1982
[24] *Bishop, I.; Simko, A.*: A new concept of stratified charge combustion – the Ford Combustion Process (FCP). SAE Technical Paper Series No. 680041, 1968
[25] *Hiller, B.; Wade, W.*: Single-cylinder PROCO engine studies – Fuel and engine calibration effects on emissions, fuel economy and octane number requirements. SAE Technical Paper Series No. 780593, 1978
[26] *Simko, A.; Choma, M.; Repko, L.*: Exhaust emission control by the Ford programmed combustion process – PROCO. SAE Technical Paper Series No. 720052, 1972
[27] *Choma, M.; Havstad, P.; Simko, A.; Stockhausen, W.*: Fuel tolerance tests with the Ford PROCO engine. SAE Technical Paper Series No. 810439, 1981
[28] *Finsterwalder, G.*: A new Deutz multifuel system. SAE Technical Paper Series No. 720103, 1972
[29] *Finsterwalder, G.; Küpper, H.*: Emissionsarmer Fahrzeugmotor auf der Basis des AD-Vielstoffverfahrens A2.5. 2nd Symposium on Low Pollution Power Systems Development; Düsseldorf, 1974
[30] *Miyake, M.* et al.: A new stratified charge combustion system (MCP) for reducing exhaust emissions, Combustion Science and Technology. 1976, Vol. 12
[31] *Kuck, H. A.*: Forschungs- und Entwicklungsergebnisse von Schichtlademotoren. In: MTZ 38 (1977), Nr. 6, S. 269–274
[32] *Kato, S.; Onishi, S.*: New mixture formation technology of direct fuel injection stratified combustion SI engine (OSKA). SAE Technical Paper Series No. 871689, 1987
[33] *Emmenthal, K.-D.; Grave, H.-J.; Oppermann, W.; Schäppertöns, H.*: Motor mit Benzin-Direkteinspritzung und Verdampfungskühlung für das VW-Forschungsauto IRVW-Futura. In: MTZ 50 (1989), Nr. 9
[34] *Schlunke, K.*: Der Orbital Verbrennungsprozess des Zweitaktmotors. Vortrag beim 10. Interna-

tionalen Wiener Motorensymposium, 27./28. April 1989. Fortschritt Berichte VDI Reihe 12 Nr. 122, Düsseldorf: VDI-Verlag 1989
[35] *N. N.*: Neuer Subaru – Zweitaktmotor im Versuch. In: MTZ 52 (1991), Nr. 1, S. 15
[36] *N.N.*: Technik Zweitaktmotoren. In: Auto Motor Sport (1989), Heft 25, S. 46
[37] *Plohberger, D.; Miculic, L. A.*: Der Zweitaktottomotor mit innerer Gemischbildung. Vortrag beim 9. Internationalen Wiener Motorensymposium, 1988. Fortschritt Berichte VDI Reihe 12 Nr. 99. Düsseldorf: VDI-Verlag 1988
[38] *Plohberger, D.; Miculic, L. A.*: Der Zweitaktmotor als PKW-Antriebskonzept – Anforderungen und Lösungsansätze. Vortrag beim 10. Internationalen Wiener Motorensymposium, 27./28. April 1989. Fortschritt Berichte VDI Reihe 12, Nr. 122. Düsseldorf: VDI-Verlag 1989
[39] *N.N.*: Mitsubishi Benzinmotor mit Direkteinspritzung. Technisches Dossier, 1995
[40] *Kato, S.*: Toyota's approach to environmental protection towards the 21st century. 19. Internationales Wiener Motorensymposium, 1997
[41] *Krebs, R.; Stiebels, B.; Spiegel, L.; Pott, E.*: FSI – Ottomotor mit Direkteinspritzung im Volkswagen Lupo. Vortrag beim 21. Internationalen Wiener Motorensymposium 4./5. Mai 2000. In: VDI Fortschrittberichte Reihe 12, Nr. 420. Düsseldorf: VDI-Verlag 2000
[42] *Büchling, J.*: Der neue 2,0-l-FSI-Motor von Audi. In: MTZ (2002), Nr. 10
[43] *Stutzenberger, H.: Preussner, C.; Gerhardt, J.*: Benzin-Direkteinspritzung für Ottomotoren – Entwicklungsstand und Ausblick. 17. Internationales Wiener Motorensymposium. VDI Fortschritt-Berichte, Reihe 12, Nr. 267. Düsseldorf: VDI-Verlag, 1996
[44] *Welter, A.; Unger, H.; Hoyer, U.; Brüner, T.; Kiefer, W.*: Der neue aufgeladene BMW Reihensechszylinder Ottomotor. 15. Aachener Kolloquium Fahrzeug- und Motorentechnik 2006
[45] *Lückert, P.; Frey, J.; Kemmler, R.; Schaupp, U.; Vent, G.; Waltner, A.*: Kunden und zukunftsorientierte Technologien am Ottomotor heute und morgen. Vortrag beim 26. Internationalen Wiener Motorensymposium 28./29. April 2005. VDI Fortschritt-Berichte, Reihe 12, Nr. 595; Düsseldorf: VDI-Verlag, 2005

# 3 Mixture Formation and Combustion Processes

Recent decades have seen significant development in both gasoline and diesel engines. While passenger car diesel engines have shifted towards direct injection, with Audi [1] making a successful move in 1989 based on experience gained from their commercial vehicle operations, the gasoline engine has remained an engine with load control through quantity governing using a throttle valve. This results in significant losses in efficiency, especially in part load operation, resulting in high specific fuel consumption. This is particularly important given that today's traffic problems mean that vehicles are driven predominantly in part load operation. Therefore, the reduction of throttle losses in the charge exchange phase has considerable potential for minimizing consumption and thus conserving resources. Besides the different approaches to reducing gas exchange losses using new concepts like variable valve control, efforts have been made for more than a hundred years to operate the gasoline engine throttle-free, through direct fuel injection into the combustion chamber, similar to the diesel engine, The idea behind modern direct injection in gasoline engines is to combine the advantages of the diesel engine with those of the gasoline engine. The advantages of diesel engines are minimal losses during gas exchange, quality control (load control only via the supplied fuel quantity) and minimal heat losses through the combustion chamber wall. What distinguishes the gasoline engine is its small size, good power performance and low emissions.

There are two different operating modes (operating strategies) for gasoline engines with direct injection, with the essential differences being in injection timing, throttle valve position and mixture composition. In the higher load range, the engine is operated with a homogeneous mixture composition like a gasoline engine with external mixture formation. In part load operation, it is operated with charge stratification. This is necessary in order to implement unthrottled operation in part load mode with homogeneous mixture formation. The charge in the engine combustion chamber is made leaner in such a way that ignition can no longer be reliably initiated due to the ignition limit for air/fuel ratios $\lambda$ of approximately 1.5. To ensure reliable ignition and inflammation of the air/fuel mixture, the mixture formation process must provide an ignitable, combustible mixture in the vicinity of the spark plug at ignition start. **Figure 3.1** shows the different operating modes of direct injection for the implementation of homogeneous mixture formation and mixture formation with stratified charge.

In intake stroke injection, at higher loads and in full load operation, the fuel is introduced into the combustion chamber during the downward movement of the piston; **figure 3.1** right. As for fuel preparation in gasoline engines with external mixture formation (intake manifold injection), the fuel is injected during the intake process and is mixed early with air coming in at high speed.

In the subsequent compression stroke, this leads to homogeneous mixture distribution at the ignition point, with an almost constant air/fuel ratio throughout the combustion cham-

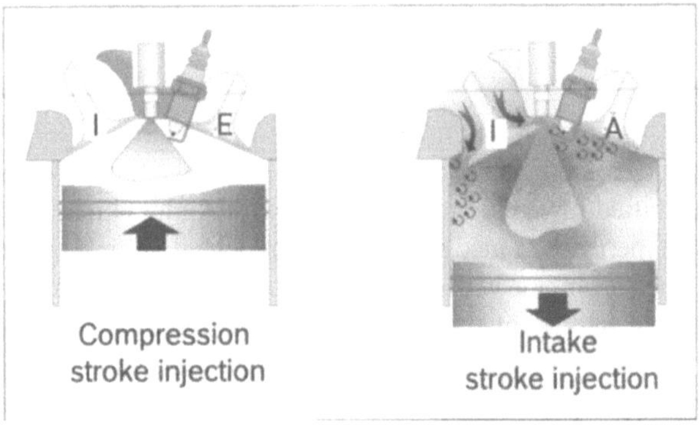


**Figure 3.1:**
Mixture composition and operating modes in the gasoline engine with direct injection:
L: stratified operating mode
R: homogeneous operating mode

ber. At full load, almost identical boundary conditions with respect to volumetric efficiency exist for manifold injection (throttle valve fully open) and direct injection (throttle-free operation). Nevertheless, gasoline engines with direct injection have higher volumetric efficiencies. This can be attributed to the heat extraction in the combustion chamber that is due to the almost total evaporation of the fuel during the intake process, which can also improve knocking behavior [2]. Moreover, there are no further dynamics of the fuel wall film in the intake manifold, since the fuel is injected directly into the combustion chamber. This results in a better response during cold start, better operational behavior during load changes and more favorable emissions.

In the stratified operating mode during part load, a gasoline engine with direct injection should ideally be operated with a fully opened throttle valve. As in diesel engines, the load is then controlled solely via the injected fuel quantity (quality control), which is directly introduced into the combustion chamber during the upward movement of the piston only immediately before ignition. This process is known as compression stroke injection; **figure 3.1** left. Ideally an ignitable mixture cloud should be present in the center of the combustion chamber, surrounded by air not required for combustion. Therefore the charge is stratified in the combustion chamber and the heat losses to the combustion chamber wall in the subsequent combustion process can be reduced by the insulating air layer. In principle, engine operation is very lean and at the point of ignition, large gradients exist in the air/fuel ratio in the combustion chamber. This is how the principle of direct injection in gasoline engines, which is simple but not easy to implement, manifests itself: by making available an ignitable mixture in the vicinity of the spark plug, at ignition point. **Figure 3.2** shows the application of the operating strategies mentioned above for the complete engine map [3].

It is imperative to delve into theoretical principles during the development of gasoline direct injection. Many effects can be analyzed only through a profound overall understanding of the individual processes. Mixture formation, especially, is of vital importance for the practical implementation of this combustion process.

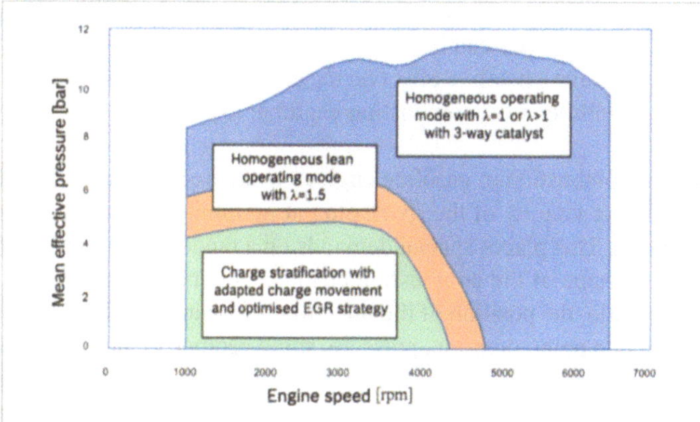


**Figure 3.2:** Operating strategies in the map [3]

## 3.1 Basics of mixture formation

Mixture formation calibration for the entire operating range is an essential basic prerequisite for achieving reliable, good combustion with high efficiency and minimal emissions. This applies to both modes: operating with homogeneous mixture and operating with stratified charge. The influences on the mixture formation process are extremely diverse. Good mixture formation reproducible in wide map ranges is a proven recipe for economical engine operation. Similarly, good drivability across a wide engine speed range depends directly on a stable mixture formation. The mixture formation is basically determined by the influences presented in **figure 3.3** [4]. The operational parameters can usually be con-

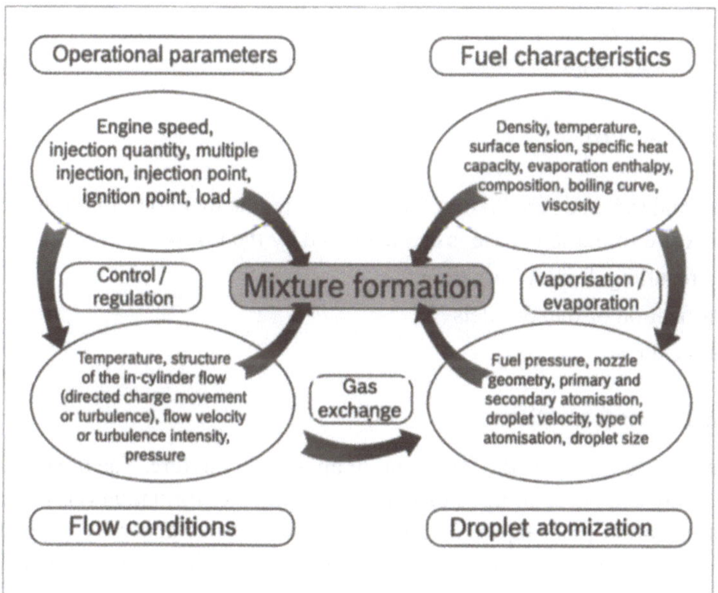


**Figure 3.3:** Parameters that influence the mixture formation process [4]

trolled without any problem by a modern engine control system. Similarly, in the age of highly developed fuels, fuel characteristics can be specifically influenced in modern refineries. The fuels, so modified, can have a positive effect on the combustion process and most importantly on the tendency to pollute in the combustion chamber [5].

It is particularly important for combustion in gasoline engines with direct injection that an ignitable mixture is present in the vicinity of the spark plug at ignition point, irrespective of engine load and engine speed. This places special demands on a range of parameters. In particular, the intake flow, the shape of the combustion chamber (cylinder head type and piston type), the compression ratio, the position of the injection nozzle and spark plug, and the injection parameters (injection point, duration, pressure, spray geometry) must be taken into consideration. A clear understanding of the processes taking place during gasoline direct injection and the related precise coordination of injection, mixture formation and combustion requires extensive research. Therefore, the development of gasoline engines with direct injection which are ready for series production and capable of being operated in stratified charge mode has only just begun.

### 3.1.1 Fuel preparation

Extremely quick mixture formation is one of the most important prerequisites for the implementation of direct injection in gasoline engines, especially in the stratified charge mode. Fuel atomization and fuel evaporation are the basis for adequate mixture formation under all operating conditions. Small droplet sizes that affect evaporation favorably are a result of:

- Small nozzle hole diameters
- High fuel delivery speed (high injection pressure)
- High air density (high compression ratio or super/turbocharging)
- Low viscosity of the fuel
- Low surface tension of the fuel.

Fuel preparation has a vital influence on mixture formation. After the fuel flows through the injection valve, the mixture formation phase is characterized by the following events [6]:

- Spray break-up and resulting droplet formation
- Evaporation or vaporization of fuel droplets
- Mixing with the intake air.

**Spray break-up and droplet formation**

Spray break-up is the first step towards the formation of an air/fuel mixture. The break-up of contiguous liquid spray can occur in three different ways: droplet formation (I), wave formation (II) and atomization (III); **figure 3.4**. These break-up types generally depend on the fuel delivery speed. At lower speeds, the surface forces of the liquid predominate, which results in droplet formation (I, also known as Rayleigh break-up). If the fuel delivery speed increases, the aerodynamic forces acting on the fuel jet also increase. Under the influence of the aerodynamic forces, the spray is initially formed into waves (II, also known as primary and secondary flow-induced break-up) and is subsequently completely atomized when the delivery speed continues to increase (III) [8]. The resulting droplets during atomization are much smaller than in the other two break-up types and the relevant area for mixture formation in the gasoline engine can be represented solely by the atomization process. The atomization of the fuel takes place when the external forces acting on the droplet become greater than the surface tension forces [9]. If a fuel droplet moves in a gaseous medium, a number of different forces and mechanisms act on it. These include, for example, inertial forces, viscous forces, gravitational force and surface force.

The Weber number *We*, which indicates the ratio of dynamic pressure to internal pressure (surface tension), is used as the characteristic number for the break-up of a liquid with low viscosity, for example gasoline:

$$We = \frac{\varrho \cdot c_{\mathrm{rel}}^2 \cdot d_{\mathrm{T}}}{\sigma}$$

where:

| | |
|---|---|
| $\varrho$ | density (air or fuel) [kg/m$^3$] |
| $c_{\mathrm{rel}}$ | relative velocity (air/fuel) [m/s] |
| $d_{\mathrm{T}}$ | diameter of the droplet [m] |
| $\sigma$ | surface tension of the liquid [N/m] |

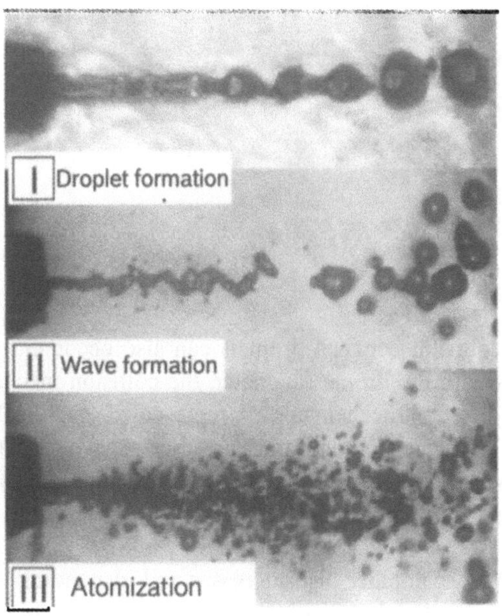


**Figure 3.4:**
Break-up types of a liquid spray [7]

If the break-up of fuel droplets takes place primarily under the influence of airflow, the break-up criterion is indicated by exceeding a critical value for the Weber number. In practice, the critical Weber number for gasoline is:

$$We \approx 10 - 12$$

i.e. droplets break up for Weber numbers greater than 10 to 12.

The spray break-up process during fuel injection and atomization can be divided into two different areas. The area in which the still contiguous liquid is broken down into droplets for the first time is described as primary break-up. This area is usually still within the nozzle [10, 11] or immediately after the exit from the injection hole [12]. In the area farther away from the nozzle, secondary break-up takes place with further droplet break-up. The individual areas are marked in **figure 3.5**. During the primary break-up near the nozzle, the still compact liquid spray core breaks down into larger droplets and ligaments [13]. In this zone, a large amount of liquid fuel is still present. Only later, farther away from the nozzle, does the ambient air supplied from outside lead to a dilution of the liquid phase, and the increasing aerodynamic forces finally bring about the secondary break-up of the droplets [13]. In addition, evaporation starts, due to the heat supplied by the ambient air. The drag at the liquid phase reduces the spray velocity.

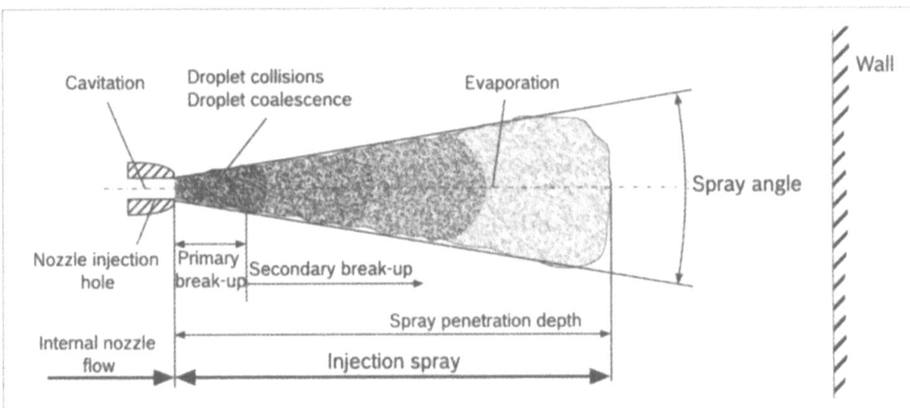


**Figure 3.5:** Spray break-up zones according to [8]

In addition to single droplet break-up, collision between droplets can also occur in the spray. Since the Weber number changes with the droplet diameter, the collision mechanisms can also change accordingly. The different mechanisms are represented graphically in **figure 3.6**. Droplet collisions lead to coalescence or temporary coalescence even at Weber numbers slightly above the critical Weber number ($We$ = 10 to 12 for gasoline). In this process, two separate droplets become one common droplet.

As the droplet diameter increases, the Weber number also increases and, depending on its diameter, the now larger droplet can again break down into smaller droplets or break down explosively [14]. The droplet size within a spray is therefore difficult to describe and characterize, so usually only the atomization quality of a spray is described in technical applications.

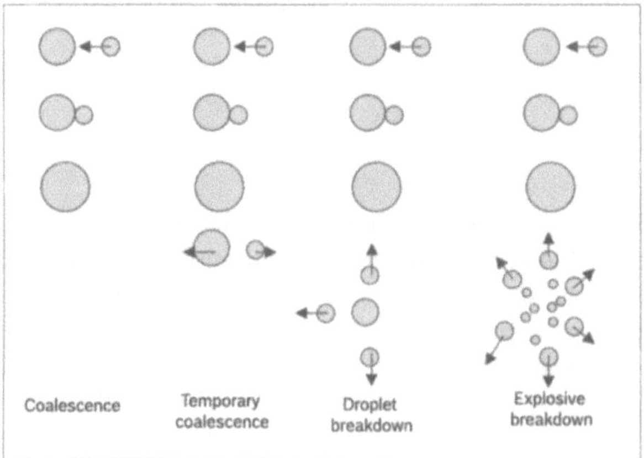


**Figure 3.6**: Collision mechanisms of droplets [14]

### Evaporation and mixing

The processes of spray break-up and droplet formation are followed by the evaporation and mixing of fuel with air. During the evaporation process, the transition from the liquid to the gaseous phase takes place within a short time [9]. The evaporation process is influenced by the following factors:

- Potential for evaporation to the environment (temperature difference)
- Atomization quality
- Mass transfer to the environment.

Evaporation potential and atomization quality have the biggest influence. Thus, the evaporation process mainly depends on the thermodynamic boundary conditions of pressure and temperature [13]. The enthalpy required for evaporation is taken from the hot environment and transferred to the fuel spray. In the process, the fuel is mixed with the ambient air. If this process takes place within the combustion chamber, the air contained in it cools down and the cylinder charge increases. In gasoline engines with direct injection, this effect is used to increase the charge in the intake stroke injection operating mode and to increase performance [15]. A flow in the engine combustion chamber superimposed on this process has a positive effect on the evaporation process and also improves the mixing and homogenization of the mixture [16]. Small droplet diameters also have a positive effect on evaporation, as do high relative velocities between the droplet and the gas [9]. An increase in the injection pressure in gasoline engines with direct injection can therefore positively influence evaporation in many respects [4, 17]. For example, the relative velocity between the droplets and the environment increases. In addition, the entrainment flow of the injection spray intensifies, through which air is also entrained in the spray (see section 3.1.2).

The complete process of evaporation is very complex, similar to that of spray break-up. Many more theoretical analyses of this subject and of the relationships explained so far in this chapter can be found in the works of Lefebvre [18], Troesch [19] and Weber [20]. In summary, the optimized atomization and evaporation of fuel are seen as the basis of good mixture formation.

### 3.1.2 Flow characteristics

Additional air movement in the combustion chamber to increase the relative velocity to the injection spray (swirl, tumble or squish flow) can improve the atomization quality and the evaporation of the fuel droplets. The formation of flow vortexes in the combustion chamber during the intake phase, and their subsequent behavior in the compression phase, are especially crucial for the subsequent combustion [21]. The flow vortexes have significant influence on heat transfer through the walls, mass transport, fuel conversion, and the propagation velocity of the combustion [4, 21, 22]. Moreover, directed flows take over the function of mixture guidance and mixture transport in gasoline engines with direct injection. Based on the piston movement, two different formation processes can be distinguished for the formation of flow vortexes; **figure 3.7**.

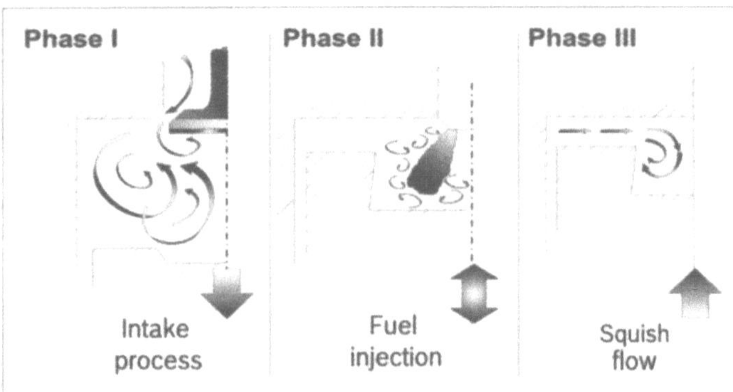


**Figure 3.7:** Different mechanisms of turbulence formation in the combustion chamber of a combustion engine [23]

During the downward movement of the piston in the intake phase (phase I), very high flow velocities develop at the valve gap of the inlet valves, which can be up to 10 times the actual piston velocity [24]. Turbulences are generated that result in the formation of recirculation areas with small-scale vortex structures in the vicinity of the inlet valve flow. Large-scale, directed flows are created in the combustion chamber [23] through interaction with the combustion chamber walls and the downward moving piston. According to [22], a charge movement generated at the inlet serves particularly to improve the ignition conditions. During the upward movement of the piston (phase II), the effect of the decreasing combustion chamber volume can be seen. The available vortexes are compressed during the subsequent compression phase. This leads to a reduction of the diameters of the vortexes and an increase of their rotation velocity. As a result, the turbulence becomes more intense and fine- grained [21]. At the end of the compression stroke (phase III), the squish flow appears as an additional acceleration of the flow in the combustion chamber. This flow is exclusively dependent on the geometry of the combustion chamber and produces especially positive characteristics with regard to the inflammation of lean mixtures [25, 26]. According to [22], increased turbulence in the compression phase significantly influences the combustion process, in particular the combustion speed and the burn-through [27].

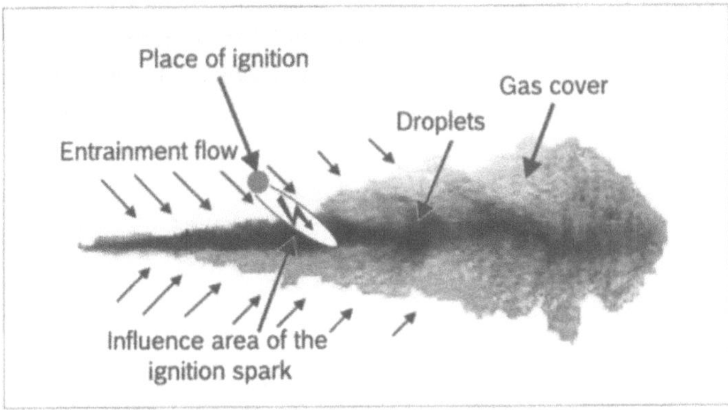


**Figure 3.8:** Entrainment flow and ignition spark influence at a single spray cone, schematically [30]

In gasoline engines with direct injection, the fuel injected under high pressure enters the combustion chamber at high velocity. This induces flow influences on the ambient gas in the periphery of the injection spray during the injection process [28, 29]. Such an entrainment flow provides further support to the mixture formation process in phase II. The entrainment flow is generated by drawing in the air from the environment of the spray. This is a result of the vacuum that is created in the spray because of vaporization. The vacuum is attributed to the velocity distribution of the droplets and the fact that the environment is cooled by the spray, as the evaporation enthalpy is taken from the environment of the spray [28, 30]. Given that very little time is available for mixture formation in gasoline engines with direct injection, the entrainment flow makes a vital contribution to the mixture formation process and is especially important for this combustion process [30]. An ignition spark already available during injection can be drawn into the spray by the air, e.g. during the spray-guided mixture formation process. This process is represented schematically in **figure 3.8**.

This additional effect can have a positive impact on inflammation, in particular in gasoline engines with direct injection having with a spray-guided mixture formation process, as the mixture volume accessed by the spark becomes larger. Depending on the direction of the piston movement at the point of injection, the air movement induced by the spray is superimposed on the respective in-cylinder flow from phases I and II. The degree of dethrottling and the engine speed have a considerable influence on the charge velocity. The inertia of the flow during the intake phase can be also used in many ways in internal combustion engines. A prolonged opening time of the inlet valves at high engine speeds, for example, leads to an increased charge in the cylinder. In addition, the amount of exhaust gas recirculated internally can be influenced by using the time window of the valve overlapping [31]. A corresponding throttle valve actuation makes it possible to re-induct residual gas from the cylinder into the inlet port.

Fluid movements are generally either directed or undirected flows. Directed flow is relatively stable and mainly characterized by its momentum. The higher the momentum, the longer and more stable the flow remains. Directed flows are squished during compression and diverted by the combustion chamber walls, because of which large-scale vortex structures break down into smaller vortexes. They become undirected flows or turbulences and no longer have any specific flow direction. Turbulent flows are irregular and their characteristic parameters change with location and time [32]. Turbulent flows can be characterized according to [32] as follows:

- Turbulent flows are irregular
- Turbulent flows are vortex flows
- Turbulent flows are three-dimensional
- Turbulent flows are unsteady.

Detailed mathematical explanations of the flow mechanics in combustion engines are available in the publications of Heywood [24], Sullivan [33] and Raposo [34], among others. Charge movements generated at the inlet can be divided into two main categories; **figure 3.9**. In both cases, the flow direction is a rotation around an axis. In case of a swirl flow, the rotational axis is in the cylinder axis. For a tumble flow, the rotational axis is perpendicular to the cylinder axis.

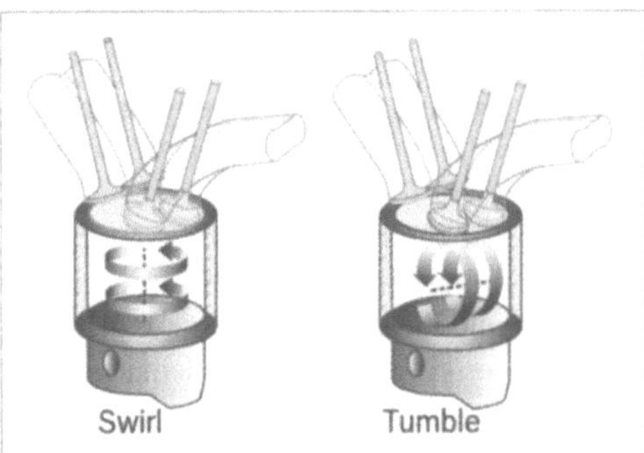


**Figure 3.9:**
Different charge movements generated by the intake port [23]

### Swirl flow

So far, swirl flows have been used primarily in diesel engines. Since the introduction of direct injection with stratified charge, the generation of swirl flows has increasingly been used in gasoline engines [16, 35] as well. To minimize frictional losses due to the flow, this type of charge movement is usually implemented with a geometrical modification of the inlet port. To do this, an asymmetric intake configuration is applied. It is mostly implemented by changing an inlet port to a pure charge port. Depending on the operating mode, the charge port can be closed with a flap. The other inlet port is used as a swirl-generating tangential port; **figure 3.10**. The flap remains closed in large ranges of the map so that the positive effect of the swirl can be used. This makes the air flow asymmetrically into the combustion chamber and generates the flow pattern that rotates around the vertical axis of the cylinder. The charge port is exposed to increase the charge only at high engine speeds and during full load in order to enable maximum volumetric efficiency [16].

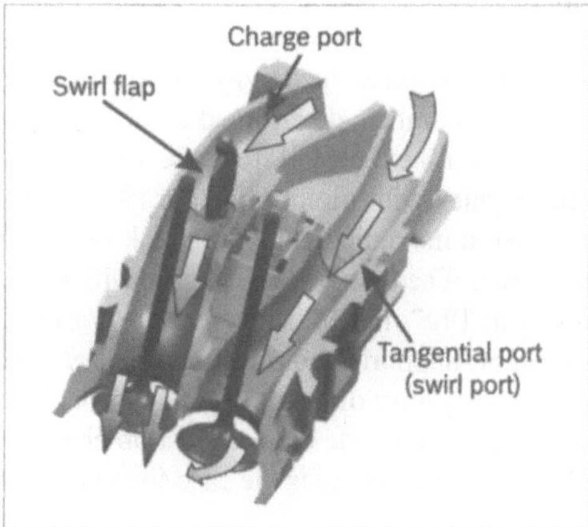


**Figure 3.10:**
Swirl port for a gasoline engine with direct injection [16]

The swirl flow can also be supported by an appropriately designed piston (bowl type piston); **figure 3.11**. The characteristic number that describes the intensity of the swirl flow can be determined by measurements on a stationary flow rate test bench and is known as the swirl number. It is defined by the rotational impulse flow to axial impulse flow ratio [36]. Swirl flows are also maintained in the compression phase [4, 36] as far as possible.

**Figure 3.11:**
Pistons for swirl support (left) and tumble support (right) [16]

### Tumble flow

Tumble flow that rotates around a horizontal axis is generated if most of the air that is drawn in by the engine flows into the combustion chamber via the upper half of the valve heads. This behavior is primarily generated by the fact that the airflow is forced to enter via the upper part of the inlet port by measures implemented at the inlet. For this purpose, either the inlet port is bent more and therefore changed to what is called a tumble port or the lower part of the inlet port is closed with a flap, thus guiding the airflow in the upper part directly to the valve. Due to the resulting reduction in cross-section, the flow in the constricted area is highly accelerated and pushes itself against the wall in the upper part of the inlet port. As a result, most of the intake air is guided via the upper valve area and flows into the combustion chamber with high kinetic energy. This results in a flow pattern that rotates perpendicular to the cylinder's vertical axis in the combustion chamber. **Figure 3.12** illustrates the principle of tumble generation.

The tumble flow is characterized by the tumble number. Since the downward moving piston is not part of a stationary flow test bench, tumble numbers determined on a stationary test bench are difficult to correlate with engine results. Like the swirl flow, the tumble flow is widely used in the development of gasoline engines with direct injection [3, 37, 38]. Tumble flows are very much dependent on the piston movement and can break down almost completely into turbulence at the top dead center. The first gasoline engine with direct injection introduced in a mass production model in 1997, the Mitsubishi GDI (*Gasoline Direct Injection*), was already equipped with a tumble supported combustion process [39]. Mitsubishi, however, used a "*Reverse Tumble*" in which most of the flow enters through an extremely steep inlet port via the lower valve area and alongside the combustion chamber wall. The change of direction takes place on the piston surface, as is the case for a conventional tumble [39].

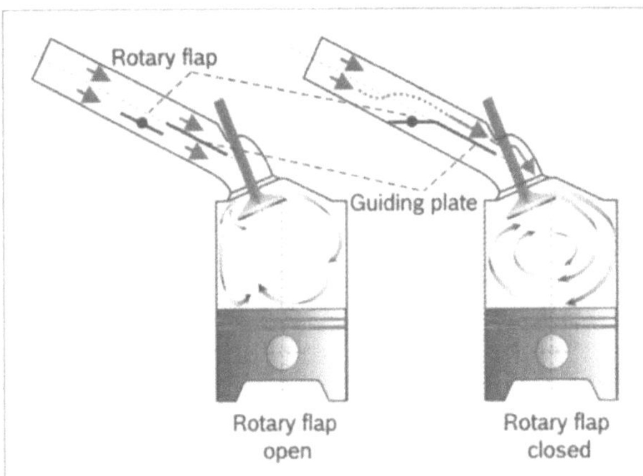


**Figure 3.12:**
Principle of tumble generation [23]

### Conclusion

When applying these findings in practice, it can be said that, in summary, additional air movement in the combustion chamber, which is generated to increase the velocity relative to the injection spray (swirl, tumble or squish flow), improves the quality of atomization and increases the evaporation speed considerably. When it comes to the aim of achieving fast mixture formation with stratified charge, this is of particular importance. It is not only the airflow and the delivery speed of the fuel from the injection nozzle, however, that influence the life of the droplet and with it the time of mixture formation, but also the air temperature that prevails in the combustion chamber when the injection process takes place. Generally speaking, the smaller the droplet size and the higher the air temperature, the shorter the life of the droplet. **Figure 3.13** shows the influence of droplet size and air temperature on the life of the droplet [40].

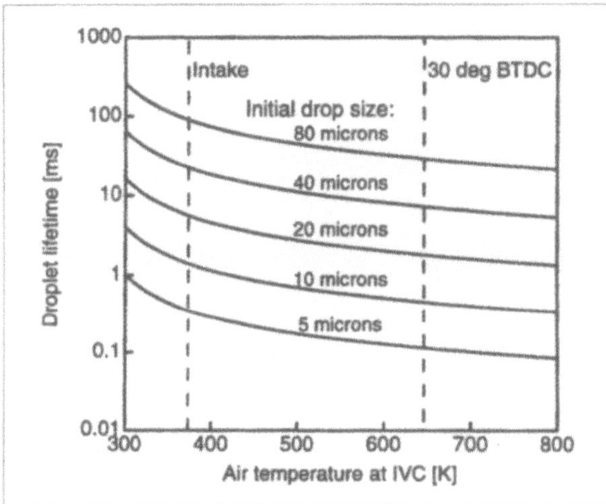


**Figure 3.13:**
Influence of droplet size and air temperature on droplet lifetime [40]

As can be seen, the lifetime of the droplet decreases with the size of the injected fuel droplet, and this increases the evaporation speed. The smallest possible droplet diameters are therefore necessary for rapid mixture formation. This is supported on the one hand by high injection pressure and small nozzle cross-sections and on the other hand by increased air movement and high air temperature. The adjustment of these boundary conditions for the injection and the necessary fast and appropriate mixture formation, which must be in line with the respective operating conditions, then leads to better combustion with fewer emissions and more favorable fuel consumption.

## 3.2 Direct injection with homogeneous mixtures

When gasoline direct injection was initially developed in four-stroke engines for passenger cars (Mercedes 300 SL), the power throughout the entire operating range was controlled by adjusting the quantity, i.e. by means of a throttle valve, as is the case in carburetor engines and engines with intake manifold injection. The injected fuel was adjusted to the intake air mass in such a way that the air/fuel ratio λ was almost the same in all load ranges. The fuel is injected directly into the combustion chamber early in the intake process; **figure 3.14**. Thus, there is sufficient time available for mixture formation and mixture homogenization, even at high engine speeds. This facilitates mixture formation, which is much easier than in part load operation with stratified charge. Until a few years ago, part load operation with stratified charge could not be mass-produced because of their purely mechanical injection systems and dynamic engine operation.

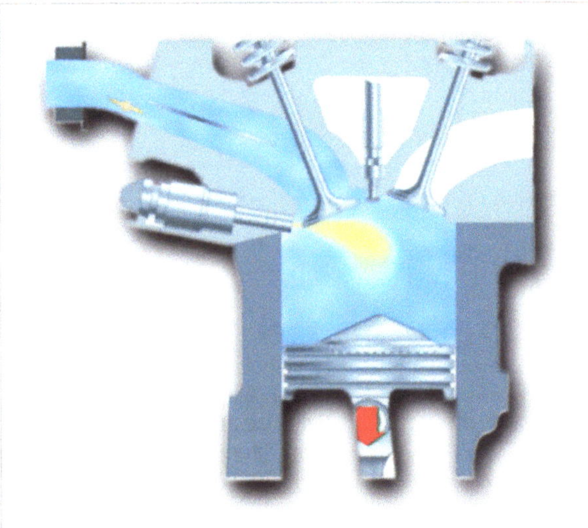

**Figure 3.14:**
Fuel injection in the intake stroke in the VW FSI [Source: VW]

While gasoline engines with direct injection that operate in the stratified charge mode in part load have emerged recently in the market, obtaining a reliable stratified charge mode is still seen as a challenge, associated with high investment costs for the electronics and injection systems. Intensive development work is therefore currently being carried out for gasoline direct injection with homogeneous mixture formation. The concept is increasingly being combined with supercharging in order to utilize the advantages of gasoline direct injection in the high load range as compared to intake manifold injection on the one hand, and on the other, to increase the specific power through supercharging with simultaneous reduction of the engine size (downsizing). The advantage here is that the frictional loss is reduced, by which the fuel consumption can be reduced by up to 10% as compared to a reference engine with intake manifold injection without supercharging and the same power. Another advantage, of course, is the experience gained by working with direct injection in series production, which promotes the further development of this combustion process for future requirements.

As a rule, gasoline engines with direct injection are less sensitive to knocking combustion than engines with intake manifold injection, due to the internal cooling caused by having the fuel directly injected into the combustion chamber. If the improved knock behavior is used for increasing the compression ratio, a rise in $\Delta\varepsilon$ from 1.5 to 2 can be assumed.

Thus, an advantage of approximately 5% in fuel consumption can be achieved according to a theoretical analysis of the ideal engine process. The compression ratio is normally increased by geometrically reducing the compression volume against the piston displacement. However, this involves an increase in the ratio of wall surface area to combustion chamber volume, or surface/volume ratio, which leads to an increase in unburnt hydrocarbon emissions due to the increasing wall influence. The positive effect of compression ratio increase can also be used through super or turbocharging, either by achieving more power through forced induction and the related charge improvement with the same engine size or, as mentioned, by reducing the engine size, i.e. downsizing, while maintaining the same engine power. This in turn means a decrease in frictional loss and a reduction in fuel consumption. Nowadays both effects are used for the development of engines with gasoline direct injection and homogeneous mixture formation.

At full load, almost identical boundary conditions with respect to volumetric efficiency exist for manifold injection (throttle valve fully open) and direct injection (throttle-free operation). Nevertheless, gasoline engines with direct injection have improved charge due to higher volumetric efficiencies. This can be attributed to the heat extraction from the combustion chamber through fuel evaporation in the combustion chamber, which is also responsible for the improved knocking behavior with direct injection. A rough estimate at standard conditions and full load (stoichiometric operation) confirms that a 9% higher theoretical volumetric efficiency can be achieved. However, it must be taken into consideration that mixture cooling in the intake manifold due to fuel evaporation takes place even for intake manifold injection, which also results in a charge increase. The fact that the volumetric efficiency is still higher for direct injection can be explained according to [15] as follows:

- In intake manifold injection, the low to medium relative velocities between fuel and air (fuel/air pressure ratio: approximately 3...5/1) lead to a combination of pressure atomization and compressed air atomization. In the case of direct injection, only pressure atomization takes place due to the high relative velocities between fuel and air (fuel/air pressure ratio approximately 50...200/1). As shown in **figure 3.15**, the diameters of fuel droplets formed at high relative velocities are considerably smaller than those formed at low relative velocities. This results in a significant increase in the number of fuel droplets and thus to an increase in the total fuel surface area of all the droplets. Heat exchange between the fuel droplets and the surrounding air and evaporation is subsequently improved, which in turn results in faster mixture formation.
- The fuel receives the evaporation enthalpy it requires from the environment during the evaporation that follows atomization; the environment provides the fuel the required. In this process, there are two aspects that positively affect fuel evaporation for direct injection. The evaporation potential is significantly higher than in the intake manifold due to the hot combustion chamber walls and the related high temperatures. This leads to considerably faster evaporation of the fuel and thus mixture cooling with a largely open inlet valve. During intake manifold injection, in contrast, some of the fuel reaches the relatively cold intake manifold walls (intake manifold wall film) and does not fully participate in the evaporation. In particular the opening and closing of the inlet valve leads to a "tearing off" of the intake manifold wall film. This tearing off and the larger fuel

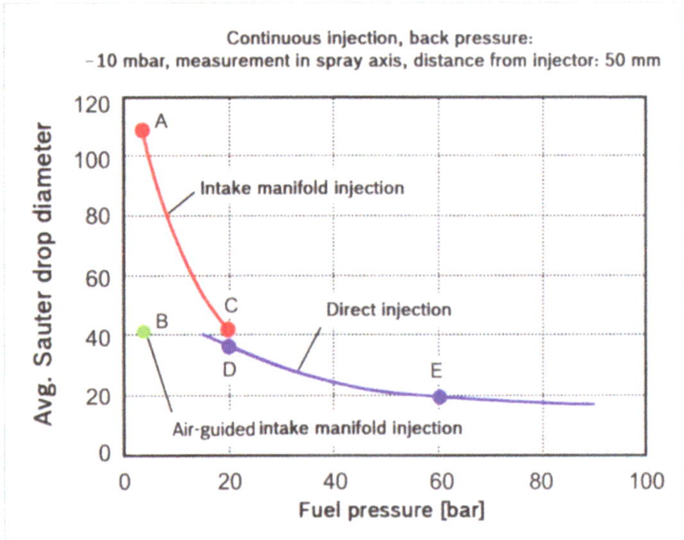


**Figure 3.15:** Droplet diameter dependent on fuel pressure [41]

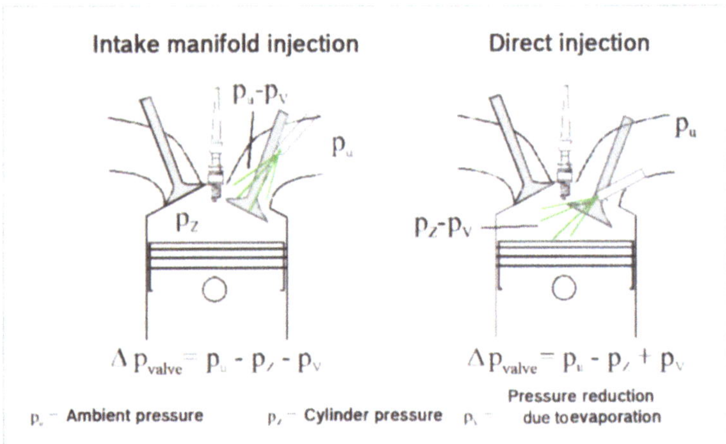


**Figure 3.16:** Pressure difference at the inlet valve for intake manifold injection and direct injection: $p_u$ ambient pressure, $p_Z$ cylinder pressure, $p_V$ pressure reduction through evaporation

droplets present in intake manifold injection need considerably more time to evaporate. Some of the evaporation therefore does not occur until the compression phase after the inlet valve is closed. The mixture cooling that results from the fuel evaporation is less substantial than for direct injection and thus produces a smaller increase of the charge than the direct introduction of fuel into the combustion chamber.

- The fuel still evaporating in the intake manifold during intake manifold injection produces a drop in temperature and pressure. In comparison to direct injection, this drop leads to a lower pressure difference at the inlet valve, which is unfavorable with regard to the charge; **figure 3.16**.

- The volume of liquid fuel injected is substantially smaller (factor approximately 700) than the volume of gaseous fuel. Based on these considerations, it can be assumed that for direct injection in the intake stroke, as compared to the intake manifold injection, a charge increase and thus a power increase of 5 to 8% is realistic, depending on the prevailing boundary conditions (flow conditions, injection parameters). This corresponds to the improvements (knocking behavior, compression ratio) theoretically estimated in different studies [15].

The 2.2 DIRECT ECOTEC engine by Opel with 114 kW is typical of engines with homogeneous gasoline direct injection [35]. This power unit is based on the 2.2 ECOTEC full aluminum engine with intake manifold injection. Gasoline direct injection works with variable injection pressure (40 to 120 bar) and variable charge movement which is generated by using electrical swirl control valves depending on the load. **Figure 3.17** shows the combustion chamber with the arrangements of the injection valve, spark plug and inlet and exhaust valves. Each cylinder has two separate inlet valves; when one of the ports is closed, an air/fuel mixture swirl is generated in the combustion chamber. This leads to extremely efficient combustion. This way, high exhaust-gas recirculation rates can be achieved that result in less fuel consumption during part load operation. Of the two inlet ports, the right port is designed as a swirl port, whereas the left inlet port is designed as a charge port and provided with a shut-off valve. The higher compression ratio (12:1 as compared to 10:1 in the 2.2 ECOTEC engine) also has a positive effect, which results in better power development. The fuel consumption lies around 6% below that of the 2.2 l intake manifold injection engine, **figure 3.18**, with a higher exhaust-gas recirculation rate being possible through direct injection [42]. At the same time, the engine with direction injection also offers up to 10% more torque at low engine speeds (maximum: 220 Nm at 3800 rpm) and 6% more power.

The Audi/VW 2.0 TFSI engine, in which direct injection and turbocharging are combined, has a displacement of 2.0 l and reaches a maximum torque of 280 Nm and a maximum power of 147 kW. This engine is operated exclusively in the homogeneous operating mode throughout the operating range. Tumble flaps in the inlet port generate an intensified tumble flow and thus increase turbulence in the combustion chamber, which improves the mixture formation. The velocity vectors calculated for the 2.0 FSI (left) and the 2.0 TFSI (right) are illustrated in **figure 3.19**. The supercharged engine achieves a significantly more intensive tumble flow through a flatter port layout and additional masking in the lower port area.

The tumble intensity of the TFSI engine as compared to the tumble intensity of the FSI engine without turbocharging is illustrated in **figure 3.20**. As in the VW FSI engine, the tumble flaps close the lower half of the inlet ports up to an engine speed of approximately 3000 rpm, by which high charge movement and thus a high combustion speed are achieved even in the low engine speed range. These flaps are opened at high engine speeds in order to achieve the required volumetric efficiency.

The fuel is injected using a pressure swirl injector, the maximum injection pressure being 110 bar. Since the stratified operating mode did not need to be taken into consideration due

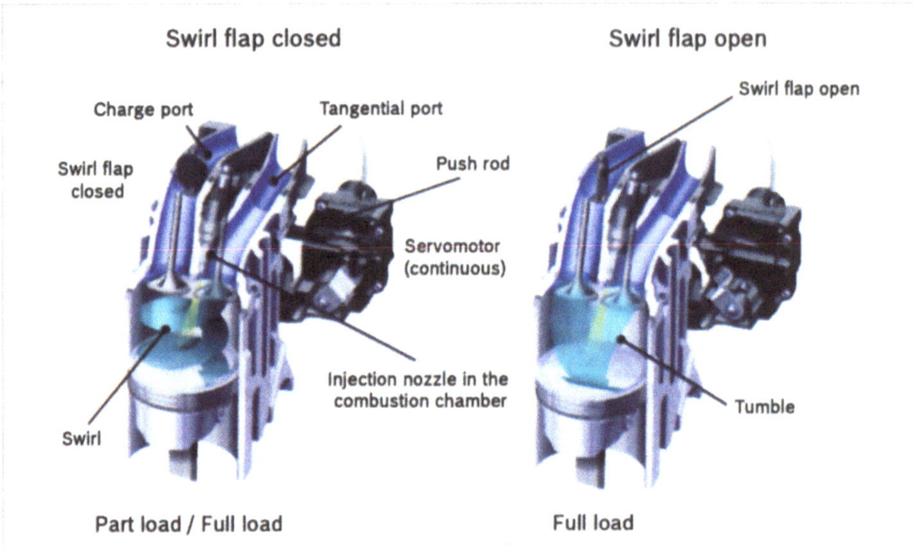


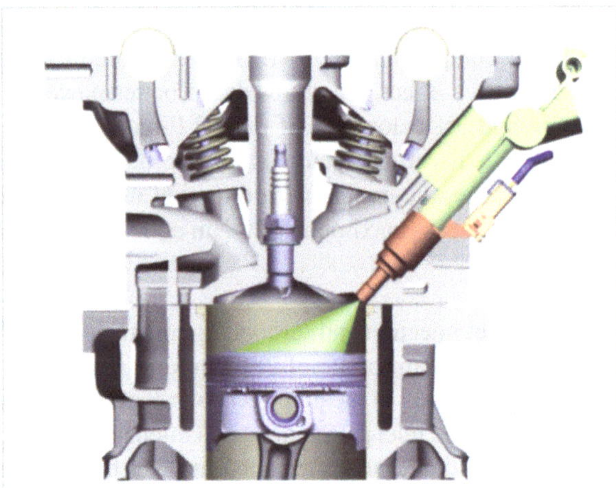

**Figure 3.17:**
Position of injection nozzle and spark plug with valve system in the intake port of the Opel 2.2 DIRECT ECOTEC engine [35]

to the homogeneous concept, an injection spray configuration that enables the best possible homogenization of the air/fuel mixture was selected. This ensures low hydrocarbon emissions during part load operation, high knock resistance at full load and fast complete fuel combustion. The distinct piston recess and lateral piston shoulders stabilize the tumble flow and thus positively influence the combustion [43]. Due to the internal mixture formation through direct injection, the compression ratio of 10.5:1 is relatively high, due to which the efficiency of the engine is also favorable. **Figure 3.21** shows the consumption map as compared to the predecessor engine with turbocharging and intake manifold injection.

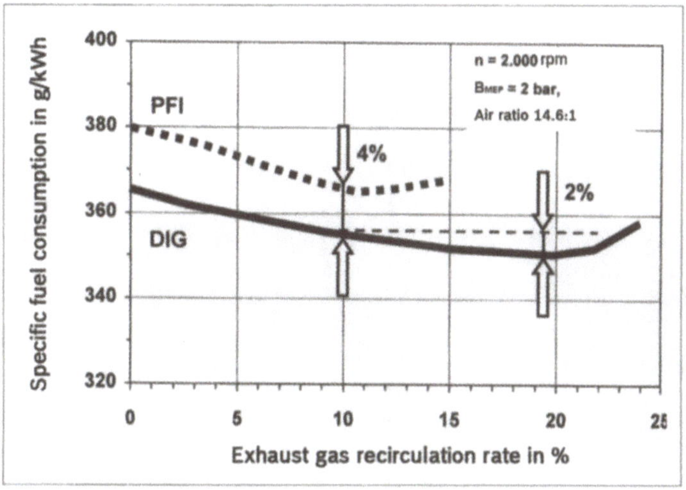


**Figure 3.18:** Opel 2.2 DIRECT ECOTEC: influence of the exhaust-gas recirculation rate on specific consumption in a gasoline engine with direct injection and intake manifold injection [42]

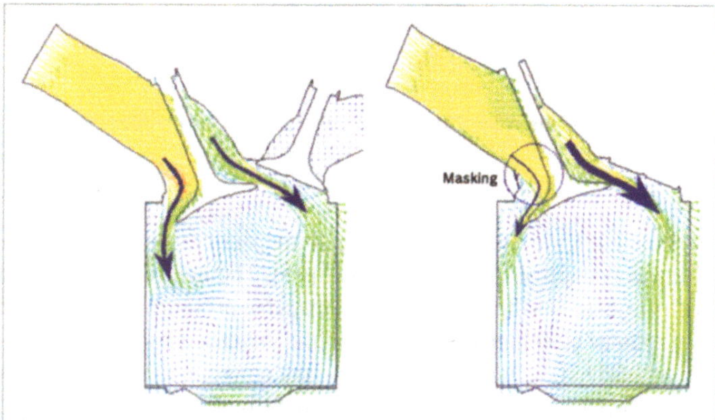


**Figure 3.19:** Velocity vectors of the intake flow for Audi 2.0 FSI (left) and 2.0 TFSI (right) [43]

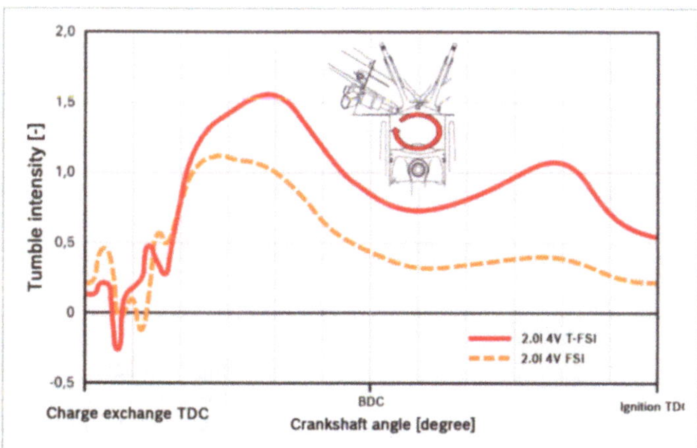


**Figure 3.20:** Unsteady tumble intensity of the TFSI engine as compared to the FSI engine [44]

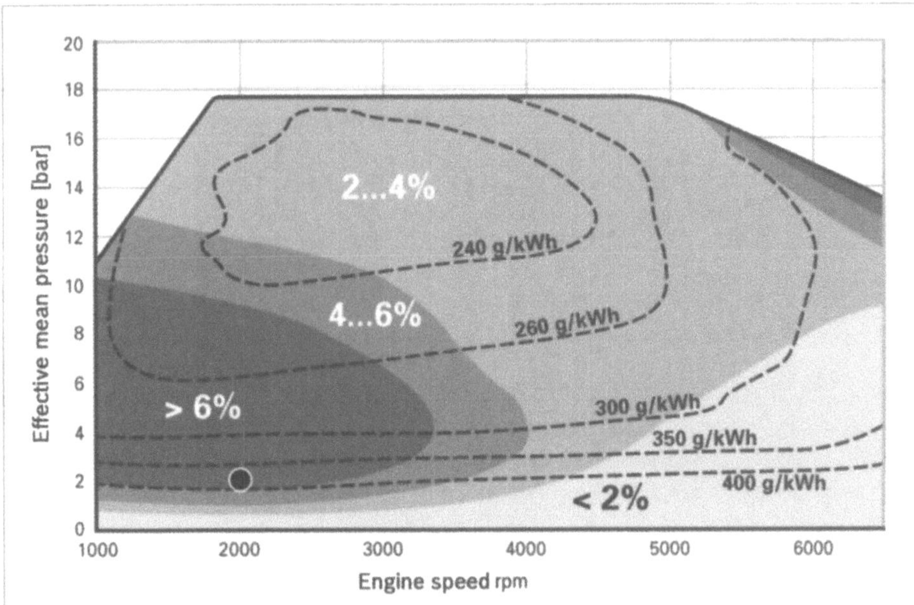


**Figure 3.21:** Specific fuel consumption of the Audi 2.0 TFSI as compared to the predecessor engine 1.8 T (132 kW) [43]

In the present BMW 12-cylinder engine, homogeneous gasoline direct injection is combined with the fully variable valve control VALVETRONIC [45]. The maximum injection pressure is 120 bar. The fuel is injected in the intake stroke through a pressure swirl injector (one-hole nozzle). This way, power is enhanced by 3%, torque increased by 5% and consumption reduced by approximately 10%, while maintaining the conventional exhaust-gas aftertreatment measures and without a requirement for sulfur-free fuels; **figure 3.22.**

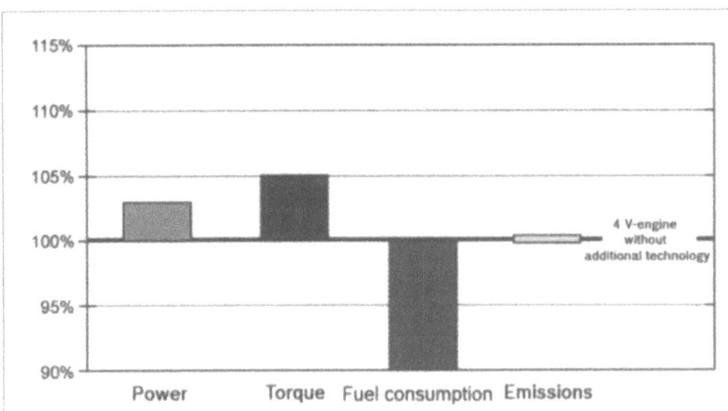


**Figure 3.22:** Comparison of the BMW 6.0 l V12 engine with the predecessor engine [45]

The new Lexus 2GR-FSE V6 engine with the D4-S dual injection concept represents a new development [46]. The engine has a displacement of 3456 ccm, a maximum power of 228 kW at 6400 rpm and a maximum torque of 376 Nm at 4800 rpm[1]. **Figure 3.23** shows a cross-sectional drawing of this engine. The concept is characterized by two injectors. One injector injects the fuel directly into the combustion chamber and the other into the intake manifold. The injection quantity is distributed between the two injectors depending on the load. If 30 to 40% of the injection quantity is directly injected into the combustion chamber, the mixture formation in the part load range is improved. The amount of fuel quantity that is directly injected is increased to 50 to 60% in the upper load range. At full load, 100% of the fuel is directly injected into the combustion chamber. This leads to a higher maximum torque and reduces the risk of knocking combustion. During the exhaust stroke, the intake manifold injector nozzles inject the fuel into the inlet ports as long as the inlet valves are closed. As soon as the inlet valves open, a homogeneous mixture is drawn into the combustion chamber. During the compression phase, the fuel is injected in the direction of the combustion chamber recess and thus a region with high fuel concentration is formed near the spark plug. With a retarded ignition point, the combustion primarily takes place in the expansion phase with correspondingly high exhaust temperatures. This measure also ensures faster heating of the catalytic converter in a shorter period of time.

The main reasons for the development of engines with direct injection and homogeneous mixture formation are, on the one hand, improved power performance and better knocking behavior than with intake manifold injection and, on the other hand, the possibility of meeting international exhaust emissions standards using the conventional exhaust-gas

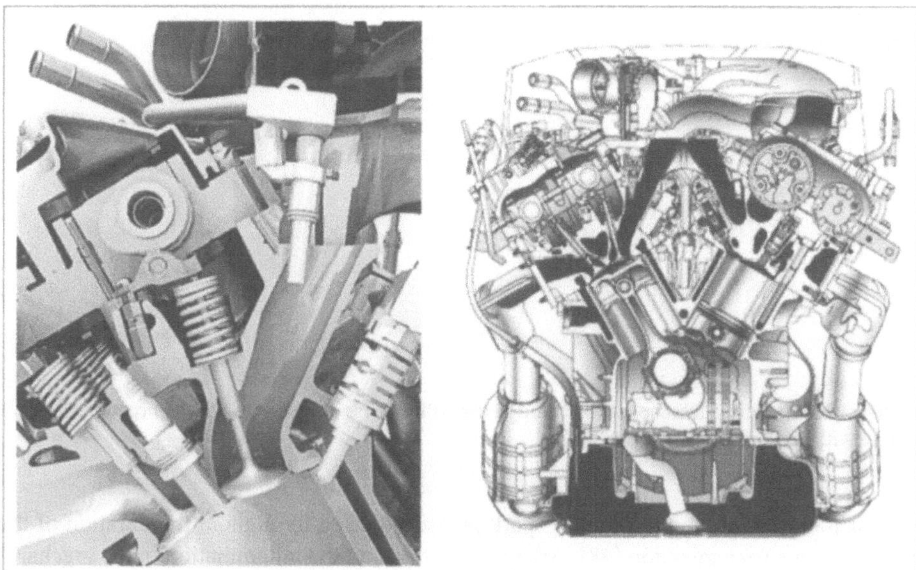

**Figure 3.23:** The Lexus 2GR FSE V6 engine with direct injection and intake manifold injection [Source: Lexus]

aftertreatment technology of the three-way catalytic converter. Thus, costs are reduced because it is not necessary to use the $NO_x$ storage catalyst required for lean operation. At the same time, it is therefore not necessary to have sulfur-free fuel available worldwide.

### BPI Process (*Bowl Prechamber Ignition*)

The BPI process (*Bowl Prechamber Ignition*) is a homogeneous lean combustion process with gasoline direct injection. In this process, stable inflammation is ensured by means of minimum fuel enrichment at the ignition spot (prechamber spark plug) at high global air/fuel ratios [47]. The enrichment is realized by injecting a minimum quantity of fuel directly into the main combustion chamber (less than 5% of the total injected fuel quantity) during the compression stroke. **Figure 3.24** shows the injection method of the BPI process. The fuel is injected into a piston recess through an injection valve at a narrow taper angle. Due to the piston movement, the prechamber spark plug penetrates into the piston recess and takes up the mixture in the recess due to the pressure ratio at the prechamber spark plug.

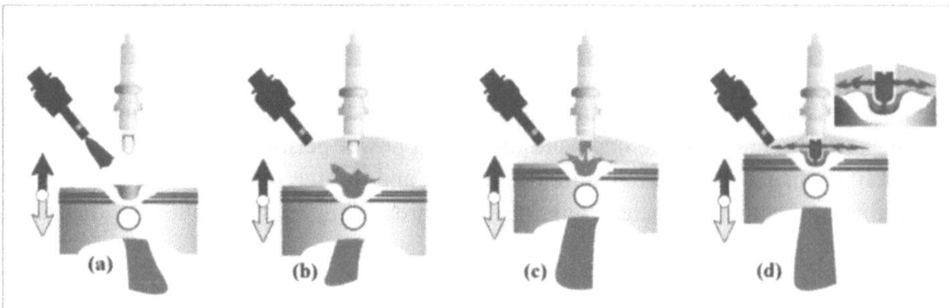


**Figure 3.24:** BPI process for the inflammation of lean mixtures [47]

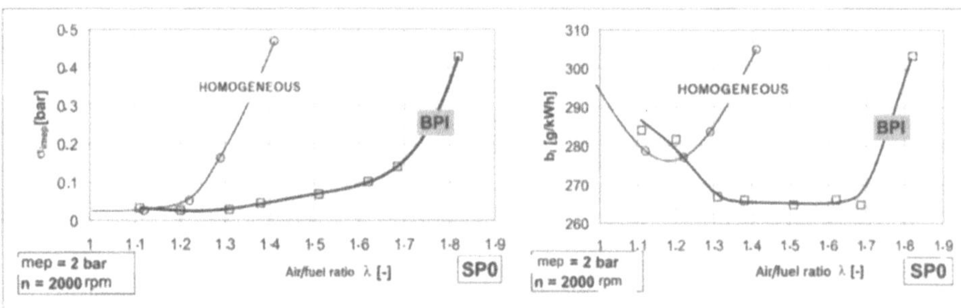


**Figure 3.25:** Extended lean operation through the BPI process as compared to the homogeneous engine with prechamber spark plug [47]

The air/fuel ratio at the prechamber spark plug is adjusted throughout the duration of the compression stroke injection [48]. After the ignition and the inflammation in the prechamber spark plug, the jet flames penetrate the main combustion chamber and ignite the lean mixture. Tests have been carried out in [47] on the inflammation of lean mixtures in the combustion chamber using the BPI process. **Figure 3.25** shows a comparison between extended lean operation in the BPI process and purely homogeneous lean operation with a prechamber spark plug. The BPI process helps separate the inflammation and main combustion whilst using one injection system.

## 3.3 Direct injection with stratified mixtures

As described in the basics of mixture formation, part load operation of a gasoline engine with direct injection with stratified mixture is a useful way of improving the process efficiency. Here, the fuel is injected during or only towards the end of the compression phase. This makes it possible to ensure stratification of the charge in the combustion chamber. An ignitable mixture is required in the vicinity of the spark plug at the point of ignition. In the area farther away from the spark plug, either a very lean but still combustible mixture or pure air or pure exhaust gas through systematic exhaust-gas recirculation should be available.

A relatively rich mixture is present in the vicinity of the spark plug ($\lambda \approx 1.0$) and a very lean mixture can be found in the wall area of the combustion chamber. Graded air/fuel ratios $\lambda$ in the range from 3.0 to 5.0 are generated for an ideal stratified charge in the test-relevant low part load. If a realistic air/fuel ratio $\lambda = 4.0$ is assumed for a typical part load point, significant efficiency advantages can be seen as compared to operation with stoichiometric air/fuel ratio and throttle control. These advantages can be illustrated by means of an ideal cycle model for the gasoline engine (constant volume cycle); figure 3.26. The continuous state characteristic curves shown in the figure result from the state conditions in the engine with intake manifold injection ($\lambda = 1$) for throttled operation in the part load. For the unthrottled mode with direct injection ($\lambda > 1$), the broken state characteristic curves with steep compression and expansion lines are obtained for an identical load point. A comparison of the heat quantities in the T-s diagram clearly shows that the heat loss during intake manifold injection is significantly higher than that during direct injection for the same heat quantity supplied (variables of the hatched areas in the *T-s* diagram). This shows that, due to the capability of achieving leaner mixtures, the theoretical efficiency of direct injection should be significantly higher than that of intake manifold injection.

The effects of these advantages of direct injection in gasoline engines as compared to the conventional concepts can be summarized thermodynamically as follows [49, 50, 51]. The efficiency $\eta_e$ of the actual engine process is obtained from the internal efficiency of the ideal process $\eta_V$ (constant volume combustion) and the different individual parasitic losses:

- Losses due to non-ideal combustion ($\Delta\eta_{BV}$)
- Losses due to leakage ($\Delta\eta_U$)
- Heat losses ($\Delta\eta_K$)

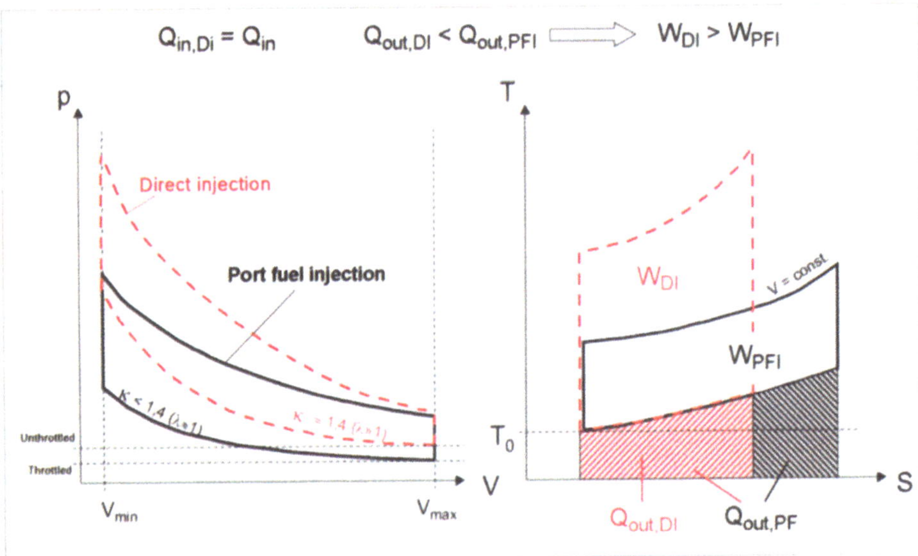


**Figure 3.26:** Efficiency increase through direct injection (constant volume cycle): $Q_{in,DI}$: supplied heat, direct injection, $Q_{in}$: supplied heat, intake manifold injection, $Q_{out,DI}$: dissipated heat, direct injection, $Q_{out,DI}$: dissipated heat, intake manifold injection, $W_{DI}$: work output, direct injection, $W_{PFI}$: work output, port fuel injection

- Gas exchange losses ($\Delta\eta_{LW}$)
- Frictional losses ($\Delta\eta_R$).

Thus, the efficiency $\eta_e$ of the actual engine process is:

$$\eta_e = \eta_V - \Delta\eta_{BV} - \Delta\eta_U - \Delta\eta_K - \Delta\eta_{LW} - \Delta\eta_R$$

Direct injection strongly influences as many as four of the variables mentioned in the equation ($\eta_V$, $\Delta\eta_{LW}$, $\Delta\eta_K$, $\Delta\eta_R$). The heat from the ambient air is absorbed due to the evaporation of fuel in the combustion chamber. This leads to a drop in temperature and pressure. With the increasing capability of achieving leaner mixtures of the mixture, the isentropic exponent also rises due to the increased amount of inert gas and the consequent drop in temperature, and approximates the value for pure air.

The gas exchange losses $\Delta\eta_{LW}$ can be reduced considerably by the ideally unthrottled intake process. **Figure 3.27** clearly shows the decrease of the gas exchange loop in the *p-V* diagram in a comparison between a conventional gasoline engine and a gasoline engine with direct injection [50, 52].

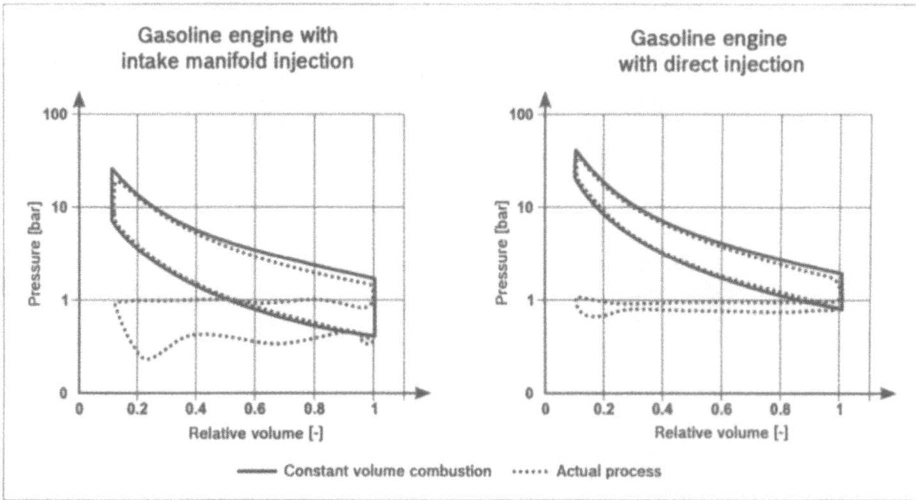


**Figure 3.27:** Comparison of combustion in the actual engine process between external and internal mixture formation [50, 52]

As mentioned earlier, the wall heat losses $\Delta\eta_K$ can be reduced in the ideal stratified operating mode by means of an isolating layer of inert gas, irrespective of whether the layer is made of residual gas, recirculated exhaust gas or air.

In addition to the positive characteristics, there are the disadvantages of increased mechanical losses $\Delta\eta_R$ through the additional operation of a high pressure pump as well as the increased piston ring friction resulting from the increased charge mass in the unthrottled mode and the related higher cylinder pressures. Furthermore, a highly complicated exhaust-gas aftertreatment strategy is required, as the conventional three-way catalytic converters lose their effect due to the permanent excess air in the stratified operating mode. This last point makes it extremely difficult to realize the thermodynamic advantages in practice.

It is particularly important for combustion in gasoline engines with direct injection that an ignitable mixture is present in the vicinity of the spark plug at ignition point irrespective of the engine load and speed. This imposes special requirements on a range of parameters. In particular the intake flow, the shape of the combustion chamber (cylinder head type and piston type), the compression ratio, the position of the injection nozzle and the spark plug, and the injection parameters (injection point, duration, pressure, spray geometry) must be considered. A clear understanding of the processes taking place during gasoline direct injection and the related precise coordination of mixture formation, ignition and combustion requires an enormous research effort. Therefore, the development of gasoline engines with direct injection which are ready for series production has only just begun.

There are various methods of fuel insertion and the related mixture formation to implement the stratified charge mode in a gasoline engine with direct injection. On examining the different concepts, three types of mixture formation processes emerge, which differ in their characteristics with regard to how the fuel is guided from the injection nozzle to the spark plug; **figure 3.28.**

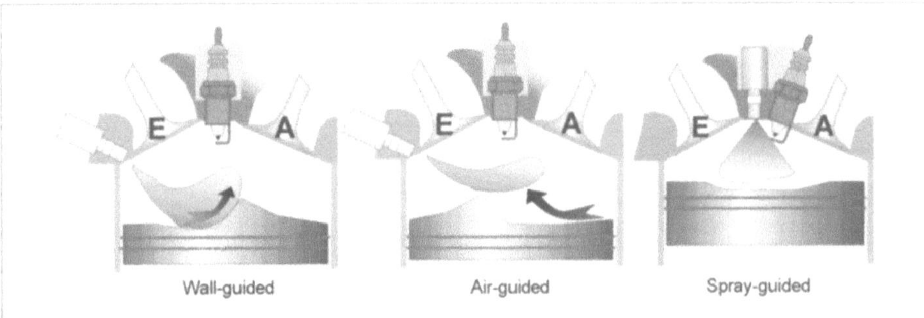


**Figure 3.28:** Classification of the combustion processes for gasoline engines with direct injection [53]

### 3.3.1 Wall-guided combustion process

Most of the stratified charge engines of the first generation currently in the market are based on the wall-guided combustion process, which features a relatively large spatial distance between the spark plug and the injection valve. The mixture formation process and transportation of the mixture to the spark plug take place through an interaction between the injected spray and the combustion chamber wall. The fuel spray is usually guided towards the spark plug, supported by a swirl or tumble flow and by specifically designing the piston surface. As the injected spray is guided directly to the piston in this combustion process, there is increased fuel accumulation and increased emissions of uncombusted hydrocarbons [51]. In addition, the injection timing is directly related to the piston movement and thus to the engine speed. The transport of the mixture cloud across the relatively long distance from the injection valve to the spark plug requires specific, stable charge movement at each engine speed and precise determination of the injection and ignition points for stabilizing the fuel vapor. The wall-guided processes realized so far, however, were not able to fully exploit the theoretical potential of direct injection for reducing fuel consumption. The thermodynamic advantages explained in the previous subchapter are not used consistently in wall-guided processes, for a variety of reasons:

- Distinct intake flow required: due to the large distance between the injection nozzle and the spark plug, the intake flow has to transport the mixture cloud over a long distance and therefore for a long period of time. A strong swirl or tumble movement at each engine speed is therefore as indispensable as the precise determination of injection and ignition points.

- Hydrocarbon deposits on the cylinder wall: by systematically guiding the fuel along the cylinder wall, it is not just at cold start and in the warm-up phase that a pronounced wall film may be formed which evaporates only partially and participates in the combustion. Under unfavorable conditions, for example, (cold engine, cold intake air), the injected fuel may not evaporate completely until the ignition point is reached and may be present in the form of fuel droplets. Subsequently, rich combustion takes place in these zones. This can result in the increased emissions of unburnt hydrocarbons. On the other hand, the injected fuel can condense on the cold combustion chamber walls or fuel droplets can form a wall film on impact with the combustion chamber walls. Due to the cold walls or the significantly reduced fuel surface, this fuel evaporates only after the combustion has already started. It is most obvious that the combustion first takes place in the lean range and then in the rich range once the wetted combustion chamber walls are reached and as a result, additional pollutants are generated.
- Hydrocarbon deposits on the piston crown: the direct accumulation of fuel on the piston leads to rich combustion in the piston crown area. The incomplete combustion results in soot formation and deposits on the piston, which are not completely burnt, due to the low combustion chamber temperatures present during part load. During the subsequent injection, these residues absorb the injected fuel like a sponge. The result is increased emissions of unburnt hydrocarbons.
- Fuel in squish area: during the transition from low to medium part load, more fuel is injected. In terms of an optimum ignition point in terms of efficiency, this means an advancement of the injection point. However, it can happen that the injected fuel seeps into the squish band through the recess rim. As the fuel here does not take part in the combustion, it causes increased emissions of unburnt hydrocarbons.
- Sooting of the inlet valves during part load, since the cleansing effect of the fuel injected in the intake manifold in gasoline engines with external mixture formation is absent.

**Figure 3.29** shows the operating map and the control behavior of a gasoline engine with direct injection and a wall-guided combustion process [53]. The map range in which a stable stratified operating mode is possible is limited to low loads and low engine speeds by the mixture formation characteristics of the injection valves used and the dependency of the injection on the piston movement. Thus the fuel saving potential is limited to these map areas which, moreover, are not necessarily relevant to the customer. Due to the high NOx raw emissions in the highly enleaned stratified charge mode, this combustion process does not exploit the possible stratification potential either and cannot prevent heat losses through the walls. In addition, to achieve stable combustion without misfires, unthrottled operation cannot always be implemented. The efficiency of the high pressure process is far from the theoretical ideal. This is due to the considerable losses caused by incomplete combustion of the fuel film formed on the piston and due to the overly advanced position of the point with 50% converted fuel mass (thermodynamic optimum is approximately 6 to 8° crankshaft angle after the top dead center). Furthermore, in the actual driving operation, there are additional losses from the exhaust-gas aftertreatment in the stratified operating mode, so that throttling has to take place at the relatively low load of approximately 3.5 bar *mep*; **figure 3.29** right. For this reason, the stratified operating mode has to be ended and

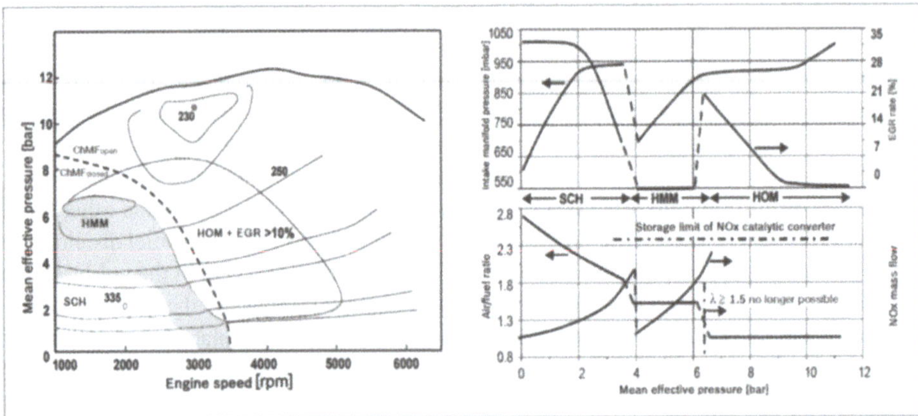


**Figure 3.29:** Map of operating modes (left) and control behavior at 2000 rpm (right) of a gasoline engine with direct injection and wall-air-guided combustion process [3]; $ChMF_{closed}$: charge movement flap open, $ChMF_{closed}$: charge movement flap closed, SCH: stratified operating mode, HMM: homogenous lean operating mode, HOM: homogeneous operating mode, EGR: with exhaust-gas recirculation

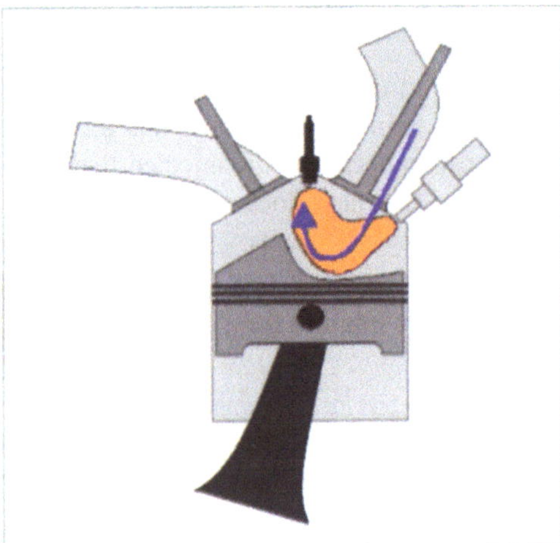

**Figure 3.30:**
Schematic diagram of the Mitsubishi GDI process [55]

a homogeneous lean operation with high exhaust-gas recirculation with the related higher fuel consumption is required.

The Mitsubishi GDI process (*Gasoline Direct Injection*), **figure 3.30**, the HPI process (*High Pressure Direct Injection*) by PSA Peugeot Citroën and the FSI process (*Fuel Stratified Injection*) by Volkswagen are examples of the wall-guided combustion processes currently available on the market. The combustion chamber of the Mitsubishi GDI engine, which was presented at the International Motor Show in Frankfurt in 1995, has been designed as a roof shaped combustion chamber with four valves [55]. An injection nozzle positioned on the cylinder rim between the two inlet valves injects the fuel centrally at a pressure of 50 bar (liquid high pressure injection) into a specially modeled recess of a deflector type/bowl type piston. At part load (mean air/fuel equivalence ratio $\lambda = 2.0$ to 2.8), where the fuel is injected during the compression phase (at the latest at a 90° CA before top dead center), a reversed tumble flow is generated using a steeply downward sloping inlet port and the special piston shape (opposite to the usual tumble flows). It stratifies the fuel near the spark plug, which is mounted centrally in the combustion chamber. At full load, injection takes place early, in the intake stroke; the mixture ($\lambda = 0.9$ to 1.0) is almost homogeneous. The compression ratio $\varepsilon$ is defined as 12.0.

The components relevant to the VW FSI combustion process are illustrated in **figure 3.31** left [3]. It includes a bowl type piston which has a "fuel recess" on the intake side and a "flow recess" on the exhaust side (not visible in the cross-sectional view). The inlet port is a charge port with tumble characteristics in its base geometry and is divided into an upper and a lower half by a cast-in plate; **figure 3.32**. The lower half can be closed by means of a flap depending on the operating mode [56]. Higher flow velocities and thus intensified tumble flow are achieved by closing the lower half of the port, e.g. in the stratified charge mode. Since the fuel spray is transported and stabilized partly by the tumble flow and partly by the piston recess, Volkswagen talks of a combined wall/air-guided process. The effect of the individual measures on fuel consumption as compared to the basic engine with intake manifold injection is

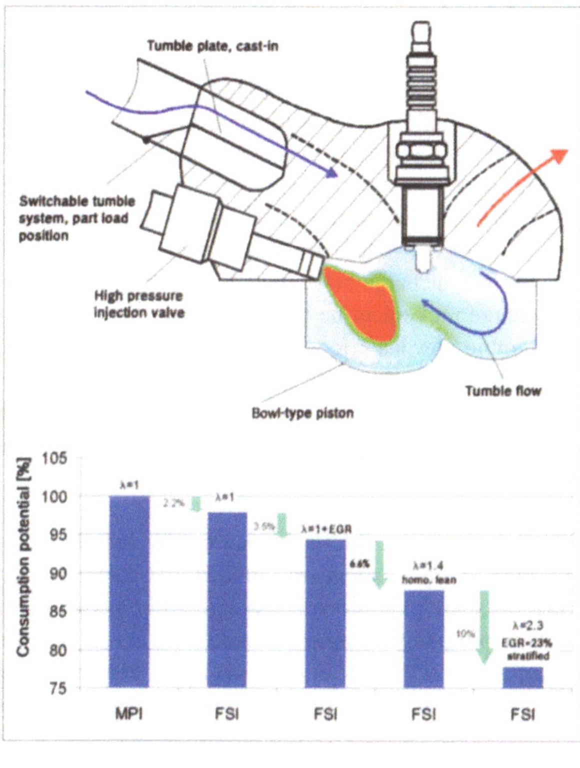


**Figure 3.31:**
Schematic diagram of the VW FSI combustion process and the fuel saving at n = 2000 rpm, *mep* = 2 bar [source: VW]

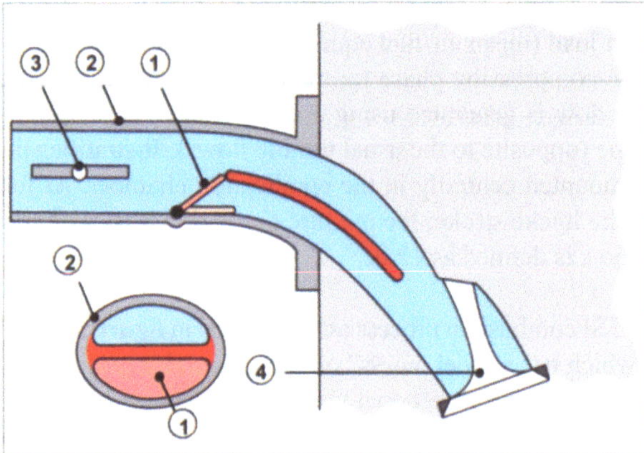


**Figure 3.32:**
Controllable charge movement in the intake manifold
1 Charge movement flap
2 Intake manifold
3 Throttle valve
4 Intake valve

shown in **figure 3.31** right for an operating point at 2000 rpm and a mean effective pressure of 2 bar. The sum of all measures results in fuel savings of 21% at this operating point.

## 3.3.2 Air-guided combustion processes

In air-guided processes, the fuel is transported to the spark plug solely by the charge movement generated on the intake side, air being simultaneously mixed with the spray. The charge movement is also supported by a specifically shaped piston. In contrast to wall-guided combustion processes, contact between the fuel and the combustion chamber walls should be avoided. Therefore, there is ideally no fuel accumulation on the combustion chamber wall. A thorough mixing of fuel and intake air is intended in parallel with charge transport. The successful implementation of this process depends on the alignment of the injection spray and the generation of a directed charge movement. It is particularly important in this context that the directed charge movement be maintained well into the compression phase. The required swirl or tumble flow leads to volumetric efficiency losses and thereby to disadvantages in the performance. Currently, only one purely air-guided process is available on the market [57, 58], though all wall-guided processes more or less make pronounced use of airflows for mixture guidance and are therefore classified as a mixture of processes, i.e. as wall/air-guided processes.

Air-guided processes, as dealt with in more recent studies carried out by individual research centers, are a kind of compromise solution. As in wall-guided processes, the distance between the injection nozzle and the spark plug is relatively large. Attempts are made to transport the injected fuel to the spark plug only by means of a directed intake flow. In contrast to a wall-guided process, there is therefore ideally no fuel accumulation on the combustion chamber wall. A thorough mixing of fuel and intake air is intended in parallel with the charge transport.

The development of an air-guided process proved to be especially difficult, as both the problems of the spray-guided process and those of the wall-guided process are to be solved at the same time. **Figure 3.33** shows the 2.0 l FSI engine of Audi which, due to its tendency, can be classified as an air-guided process, with an open charge flap in the homogeneous operating mode and a closed charge flap in the stratified charge mode [58]. The operating ranges in the map are illustrated in **figure 3.34**. This Audi engine shows a consumption

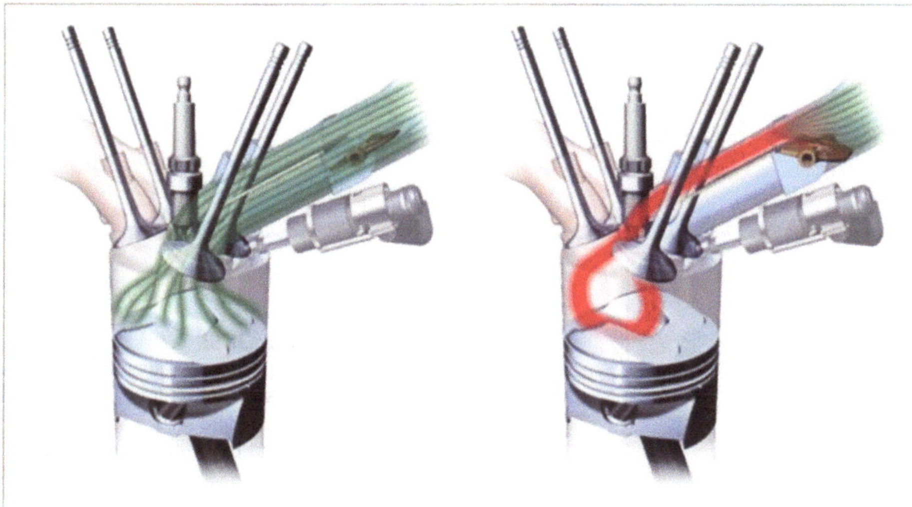

**Figure 3.33:** Operating modes of the Audi 2.0 l FSI engine: left: homogeneous operating mode; right: stratified charge mode [Source: Audi AG]

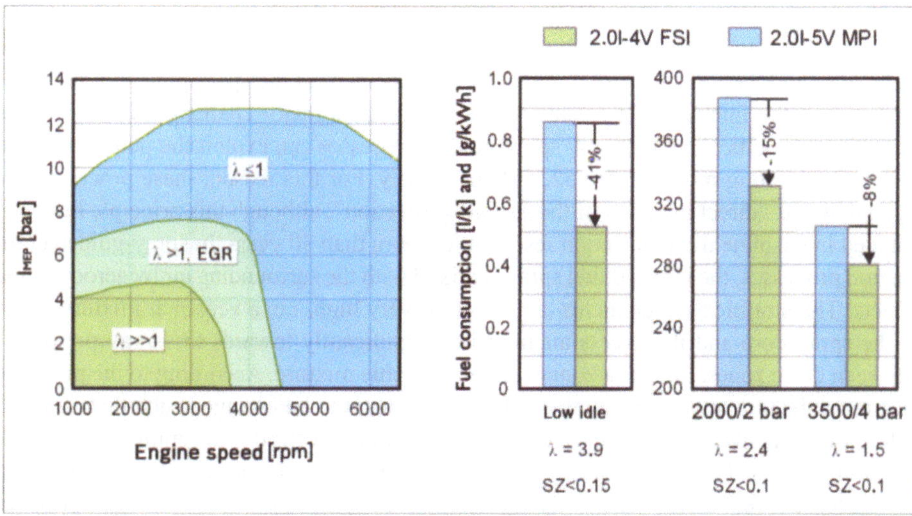


**Figure 3.34:** Operating ranges in the map and consumption potential of the Audi 2.0 l FSI engine

advantage of 41% during low idle as compared to its predecessor with intake manifold injection (2.0 l, 5V engine) with an enleanment to the air/fuel ratio $\lambda = 3.9$.

**Figure 3.35** shows a comparison of the cumulative heat release for the Audi FSI engine and the engine with intake manifold injection. What can be clearly seen for the FSI engine is the rather advanced and thus thermodynamically unfavorable center of combustion (50% conversion) at 5° CA before the top dead center. The center of combustion of the FSI engine can be shifted towards the top dead center (approximately 0° CA) by means of exhaust-gas recirculation.

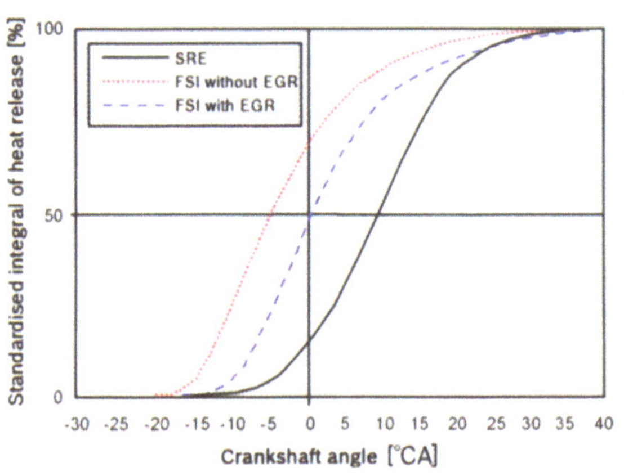


**Figure 3.35:**
Comparison of the cumulative heat release: Audi with intake manifold injection and Audi FSI with and without EGR (exhaust-gas recirculation)

### 3.3.3 Spray-guided combustion processes

Spray-guided combustion processes are characterized by smaller distances between the spark plug and the injection valve. The close arrangement of the spark plug and the injector between the inlet and the exhaust valve represents a major challenge [2, 49]. At the same time, experts agree that only spray-guided processes can exploit the potential of the stratified operating mode [2, 50, 52, 59] completely. For this reason, these processes are also known as "direct injection of the second generation", although this principle has been pursued in gasoline direct injection research for more than 30 years. In spray-guided combustion processes, the injected fuel spray is mixed with the surrounding air by aerodynamic effects. The stratification gradients are therefore very high, i.e. a very rich mixture exists in the spray core and the excess air increases significantly towards the spray periphery. Between these zones, there is a zone with the ignitable mixture. According to the principle of direct injection in gasoline engines, the spark plug is to be arranged such that an ignitable mixture is available in the vicinity of the ignition electrodes at ignition point at all operating points. The process largely depends on the spray characteristic of the injection valve used and reacts very sensitively to interferences or fluctuations in this regard. Known problems of the spray-guided process are:

- Carbon deposit formation on the injection nozzle (effect on spray pattern): due to the low combustion temperatures at part load and low idle, carbon deposits can be formed on the injection nozzle and can influence the spray pattern considerably.
- Tolerances within the spray pattern of the injection nozzles: manufacturing tolerances and operating states, for certain injection valves, strongly affect the spray pattern and thus the combustion process of the engine.
- Carbon deposits on the spark plug (ignition misfires): the low combustion temperatures at part load and low idle mode can also lead to carbon deposit formation on the spark plug. Ignition misfires are a direct consequence.
- High thermal load on the spark plug: during injection, liquid fuel comes in direct contact with the spark plug still hot from the previous combustion and evaporates quickly. This results in a very strong cooling of the spark plug (thermal shock resistance).
- Inadequate mixture formation as a result of retarded injection: in this process, injection and ignition are strongly bound to each other chronologically and are only a few ° CA away from each other. In order to have a favorable ignition point in terms of efficiency, the injection is retarded; as a consequence, there is very little time left for mixture formation.
- Inadequate mixture formation at low engine speeds: at low engine speeds, the relative velocity between the intake flow and the injection spray can be low. The fuel droplets are bigger and therefore evaporate with more difficulty. Together with the retarded injection (see previous point), an altogether too rich mixture results.
- Mixture cloud drifts at higher engine speeds: at higher engine speeds, the velocity of the intake flow and thus the relative velocity between the fresh charge and the fuel spray increases. The fuel droplets are smaller and therefore lighter. With the higher flow velocity of the fresh charge, stable stratification in the vicinity of the spark plug may not be achieved, due to drifting of the fuel droplets.

The requirements for a future stratified combustion process (mentioned below) arise from the problems mentioned and the basics of thermodynamics, keeping in mind the optimal utilization of the theoretically possible efficiency potentials [2, 50, 59]:

- A high, steadily reproducible spray quality of the injection valve in a large engine speed and load range in order to be able to achieve complete combustion with low CO and hydrocarbon emissions. The stratification range to be achieved must be completely congruent with the customer map range for the relevant customer. The technologies currently available on the market have different consumption potentials, some of which are at the threshold that can be experienced in practice [60].
- While designing the combustion chamber, the focus of development should be on wetting the combustion chamber surface with fuel as little as possible.
- High stability of the combustion process with suitable mixture quality even for retarded injection points, in order to ensure that the center of combustion can be shifted to the most favorable position in terms of thermodynamics. The NOx raw emissions must be taken into consideration.

- High flexibility of the injection system used, with regard to the injection timings and the possibility of multiple injections. This is necessary in order to be able to influence the mixture composition near the spark plug. Furthermore, the injector should be non-sensitive to contamination and have a high long-term stability with regard to spray behavior.
- Use of the highest possible injection pressures in order to keep the mixture formation duration as short as possible during retarded injection. Thermodynamically optimal centers of combustion can be achieved only by the use of increased injection pressures together with single injection or, if required, with multiple injection. In addition, a reduction of the particle emissions can be achieved by faster evaporation of the smaller fuel droplets. In this process, the stratification range is no longer limited by emissions values (especially hydrocarbon and particle emissions), but is determined by the resultant air/fuel equivalence ratio $\lambda$ and thus by the consumption advantage as compared to the homogeneous operating mode [2, 4 and 60].
- Application of a powerful, robust, variable ignition system which meets the special requirements of the spray-guided process. The spark plug should be resistant to wear and thermal shock.

Some of the requirements mentioned above need to be taken into consideration early on, while deciding on the position of the spark plug and injection valve. Modern gasoline engines are mostly designed as four-valve engines in order to avoid flow losses and to increase the charge at high loads. This places certain demands on the arrangement of the spark plug and the injection valve. **Figure 3.36** gives an evaluated overview of the possible mounting positions [53]. The aim of the following list is to explain in detail the specific advantages and disadvantages of the arrangements.

### Transverse arrangement

In this arrangement, which is derived from the wall-guided and wall/air-guided processes, the injection valve is located above the inlet ports in the cylinder head. It is quite easy to implement, as both the spark plug and the injection valve are accessible from one of the longitudinal sides of the engine. The advantage of this variant is that the injector is on the cooler intake side and thus the thermal load on it is significantly lower. Furthermore, this arrangement prevents the formation of a fuel vapor lock at extreme thermal conditions. Accordingly, the spark plug which is already thermally highly loaded, must be oriented in the direction of the exhaust valves, which is the first disadvantage of this arrangement in view of component loading. The positioning of the spark plug close to the exhaust side, however, reduces the knocking tendency at full load. Due to the limited space in modern multiple-valve engines, it is not easily possible to scale the cooling ducts to the center of the combustion chamber. This leads to a further increase of the thermal loading on the components. Furthermore, there could be more restrictions in the inlets and exhaust port areas, as the components must be accessible from the outside.

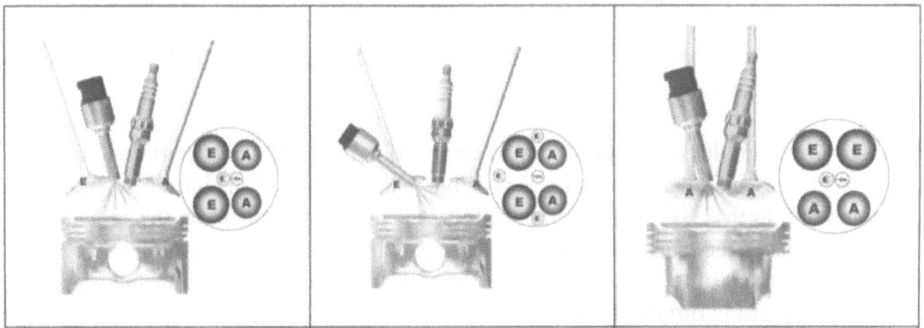

| Transverse arrangement | Distant arrangement | Longitudinal arrangement |
|---|---|---|
| + good injector cooling | + good spark plug cooling | + good compromise with respect to thermal load on spark plug and injector |
| + simple arrangement | + simple construction | + no influence on the construction of the inlet port |
| – longitudinal water cooling difficult | – longitudinal water cooling difficult | + longitudinal as well as transverse water cooling possible |
| – restricted inlet or exhaust port | – restricted inlet or exhaust port | – arrangement difficult (fuel rail / ignition coils) |
| – high thermal load on spark plug | – high thermal load on injector (rapid formation of deposits) | = Injector<br>= Spark plug<br>E = Inlet valve<br>A = Exhaust valve |

+ very good, – less good

**Figure 3.36:** Respective advantages and disadvantages of spark plug and injection valve mounting options in the combustion chamber of gasoline engines with direct injection and spray-guided combustion process [53]

### Distant arrangement

This arrangement is a rather exotic variant and not actually a real spray-guided process. Nevertheless it is mentioned in literature and is termed as such. It is a relatively simple design with the cylinder heads used for air-guided and wall-guided processes converted into a spray-guided process by using a modified injection valve. Due to the great distance between the spark plug and the injector, this arrangement does not involve any significant advantages for the process. It is questionable whether such an arrangement would actually work in a spray-guided process.

### Longitudinal arrangement

Longitudinal arrangement is the best compromise with respect to the thermal load. The load on spark plug and injector is balanced equally without the risk of overloading either.

Moreover, no restrictions are placed on the inlet or exhaust ports. However, the arrangement becomes quite challenging when the packaging of ignition coils and fuel supplies are taken into consideration.

To successfully implement a spray-guided combustion process, it is important to create an ignitable mixture cloud as compactly as possible in the vicinity of the ignition electrodes at ignition point; **figure 3.37**. However, this process is highly dependent on the spray characteristic and reacts very sensitively to disturbances, which can lead to ignition misfires. The spatial arrangement of the injection nozzle, the spark plug and the injection spray propagation is therefore of enormous importance.

**Figure 3.38** shows a comparison between intake manifold injection and spray-guided gasoline direct injection in stratified charge mode at an operating point with the engine speed $n = 2000$ rpm and an induced mean effective pressure *imep* = 3 bar [15]. For direct injection, the fuel consumption and the $NO_x$ emissions are better by approximately 23% or approximately 40% than in the case of intake manifold injection. The hydrocarbon emissions, in contrast, are up to 65% higher than the values for the engine with intake manifold injection.

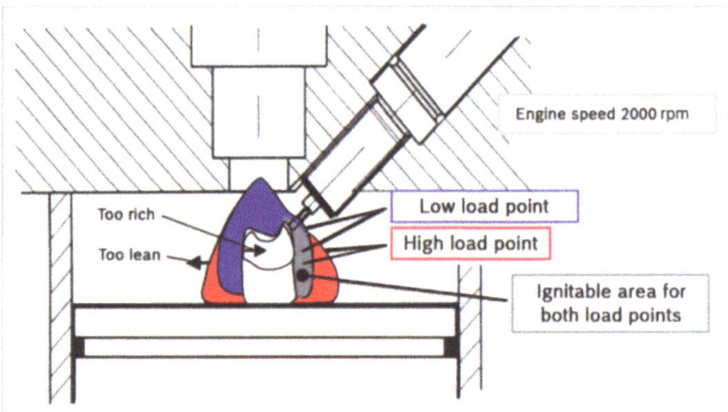


**Figure 3.37:** Spray-guided process, ignitable area

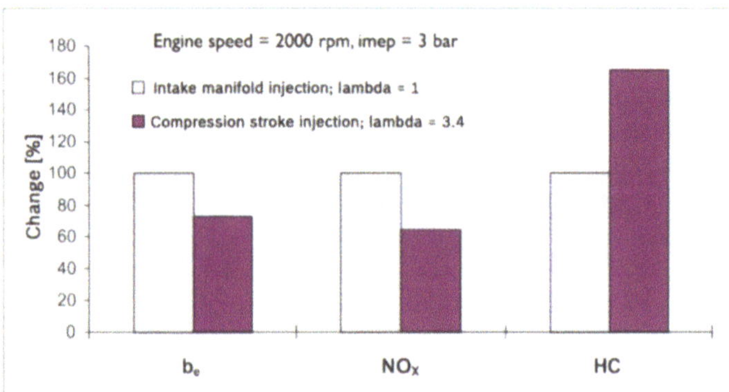


**Figure 3.38:** Comparison between processes with direct injection and intake manifold injection [15]

**Figure 3.39** shows the relevant cylinder pressure curves obtained by indexing, in a *p-V* diagram and the resultant energy conversions (cumulative heat release). Due to the higher air mass (throttle-free operation), the gasoline engine with direct injection achieves significantly higher peak pressures as compared to intake manifold injection. The low gas exchange losses in the operation mode with direct injection are hardly noticeable in the figure. While the heat release curve of intake manifold injection demonstrates characteristic behavior with an optimal 50% conversion point at approximately 8° CA after the top dead center in terms of efficiency, direct injection shows a completely different behavior. Due to the relatively early injection in the compression phase, the ignition point is also advanced as opposed to intake manifold injection, in spite of a clearly shorter combustion delay (caused by the rich zones in the vicinity of the spark plug during stratified charge). This leads to a shift of the 50% conversion point to a position unfavorable for efficiency. Retarded injection is not possible due to the strongly increasing soot formation as a result of inadequate fuel atomization and evaporation. In addition, thermodynamically unfavorable, slow fuel conversion also occurs towards the end of combustion (caused by the lean zones during stratified charge in the end gas area) [15].

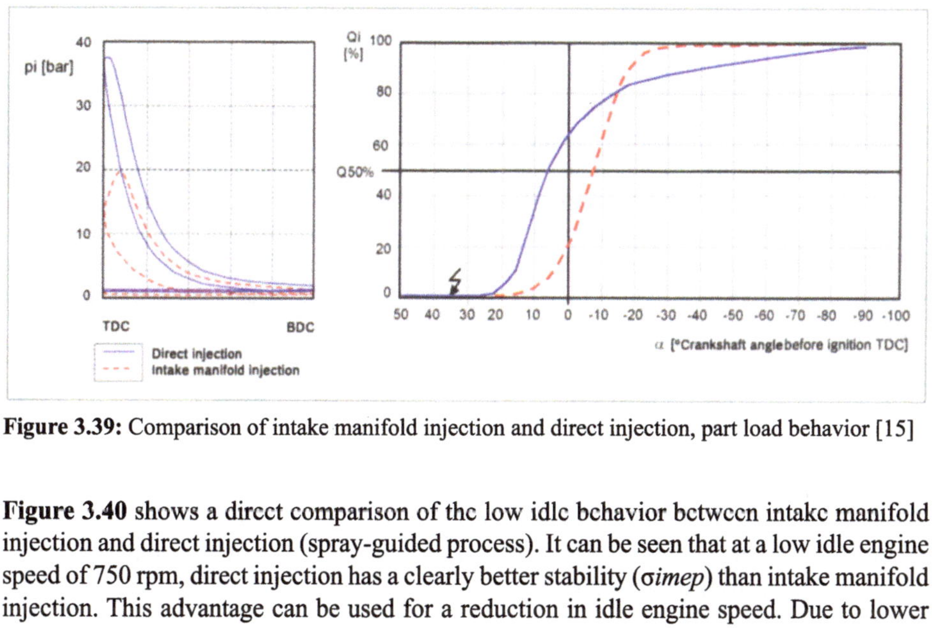


**Figure 3.39:** Comparison of intake manifold injection and direct injection, part load behavior [15]

**Figure 3.40** shows a dircct comparison of thc low idlc bchavior bctwccn intakc manifold injection and direct injection (spray-guided process). It can be seen that at a low idle engine speed of 750 rpm, direct injection has a clearly better stability ($\sigma imep$) than intake manifold injection. This advantage can be used for a reduction in idle engine speed. Due to lower heat losses through the wall, but in particular due to the elimination of high gas exchange losses, a fuel consumption advantage of 58%, as compared to intake manifold injection, is obtained together with the minimization of frictional losses when the engine speed is reduced ($\lambda$ increases). Since only a little fuel needs to be injected in the low idle operation, the injection is very retarded. Together with the low flow velocities of the fresh charge, good stratification can be achieved and hydrocarbon emissions are reduced too.

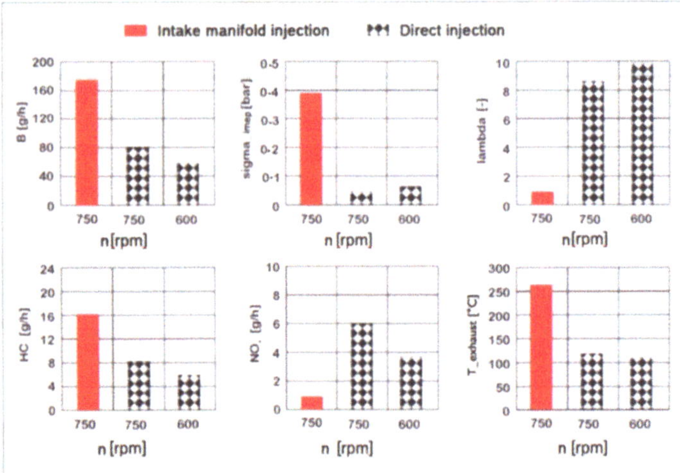


**Figure 3.40:** Comparison of intake manifold injection and direct injection, low idle behavior [15]

Research into the lean homogeneous operating mode and the stratified charge mode shows that the total efficiency gain in the stratified charge mode is achieved in today's engines with direct injection largely due to the reduced throttle losses during the intake phase. During the combustion phase, the homogeneously operated engine with intake manifold injection can even be preferable as the center of combustion is more favorable. As in the case of intake manifold injection, the adjustment of the ignition point plays a crucial role in direct injection, especially in the stratified charge mode. The injection point adjustment is of greater importance for a spray-guided combustion process than for a wall-guided combustion process, as the mixture cloud is usually much more compact in the spray-guided process than in the wall-guided process. However, the same general principles apply for both types.

Potential fuel savings for future concepts with direct injection in comparison with intake manifold injection depend on load and engine speed due to the different operating strategies (stratified charge, homogenization) required in the engine map. **Figure 3.41** shows the extent of fuel consumption reduction with optimum utilization of the possibilities offered by direct injection in gasoline engines, which can be achieved using an optimized spray-guided or wall-guided process (values in brackets) in comparison with a conventional engine with intake manifold injection.

Fuel savings of up to 5% are possible in full load operation. This is mainly due to the charge increase due to internal evaporation. In low idle operation, fuel savings with extremely stratified charge, which is realized by using the spray-guided process, can exceed 50%. According to estimations or comparative studies on engines with intake manifold injection, these fuel savings are mainly achieved through the reduction of the gas exchange losses as a result of the unthrottled operation and the large amount of excess air. In the medium engine speed range, fuel savings of 10 to 25% seem realistic, depending on the combustion process selected.

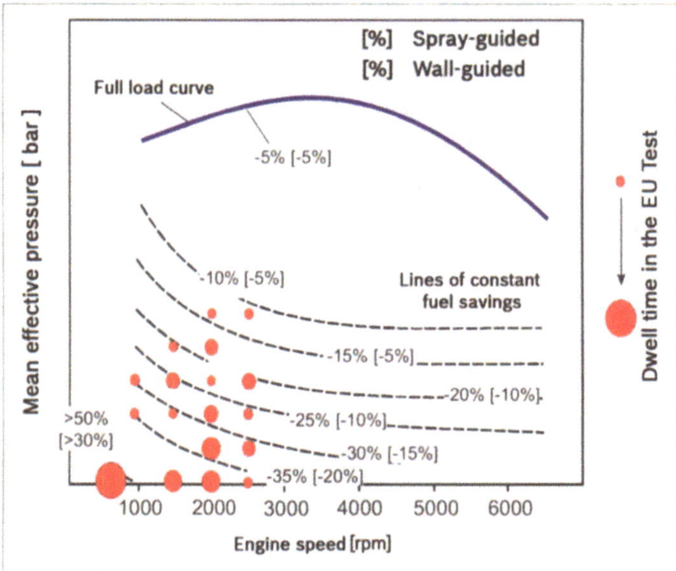


**Figure 3.41:** Fuel savings through gasoline direct injection in the engine map [61]

**Emissions**

The distinct disadvantages of direct injection are the hydrocarbon and $NO_x$ emissions and the very low exhaust gas temperatures in the unthrottled stratified charge mode, which imply major problems with regard to exhaust-gas aftertreatment (response time, response temperature and conversion rate of the catalytic converter). In the range close to low idle, the exhaust gas temperatures lie below the light-off temperature for oxidation catalytic converters. The temperature of the exhaust gas can be increased by reducing the fresh gas mass flow through partial throttling, due to which, however, the gas exchange losses are increased again. The exhaust gas temperature can also be increased by secondary air injection or exhaust-gas recirculation. Exhaust-gas recirculation offers the additional advantage that the $NO_x$ emissions can be reduced during lean operation.

In general, design-related measures (combustion chamber design, displacement, stroke-to-bore ratio, compression ratio) and operational measures (ignition point, valve timings, air/fuel ratio, exhaust-gas recirculation, overrun fuel cut-off) alone are not sufficient to meet the current and future emissions standards. Additional exhaust-gas aftertreatment, especially for reducing the unburnt hydrocarbons and nitrogen oxides, is thus indispensable. Soot emissions have to be taken into consideration, too. Since gasoline engines with direct injection are operated with global air/fuel equivalence ratios greater than 1 (up to $\lambda > 8$) in the stratified charge mode, the nitrogen oxides cannot be reduced with the conventional three-way catalytic converter. In homogeneous operation mode ($\lambda = 1$), in contrast, conventional three-way catalytic converters are another effective measure for reducing the pollutants in gasoline engines with direct injection.

To increase the exhaust gas temperature, it is possible to achieve an insulating effect in the external area of the combustion chamber through accordingly high exhaust-gas recirculation rates with a partial stratification of the recirculated exhaust gas; **figure 3.42**. On the other hand, the engine can be operated with low air/fuel ratios and therefore with higher combustion and exhaust-gas temperatures by replacing the air that is not required for combustion with recirculated exhaust gas.

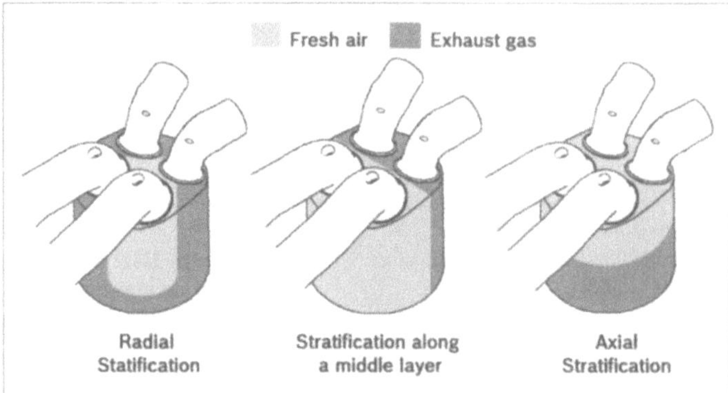


**Figure 3.42:**
Ideal exhaust-gas stratification concepts [62]

**Figure 3.43** shows the results of the investigation of an engine with stratified and homogeneous exhaust-gas recirculation [62]. Due to the poorer inflammation conditions, an increase of the specific fuel consumption can be seen for an exhaust-gas recirculation rate above 20%.

Combustion fluctuations strongly increase during homogeneous recirculation from an exhaust-gas recirculation rate of 15% onwards. From 20% onwards, the first ignition misfires occur. Fluctuations also increase in the stratified exhaust-gas recirculation, but there are no misfires even at exhaust-gas recirculation rates up to 40%.

Due to the increased amount of exhaust gas of up to 40%, nitrogen oxides can be reduced by more than 90%. Hydrocarbon emissions increase due to the mixing of exhaust gas and fuel and the related flame-quenching effects but can easily be reduced with oxidation catalytic converters.

Of the injection processes, high pressure injection with spray-guided process has the potential for the lowest fuel consumption. In comparison to the wall-guided process, there is less friction due to the simpler piston shape and the resultant lower piston weight. The short distance between the injector and the spark plug facilitates the achievement of short time intervals between injection and ignition, which is favorable for a stratified charge.

However, direct injection is possible only through common rail injection, which is highly variable. Common rail systems for engines with direct injection offer high flexibility for adapting the injection to the engine operation.

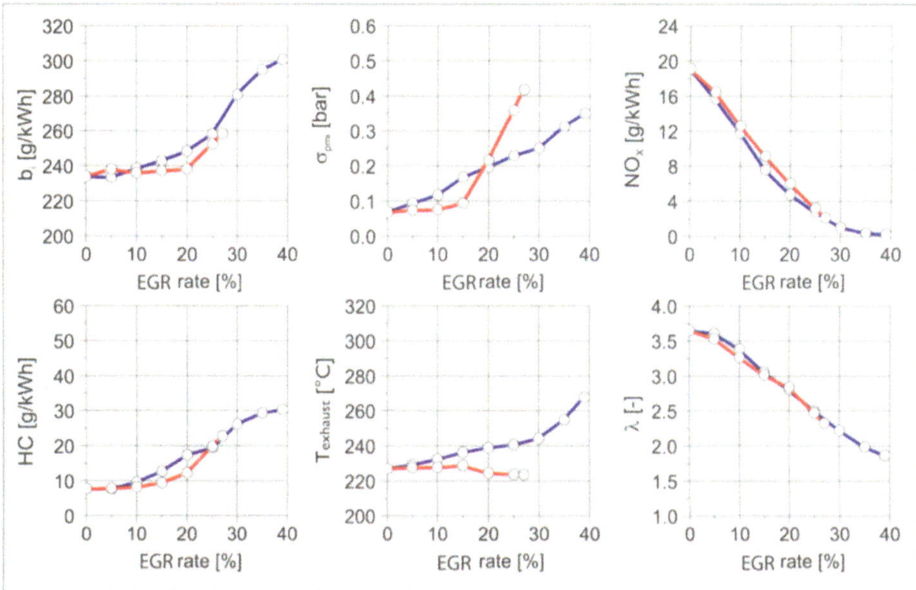


**Figure 3.43:** Comparison of stratified (red) and homogeneous (blue) exhaust-gas recirculation at $n = 200$ rpm and $imep = 3$ bar [62]

The difficulty in the development of this process, however, is to achieve stable and good mixture formation in spite of the close placement of the injection nozzle and the spark plug and the related extremely short time interval between injection and ignition. If this cannot be optimized, especially in view of the fact that the injection is carried out against a high combustion chamber pressure at a very late point in time in the compression phase to ensure stable stratification of the mixture, combustion misfires and increased soot formation could result. High injection pressures, which permit very good fuel atomization and therefore extremely fast evaporation of the fuel, are required for this purpose. The recently developed injection systems [63, 64] with increased maximum injection pressures of up to 200 bar are the first step in this direction. Further pressure increases that can be implemented in standard production can be expected in the future.

### Research results on increase of injection pressure

Currently, it is not possible to increase the injection pressure to more than 200 bar, due to the endurance limit of injection systems. Since gasoline, unlike diesel, does not have any lubricating effect, the wear and tear of both the injection pump and the injector is still too high for series application at high pressures. [4] has carried out initial basic investigations into increasing the injection pressure to approximately 600 bar using a research unit.

For this, the average droplet diameters were first visually determined in an injection pressure chamber with different injection pressures at different chamber pressures between 4 and 16 bar. The results are shown in **figure 3.44**. If the injection pressure is increased from

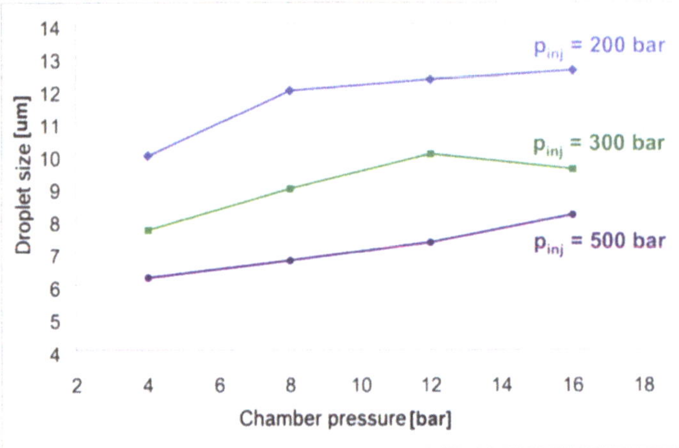


**Figure 3.44:** Average droplet size for injection pressure variation [4]

200 to 500 bar, the average diameter of the fuel droplets decreases by approximately 36% at a back pressure of 16 bar. This is the first indication that increased injection pressure improves mixture formation.

The increased pressure at the injection nozzle therefore leads to finer atomization of the fuel when it is discharged. The small droplets have a greater total surface area and can thus absorb the heat much faster and evaporate. The mixture formation time can be considerably reduced. The mixture quality is improved.

**Figure 3.45** sshows the specific fuel consumption measured on a single-cylinder research engine at injection pressures of 200, 300 and 590 bar at an engine speed of $n$ = 2000 rpm and an indicated mean effective pressure of *imep* = 3 bar (corresponds to *mep* = 2 bar). At

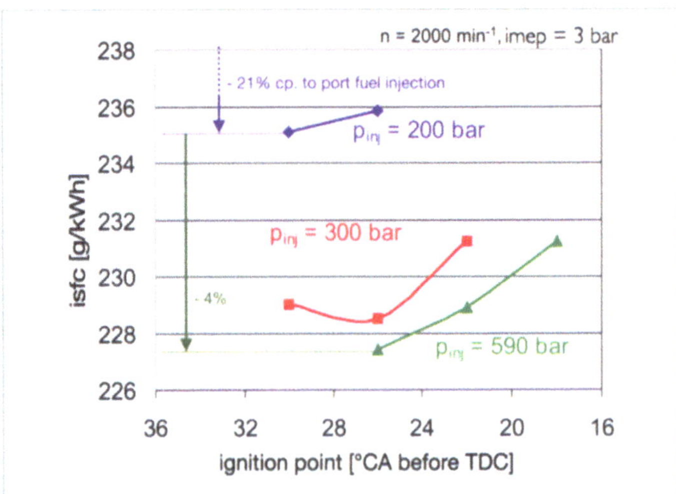


**Figure 3.45:** Fuel consumption with injection pressure variation [4]

an injection pressure of 200 bar, a fuel consumption advantage of 21% as compared to intake manifold injection is achieved. An increase of the injection pressure to 590 bar results in a further decrease of fuel consumption by approximately 4%. In this context, it should be noted that the injection nozzle used for pressures above 200 bar was developed for diesel fuel and is therefore not particularly suitable for gasoline direct injection. Injection nozzles suitable for these high pressures are not yet available for gasoline direct injection.

This consumption advantage alone, which was determined in the first investigations, would hardly justify the additional effort for a suitable injection system. However, the positive effect on the mixture formation at extremely high injection pressures becomes evident when the dependency of soot formation on the injection pressure is considered; **figure 3.46**.

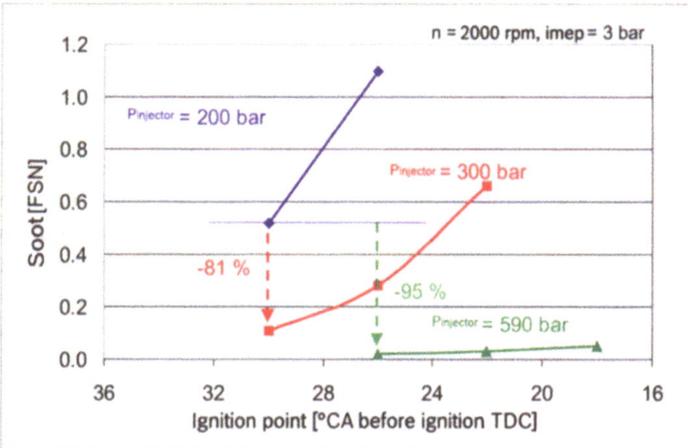


**Figure 3.46:** Influence of injection pressure on soot emissions [4]

Starting from an injection pressure of 200 bar, a reduction of soot emissions by 81% is achieved by increasing the injection pressure by 100 bar. However, the soot formation increases significantly when the ignition angle is changed. At an injection pressure of 590 bar, a reduction of the soot emissions by 95% is achieved. A soot level as low as this is maintained even at an ignition point variation of more than 8° CA, where the fact that the injection point, like the ignition point, can be retarded is taken into account. The high injection pressure therefore not only results in enormous reductions in the soot level, but also in degrees of freedom when selecting the injection and ignition points. This is important in order to be able to optimize the center of combustion as far as possible in terms of efficiency. Often, this is not successful due to injection timing that is too early and the possible ignition points related to them, both of which are too far ahead of the top dead center as the injector and the spark plug are positioned too close together. High injection pressure therefore improves the mixture preparation considerably.

In addition to the improvement in the soot formation and the possible injection and ignition window, the improved mixture formation also influences the burn-out behavior. Optical examinations with a rapid photo multiplier camera have shown that, at identical engine parameters, the combustion starts earlier, occurs faster and has less soot luminescence. Now, if the center of combustion can be optimally adjusted with the high injection pressure, faster combustion enables a thermodynamically favorable process cycle with approximation to the desired constant volume combustion. In order to achieve this visionary target of optimum spray-guided direct injection, two more problems, which should not be underestimated, are to be solved in addition to ensuring fast mixture preparation at ideal stratification. These problems are the development of an injection system suitable for the required high injection pressures and the reduction and treatment of the increased nitrogen oxide emissions related to stratified charge.

## 3.4 Ignition and combustion

The combustion of an optimally prepared air/fuel mixture in a conventional gasoline engine requires ignition. The process of efficient ignition is by no means trivial. Literature talks of spark ignition when a mixture that would not normally auto-ignite is locally brought to ignition by an ignition source. Inflammation of the mixture takes place within the ignition volume of the ignition source, followed by unsteady flame propagation into the unburnt gas. Spark ignitions can be caused by highly energetic plasmas, which are still usually generated by electrical sparks between the electrodes of a spark plug [65]. **Figure 3.47** shows the influences on the spark ignition process schematically.

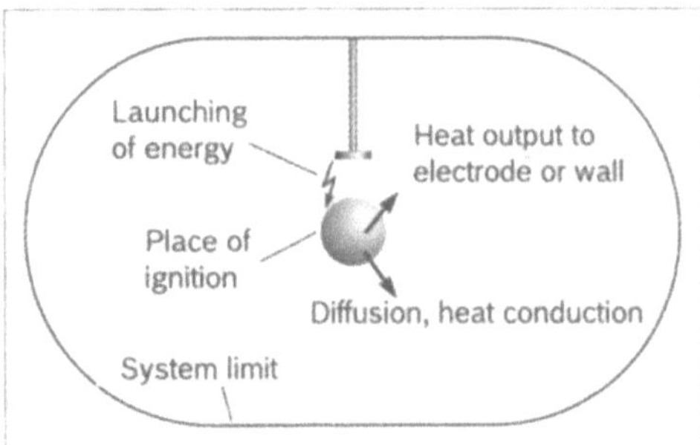


**Figure 3.47:**
Schematic representation of spark ignition [66]

Spark ignition can be influenced by the type of energy induction, the mixture composition with regard to ignition quality and its transport properties – diffusion and heat conduction. Diffusion and heat conduction are particularly important for the complete ignition of the mixture [66]. For current gasoline engines, electrical spark ignition is the variant most commonly used as the classical coil ignition system. An ignition spark initiates the combustion by igniting the air/fuel mixture compressed in the combustion chamber. The ignition spark is generated by a sparkover between the electrodes of the spark plug, which extend into the combustion chamber. The ignition system has to generate the high voltage required for the sparkover at the spark plug and ensures the timely triggering of the ignition spark. For more detailed explanations of this and other ignition systems, please refer to section 11.1. **Figure 3.48** shows an example of a current and voltage characteristic of the ignition process in a gasoline engine with direct injection and a coil ignition system.

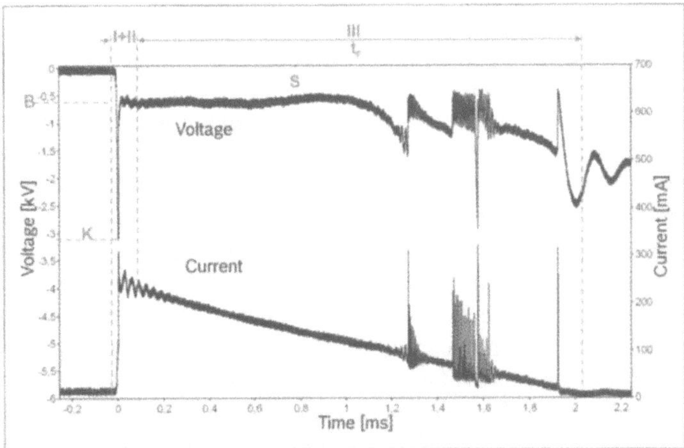


**Figure 3.48:** Time characteristic of voltage and current, measured at the electrodes of a spark plug [53]

The overall discharge can be divided into three phases: breakdown phase (I), arc phase (II) and glow discharge phase (III) [66, 67, 68]. The capacitances charged to ignition voltage in the secondary circuit of the spark plug discharge suddenly in a spark head (K in **figure 3.48**, also called flashover). This electrical breakdown takes place in the nanosecond range and is characterized by a voltage of several kV [69].

### Specific characteristics of the ignition of lean mixtures

For gasoline engines with external mixture formation in stoichiometric operation with high volumetric efficiency, with almost ideal mixture formation conditions and ignitable mixture composition, the requirements of the ignition system with regard to inflammation are relatively low. However, the requirements of the ignition system and the spark plugs for the ignition of lean mixtures, especially for gasoline engines with direct injection, are comparatively high. The objective of spark ignition for gasoline engines with direct injection is the reliable inflammation of non-homogeneous mixtures for a lean, global air/fuel ratio but with a locally very limited area of ignitable air/fuel mixture at the same time. The high flow velocities induced by the charge movement in the vicinity of the spark plug also complicate the flame kernel formation [70]. Furthermore, a wetting of the spark plug with liquid fuel cannot be ruled out due to the short mixture formation time in the stratified operating mode. As a result of these special characteristics, an ignition system suitable for gasoline engines with direct injection should have the following features:

- Reliable ignition in a large range of the air/fuel ratio
- Insensitivity to the liquid phase

- Robustness against the prevailing flow conditions
- Insensitivity to deposits
- Stable location of ignition
- Variability of the ignition parameters (e.g. spark duration, spark energy) for adjustment to the respective operating and load points.

When compared to the wall-guided and air-guided processes [69, 71], the proximity of the injection valve to the spark plug and the retarded injection for gasoline engines with direct injection and spray-guided injection processes add another procedural influence on the inflammation conditions. Because of the strong spray momentum in the ignition gap due to the injection, high flow velocities are generated in the spark gap. In comparison to stoichiometric operation with an identical charge movement, up to five times higher values for the flow velocities in the spark gap and significantly greater ranges of variation [71] result for compression stroke injection. Due to the greater penetration depth of the fuel spray of a multi-hole valve, in comparison to an injection valve, which opens outwards (c.f. section 4.2.3, A-nozzle), this effect is especially pronounced when multi-hole valves are used. In the vicinity of the spark plug, flow velocities of up to 30 m/s can occur at the point of ignition [72, 73, 74]. **Figure 3.49** shows the dependencies of the lean inflammation limit on the flow velocities perpendicular to the spark axis. The ignitability increases up to a velocity of approximately 5 m/s.

The reason for this is the deflection of the ignition plasma and the corresponding increase of the mixture volume covered by the plasma, without resulting in a spark breakaway. Furthermore, the deflection of the spark without breakaway reduces the contact areas of

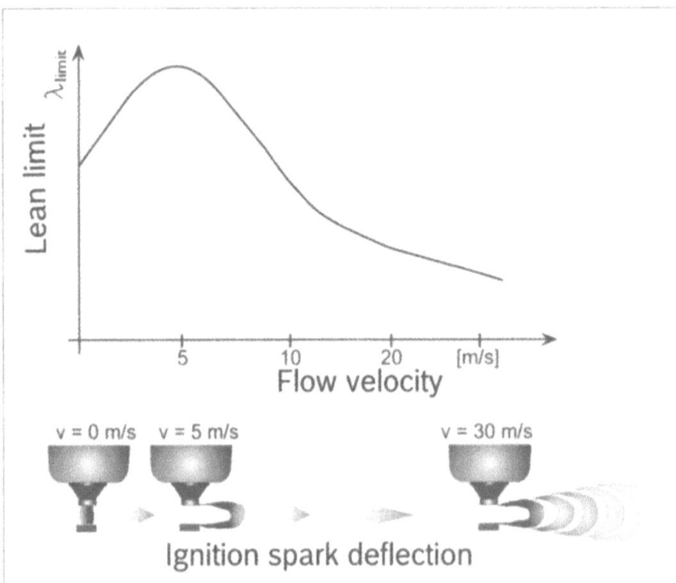


**Figure 3.49:** Lean operation limit depending on flow velocity perpendicular to spark axis [23, 74]

the plasma to the electrodes and thus also reduces the loss of heat. At the same time, the energy input into the mixture is increased; **figure 3.47**. For higher flow velocities with spark breakaway, the lean operation limit decreases. The spark breakaway causes more favorable re-ignition for the ignition spark in terms of energy, through a shorter route. This new flashover and ignition process costs additional energy, **figure 3.47**, and the accumulation of ignition energy occurs in multiple flame kernels. This leads to a strong dependency of the inflammation conditions on location and time [69].

As an example, **Figure 3.50** shows the correlation of the ignition voltage characteristic to the spark breakaway with optical tests. Since these tests were carried out at an engine speed of 1500 rpm at motored operation, the deflection of the spark is relatively slow, the ignition voltage increases evenly and the process can be observed clearly.

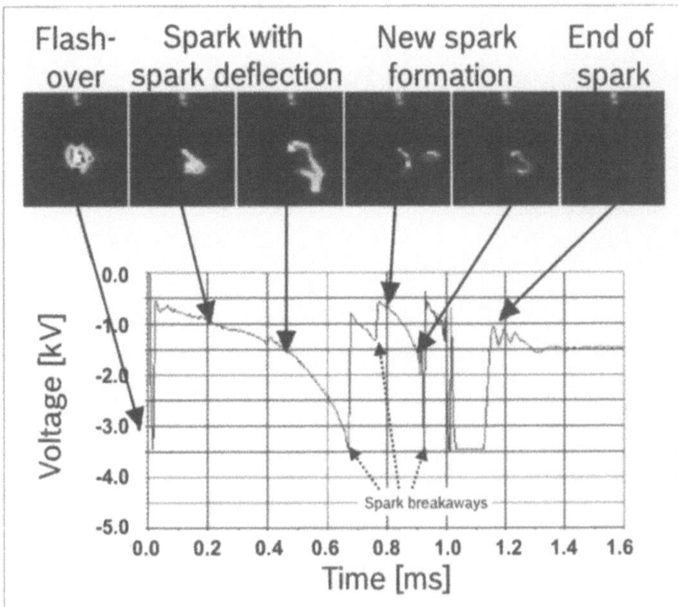


**Figure 3.50:** Ignition voltage characteristic during a spark breakaway in the engine at trailing throttle [69]

The ignition spark resulting from the first voltage breakdown is deflected up to approximately 0.65 ms after the breakdown. Subsequently, due to the high flow velocities, the first spark breakaway takes place [68], which can be seen clearly from the sudden drop in the ignition voltage. Using the residual energy of the ignition coil still available, two other sparks are generated until no more energy is available in the ignition coil. This leads to the ignition voltage characteristic typical of high flow velocities in the vicinity of the spark plug. **Figure 3.51** shows the ignition voltage characteristic in different operating modes of a gasoline engine with direct injection. No deflection of the spark takes place at highly throttled operation in low idle mode during stoichiometric operation. At 2.4 ms, the combustion duration is relatively long. In the partially dethrottled homogeneous operating mode at 3 bar indicated mean effective pressure (*imep*), higher flow velocities are already present at the point of ignition, and these lead to frequent deflections of the ignition voltage signal (saw tooth profile) and reduce the spark duration to approximately 2.0 ms. If the injection is switched off at this operating point, the signal only changes insignificantly. The deflection of the spark has become greater; however, this is attributed to the higher breakdown voltage for pure air and the lack of an ionization effect of the combustion prereactions. This does not affect the spark duration, however, from which it can be concluded that the spark breakaways result only from charge movement. At an almost dethrottled stratified charge operating point, the charge movement and the high spray momentum interfere with each other during the injection close to the point of ignition in the compression stroke.

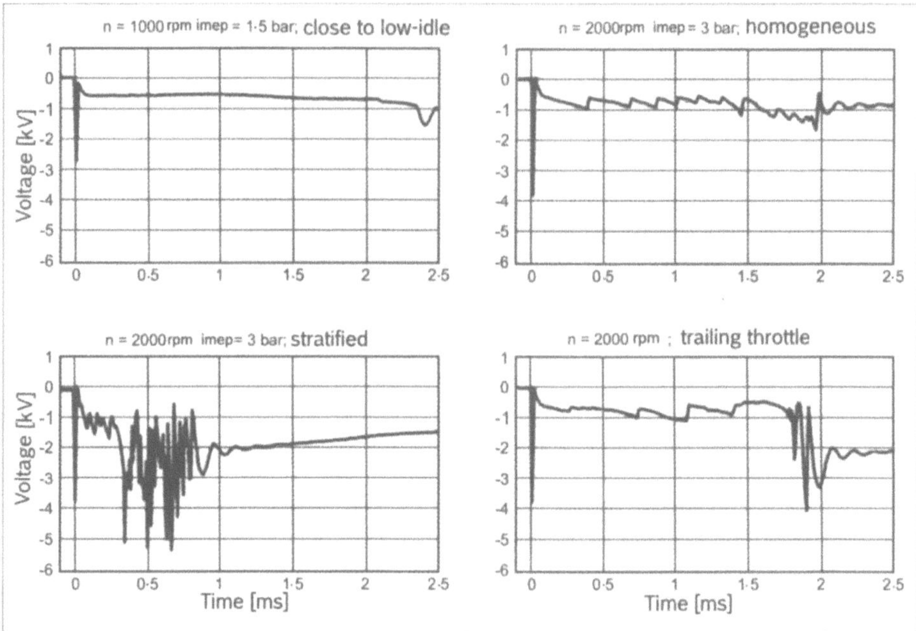


**Figure 3.51:** Ignition voltage characteristic for different operating modes of a gasoline engine with direct injection [53]

The great increase of the flow velocities in the spark plug gap is evident because of the extremely reduced spark duration of less than 1 ms and the steep voltage increases [69, 73]. The immediate conclusion is that having the spark plug and injection valve positioned very close to each other in a gasoline engine with direct injection and spray-guided combustion process leads to a significant increase of the flow velocities in the spark plug gap. This reduces the spark duration to such an extent that sufficient energy for inflammation can no longer be transferred to the mixture. Tests that have been carried out in [72] confirm this assumption.

### Abnormal ignition in gasoline engines with direct injection

The high stratification gradients of the air/fuel mixture and the possible wetting of the spark plug with liquid fuel, which are common for gasoline engines with direct injection and a spray-guided combustion process, lead to heavy and persistent contamination in the insulator part of the spark plug. These carbonaceous residues from combustion, such as soot or oil deposits, are very conductive in comparison to the ambient air. Together with the high ignition energy (up to 140 mJ [69, 71, 75]) required for a steady stratified operating mode, the increased conductivity of the insulator nose leads first of all to the spark sliding into the breathing space of the spark plug. The air gap to be overcome is smaller in the breathing space of the plug than in the actual spark plug gap, causing spark flashover. Due to the strong mixture stratification and the corresponding sensitivity of the process to changes where ignition takes place, spark failures and misfires can result when surface type sparks occur.

**Figure 3.52** compares the voltage characteristic in the spark plug during an ignition process with air spark ignition and complete combustion with a combustion misfire due to surface type spark ignition.

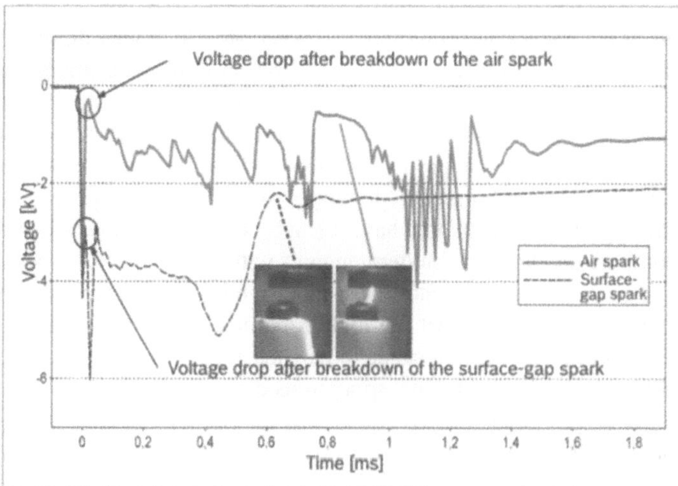


**Figure 3.52:** Difference between air sparks and surface type sparks in the ignition voltage characteristic [75, 76]

The main characteristic of cycles with surface type spark ignition is the completely different history of the voltage, **figure 3.48**, immediately after the voltage breakdown, **figure 3.52**, which is higher by approximately 3 kV in this case. Due to the related higher electrical power during the combustion phase, the combustion duration of a surface type spark is significantly shorter than that of an air spark. The previous analysis also shows how useful ignition voltage detection can be when a combustion process is adjusted. By means of voltage analysis, a clear distinction can be made between misfires due to mixture composition and misfires due to ignition. The tendency to surface type sparks can be minimized both by design changes in the spark plug (annular orifice in the electrode area) and by electrical measures (positive polarity of the spark plug) [75]. The publications by Herden [77], Fansler [78, 79] and Maly [67] deal with the theory of spark ignition in detail.

### Combustion during direct injection and homogeneous operating mode

As already described, engines with gasoline direct injection can be operated in two different ways: homogeneously in higher engine loads and engine speeds or with charge stratification in the lower part load range. In the homogeneous operating mode, the combustion process is very similar to that of engines with intake manifold injection: starting from the spark plug, after the inflammation of the air/fuel mixture, an almost spherical flame front propagates into the combustion chamber, which can be more or less strongly folded and clefted in interaction with the in-cylinder flow and through the turbulence interaction. When the mixture homogenization is good, combustion takes place almost fully premixed. Combustion can then be characterized, then, by an ideal premixed flame. However, it is not usual, either for direct injection or for conventional gasoline engines with external mixture formation, for ideal mixture homogenization to be achieved, so that at the point of ignition cyclic fluctuations of the mixture state occur at the spark plug in interaction with the internal flow in the combustion chamber, which negatively affect the stability of inflammation. During gasoline direct injection, the fuel is injected into the combustion chamber only during the intake stroke which, in case of unfavorable injection spray characteristic and bad timing, can result in the spray contacting the piston surface. Such wetting of the piston crown leads inevitably to a fuel film that burns poorly because of bad evaporation conditions. The fuel film burns off slowly and moreover incompletely in a diffusion flame and therefore leads to increased emissions of soot and unburnt hydrocarbons. This is a circumstance that must be taken into account, particularly for combustion processes with a vertical injector position. Such piston wetting in the homogeneous operating mode can be prevented by using suitable injection systems that have a small vertical penetration depth of the fuel spray in the combustion chamber. Modern engines with direct injection therefore work with much more complex and thus more cost-intensive injection systems than conventional gasoline engines, in order to meet the higher requirements during mixture preparation. Furthermore, the point of injection needs to be chosen in such a way that the piston is positioned at sufficient distance to the fuel spray in its downward stroke.

### Flame propagation in the stratified charge mode

As already mentioned, the stratification of the air/fuel mixture in the combustion chamber enables the reduction of the flow losses during gas exchange by dethrottling the intake process. Moreover, the insulating gas layer of air or residual gas should serve to reduce the heat losses through the wall and therefore to achieve an additional increase in thermal efficiency. This does, however, depend on the fact that, unlike homogeneous combustion, combustion in the stratified charge mode does not reach the walls of the combustion chamber but comes to a halt at some distance from the cylinder wall in very lean mixture ranges [80]. For this reason, the development of combustion processes with gasoline direct injection and charge stratification, in addition to ensuring steady inflammation, also takes into account the optimization of those parameters that ensure such stratified charge combustion behavior. In this context, measurement procedures are required for a systematic analysis of the variables that influence the local flame propagation, which go beyond the global measured variables such as combustion chamber pressure or exhaust-gas values. It therefore seems to be reasonable to supplement standard measuring methods such as pressure indication with appropriate special measuring methods in order to obtain information on local flame development. In this regard, modern optical measuring methods are especially suitable. Different measurement procedures based on fiber optics have proven to be successful for the examination of flame propagation in combustion engines in particular [81, 82, 83].

By means of a multiple optical fiber measuring method, the combustion radiation in the combustion chamber of an engine can be detected by fiber optics endoscopes in both the ultraviolet and the visible wavelength range by means of fast photo multiplier modules with a sampling frequency of up to 200 kHz [4, 84].

Among other things, the measuring system, therefore, enables a direct comparison of the combustion in the homogeneous and stratified charge mode in engines with direct injection. **Figure 3.53** shows an exemplary arrangement of a fiber optics endoscope for an analysis of flame development at different places. Starting on the intake side between the two inlet valves, a section of the combustion chamber is visualized. The detection range of the endoscope has an overall cone angle of 80°, the direction of detection being tilted by 30° to the horizontal.

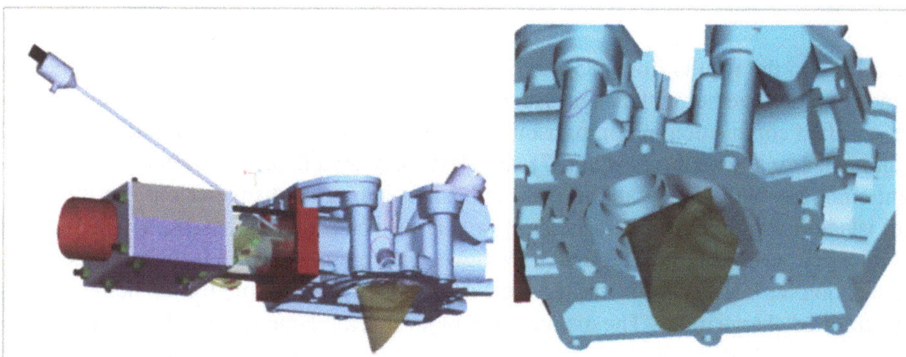

**Figure 3.53:** Endoscope position in the cylinder head; detection cone

**Figure 3.54** shows a series of images of the flame development for a part load operating point of a single-cylinder gasoline engine with spray-guided gasoline direct injection using a multi-hole injector. The upper row shows the flame propagation in the homogeneous operating mode through fuel injection into the intake stroke. For comparison, the lower image sequence depicts the combustion of a stratified air/fuel mixture at identical engine load and engine speed as well as under identical boundary conditions for the intake flow. Whereas for a largely homogeneous mixture state the flame front, starting from the spark plug, propagates through the entire detection range of the visualization system, stratified combustion behaves quite differently. After the inflammation of the mixture at the spark plug, the combustion is limited locally. The flame hardly reaches the walls of the combustion chamber on the intake side (in the lower part of the pictures). The demand for an isolated, compact flame would therefore be met if the same statement was also valid for the other propagation directions. However, this cannot be clarified sufficiently with two-dimensional measuring methods.

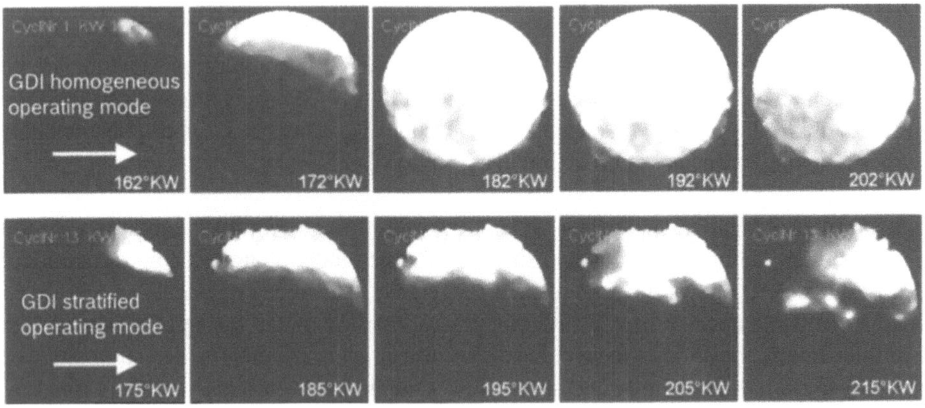


**Figure 3.54:** Comparison of homogeneous operating mode and stratified charge mode; representative individual cycles; top dead center position at 180° CA

According to the results from [85] and [86], much stronger deviation from the spherical flame contour, together with a significant deformation of the reaction zone, is detected for the stratified combustion in comparison to the homogeneous operating mode. In the case examined, as was also the case in [87], the partly premixed flame that occurs under strong light emission can disintegrate during combustion into several smaller regions (combustion islands) which, due to relatively slow mixing controlled processes, are maintained for a longer period of time. **Figure 3.55** shows, for comparison, the isolines of the flame propagation during flame growth and the extinguishing of the flame for the homogeneous and stratified charge mode. The disintegration into several flame regions during the largely diffusion controlled burn-out is associated with the steep fuel concentration gradient due to the high mixture non-homogeneity in the combustion chamber and reinforced by the influence of the in-cylinder flow. Furthermore, depending on the combustion chamber flow, it is observed that the flame is blown over. The worst case scenario is that this causes the flame to be expelled and concentrated in one area of the combustion chamber. It follows that the complex interaction between in-cylinder flow, mixture formation and ignition in the stratified charge mode should be adjusted not only with regard to the inflammation phase but also taking the subsequent flame development into consideration.

To characterize the spatial flame propagation and complement two-dimensional measuring methods, an advanced 3D measuring procedure is used in [84]; **figure 3.56**. In this process, up to three fast photo multiplier measuring systems are connected to visualize the combustion

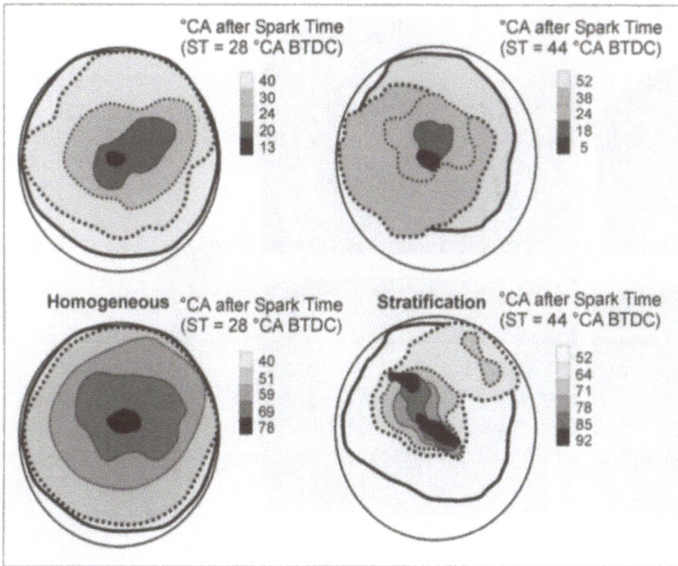


**Figure 3.55:** Isolines of flame propagation for homogeneous (left) and stratified charge mode (right) [87]

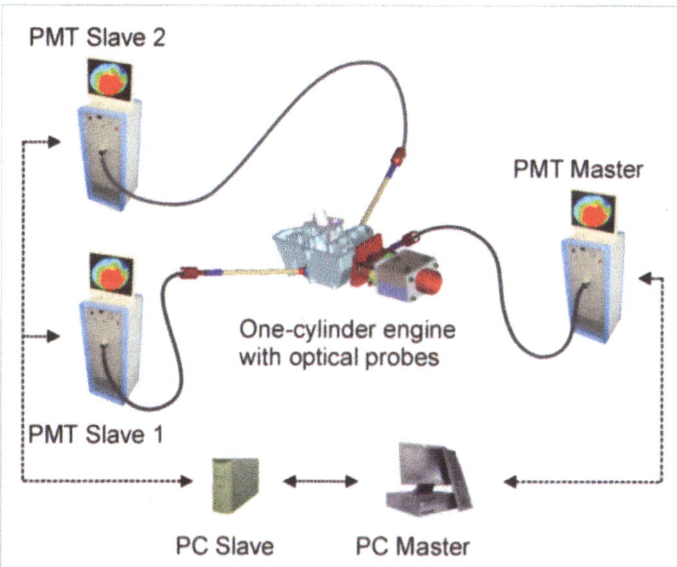


**Figure 3.56:** Experimental set-up of "3D flame" measurement procedure [84]

synchronously from different directions of detection. The flame images obtained this way are reconstructed into a three-dimensional model in the subsequent evaluation process using special software, from which information about the development of the flame volume can be derived in terms of time as well as the location of the combustion center; **figure 3.57**.

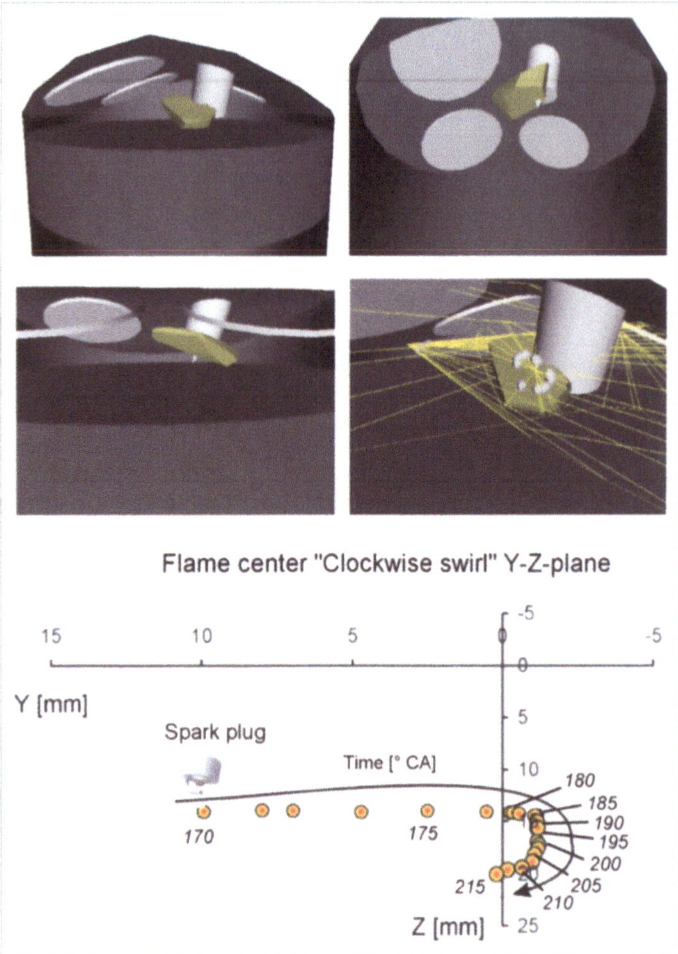


**Figure 3.57:**
3D reconstruction of flame development; center analysis [84]

By doing so, it is possible to systematically optimize the relevant variables from mixture formation, combustion chamber flow and ignition to improve combustion.

Spray-guided combustion processes, in particular, should work without special air guidance due to the close arrangement of the spark plug and the injection valve. A systematic charge movement, however, can support the mixture formation and combustion even for spray-guided combustion processes. A charge movement can improve the heat input into the spray and thus support the mixture formation process, which has to take place in a very short period of time. Insufficient mixture preparation with high gradients of the air ratio becomes apparent through the long burn-out duration in the stratified charge mode, combined with diffusion combustion. A systematic charge movement through a global flow structure generates turbulences in the combustion chamber that can be used to accelerate the flame front at the point of ignition. This way, faster and more complete combustion with improved overall efficiency can be achieved [84]. Different characteristic charge movements can differ in their effects on combustion. A swirl motion is usually maintained for a relatively long time and therefore has a lot of kinetic energy at the point of ignition and during combustion, which can accelerate the combustion speed. A tumble flow can influence the fuel transport systematically and prevent an adsorption of fuel on the piston and thus can positively influence the hydrocarbon emissions.

The kinetic energy brought into the combustion chamber through a charge movement primarily generates a turbulent fluctuation velocity and therefore leads to the increase in the speed of combustion. By means of a global combustion chamber flow, the flame can be guided into the center of the combustion chamber and can be stabilized there [4, 23].

## Knocking

Modern engines have a high compression ratio with regard to good fuel consumption under part load. At full load, critical states can arise in the combustion chamber due to this, together with the occurrence of "knocking". In contrast to normal combustion, knocking combustion is characterized by auto-ignition at one or several locations in the mixture still unburnt toward the end of compression; **figure 3.58**.

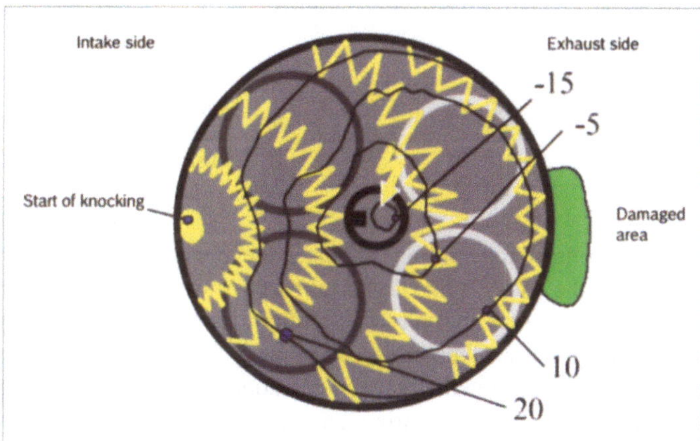


**Figure 3.58:** Flame propagation during knocking

As can be seen from the flame contours, the normal flame, starting from the spark plug, propagates through the combustion chamber at normal flame speed. Towards the end of combustion, spontaneous ignition occurs in the still unburnt mixture (end gas area) due to critical states (pressure, temperature, prereactions).

This auto-ignition results in extremely fast fuel conversion, together with a strong pressure wave. These pressure waves can cause material damage (e.g. in the top land area) especially in places where they are reflected in the combustion chamber. Furthermore, knocking also leads to thermal overloading and can therefore cause glow ignitions. In the pressure characteristic, knocking is visible toward the end of combustion; **figure 3.59**. Only then, starting with a steep pressure increase as a consequence of the fast conversion of the end gas, do high frequency oscillations superimpose the normal pressure characteristic. They are caused by gas oscillations in the combustion chamber (local pressure differences). They are also transmitted to the engine components and are responsible for the knocking noise, which is also called "pinging". Extended periods of knocking operation usually lead to engine damage. Therefore knocking should always be avoided.

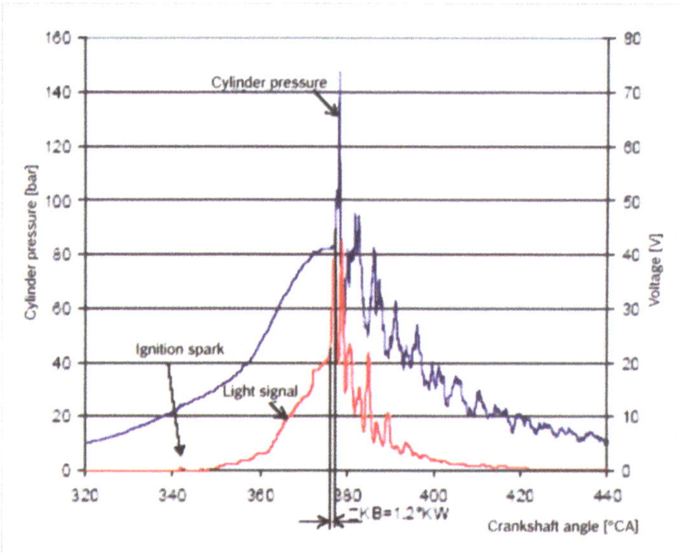


**Figure 3.59:** Pressure characteristic and optical fiber signal during knocking combustion

For older engines without knock control, the ignition timings in the higher load range were always set with a safety margin to the knock limit (first occurrence of working cycles with knocking). This results in efficiency losses. Usually the ignition timing was retarded by approximately 5° CA; the efficiency losses then are approximately 2 to 3%. For engines with knock control, the knock limit is detected during each operation and the ignition angle is adjusted to it. When knocking occurs (detected through a structure-borne sound sensor), the ignition timing is retarded by a particular value. After a specified time, the ignition timing is advanced again until new knocking is detected or the optimum power is reached.

Today's gasoline engines show a special knocking behavior during full load at high engine speed, which is called "extreme knocking" or "mega knocking" and which is probably due to the strong exploitation of the engine. This phenomenon is characterized by stochastically extreme knocking working cycles with damage related pressure amplitudes. The anti-knock control, when close to the knock limit and with advancing of the ignition timing, shows no continuously increasing knock intensity with superimposed, high frequency pressure amplitudes. On the contrary, individually occurring "extreme knockers" can be seen at the knock limit itself, **figure 3.60**. These are many times higher than the pressure amplitudes during normal knocking combustion. This behavior of individual working cycles cannot be foreseen at all in the previous working cycles, irrespective of whether the engine is operated with a steady ignition angle or with anti-knock control. Furthermore, this phenomenon occurs irrespective of the mixture formation process both for naturally aspirated engines and forced induction engines as well as for engines with direct injection. The event-controlled knock control systems generally used nowadays can react with a retardation of the ignition angle only after knocking combustion has occurred. Since the extreme knockers can be already seen in a narrow ignition angle range at the knock limit and today's knock control systems are designed close to the knock limit, an ideal design in terms of power and consumption can no longer be implemented without the risk of sudden destruction of the engine. To be able to continue to increase the efficiency of today's and future engine generations, it is therefore necessary to find new approaches to prevent this extreme knock behavior by means of a systematic causal analysis.

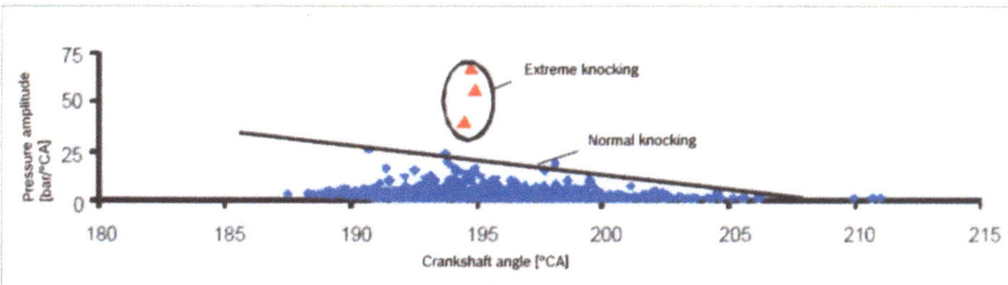


**Figure 3.60:** Extreme knocking

The operating conditions, the fuel and the shape of the combustion chamber have significant influence on knocking. The knock tendency of an engine always increases under the following boundary conditions:

- High pressure and high temperature in the residual unburnt mixture (end gas)
- Approximation to stoichiometric air/fuel ratio ($\lambda = 1$)
- Slow process sequence (prereaction times are long)
- Low octane number of the fuel
- High compression through high compression ratio and/or super/turbocharging of the intake mixture.

For engine operation, the following can be concluded: the knock tendency of a gasoline engine increases when the compression ratio $\varepsilon$, the ignition angle $\alpha_z$ (advanced timing), the mixture temperature, the engine temperature, the mean effective pressure and the engine dimensions (cylinder bore) are increased. Higher engine speeds generally reduce the knock tendency because the combustion duration becomes shorter. Depending on the engine operating point, two types of knocking are identified:

- **Knocking under acceleration:** knocking under acceleration occurs during full load acceleration at low engine speeds. It is audible as a clear "pinging" and seldom leads to damage.
- **High-speed knocking:** high speed knocking occurs at high engine speeds in the upper load range. Due to the accompanying driving noise, it is seldom audible and almost always leads to severe engine damage. Persistently knocking engine operation must be avoided, as otherwise it will cause significant, irreparable knock damage, usually in the piston, in the top land area.

The shape of the combustion chamber can influence combustion and knocking considerably. Combustion chambers with low knock tendency (anti-knock combustion chambers) have to meet the following basic requirements:

- Short flame paths: compact combustion chamber, central spark plug position
- Avoidance of hot spots at the end of the flame path: spark plugs close to the exhaust valve
- High flow velocities: tumble and/or swirl motion and/or squish flow.

Combustion chambers with high concentration of the charge close to the spark plug are more advantageous than expanded combustion chambers (long flame paths). Engines with large cylinder dimensions tend more to knocking due to their long flame paths than engines with smaller cylinder dimensions. Today's combustion chamber shapes are designed to take these boundary conditions into account, at least partially. The demand for low emissions must, however, not be ignored.

## 3.5 Gasoline direct injection with homogeneous charge compression ignition

### 3.5.1 Thermodynamics

Ever since Nikolaus August Otto conducted the first tests on the spark ignition engine that he developed [90], there has been a basic understanding of homogeneous charge compression ignition and of the fact that that auto-ignition can occur even in homogeneous mixtures if the prereaction time is long enough and the pressure and temperature profiles are suitable. Interestingly, auto-ignition wasn't the objective but an undesired side effect of the gasoline engine operating process, taking the form of the knocking combustion phenomenon. In the gasoline engine, knocking occurs after spark ignition because of auto-ignition in the cylinder charge areas that have not yet been reached by the flame (the end gas). This sudden simultaneous conversion of a locally limited part of the cylinder charge results in the formation of a pressure wave, which is reflected at the combustion chamber wall and results in great pressure vibrations. This causes damage because of the mechanical load but, in particular, also because of the greatly increased convective heat transmission at the combustion chamber walls and the piston during the high-temperature phase of combustion. For these reasons, the primary objective of engineers in the course of the development of the gasoline engine has always been to prevent auto-ignition of the cylinder charge.

Auto-ignition during homogeneous charge compression ignition, however, differs from knocking combustion in conventional gasoline engines: on the one hand, the mixture is considerably more diluted (lower fuel concentration), due to which local heat release is reduced, and on the other, the almost homogeneous distribution of temperature and reaction progress across the entire combustion chamber leads to auto-ignition occurring evenly in the entire cylinder charge and not, as is the case for knocking, locally limited in the end gas area. The formation of a harmful detonation wave is therefore avoided in homogeneous charge compression ignition.

The different forms of the reaction process depending on the available temperature gradient can be easily explained by means of a very simplified, one-dimensional model as per Zeldovich [91]; **figure 3.61**. As a simple assumption, a gas mixture with the properties of an ideal gas whose local reaction intensity is seen as dependent only on the temperature was taken as a basis. The local mixture temperature drops linearly starting from the origin ("exothermal center"). Based on the increase of the temperature gradient, combustion can be classified into different types:

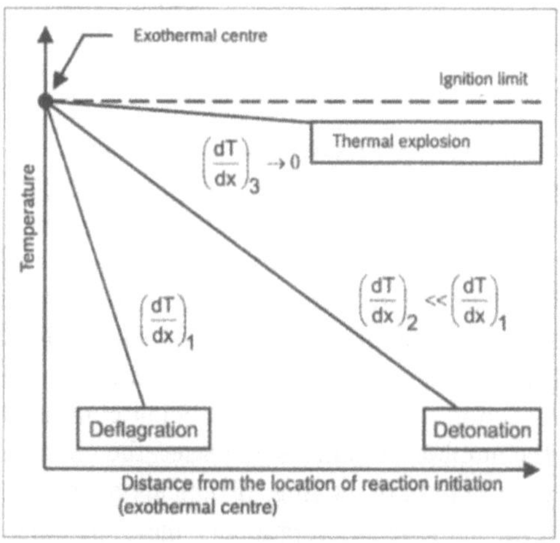


**Figure 3.61:**
Simplified model of energy conversion based on the temperature gradient as a function of the distance from the place of reaction initiation ("exothermal center") in a one-dimensional system [91]

In case of a steep temperature gradient, deflagrative combustion commences from the exothermal center where the ignition conditions are reached first. A flame front is formed as a local reaction zone that propagates in the direction of the falling temperature gradient by heating adjacent mixture areas to ignition temperature so that energy conversion is triggered there too. This case can be seen as an exemplary approximation for classic gasoline engine flame front combustion, during which combustion propagates into the entire combustion chamber starting from the spark plug.

The propagation rate of the reaction zone increases with decreasing gradients. The mean temperature gradient depicted in the diagram represents what is known as detonative combustion. In contrast to the deflagration where the flame propagation is facilitated by chemical reaction and molecular transport processes, propagation during detonation is sustained by a pressure wave, which is maintained by the chemical reaction and the corresponding heat release [92]. This state, which is characterized by strongly accelerated conversion, is typical of the conditions during knocking combustion where parts of the end gas have a slightly higher temperature than the rest of the mixture, which leads to the triggering of auto-ignition. At high temperatures and pressures in the end gas, the locally similar ignition conditions lead to the formation of a pressure wave and subsequently to harmful pressure vibrations.

If the temperature gradient is decreased further, a thermal explosion takes place when gradients almost disappear. During the process, an approximately simultaneous energy conversion takes place in the entire gas mixture with no need for flame propagation. This combustion mode represents the ideal for homogeneous charge compression ignition. The start of ignition then depends only on the local reaction conditions, which are determined by pressure, fuel concentration and prereaction state, in addition to the temperature. Minor non-homogeneities of the local reaction conditions are compensated for, with only minor time delay, by the energy release (pressure and temperature increase) of the first igniting mixture areas, so that the entire mixture ignites in rapid succession and the reaction therefore intensifies automatically as soon as the first auto-ignitions have started ("thermal feedback").

In reality, in addition to the temperature, chemical reaction kinetics also plays an important role, as the ignition processes in hydrocarbon/air mixtures are radical chain reactions [92]. The release of highly reactive intermediate products (radicals) that react with the chemical species of the fresh mixture or other intermediate products to form other radicals is of crucial importance for the reaction process. The prereactions occurring both in the mixture and in the actual conversion reactions during combustion are determined by such radical chain reactions. If the overall molar mass of radicals does not change during the reaction process, it is known as a chain propagation reaction, the chain reaction being continued without intensification or attenuation. When the number of radicals increases, the reaction intensifies automatically; such a reaction is called chain branching. If radicals react with each other or with intermediate products to form stable reaction products, these are chain breaking reactions that inhibit the reaction progress.

In contrast to the purely thermal ignition described above, the temperature does not increase immediately for radical chain explosions; instead, the increase occurs only after an ignition delay or induction period. This can be attributed to the fact that a sufficient number of reactive radicals must first be generated through chain branching reactions, which consequently trigger the fuel conversion with a significant temperature increase. The prereactions that take place during the ignition delay time do not lead to a noticeable temperature increase [92]. However, since the speed and equilibrium position of the prereactions depend on the temperature, there are certain conditions that can cause the equilibrium of the chain triggering reactions to be displaced if temperature increases during the prereaction period, thus eliminating the grounds for chain branching ("degenerate chain branching"). The ignition delay then increases in spite of the progressing reaction. In this case, a "negative temperature coefficient" of the ignition delay can be observed. This phenomenon is especially pronounced for the auto-ignition of long-chain hydrocarbons, as found in diesel fuel in particular. The actual auto-ignition is then preceded by a low temperature conversion with a faintly bluish combustion radiation ("cold flames"). This first conversion leads to an increase in pressure and temperature, through which the reaction reaches the range of the negative temperature coefficient where combustion radiation can no longer be detected. After a certain period of time, the main conversion reaction starts with a clearly noticeable flame. Such multi-stage ignition, however, is not observed in gasoline; instead, the ignition there is immediately followed by the main conversion reaction (one-stage high temperature ignition).

In conclusion, it can be deduced that for the practical implementation of homogeneous charge compression ignition at the desired point of ignition, the mixture should be as close as possible to the ignition conditions in the entire combustion chamber. Ignition then starts first in the areas which reach the required prereaction level first and subsequently propagates quickly into the entire combustion chamber. In contrast to the simplified approach based on Zeldovich used above, the ignition conditions are determined not only by temperature but also by temperature history, pressure and kinetic prereaction level. As in practice, these parameters can sometimes vary very strongly locally, the local ignition delay being the result of numerous individual parameters and their temporal development. Even if the locally different prereaction level means a deviation from the ideal of the perfect homogeneous charge compression ignition, these deviations can well be desired under engine conditions, as – when strategically used – they offer a control mechanism to reduce the conversion rate, through which, for example, controlled conversion with moderate pressure gradients can be achieved at higher engine loads in spite of a rich mixture.

Based on a model developed in [93], figure 3.62 illustrates the differences between conventional spark ignited combustion and homogeneous charge compression ignition. Whereas for spark ignition the flame front gradually converts the cylinder charge (integration via mass increment dm), the entire cylinder charge is almost simultaneously converted during homogeneous charge compression ignition (locally different reaction progress is indicated by the inclination of the reaction zone in the diagram), the individual reactions taking place over an expanded period of time in comparison to flame front combustion (integration via differential heat part d$q$).

In spite of the conversion reactions temporally expanded with regard to the individual fuel molecule, the global conversion rate increases due to the almost simultaneous conversion of the entire cylinder charge in comparison to flame front combustion (with sequential

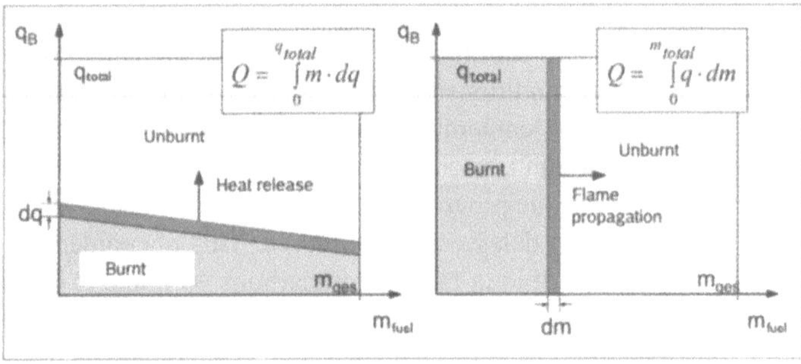


**Figure 3.62:** Idealized models of spark ignited (right) and homogeneous charge compression ignition (left) [93]

conversion). The fast heat release close to the top dead center leads to the fact that with homogeneous charge compression ignition an almost ideal constant volume cycle can be achieved. It brings a significant thermodynamic efficiency advantage., in contrast to processes with smaller constant volume levels – especially the gasoline engine with direct injection in the stratified charge mode, which due to the slow burn-out is often to be operated with centers before top dead center [4, 15, 94] – HCCI (*Homogenous Charge Compression Ignition*) also involves other efficiency advantages due to the strong mixture dilution and the frequent increase in compression.

## 3.5.2 Concepts and operating strategies for gasoline auto-ignition

In HCCI, it is not possible to directly influence combustion phasing and thereby control the combustion start via an individual parameter, such as the ignition timing for spark ignition or the injection start for the diffusion combustion of diesel engines. Instead, the start of combustion is influenced by the auto-ignition properties of the air/fuel mixture. The main influencing variables are mixture properties (auto-ignition tendency) and the temporal temperature profile before ignition.

Combustion phasing is determined in particular by the following parameters:

- Auto-ignition properties of the fuel
- Local fuel concentration (air ratio and mixture homogeneity)
- Residual gas amount and composition of the residual gas, if required
- Temporal pressure and temperature profile in the mixture from the moment of fuel insertion (intake air temperature and residual gas temperature, intake manifold pressure, compression)
- Other engine specific characteristics.

**Figure 3.63** summarizes the external intervention possibilities already used in several tests for influencing the combustion phasing; a division into the two categories mentioned above – temporal temperature profile in the mixture and variation of the auto-ignition tendency

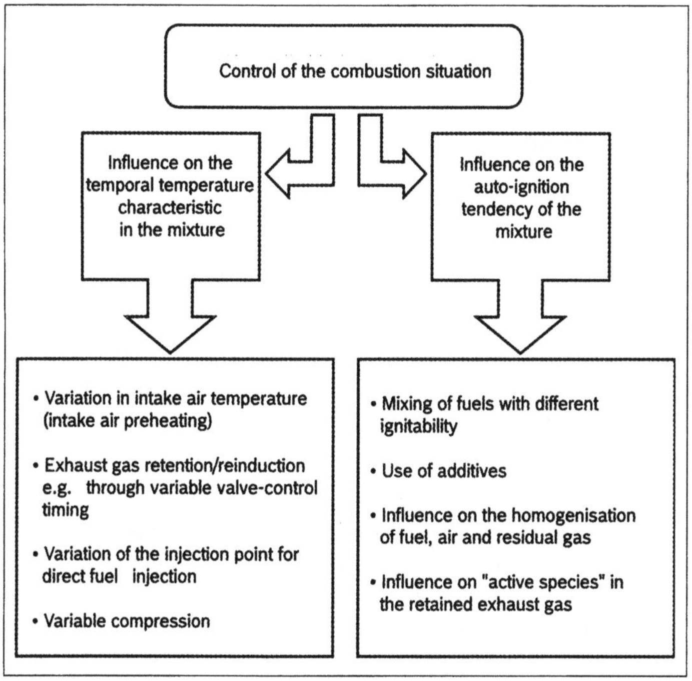


**Figure 3.63:** Parameters for HCCI control

of the mixture – is made. Below, the influence of the individual parameters is explained and their suitability to control combustion phasing at different operating conditions is analyzed.

### Intake air preheating

The easiest way to control HCCI via the temperature profile is through variation of the intake air temperature, which has been done in numerous tests on four-stroke engines [95, 96]. However, the easy handling (even transition from spark ignition combustion or trailing throttle through continuous increase of the intake air temperature) and the size of the map achieved this way (limitation only through volumetric efficiency, pressure gradient and available heating power), are accompanied by major disadvantages during transient engine operation as in reality, for reasons of efficiency, exhaust-gas/fresh air heat exchangers can be used to provide thermal heat. The reaction times of such heat exchangers are too long when a rapid change of the intake air temperature is required, e.g. for a fast increase or decrease of power demand (change from low idle or low load to full load and vice versa, e.g. when accelerating a vehicle, when braking or during gear shift).

### Increase of the compression ratio

Another parameter that influences the combustion phasing is the compression ratio. A rigid increase of the compression without variation possibility, however, has the disadvantage that full load operation with spark ignition process may be possible only with huge limitations in addition to the lack of any way of intervening in the HCCI operation. For compression ratios greater than approximately 13:1, the danger of knock during the operation with gasoline increases considerably and seriously limits the permissible load range. A solution could be to use variable compression, with high demands being made on the accuracy and speed for the setting of a particular compression ratio [97, 98].

### Variation of the fuel composition

The fuel properties are an important factor for the auto-ignition tendency of the mixture. The control of combustion phasing on the basis of a variable mixture of two fuels with different ignition quality has already been tested successfully [99, 100]. Another option could be the addition of additives to the fuel, with which the auto-ignition tendency of the mixture can be influenced [101, 102]. For mobile use in a vehicle, however, an operating procedure that requires refueling with two different fuels or additives is not convenient.

### Variation of the residual gas amount

In comparison to intake air preheating, the variation of the residual gas percentage is a much better solution for transient applications. As in preheating, the temperature is augmented by an increased residual gas amount at the beginning of compression to such an extent that auto-ignition conditions are reached at the end of compression. Different strategies can be applied to increase the residual gas amount in the cylinder; **figure 3.64**.

The external exhaust-gas recirculation is the easiest and most economic solution, and is already available in most standard engines. This option, however, has the disadvantage that the exhaust gas cools down strongly even for short return lines. Moreover, conventional EGR valves allow neither exact measuring of the EGR exhaust-gas recirculation rate for each cycle nor precise control of the actual exhaust-gas recirculation amount under all operating conditions. Both processes, however, are basic conditions for residual gas controlled HCCI control in as wide a map range as possible. For these reasons, external exhaust-gas recirculation is not suitable in gasoline engines for reaching the desired temperature increase in the cylinder via an increase of the residual gas amount.

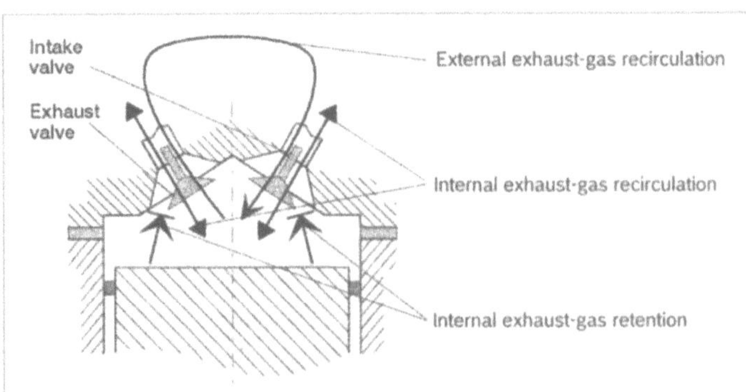


**Figure 3.64:** Strategies for increasing the amount of residual gas in the cylinder

During internal EGR (or exhaust-gas re-induction), however, the exhaust gas is first pushed into the intake or exhaust tract and is subsequently re-inducted. Internal EGR via the exhaust requires a second opening of the exhaust valves during the intake cycle. Internal EGR via intake can in contrast be implemented by means of a phase adjuster by advancing the “inlet open” timing. As a disadvantage, however, it is to be taken into account that the exhaust gas pushed back into the intake inlet cools down relatively quickly while also heating up the intake area and the fresh mixture in it. Such a strategy therefore leads on the one hand to an increased demand for residual gas to implement the desired temperature increase in the cylinder, whereas on the other hand the actual volumetric efficiency is reduced due to the heating of the fresh mixture. It is therefore a less efficient, but comparatively easily handled procedure to increase the amount of residual gas in the cylinder.

To implement internal EGR via exhaust, normally at least one simple, fully variable valvetrain is required, which allows a second opening of the exhaust valves when the corresponding cam is already on the base circle [103, 104]. The advantage of exhaust-gas re-induction from the exhaust is the relatively good controllability and less cooling in comparison to the processes discussed above. An easier solution is to shift the closing timing of the exhaust valve to the intake cycle, but this is possible only for moderate compression or pronounced valve seat pockets. In this case, the intake control timing has to be adjusted (e.g. by means of a phase adjuster) such that no fresh mixture reaches the exhaust tract [105].

Exhaust-gas retention is the third option. Here the exhaust-gas mass pushed out is directly reduced by shortening the “exhaust opens” time and by advancing the “exhaust closes” time, so that a considerable amount of residual gas remains in the combustion chamber after the gas exchange is completed. The advantage of this option is that the residual gas cools down only a little, which takes place via heat losses through the combustion chamber wall. To set the desired internal temperature of the cylinder at compression start, this process requires smaller residual gas percentages than the processes described above. For an optimized valve lift curve (short opening time for comparatively big valve lift), it is also possible to reduce flow losses, as the retained amount of residual gas does not need to pass the valves (multiple times). By adjusting the “exhaust closes” time, it is possible to control the amount of residual gas directly and continuously. If required, this can be determined precisely for each cycle depending on the adjustment mechanism. Moreover, it is even possible to achieve extremely high exhaust-gas retention rates and therefore high residual gas percentages, in combination with corresponding valve lift curves.

The special suitability of exhaust-gas retention for operating points with low load, i.e. a low amount of converted fuel and therefore high heating demand, has also been confirmed in experiments [106].

Apart from purely heating the mixture, residual gas also has a number of other effects, which have been examined in different analytical and experimental studies [107, 108]. These include the effect of charge dilution (reduction of the molar fuel amount in the mixture by increasing the residual gas amount instead of throttling), the effect on the heat capacity of the mixture (increased $CO_2$ and $H_2O$ amounts by replacing excess air with residual gas) and chemical influences (dissociation reactions).

An influence to be considered separately from the other factors is the effect of non-homogeneities in charge distribution, which can be caused by incomplete mixing of fresh mixture and residual gas [108]. This makes it more likely that even for smaller amounts of residual gas, high-temperature zones will be close to areas of high fuel concentration [106], by which the initiation of ignition is enhanced.

### 3.5.3 Intervention options for direct injection

The use of direct injection instead of external mixture formation enables direct access to the temperature history of the fuel molecules and therefore also to the temporal temperature profile in the mixture and its willingness to react. The injection point determines the time that the fuel has to prereact until the end of compression is reached. Furthermore, the internal evaporation of the fuel leads to a cooling of the cylinder charge, which has different effects on the gas exchange, on the temperature profile and finally also on the composition of the air/fuel/residual gas mixture [109, 110], depending on the injection point. Sophisticated injection strategies involve multiple injections per working cycle, with which a portion of the injection amount is inserted in the low load range at a very early point in time. For high loads, the reaction is limited through retarded injection when the intake is finished. in order to prevent combustion phasing to take place too early [111]. If the injection is too late, though, locally increased fuel concentrations cause the combustion point to advance and the exhaust-gas raw emissions to increase [112, 113].

A suitable injection strategy also makes it possible to control combustion phasing. By means of corresponding valve timing, high residual gas amounts can be achieved in the combustion chamber. This leads to an intermediate compression during gas exchange. The exhaust gas is therefore compressed again and temperatures above 1000 K can be attained. These conditions can specifically be employed for direct fuel injection. The residual oxygen still present in the exhaust gas can then be used for minor combustion within the intermediate compression when the composition of the mixture is too stoichiometric during the main combustion at the top dead center of the ignition in gas exchange. This requires good mixture formation for advanced injection directly after "exhaust closes" [114]. **Figure 3.65** shows the pressure and valve increase curves for the double injection strategy for HCCI and compares them to conventional gasoline engine operation with spark ignition.

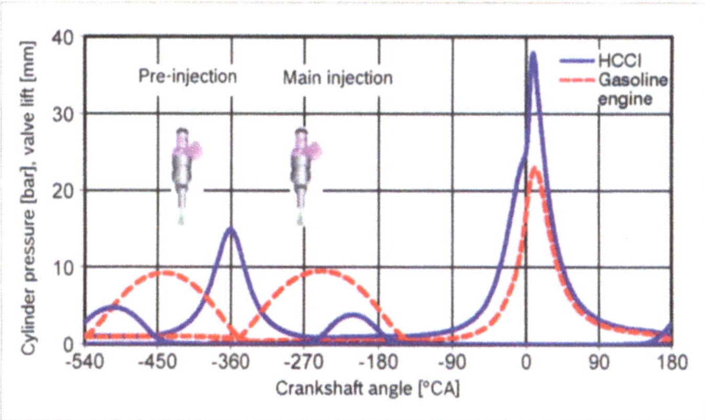


**Figure 3.65:** Injection strategy and pressure characteristic for controlled auto-ignition with exhaust-gas retention through corresponding valve timing [114]

## 3.5.4 Potential with regard to consumption and pollutant emissions

Initial comparisons on the basis of loss distribution calculations confirm the high efficiency potential of HCCI and especially the comparatively low heat losses through the wall that can be achieved for this combustion process [115]; **figure 3.66**. Overall, the gasoline engine with spray-guided direct injection shows the best results in the test; however, with $\varepsilon = 12.0$, this engine has a higher compression than the reference power engines with conventional manifold injection, variable valve control and hydrogen combustion ($\varepsilon = 10.5$ in each case) as well as homogeneous charge compression ignition (CAI, $\varepsilon = 11.0$).

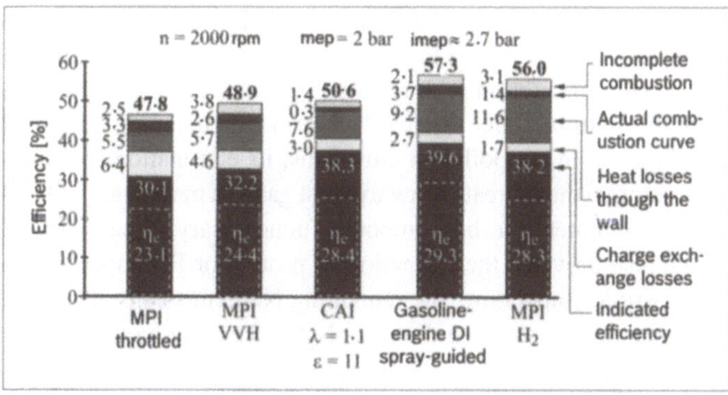


**Figure 3.66:** Comparison of the loss distributions of different combustion and load control processes [115]

In [116], Lavy et al. describe the work on HCCI at the Institut Français du Pétrole (IFP). Here too, a concept with gasoline direct injection has been examined. **Figure 3.67** summarizes the results of this investigation in comparison to a gasoline engine with direct injection and spark ignition in the homogeneous operating mode at $\lambda = 1$ and in the lean stratified operating mode.

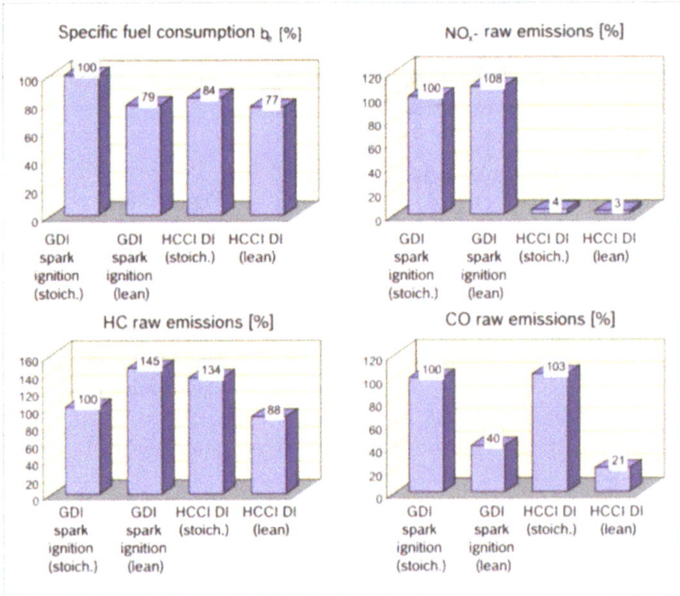


**Figure 3.67:** Potential of HCCI compared to gasoline direct injection (GDI), $n = 2000$ rpm, *bmep* = 2 bar [116]

It becomes clear that in the operating point observed, HCCI with direct injection has the best values for all limited exhaust pollutants and for actual consumption during lean operation. Although only one individual operating point has been depicted, it clearly shows the potential of HCCI with direct injection of gasoline. Moreover, the low conversion temperature for HCCI allows the prevention of the thermal NO formation to a large extent, because the combustion peak temperature can generally be kept below the NO formation temperature of approximately 2000 K. This is due to the homogeneous conversion and the strong mixture dilution and, in case of dilution with residual gas, it is additionally due to the increased heat capacity of the mixture. Since thermal NO formation in the conventional gasoline engine contributes the most to the pollutant emissions, its elimination enables a reduction of the $NO_x$ concentration in the untreated raw exhaust gas to a magnitude of typically 10 ppm, which makes aftertreatment for this component unnecessary. Because of the low fuel concentration, which also prevents the formation of prompt or Fenimore NO in the flame – and the low combustion temperature, the remaining $NO_x$ emissions are attributed mainly to the continued reaction of nitrogen dioxides $N_2O$ formed at high pressures to NO [92]. For the conventional gasoline engine, the percentage of the nitrogen dioxide $NO_2$ emissions of the overall $NO_x$ can be ignored due to the high combustion temperatures, whereas for the conventional diesel engine, the $NO_2$ percentage can constitute up to 30% of the overall emissions [117] due to the non-homogeneous mixture composition and the overall lower temperature level. However, no concrete test results on the percentage of $NO_2$ emissions for HCCI with gasoline have been presented so far.

### 3.5.5 Acoustics

The introduction of new, thermodynamically more efficient combustion processes for automotive propulsion has always brought new tasks for the acoustician, e.g. an increased level of excitation that is due to higher peak pressures and pressure rise rates, or changed noise characteristics due to new combustion processes [118]. For gasoline engines with direct injection and controlled auto-ignition (CAI), the rapidly starting auto-ignition combined with faster combustion can lead to an increase of all acoustic characteristic variables directly connected to the combustion peak pressure, its maximum pressure gradient and its derivation, relative to the ITDC (ignition top dead center). In addition, the generation of the thermodynamic boundary conditions which lead to controlled auto-ignition can affect acoustics. Just as important is the fact that CAI can currently not be operated in the entire engine map, as shown in **figure 3.68**. This is especially the case when operating the engine at higher loads and speeds.

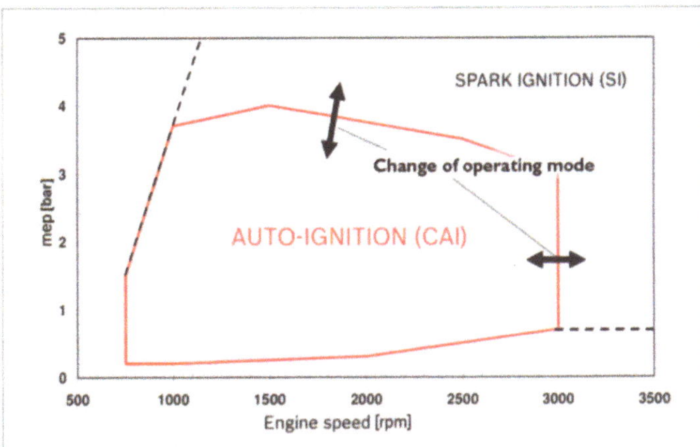


**Figure 3.68:** Operating range of CAI in the engine map

Therefore, operating mode changes between spark ignition (SI) and preferably dethrottled CAI mode are required either at constant engine speed and variable load or at constant load and variable engine speed. Below, the combustion excitation of the controlled auto-ignition is compared with conventional spark ignited homogenous combustion as well as with a modern common rail diesel engine with turbocharging (TDCI). Furthermore, the interaction of application development and sound and vibration is described while investigating the influence of different control settings doing parametric studies. Finally, the change of operating modes and potential requirements to achieve acceptable noise and vibration characteristics is laid out.

#### Combustion excitation

If the controlled auto-ignition is carried out according to the principle of residual gas compression through negative valve overlapping [119, 120], the residual gas mixture is precompressed in the top dead center of charge exchange due to the early closing of the exhaust valve in order to provide the corresponding thermodynamic boundary conditions for auto-ignition [119]. According to the first law of thermodynamics the pressure rise during combustion can be defined as follows:

$$\frac{dp}{d\alpha} = \left[\frac{(\gamma - 1)}{V}\frac{dQ_H}{d\alpha}\right] - \left[\gamma \frac{p}{V}\frac{dV}{d\alpha}\right]$$

With $V$ denoting the actual volume of the cylinder, $\alpha$ the crank-angle, $\gamma$ the ratio of specific heats, and $dQ_H/d\alpha$ denoting the heat release. However, during CAI operation the pressure rise rate $dp/d\alpha$ is dominated by the heat release, thus by $dQ_H/d\alpha$. Consequently the heat release rate during CAI operation must be controlled appropriately, to achieve acceptable noise and vibration levels.

**Figure 3.69** shows such a cylinder pressure curve and each corresponding pressure gradient in comparison with a spark ignition GDI engine and a common rail diesel engine at identical load and engine speed. The short time frequency analysis of the combustion corresponding to the pressure characteristic is shown in the right half of the figure next to the pressure characteristic for all three combustion types. The excitation level of each combustion is written above the time-frequency analysis.

It can be clearly seen that CAI can lead to high excitation levels due to the high peak pressure and high pressure gradient. However, as a result of the injection strategy and mixture

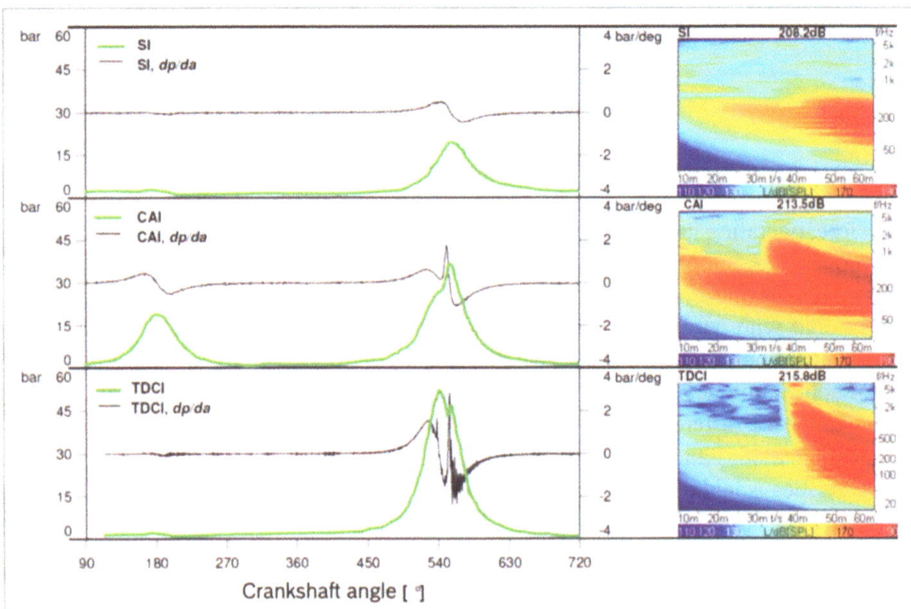


**Figure 3.69:** Analysis of combustion excitation through spark ignition (SI), controlled auto-ignition (CAI) and diesel engine with turbocharging and direct injection (TDCI), [121]

formation, the CAI process, in contrast to TDCI, shows lower pressure gradients, which are an indirect measure of the combustion excitation in the frequency range 0.5 to 5 kHz. The characteristic of the pressure gradients of the two auto-ignition processes also show the typical curve of prereactions and main reactions that start suddenly. As the short time frequency analysis of the combustion excitation shows, both auto-ignition combustion processes excite combustion chamber modes, visible due to higher narrowband excitation levels, that drift to lower frequencies over the duration of combustion due to the falling temperature in the combustion chamber as combustion progresses. An early start of combustion and fast conversion rates can therefore be assumed [122].

**Figure 3.70** depicts a calculation of the radiated noise, based on [123], of a spark ignited gasoline engine (SI) compared to controlled auto-ignition (CAI) at identical load and engine speed. An increase in the noise level of 1.6 dB(A) in the radiated noise as well as an increase of the noise level in the frequency range 0.3 to 3.5 kHz result in an examined operating point of 2000 rpm, 2 bar mean effective pressure (pme). Modulations caused by cylinder-to-cylinder deviations can lead to increased roughness in the engine's radiated noise. **Figure 3.71** shows the upper and lower scattering range for a statistically significant cycle number for both combustion processes. A comparison of stability and synchronization shows that

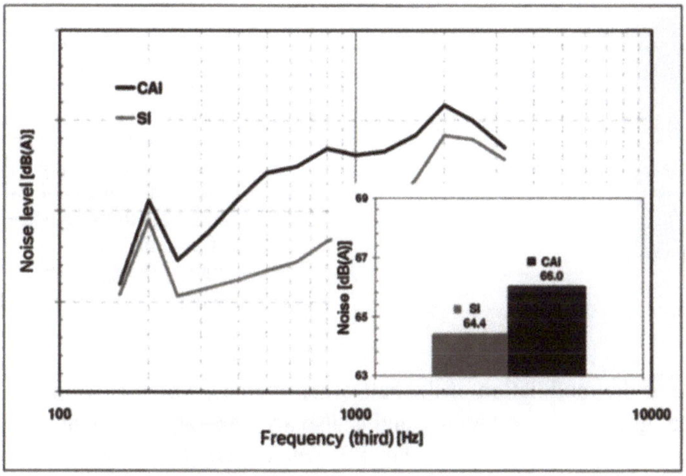


**Figure 3.70:** Radiated noise and level at 2000 rpm, $p_{me}$ 2 bar; SI and CAI in comparison

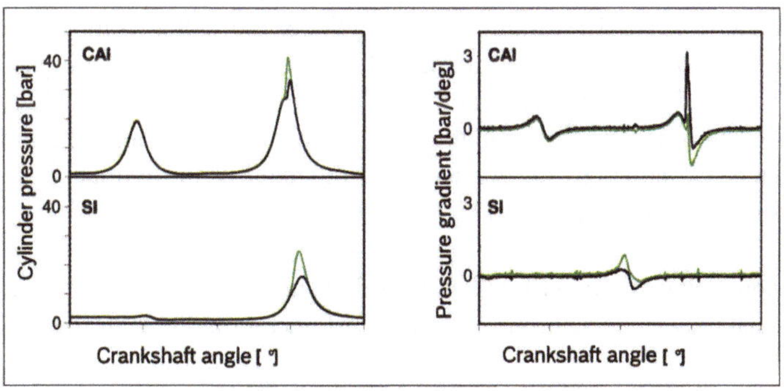


**Figure 3.71:** Scattering range for SI and CAI combustion at 2000 rpm, $p_{me}$ 2 bar in comparison: left: scattering range of the cylinder pressure; right: scattering range of the pressure gradient

although CAI has a greater variation of the peak pressure, **figure 3.71** (left), and of the pressure gradient, **figure 3.71** (right), it shows better stability with regard to its position relative to the ITDC **figure 3.71**.

It may be helpful to analyze modulation and roughness as shown in **figure 3.72** to further analyze engine noise. The left half of the figure shows the excitation level of the respective carrier frequency versus the modulation frequency $f_M$. The increased vertical modulation frequency bands can be assigned to the respective engine excitation orders. The right half of the figure shows an analysis of the roughness $R$ in asper (Latin for rough) [124] over the analyzed period of time.

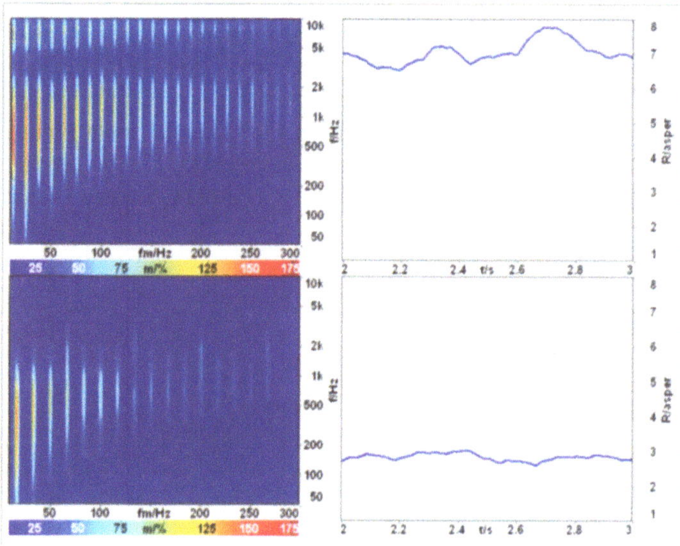


**Figure 3.72:** Analysis of modulation and roughness (right half of the figure) of the direct combustion noise of SI and CAI at 2000 rpm, $p_{me}$ 2 bar [124]

CAI leads to an increase of the modulation of the half-order engines, which is similar to that of other compression ignition diesel engines, and is also accompanied by an increase in the excitation level of these modulations at higher frequencies. The half-order characteristics feature better cylinder individual combustion stability during CAI operation, but higher spread of the cylinder-to-cylinder variation of the cylinder individual combustion characteristics. This can be attributed to harder running of the engine as well as a higher cylinder-to-cylinder deviation of the peak pressures for CAI. The roughness analysis [124] shows that the engine radiated noise becomes rougher in CAI operation. Furthermore, the roughness varies. This can be explained by the greater scattering range of the cylinder peak pressure and of the pressure gradients; **figure 3.71**.

A comparison of the combustion excitation SI with CAI with regard to the resulting crankshaft drive forces shows that they can also increase during CAI operation due to peak pressure and peak position; **figure 3.73**.

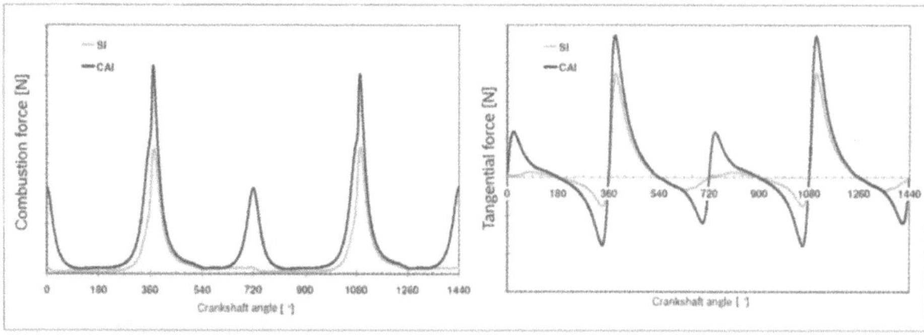


**Figure 3.73:** Combustion and tangential forces for SI and CAI operation at 2000 rpm, $p_{me}$ 2 bar in comparison; left: combustion forces for SI and CAI operation; right: tangential forces for SI and CAI operation

Combustion excitation not only results in an increase of the indirect combustion noise, which is due to the increased tangential force level during operation in the CAI combustion process. It should also be taken into account that in addition to potentially reinforcing the belt and auxiliaries at the crankshaft drive, combustion excitation also either attenuates or decouples the entire powertrain with an adapted flywheel or decoupling design and that the powertrain mounting system also needs to be altered accordingly.

### Interaction of application development and acoustics

Efficient development of the combustion system on the test bench requires early integration of application and noise development [125]. The optimal compromise of fuel consumption, emissions and engine noise should be found. In doing so, systematic parametric studies and an analysis of the main influencing factors of different control parameters of the engine/ combustion system are useful. **Figure 3.74** shows the influence of the intake control timing on the noise level during CAI operation, for example.

Here it becomes apparent that, due to the changing control timing strategy, increased attention should be paid to the acoustic behavior of the intake system in order to detect, for example, target conflicts between level and noise quality at an early point in time. Due to the diesel-like load regulation during CAI operation, it is reasonable to thoroughly examine the influence of injection start as the main parameter influencing the combustion characteristic and therefore the radiated noise. The injection rate and injection timing can have a significant effect on the combustion noise characteristic. As an example, **Figure 3.75** shows the dependency of the noise level on the injection start, whereby the injection point is referring to the point in time at which the main injection starts (as is already known from diesel engines, flexible injection strategies can be implemented employing solenoid or piezoelectric injection nozzles). A very advanced injection point leads to an increased noise level due to the longer mixture formation period. It is accompanied by an intensifying, diesel-like knocking noise.

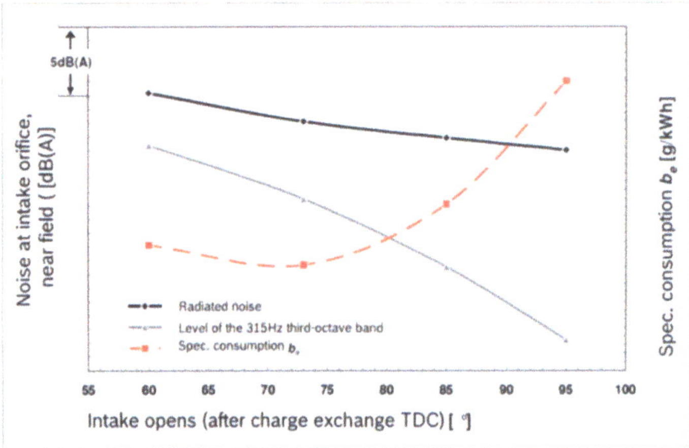


**Figure 3.74:** Influence of opening point of the intake valve (start of activation) on noise during CAI

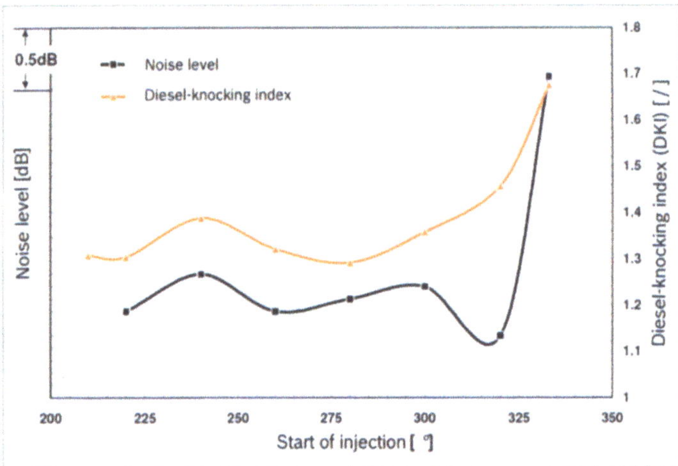


**Figure 3.75:** Interaction of main injection start, noise and noise quality (DKI)

The intensity of the knocking noise can be characterized and quantified by means of a narrowband modulation analysis, so that limit values can be targeted here too. An analysis of the noise by means of this method based on [126] for different injection points shows a progressive increase of diesel knocking intensity for a very advanced injection point, as is illustrated by the increase of the diesel knocking index (DKI) in **figure 3.75**. For advanced injection, the time for mixture formation is longer (ignition delay) and therefore leads to enhanced intermixing and finally to faster and earlier combustion [121].

During CAI operation, the engine should preferably be operated in dethrottled mode to achieve the maximum possible fuel consumption advantage. A possible increase in intake orifice noise can be reduced by slight throttling in combination with the negative valve overlapping strategy. **Figure 3.76** shows the airborne noise level at the intake orifice de-

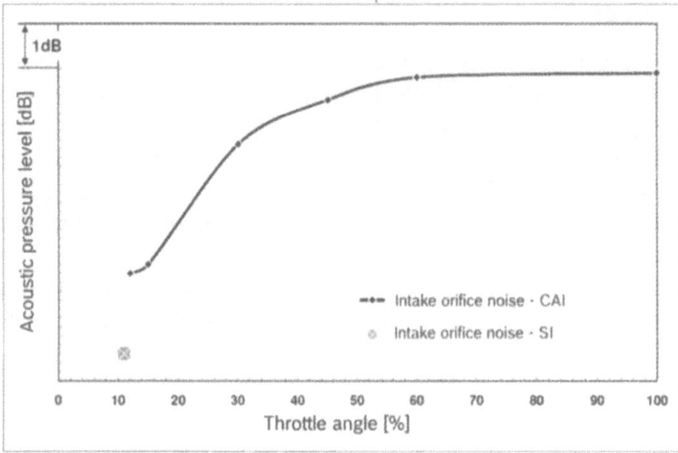


**Figure 3.76:**
Noise level at intake orifice depending on throttle valve position

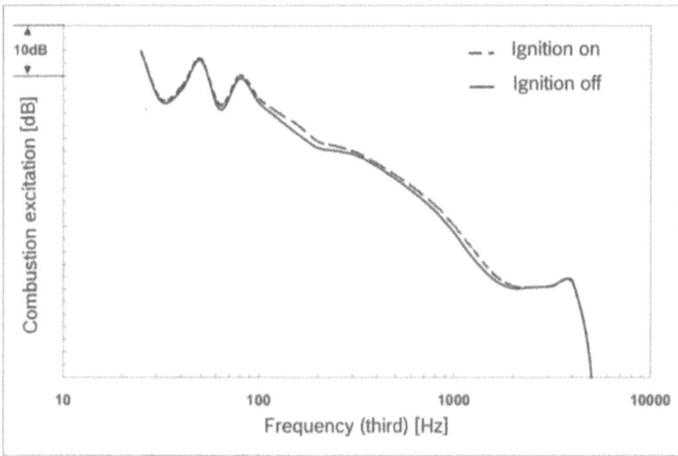


**Figure 3.77:**
Influence of spark-supported CAI combustion on combustion excitation

pending on the throttle valve angle, which behaves in accordance with according the fluid-dynamic laws that apply.

CAI combustion that is already taking place can be controlled by means of an additionally activated spark ignition when certain limits are adhered to. Based on an analysis of the combustion excitation and the radiated noise, it can be concluded that this has no significant influence on the noise excitation of the respective combustion, **figure 3.77**, as the excitation level for the spark supported combustion increases by only 0.5 dB. The advantage of such an operating strategy is that the combustion and its center can be shifted for constant duration relative to the ITDC. This strategy can be useful for the application of the operating mode changes.

## Change of the operating modes

The operating range of known CAI processes requires a change of modes between SI combustion and CAI, as is demonstrated in **figure 3.68**. This can be necessary in the low load range at increasing engine speed or at increasing load over the engine speed band relevant to CAI. On the one hand, a possible change of the radiated noise and its characteristic should be taken into account and on the other hand, the mode change should take place without vibration and jerks, i.e. without noticeable drop of the mean effective pressure $p_{me}$ and significant change of the peak pressure $p_{max}$ or the pressure gradient $dp/d\alpha$. Exact controllability of these variables is therefore the focus when developing strategies for the mode change. **Figure 3.78** shows the measured noise level of two different strategies when switching from SI to CAI.

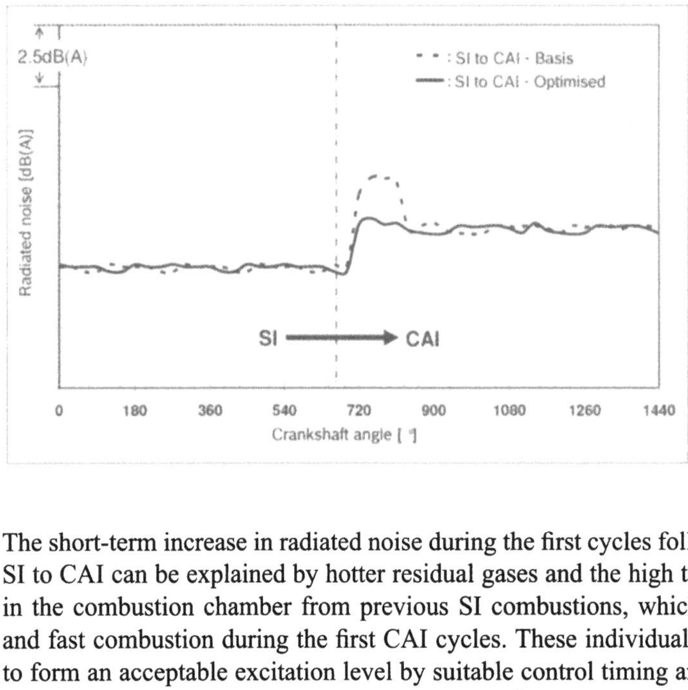


**Figure 3.78:** Radiated noise during operating mode change from SI to CAI; initial state and optimized strategy

The short-term increase in radiated noise during the first cycles following the change from SI to CAI can be explained by hotter residual gases and the high temperature still present in the combustion chamber from previous SI combustions, which lead to an early start and fast combustion during the first CAI cycles. These individual cycles can be adjusted to form an acceptable excitation level by suitable control timing and injection strategy by means of complex control models [127]. According to [123], the quality – let us call it '*NV*' – of noise and vibration, relates to the individual consecutive combustion cycles as a function of the following parameters:

$$NV \sim f[\,p_{max}\,;\,\Delta p_{max} = (p_{max,i+1} - p_{max,i})\,,\,(imep_{i+1} - imep_i)\,]$$

with $p_{max}$ denoting the maximum cylinder pressure, *imep* the indicated mean effective pressure, and the index $i$ indicating an individual engine cycle, and $i+1$ denoting the consecutive cycle of the following firing cylinder. *NV* should be minimized.

As a possible quality criterion for the creation of requirements specification with regard to low frequency powertrain vibrations (< 32 Hz), the vibration dose value (VDV), for example, can be used due to the change of the operating mode [128].

## 3.6 The combined diesel/gasoline engine combustion process

The auto-ignition of homogeneous gasoline/air mixtures also allows a reduction in fuel consumption and operation with extremely low pollutant emissions. As explained in the previous chapters, a particular amount of recirculated exhaust gas is required to ensure auto-ignition in a gasoline engine. Corresponding operation with homogeneous fuel/air mixture can also be implemented for the diesel engine, increased exhaust-gas recirculation rates also being used to control the ignition. For both engine variants (gasoline or diesel engine), direct injection is the most reasonable solution for the mixture formation in order to bring the homogeneous mixture to auto-ignition at the correct point in time in the compression phase. For the gasoline engine, the injection takes place relatively early in the intake phase to ensure auto-ignition, whereas for the diesel engine, a retarded injection is used in the compression phase so that the diesel/air mixture is not brought to auto-ignition prematurely. Outside of the possible HCCI operating range which, as described, is limited to the low load and engine speed range, the gasoline engine is conventionally operated with spark ignition and the diesel engine with heterogeneous mixture.

To initiate auto-ignition in the gasoline engine in the time available during the compression phase, an increase of the compression ratio is desirable for this engine as compared to the compression ratio for engines with manifold injection or with conventional direct injection. For the diesel engine, in contrast, the compression ratio is reduced as compared to the traditionally higher compression ratio for diesel engines in order to prevent a very advanced auto-ignition during the compression of the air/diesel fuel mixture. This way, an equalization of the compression ratios takes place in the research and development of the homogeneous combustion processes with auto-ignition for gasoline and diesel engines. As a result, the possibility of operating a combustion engine both as a gasoline engine and a diesel engine emerges.

Moreover, new alternative fuels, the designer fuels GTL (*Gas to Liquid*) and BTL (*Biomass to Liquid*), for corresponding specification in the boiling range of higher boiling gasoline components (160 to 220°C) and low boiling diesel components (200 to 260°C), allow both engine concepts to operate with homogeneous mixture formation and auto-ignition either in the compression phase or at the beginning of the expansion phase, depending on the operating condition. Volkswagen is developing this concept under the term *Combined Combustion System* (CCS), figure 3.79 [129], and Daimler under the term *Diesotto* [130]. Research in this area to describe this concept has been carried out for some time now.

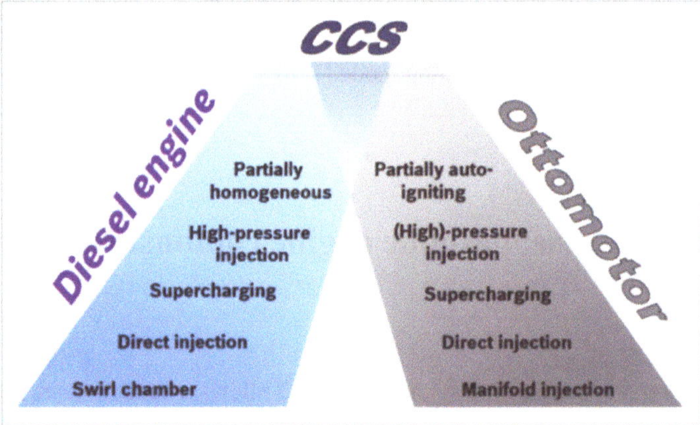


**Figure 3.79:** From diesel and gasoline engine to combined combustion system (CCS) [129]

Ongoing production and use of biogenic fuels for the combustion process with homogeneous mixture formation and auto-ignition can enable further development of the combustion engine so that both gasoline and diesel engines can be operated with the same fuel in the operating ranges where this combustion process is possible. This leads to a drastic reduction in carbon dioxide emissions; **figure 3.80** [130].

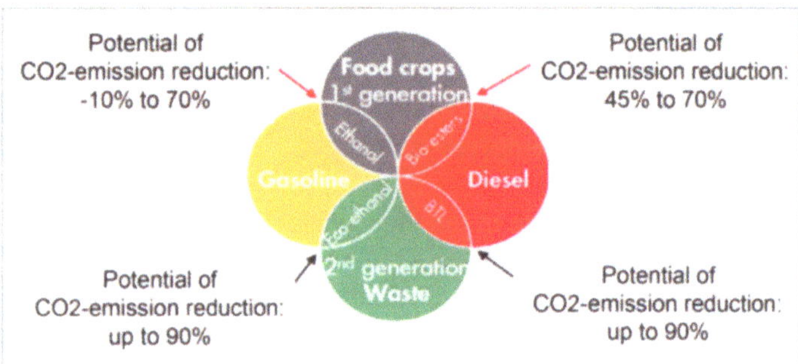


**Figure 3.80:** Fuel alternatives from biological natural products [130]

## Literature


[1] *v. Basshuysen, R.; Steinwart, J.; Staehle, H.; Bauder, A.*: Audi Turbodieselmotor mit Direkteinspritzung. In: MTZ 50 (1989), 1989

[2] *Spicher, U.; Heidenreich, Th.; Nauwerck A.*: Stand der Technik strahlgefuehrter Verbrennungssysteme. In: Strahlgefuehrte Verbrennungssysteme. Essen: Expert-Verlag, 2004

[3] *Stiebels, B.; Schweizer, M.; Ebus, F.; Pott, E.*: Die FSI-Technologie von Volkswagen – nicht nur ein Verbrauchskonzept. In: Direkteinspritzung im Ottomotor IV. Essen: Expert-Verlag, 2003

[4] *Nauwerck, A.*: Untersuchung der Gemischbildung in Ottomotoren mit Direkteinspritzung bei strahlgefuehrtem Brennverfahren. Dissertation, Universitaet Karlsruhe (TH). Logos-Verlag, 2006

[5] *Nauwerck A., Kettner M., Velji A., Spicher U.*: Gemischbildung und Verbrennung in Ottomotoren mit strahlgefuehrtem Brennverfahren. In: Direkteinspritzung im Ottomotor IV. Essen: Expert-Verlag, 2003

[6] *Maass, J.; Leyh, B.; Tschoeke, H.*: Computersimulation der Gemischbildung fuer Hochdruck-Benzindirekteinspritzung (HD-BDE). In: 12. Aachener Kolloquium Fahrzeug- und Motorentechnik", 2003

[7] *Ohnesorge, W.*: Formation of drops by nozzles and the breakup of liquid jets. In: Zeitschrift fuer Angewandte Mathematik und Mechanik, Band 16, S. 355–358, 1936

[8] *Baumgarten, C.*: Modellierung des Kavitationseinflusses auf den primaeren Strahlzerfall bei der Hochdruck-Dieseleinspritzung. Dissertation, Universitaet Hannover, Fortschritt-Bericht VDI, Reihe 12, Nr. 543, 2003

[9] *Lenz, H. P.*: Gemischbildung bei Ottomotoren. Wien, New York: Springer Verlag, 1990

[10] *Arcoumanis, C.; Gavaises, M.; French, B.*: Effect of fuel injection process on the structure of diesel sprays. SAE Technical Paper 970799, 1997

[11] *Gilles-Birth, I.; Bernhardt, S.; Spicher, U.; Rechs, M.*: A study of the in-nozzle flow characteristic of valve covered orifice nozzles for gasoline direct injection. SAE Technical Paper, 2005-01-3684, 2005

[12] *Faeth, G. M.; Hsiang, L.-P.; Wu, P.-K.*: Structure and breakup properties of sprays. In: International Journal Multiphase Flow, Vol. 21, S. 99–127, 1995

[13] *Warnatz, J.; Maas, U.; Dibble, R. W.*: Verbrennung: Physikalisch-Chemische Grundlagen, Modellierung und Simulation, Experimente, Schadstoffentstehung. 3. Auflage, Springer Verlag, 2001

[14] *Bayvel, L.; Orzechowski, Z.*: Liquid atomization. Taylor & Fransis, 1993

[15] *Kraemer, S.*: Untersuchung zur Gemischbildung, Entflammung und Verbrennung beim Ottomotor mit Direkteinspritzung. Dissertation, Universitaet Karlsruhe (TH), 1998

[16] *Kauer, R.; Oswald, H.; Pritze, S.; Rothenberger, P.; Wiedenhoeft, D.*: Benzindirekteinspritzung mit stoechiometrischer Verbrennung am 2,2l ECOTEC Aluminium-Motor fuer den Opel Signum. In: Direkteinspritzung im Ottomotor IV. Essen: Expert-Verlag, 2003

[17] *Zimmermann, D.; Kerek, Z.; Wirth, M.; Gansert, K.; Grzeszik, R.; Storch, A.; Josefsson, G.; Sandquist, H.*: Entflammung geschichteter Gemische im strahlgefuehrten Benzin-DI-Motor: Optimierung mit Hilfe von Hochgeschwindigkeitsvisualisierung. In: 6. Internationales Symposium fuer Verbrennungsdiagnostik, 2004

[18] *Lefebvre, A. H.*: Atomization and sprays. Hemisphere Publishing Corporation, 1989

[19] *Troesch, H. A.*: Zerstaeubung von Fluessigkeiten. Dissertation, Fortschritt-Berichte VDI, Reihe 3, Nr.607, 1999

[20] *Weber, C.*: Zum Zerfall eines Fluessigkeitsstrahls. In: Zeitschrift fuer angewandte Mathematik und Mechanik, Heft 11, S. 136–154, 1931

[21] *Pischinger, F.*: Motorische Verbrennung. Abschlussbericht des SFB 224, 2001

[22] *Spicher, U.*: Untersuchung ueber die raeumliche Ausbreitung und das Erloeschen der Flamme bei ottomotorischer Verbrennung. Dissertation, RWTH Aachen, 1982

[23] *Fischer, J.*: Einfluss variabler Einlassstroemung auf zyklische Schwankungen bei Benzin-Direkteinspritzung. Dissertation, Universitaet Karlsruhe (TH). Logos-Verlag, 2004

[24] *Heywood, J. B.*: Internal Combustion Engine Fundamentals. McGraw-Hill, 1988

[25] *Boewing, R.*: Der Einfluss von Zuendung und Zylinderinnenstroemung auf die Verbrennung in Ottomotoren mit hoher Ladungsverduennung. Dissertation, RWTH Aachen, 2000

[26] *Kettner, M.*: Experimentelle und numerische Untersuchungen zur Optimierung der Entflammung von mageren Gemischen bei Ottomotoren mit Direkteinspritzung. Dissertation, Universitaet Karlsruhe (TH). Logos-Verlag, 2006

[27] *Fansler, T. D.; French, D. T.*: Swirl, squish and turbulence in stratified-charge engines: Laser-Velocimetry Measurements and Implications for Combustion. SAE Technical Paper 870371, 1987

[28] *Gindele, J.*: Untersuchung zur Ladungsbewegung und Gemischbildung im Ottomotor mit Direkteinspritzung. Dissertation, Universitaet Karlsruhe (TH). Logos-Verlag, 2001
[29] *Pfeil, J.; Fischer, J.; Kettner, M.; Spicher, U.*: Influence of spray propagation on the gas phase in a direct injection gasoline engine. In: 10th International Symposium on Flow Visualization, 2002
[30] *Herden, W.; Vogel, M.*: Perspektiven alternativer Zuendsysteme. In: Diesel- und Benzindirekteinspritzung. Essen: Expert-Verlag, 2004
[31] *Pischinger, S.; Salber, W.; Dilthey, J.; Adomeit, G.; Gruenefeld, G.; Brochhaus, M.*: Ladungsbewegung und Gemischbildung bei Ottomotoren mit voll variabler Ventilsteuerung. In: MTZ 62 (2001), Nr. 11, S. 936–945
[32] *Rotta, J.*: Turbulente Stroemungen. Teubner Verlag, Stuttgart, 1972
[33] *Sullivan, P.; Ancimer, R.; Wallace J.*: Turbulence averaging within spark ignition engines. In: Experiments in Fluids, Volume 27, S. 92–101, Springer-Verlag, 1999
[34] *Raposo, J.; Hentschel, W.; Merzkirch, W.*: Analysis of the dynamical behavior of coherent structures in in-cylinder flows of internal combustion engines. In. 10th International Symposium on Applications of Laser Techniques to Fluid Mechanics, 2000
[35] *Schnittger, W.; Koenigstein, A.; Pritze, S.; Poepperl, M.; Samstag, P. R. M.*: 2.2 Direct Ecotec – Neuer Ottomotor mit Direkteinspritzung von Opel. In: MTZ 64 (2003), Nr. 12, S. 1010–1019
[36] *Glanz, R.*: Differentielle Erfassung von Tumble-Stroemungsfeldern. In: MTZ 61 (2000), Nr. 1, S. 40–45
[37] *Buechling, J.*: Der neue 2,0-l-FSI-Motor von Audi. In: MTZ 62 (2002), Nr. 10, S. 788–789
[38] *Winterkorn, M.; Spiegel, L.; Bohne, P.; Soehlke, G.*: Der Lupo FSI von Volkswagen – So sparsam ist sportlich Teil 1 (Teil 2). In: ATZ 102 (2000), Nr. 10/11, S. 832–841, 964–974
[39] *Kume, T.; Iwamoto, Y.; Iida, K.; Murakami, M.; Akishino, K.; Ando, H.*: Combustion control technologies for direct injection SI engine. SAE Technical Paper, 960600, 1996
[40] *Anderson, W.; Yang, J.; Brehob, D. D.; Vallance, J. K.; Whiteaker, R. M.*: Understanding the thermodynamics of direct injection spark ignition (DISI) combustion systems: an analytical and experimental investigation. SAE Technical Paper, 962018, 1996
[41] *Herzog, P. L.; Athenstaedt, R.*: Kraftstoffsysteme fuer DE-Ottomotoren. In: 17. Internationales Wiener Motorensymposium. VDI Forschritt-Berichte, Reihe 12, Nr. 267. Duesseldorf: VDI-Verlag, 1996
[42] *Voss, E.; Schmittger, W.; Koenigstein, A.; Scholten, I.; Poepperl, M.; Pritze, St.; Rothenberger, P.; Samstag, M.*: 2,2 l ECOTEC DIRECT – Der neue Vollaluminiummotor mit Benzindirekteinspritzung fuer den Opel Signum. 24. Internationales Wiener Motorensymposium, 2003
[43] *Wurms R.; Kuhn, M.; Zeilbeck, A.; Adam, St.; Krebs, R.; Hatz, Wolfgang, W.*: Die Audi Turbo FSI Technologie. 13. Aachener Kolloquium Fahrzeug- und Motorentechnik, 2003
[44] *Krebs, R.; Boehme, J.; Dornhoefer, R.; Wurms, R.; Friedmann, K.; Helbig, J.; Hatz, W.*: Der neue Audi 2,0T FSI Motor – Der erste direkteinspritzende Turbo-Ottomotor bei Audi. 25. Wiener Motorensymposium, 2004
[45] *Jaegerbauer, E.; Froehlich, K.; Fischer, H.*: Der neue 6,0-l-Zwoelfzylindermotor von BMW. In: MTZ 64 (2003), Nr. 7/8, S. 546–555
[46] *Tsuji, N.; Sugiyama, M.; Abe, S.*: Der neue 3.5L V6 Benzinmotor mit dem innovativen stoechiometrischen Direkteinspritzsystem D-4S. 27. Internationales Wiener Motorensymposium, 2006
[47] *Kettner, M.; Fischer, J.; Nauwerck, A.; Spicher, U.; Velji, A.; Kuhnert, D.; Latsch, R.*: Ein neues Brennverfahren mit Mehrfacheinspritzung fuer Ottomotoren mit Direkteinspritzung. 9. Tagung: Der Arbeitsprozess des Verbrennungsmotors, Graz, 2003
[48] *Kettner, M.; Fischer, J.; Nauwerck, A.; Spicher, U.; Velji, A.; Kuhnert, D.*: The BPI Flame Jet Concept to Improve the Inflammation of Lean Burn Mixtures in Spark Ignited Engines. SAE 2004-01-0035, 2004.
[49] *Herden, W.; Vogel, M.*: Visionen idealer strahlgefuehrter BDE-Brennverfahren. In: Diesel- und Benzindirekteinspritzung. Essen: Expert-Verlag, 2002

[50] *Kemmler, R.; Frommelt, A.; Kaiser, T.; Schaupp, U.; Schommers, J.; Waltner, A.*: Thermodynamischer Vergleich ottomotorischer Brennverfahren unter dem Fokus minimalen Kraftstoffverbrauchs. In: 11. Aachener Kolloquium Fahrzeug- und Motorentechnik, 2002

[51] *Pischinger, R.; Grassnig, G.; Taucar, G.; Sams, T.*: Thermodynamik der Verbrennungskraftmaschine. Berlin, New York: Springer Verlag, 1997

[52] *Lueckert, P.; Rau, E.; Schaupp, U.; Vent, G.; Waltner, A.*: Weiterentwicklung der Benzin-Direkteinspritzung bei Mercedes-Benz. DaimlerChrysler-Sonderdruck zum 13. Aachener Kolloquium Fahrzeug- und Motorentechnik, 2004

[53] *Xander, B.*: Grundlegende Untersuchungen an einem Ottomotor mit Direkteinspritzung und strahlgefuehrtem Brennverfahren. Dissertation Universitaet Karlsruhe (TH). Logos-Verlag, 2006

[54] *Steeper, R.; Stevens, E.*: Effects of piston wetting in an optical DISI engine. In: Direkteinspritzung im Ottomotor III. Essen: Expert-Verlag, 2001

[55] Mitsubishi Motors: Mitsubishi Benzinmotor mit Direkteinspritzung. Technisches Dossier, 1995

[56] Bosch: Ottomotor-Management, 3. Auflage. Wiesbaden: Vieweg Verlag, 2005

[57] *Buechling, J.*: Der neue 2,0-l-FSI-Motor von Audi. In: MTZ 63 (2002), Nr. 10, S. 788–789

[58] *Grigo, M.; Klein, H.; Lietz, O.; Kunkel, R.; Dornhoefer, R.; Erdmann, H.*: Der neue Audi 2,0l 4V FSI Motor. In: 11. Aachener Kolloquium Fahrzeug- und Motorentechnik, 2002

[59] *Huebner, W.; Witt, A.; Hoss, B.; Albrecht, T.; Durst, B.; Abdelfattah, A.*: Methodeneinsatz bei der Entwicklung eines weiterfuehrenden DI-Brennverfahren. In: Direkteinspritzung im Ottomotor IV. Essen: Expert-Verlag, 2003

[60] *Lueckert et al.*: Kunden- und zukunftsorientierte Technologien am Ottomotor – heute und morgen. 26. Internationales Wiener Motorensymposium 2005

[61] *Reissing, J.*: Spektroskopische Untersuchung an einem Ottomotor mit Benzin-Direkteinspritzung. Dissertation, Universitaet Karlsruhe (TH). Logos-Verlag, 1999

[62] *Sarikoc, F.; Kettner, M.; Velji, A.; Spicher, U.; Krause, A.; Elsaesser, A.*: Potential of reducing the NOX emissions in a spray guided DI gasoline engine by stratified exhaust gas recirculation (EGR). SAE Technical Paper 2006-01-1261

[63] *Kufferath, A.; Samenfink, W.; Gerhardt, J.*: Die neue Emissionsstrategie der Benzin-Direkteinspritzung. In: MTZ 64 (2003), Nr. 11

[64] *Achleitner, E.; Berger, S.; Frenzel, H.; Klepatsch, M.; Warnecke, V.*: Benzin-Direkteinspritzsystem mit Piezo-Injektor fuer strahlgefuehrte Brennverfahren. In: MTZ 65 (2004), Nr. 5

[65] Bosch: Zuendung im Ottomotor, Fachwissen Kfz-Technik, Motorsteuerung fuer Ottomotoren. 2002

[66] *Dreizler, A.*: Physikalisch-chemische Grundlagen der Zuendprozesse. Vorlesungsskript, Institut fuer technische Verbrennung der Universitaet Stuttgart WS 97/98

[67] *Maly, R.*: Fuel economy in road vehicles powered by spark ignition engines, chapter spark ignition: its physics and effect on the internal combustion engine. S. 91–148, Plenum Press, 1984

[68] *Maly, R.; Saggaue, B.; Spaude, H. W.; Vogel, M.; Wagner, E.; Greiner, E.; Horch, E. J.*: Die Auswirkungen einer verbesserten elektrischen Entflammung auf die Verbrennung im Ottomotor. In: Automobil Industrie, Heft 3, S. 37–41, 1978

[69] *Zimmermann, D.; Kerek, Z.; Wirth, M.; Gansert, K.; Grzeszik, R.; Storch, A.; Josefsson, G.; Sandquist, H.*: Entflammung geschichteter Gemische im strahlgefuehrten Benzin-DI-Motor: Optimierung mit Hilfe von Hochgeschwindigkeitsvisualisierung. 6. Internationales Symposium fuer Verbrennungsdiagnostik, 2004

[70] *Kaiser, T.; Hoffmann, A.*: Einfluss der Zuendkerzen auf das Entflammungsverhalten in modernen Motoren. In: MTZ 61 (2000), Nr. 10, S. 656–663

[71] *Wirth, M.; Zimmermann, D.; Friedfeldt, R.; Caine, J.; Schamel, A.; Storch, A.; Ries-Mueller, K.; Gansert, K.-P.; Pilgram, G.; Ortmann, R.; Wuerfel, G.; Gerhardt, J.*: Die naechste Generation der Benzin-Direkteinspritzung – gesteigertes Verbrauchspotenzial bei optimierten Systemkosten. 12. Aachener Kolloquium Fahrzeug- und Motorentechnik, 2003

[72] *Boewing, R.*: Der Einfluss von Zuendung und Zylinderinnenstroemung auf die Verbrennung in Ottomotoren mit hoher Ladungsverduennung. Dissertation, RWTH Aachen, 2000
[73] *Gansert, K. P.; Storch, A.*: Ignition and inflammation at spray guided combustion systems. 1st Ignition Symposium, 2004
[74] *Herden, W.*: Zuendung und Entflammung bei Benzindirekteinspritzung. Diesel- und Benzindirekteinspritzung. Essen: Expert-Verlag, 2000
[75] *Ricardo, P.; Gansert, K. P.*: Insulator nose flashover: detection method, influence on combustion. 1st Ignition Symposium, Posterpraesentation, 2004
[76] *Trachte, D.; Kaiser, T.*: Trends in spark plug engineering. 1st Ignition Symposium, 2004
[77] *Herden, W.; Vogel, M.*: Perspektiven alternativer Zuendsysteme. Diesel- und Benzindirekteinspritzung. Essen: Expert-Verlag, 2004
[78] *Fansler, T. D.; Drake, M. C.; Stoikovic, B.; Rosalik, M. E.*: Local fuel concentration, ignition and combustion in a stratified charge spark ignited direct injection engine: spectroscopic, imaging and pressure-based measurements. In: International Journal of Engine Research, Volume 4, Nr. 2, S. 61–86, 2003
[79] *Fansler, T. D.; French, D. T.*: Swirl, Squish and turbulence in stratified-charge engines: laser-velocimetry measurements and implications for combustion. SAE Technical Paper 870371, 1987
[80] *Koch, T.; Schaenzlin, K.; Boulouchos, K.*: Verbrennungsablauf im Ottomotor mit Direkteinspritzung. In: MTZ 64 (2003), Nr. 7/8
[81] *Spicher, U.; Schmitz, G.; Kollmeier, H.-P.*: Application of a new optical fiber technique for flame propagation diagnostics in IC engines. SAE Paper 881637
[82] *Ault, J. R.; Witze, P. O.*: Evaluation and optimization of measurements of flame kernel growth and motion using a fiber-optic spark plug probe. SAE Paper 980139
[83] *Philipp, H.; Plimon, A.; Fernitz, G.; Hirsch, A.; Fraidl, G.; Winklhofer, E.*: A tomographic camera system for combustion diagnostics in SI engines. SAE Paper 950681
[84] *Han, K.-M.; Velji, A., Spicher, U.*: A new approach for three-dimensional high-speed combustion diagnostics in internal combustion engines. SAE Paper 2006-01-3315
[85] *Reckers, W.; Schwab, H.; Weiten, C.; Befrui, B.; Kneer, R.*: Untersuchungen der Flammenausbreitung und der zyklischen Verbrennungsschwankungen in einem direkt einspritzenden Ottomotor mittels Zylinderdruckmessung und zeitsynchroner Hochgeschwindigkeits-Visualisierung. 5. Internationales Symposium fuer Verbrennungsdiagnostik, 2002
[86] *Kubach, H.; Mayer, K.; Spicher, U.*: Untersuchungen zur Realisierung einer russarmen Verbrennung bei Benzin-Direkteinspritzung. Abschlussbericht BWPLUS, 2001
[87] *Spicher, U.; Koelmel, A.; Kubach, H.; Toepfer, G.*: Combustion in spark ignition engines with direct injection. SAE Paper 2000-01-0649
[88] *Rothe, M.; Schubert, A.; Spicher, U.*: Thermodynamischer Ansatz zur Bewertung des Klopfverhaltens von Ottomotoren in der Volllast. 7. Internationales Symposium fuer Verbrennungsdiagnostik. Baden-Baden, 2006
[89] *Rothe, M.; Han, K.-M.; Spicher, U.*: Untersuchungen zur Ursache fuer Extremklopfen bei Ottomotoren mit Direkteinspritzung. 15. Aachener Kolloquium Fahrzeug- und Motorentechnik 2006
[90] *Sass, F.*: Die Geschichte des deutschen Verbrennungsmotorenbaus von 1860 bis 1918. Springer Verlag, 1962
[91] *Zeldovich, Y. B.; Librovich, V. B.; Makhviladze, G. M.; Sivashinsky, G. I.*: Development of detonation in a non-uniformly preheated gas. In: Astronautica Acta, Band 15; S. 313–321, Pergamon Press, 1970
[92] *Warnatz, J.; Maas, U.; Dibble, R. W.*: Verbrennung: Physikalisch-Chemische Grundlagen, Modellierung und Simulation, Experimente, Schadstoffentstehung. Berlin/Heidelberg: Springer Verlag, 2001
[93] *Onishi, S.; Jo, S. H.; Shoda, K.; Jo, P. D.; Kato, S.*: Active thermo-atmosphere combustion (ATAC) – a new combustion process for internal combustion engines. SAE Technical Paper 790501, 1979

[94] *Peters, H.*: Experimentelle und numerische Untersuchungen zur Abgasrueckfuehrung und Abgasschichtung beim Ottomotor mit Direkteinspritzung und strahlgefuehrtem Brennverfahren. Dissertation, Universitaet Karlsruhe (TH). Logos-Verlag, 2004

[95] *Najt, P. M.; Foster, D. E.*: Compression-ignited homogeneous charge combustion. SAE Technical Paper 830264, 1983

[96] *Thring, R. H.*: Homogeneous charge compression ignition (HCCI) engines. SAE Technical Paper 892068, 1989

[97] *Haraldsson, G.; Tunestål, P.; Johansson, B.; Hyvoenen, J.*: HCCI Combustion phasing with closed-loop combustion control using variable compression ratio in a multi cylinder engine. SAE Technical Paper 2003-01-1830, 2003

[98] *Wolters, P.; Salber, W.; Dilthey, J.*: Radical activated combustion: a new approach for gasoline engines. A New Generation of Engine Combustion Processes for the Future? (IFP International Congress), S. 153–162, Paris: Editions Technip, 2001

[99] *Christensen, M.; Hultqvist, A.*; Johansson; B.: Demonstrating the multi fuel capability of a homogeneous charge compression ignition engine with variable compression ratio. SAE Technical Paper 1999-01-3679, 1999

[100] *Olsson, J.-O.*: The HCCI engine – high load performance and control aspects. Dissertation, Lund Institute of Technology, 2004

[101] *Eng, J. A.; Leppard, W. R.; Sloane, T. M.*: The effect of di-tertiary butyl oxide (DTBP) addition to gasoline on HCCI combustion. SAE Technical Paper 2003-01-3170, 2003

[102] *Mulenga, M. C.; Ting, D. S-K.; Reader, G. T.; Zheng, M.*: The potential for reducing CO and NOx emissions from an HCCI engine using H2O2 addition. SAE Technical Paper 2003-01-3204, 2003

[103] *Fraidl, G. K.; Piock, W. F.; Fuerhapter, A.; Unger, E. M.; Kammerdiener, T.*: Homogene Selbstzuendung – Die Zukunft der Benzin-Direkteinspritzung? In: MTZ 63 (2002), S. 828–837

[104] *Fuerhapter, A.; Piock, W. F.; Fraidl, G. K.*: Homogene Selbstzuendung – Die praktische Umsetzung am transienten Vollmotor. In: MTZ 65 (2004), Nr. 2, S. 94–101

[105] *Kaahaaina, N. B.; Simon, A. J.; Caton, P. A.; Edwards, C. F.*: Use of dynamic valving to achieve residual-affected combustion. SAE Technical Paper 2001-01-0549, 2001

[106] *Wolters, P.; Salber, W.; Duesmann, M.*: Prozessoptimierung der kontrollierten Selbstzuendung mittels elektromechanischem Ventiltrieb. 6. Tagung Motorische Verbrennung, Muenchen 2003

[107] *Law, D.; Kemp, D.; Allen, J.; Kirkpatrick, G.; Copland, C.*: Controlled combustion in an IC-engine with a fully variable valve train. SAE Technical Paper 2000-01-0251, 2000

[108] *Zhao, H.; Peng, Z.; Williams, J.; Ladommatos, N.*: Understanding the effects of recycled burned gases on the controlled self-ignition (CAI) combustion in four-stroke gasoline engine. SAE Technical Paper 2001-01-3607, 2001

[109] *Thirouard, B.; Cherel, J.; Knop, V.*: Investigation of mixture quality effect on CAI combustion. SAE Technical Paper 2005-01-0141, 2005

[110] *Cao, L.; Zhao, H.; Jiang, X.; Kalian, N.*: Investigation into the effect of injection timing on stoichiometric and lean CAI operations in a 4-Stroke GDI engine. 6. Tagung Motorische Verbrennung, Muenchen 2003

[111] *Nieberding, R. G.*: Die Kompressionszuendung magerer Gemische als motorisches Brennverfahren. Dissertation, Universitaet Siegen, 2001

[112] *Marriott, C. D.; Reitz, R. D.*: Experimental investigation of direct-injection gasoline for premixed compression ignited combustion Phasing control. SAE Technical Paper 2002-01-0418, 2002

[113] *Guenthner, M.*: Untersuchung der Eigenschaften und Kontrollmoeglichkeiten der homogen kompressionsgezuendeten Verbrennung von Ottokraftstoff. Dissertation, Universitaet Karlsruhe (TH). Logos-Verlag, 2004

[114] *Herrmann, H.-O.; Herweg, R.; Karl, G.; Pfau, M.; Stelter, M.*: Einsatz der Benzin-Direkteinspritzung in Ottomotoren mit homogen-kompressionsgezuendeter Verbrennung. Tagung Direkteinspritzung im Ottomotor V. Essen: Expert-Verlag, 2005

[115] *Merker, G.; Schwarz, C.; Stiesch, G.; Otto, F.*: Simulation der Verbrennung und Schadstoffbildung. Wiesbaden: Teubner Verlag, 2004

[116] *Lavy, J.; Dabadie, J.-C.; Duret, P.; Angelberger, C.; Le Coz, J.-F.; Cherel, J.*: Controlled Self-ignition (CAI): A new highly efficient and near-zero NOX emission combustion process for gasoline engine application. Tagung A New Generation of Engine Combustion Processes for the Future?, Rueil-Malmaison, 2001

[117] *Pischinger, R.; Klell, M.; Sams, T.*: Thermodynamik der Verbrennungskraftmaschine. Wien: Springer Verlag, 2002

[118] *v. Basshuysen, R.; Kuipers, G.; Hollerweger, H.*: Akustik des Audi 100 mit direkteinspritzendem Turbo-Dieselmotor. In: ATZ 92 (1990), Nr.1

[119] *Koopmans, L.*: HCCI combustion by retaining residuals. PhD Thesis. Chalmers University of Technology, Goeteborg, 2005

[120] *Kontarakis, G.; Collings, N.; Ma, T.*: Demonstration of HCCI using a single cylinder four-stroke engine with modified valve-timing. SAE-Paper 2000-01-2870, Society of Automotive Engineers, Warrendale, USA, 2000

[121] *Stoffels, H.; Collings, N.*: Gasoline HCCI – a new challenge in powertrain NVH engineering. MTZ/ATZ Akustiktagung 2006. Stuttgart, 2006

[122] *Hensel, S.; Sauter, W.; Velji, A.; Spicher, U.*: Analyse der zyklischen Schwankungen bei homogen kompressionsgezuendeter Verbrennung (HCCI). In: Salber, W. et al.: Kontrollierte Selbstzuendung. Haus der Technik, Essen, 20.-21. Oktober 2005. Aachen: Shaker-Verlag, 2005

[123] *Heuer, S.*: Verbrennungsgeraeusch der direkteinspritzenden Hubkolbenmaschine. Dissertation, RWTH Aachen, 2001

[124] *Zwicker, E.; Fastl, H.*: Psychoacoustics, 2nd ed. Berlin, Heidelberg, New York: Springer Verlag, 1999

[125] *Menne, R. J.; Rechs, M.*: Optimierte Prozesse fuer die Grossserie. Berlin, Heidelberg: Springer Verlag, 2000

[126] *Bodden; M.; Heinrichs, R.*: Diesel sound quality engineering. 32. Deutsche Jahrestagung fuer Akustik DAGA'06. Berlin, Deutsche Gesellschaft fuer Akustik e.V., 2006

[127] *Karagiorgis, S.; Collings, N. et al.*: Dynamic modeling of combustion and gas exchange processes for controlled self-ignition engines. Proceedings of the American Control Conference, S. 1880-1885. Minneapolis, Minnesota, USA, 2006

[128] *Griffin, M. J.*: Handbook of human vibration. London: Academic Press, 1990

[129] *Steiger, W.; Kohnen, Ch.*: Neue Brennverfahren auf Basis einer neuen Kraftstoffspezifikation. 27. Internationales Wiener Motorensymposium, 2006

[130] *Weber, Th.*: Advanced power: Zukunftsorientierte Antriebstechnologien fuer unsere Kunden. 27. Internationales Wiener Motorensymposium, 2006

# 4 Injection Systems and System Overview

In order to tap the maximum potential of GDI (Gasoline Direct Injection) and avoid the usage of fuel gained thereby due to higher power consumption of the ancillaries, each individual component must be optimized, keeping in mind increased demand. The greatest single contributor to these measures is minimizing the power drawn by the pump for high pressure generation by regulating the fuel quantity and reducing the friction. Additionally, the injection components interacting with the engine ambient conditions are tuned to each other so that for all possible operating conditions, it is possible to achieve a mixture formation tailored to the requirements of performance, lower consumption and lower raw emissions.

As shown in **figure 4.1**, the correlation between the individual engine operating parameters and the injection parameters and their effect on mixture formation, combustion and emissions is very complex. The development and design of the injection system must take these correlations into account as well as other boundary conditions of the combustion process, such as combustion chamber geometry, injection spray spread, the geometrical arrangement of the injector nozzle and the spark plug, intake flow and air supply in the combustion chamber, and tune these to each other in the best possible way.

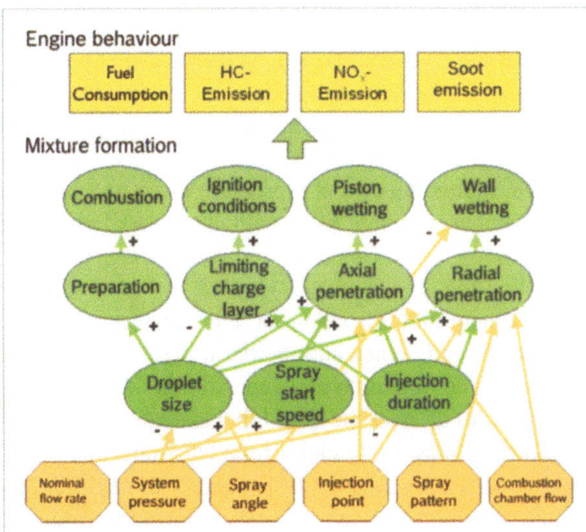


**Figure 4.1:**
Parameters and their interaction for mixture formation [1]

The high degree of parameter correlation makes adapting the injection system for the entire operating range a costly affair and impedes the development of an injection concept that accounts for the varied circumstances at different operating points and for demands related to fuel consumption, emission of pollutants, costs and driving comfort. In order to achieve good atomization of the injected fuel, as is required by all combustion processes with gasoline direct injection, the following methods of injecting the fuel directly into the combustion chamber are the two main methods used in the automobile industry:

- Low pressure air assisted direct injection
- High pressure common rail direct injection.

Low pressure direct injection was rapidly developed by Orbital [2, 3] towards the end of the 1980's and the beginning of the 1990's. The main objective then was to inject the densest possible air/fuel vapor with very small fuel droplets into the combustion chamber at a relatively low injection pressure. When this injection technology was being developed, there were no suitable high pressure injection systems for fluid fuel available that met the requirements imposed by dynamic engine operation on the mixture formation during direct injection in gasoline engines. It was only in the mid-1990's that high pressure common rail injection was developed for diesel engines [4, 5]. From then on, this injection technology was also developed for direct injection in gasoline engines and was implemented for the first time in the Mitsubishi GDI, which was released on the Japanese market in 1995 and the European market in 1997.

## 4.1 Low pressure air assisted direct injection

Air assisted low pressure direct injection, often referred to as low pressure mixture injection, features partial mixing of air and fuel before being injected into the combustion chamber. The basic set-up of a low pressure direct injection system of this type with external generation of air pressure as per the Siemens/OCP system (*Orbital Combustion Process* [3]), is shown in **figure 4.2**. In this case, fuel is injected at a pressure of up to 8 bar into the low pressure injection valve's prechamber, which is filled with compressed air (up to 5 bar).

In high pressure direct injection, the disintegration of fuel spray into small droplets and its subsequent vaporization occur due to the turbulence and inertia in the fuel spray. The penetration depth is correspondingly large. The formation and vaporization of droplets occur in air assisted injection if there is a flow of air over the liquid fuel and the tensile stress at the fuel surface is greater than the surface tension. In case of the OCP injection valve, the fuel droplets are disintegrated if they are accelerated by the expanded air that is escaping out of the nozzle. Much of the fuel disintegration and vaporization occurs within the air chamber in the injection valve. Thus, the already vaporized air/fuel mixture is injected into the combustion chamber, thereby scavenging the air chamber with the compressed air that is fed into it and injecting the fuel that has not yet vaporized in the air chamber along with the compressed air. The air is compressed to 5 bar using the externally operated air compressor. However, it is also possible to supply the compressed air through the energy released during combustion (compare e.g. DMI process of AVL [8]).

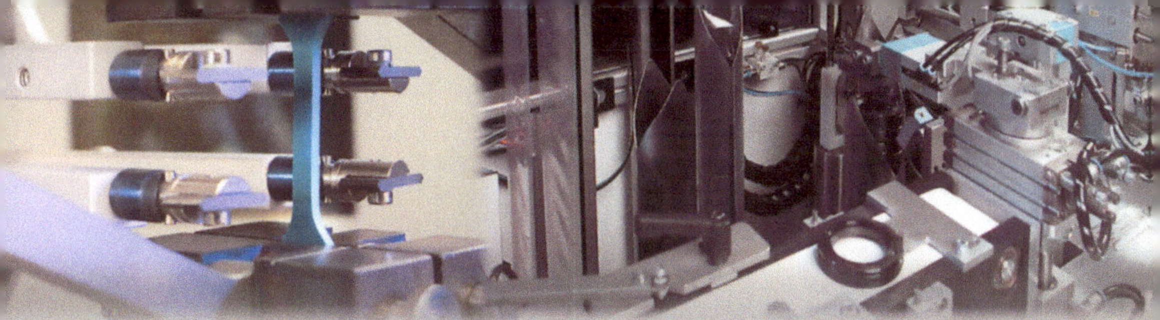

Only a small quantity of the injection spray or the injected cloud of air and fuel mixture penetrates into the combustion chamber because of very small fuel droplets formed due to partial vaporization. The mixture is then ready to the point where it can be ignited. The further mixing of air and fuel, charge transfer and mixture distribution in the combustion chamber can be additionally supported by swirl and/or tumble flow.

This process also decreases the droplet size as the injection pressure increases. **Figure 4.3** shows the dependency of the droplet diameter at two different injection pressures [9]. A high rate of exhaust-gas recirculation (EGR) can be achieved [10] only through the very small droplets in the injected mixture cloud.

The mixture cloud is typically injected with an outwards opening injection nozzle (A-nozzle). It has a ring-shaped cross-section at the opening and features a wide spray cone; **figure 4.4**. Thus, principally conducive charge stratification can be achieved with very early injection, i.e. with low back pressure in the combustion chamber. Due to low injection pressure, delayed injection towards the end of the compression stroke, i.e. at high cylinder pressure, is not possible. Although conducive charge stratification can be achieved by injecting a mixture vapor early, there is a risk of the vapor being blown away before the

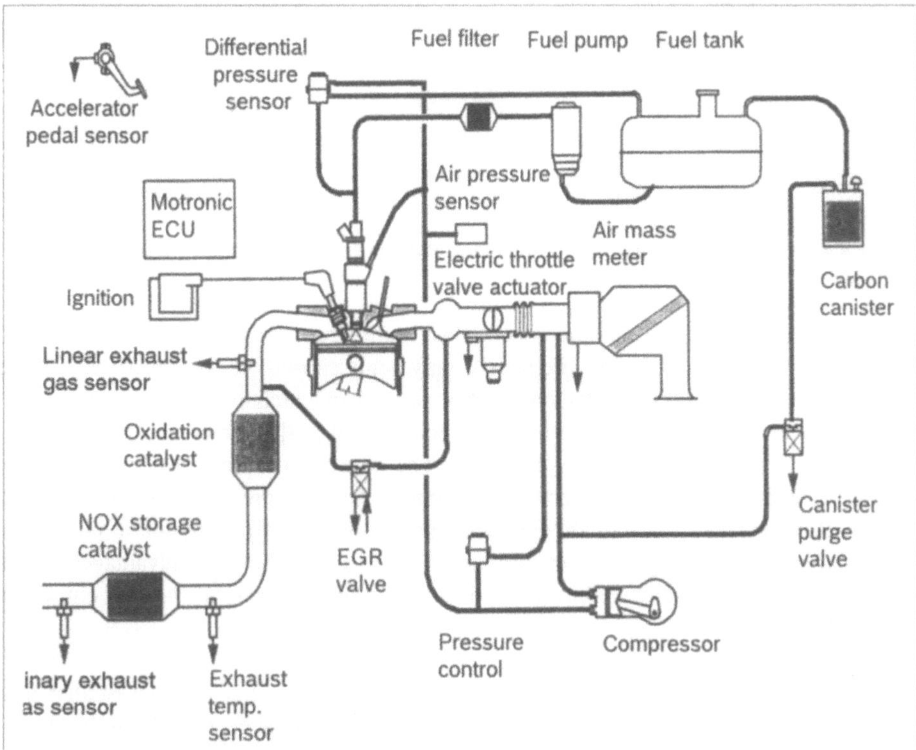


**Figure 4.2:** System representation of an air assisted injection system [3]

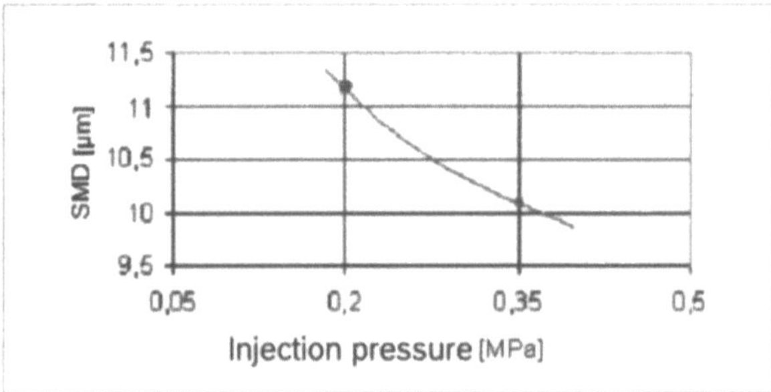


**Figure 4.3:**
Droplet size depending on injection pressure [9]

ignition point and thus not being ignited due to the cylinder-internal airflow. This is also the significant disadvantage of this process, since stable charge stratification conducive for low load operating points can seldom be generated. The loss of net output due to the need to generate compressed air is another disadvantage. Since this process is relatively easy to implement, it was tested by some automobile manufacturers despite the aforementioned disadvantages. During this time however, the main focus was on achieving homogenous operation with a stoichiometric air/fuel ratio [10].

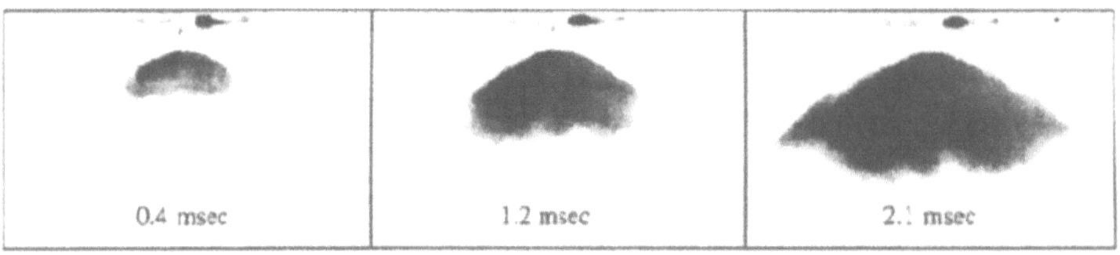


**Figure 4.4:** Spray pattern for air assisted direct injection: injection quantity = 69 $mm^2$ [11]

Until now, air assisted low pressure injection has mainly been tested in two-stroke engines [12, 13]. Tests and developments are increasingly being carried out for four-stroke engine application as well. A spray-guided combustion process in which injection valve and spark plug are positioned closely to each other is implemented [14] in practically all the applications of this process in four-stroke engines. The tests have revealed that a particularly lean mixture can be created by injecting the mixture. This and the possibility of achieving high EGR rates have favorably affected fuel consumption and the emission of pollutants. Performance with regard to hydrocarbon emissions is one of the main advantages of this process. Since liquid fuel has almost no contact with the flame front, the soot emissions are also low.

Systems for air assisted low pressure direct injection are manufactured by the Australian firm Orbital Engine Company [3], among others. For instance, injection valves from this manufacturer are in use in a Mercury Marine 200 HP V6 two-stroke outboard engine and other two-wheelers [15].

## 4.2 High pressure common rail direct injection

The common rail injection system was originally developed for diesel engines. It is based on the research carried out by the ETH Zurich from 1976 to 1992. The common rail injection system derives its name from the fuel rail or the common rail. Based on the research by ETH Zurich, a prototype of the "Unijet System" was developed by Magneti Marelli, Centro Ricerche Fiat and Elasis by the end of the 1980's. The development was transferred to Bosch for industrial series production whereby ETH Zurich and Fiat did significant pioneering work and are thus recognized as founders of the Common Rail System.

It was only in the last few years that the common rail injection technology was implemented in the gasoline engines. Mitsubishi introduced the first series passenger car with stratified gasoline direct injection, first with the Carisma GDI in 1995 in Japan and then in Europe in 1997.

The conditions for good applicability of an injection system on the entire engine map of a gasoline engine with direct injection are: the randomly selectable injection point, multiple injections when necessary within one working cycle of a cylinder, a high quality of atomization, a spray penetration depth adapted to the process of mixture formation, and variable system pressure, particularly in case of high pressure injection. The common rail system best meets these requirements. The common rail is used to store and dampen the fuel pressurized (system pressure) by the low pressure fuel pump and the high pressure pump. From this common rail, high pressure fuel is then injected into the individual cylinders via the injection valves with a fuel pressure adapted to the respective operating conditions. Unlike some systems that are known for diesel application such as rail injection pump and pump nozzle technology, high pressure is constantly available in the common rail system. Thus fuel can be injected at any time using an electrically actuated injection valve. Common rail injection systems are state-of-the-art technology for gasoline engines with direct injection [16, 17].

The design of a common rail injection assembly (high pressure circuit) is shown in **figure 4.5** [16]. It consists of a high pressure fuel pump, the common rail, a pressure sensor, a pressure control valve and one injection valve for each cylinder. Increasing injection pressures necessitate cooling of the fuel by means of a radiator.

The combination of the common rail system with modern engine control is shown in **figure 4.6** [18]. The fuel system consists mainly of a low pressure and a high pressure circuit. The low pressure circuit on the tank side, in which a low pressure of approximately 4 bar is generated, comprises an electrical fuel pump and a mechanical pressure controller and is connected to the high pressure pump inlet. The high pressure fuel pump then generates the high pressure (state-of–the-art today is a maximum of 200 bar) and pumps the fuel into the high pressure circuit (common rail) to which the injection valves are connected. In the high pressure circuit, the control circuit comprising pressure sensor and pressure control

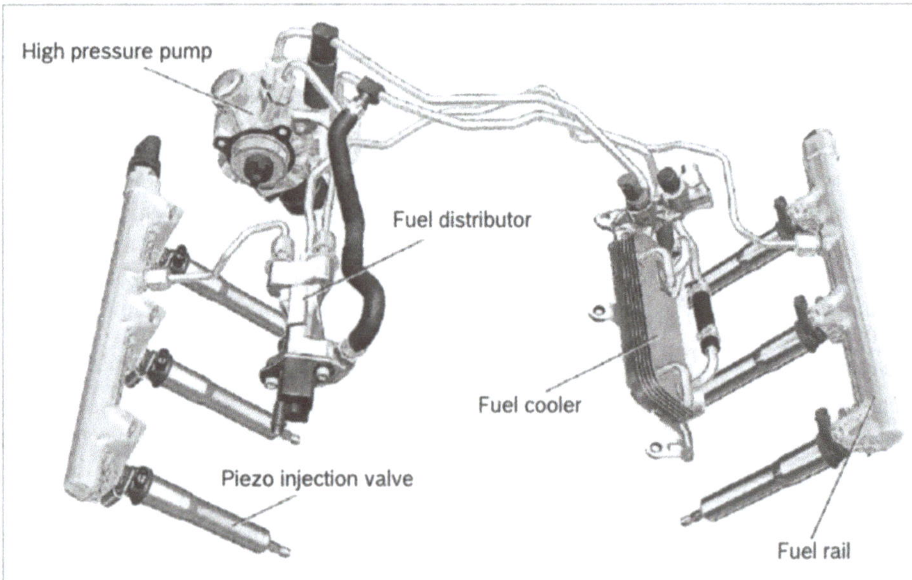


**Figure 4.5:** High pressure circuit cycle of a system for gasoline direct injection with common rail, DaimlerChrysler [16]

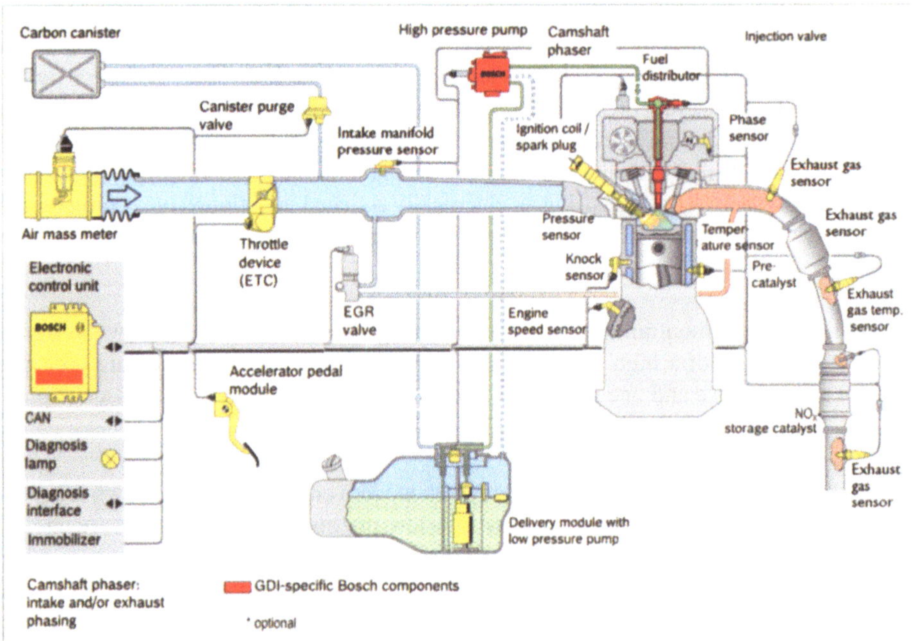


**Figure 4.6:** Overall engine control with injection system [18]

valve sets the pressure in the entire operating map – independent of the injection and pump delivery rate – as per the requirements specific to the respective operating conditions.

Several variations with respect to the injection point, injection pressure and number of injections (e.g. pilot injection, multiple injection and post injection) are possible, with the constantly available high pressure at the injection valve, the electrical actuation of the injection valves (independent of the camshaft setting) and the adjustable injection pressure. The desired injection pattern can also be achieved using modern piezo-actuated injection valves [16, 17].

External energy (pump output, compressor output) must be applied for both air assisted high pressure systems and low pressure injection systems. For a high pressure injection pump of a common rail injection system, these outputs are between roughly 0.6 kW (50 bar, $n = 2000\ \text{min}^{-1}$) and 3.0 kW (150 bar, $n = 6000\ \text{min}^{-1}$), depending on the injection pressure and the engine speed. The output for operating a compressor for air assisted low pressure injection is approximately 2.0 kW at 5 bar air pressure and $n = 6000\ \text{min}^{-1}$ (air quantity approximately 20 to 25% of the injected fuel) [19].

### 4.2.1 High pressure fuel pumps

The high pressure pump raises the pressure of the fuel coming from the low pressure pump (approximately 4 bar) in the low pressure circuit to the high pressure that is required for injection (up to 200 bar). The supply pressure depends on the respective operating point of the engine. For this the supplied fuel flow should have the least possible pulsations of pressure and volume flow to maintain constant pressure in the fuel accumulator. The supplied volume flow should be independent of the supply pressure. Therefore piston pumps are used. To achieve the necessary drive power, the high pressure fuel pump is operated using the camshaft. Since the high pressure is generated only at peak engine speed, the fuel injection in the initial working cycles of cranking occurs only at the pressure of the low pressure circuit [18].

Principally, leakages occur between pistons and cylinders of the pump. Mixing engine oil and fuel should be avoided in both directions, otherwise either the oil becomes thin, thereby reducing lubrication, or the fuel is contaminated, thus increasing the hydrocarbon emissions. Therefore, pistons and cylinders of the pump and often also the parts of the pump piston drive are sealed against the oil lubricated cylinder head and are lubricated with fuel (media lubrication). These sliding systems of the pumps are subjected to tribological stress due to media lubrication combined with the surface compressions in the sliding systems of the pump that increase constantly due to high supply pressures, and to gasoline's poor lubricating characteristics, which is worsened by the reduced sulfur content [18, 20, 21].

The high pressure fuel pump is therefore one of the significant challenges in the development of modern high pressure GDI systems. Based on the latest results in the development of the combustion process, minimizing the high pressure pump's drive power – comprising work done per minute and friction loss – and increasing the supply pressure are the main objectives of the modern high pressure GDI systems in which friction loss can be resolved only by reducing the wear of the sliding system of the pump. Modern tools of engineering ceramics [21] and DLC coatings (*Diamond-like Carbon*) are the possible solutions for this.

Other development objectives stem from general demands in automobile engineering. Thus, the high pressure fuel pump should be suitable for application worldwide, i.e. also for an ethanol component of 25% or more. This further restricts the material selection (e.g. alcohol corrosion of aluminum). Packaging requirements such as compact dimensions avoiding the use of a leakage (feedback) line to the tank without the consequent thinning of engine oil and the costs also play an important role here [22].

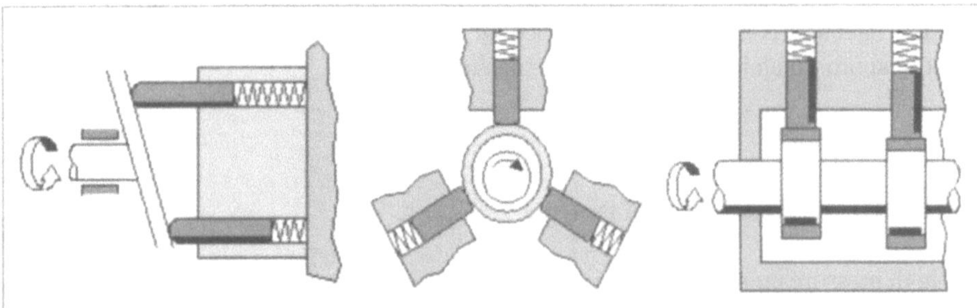

**Figure 4.7**: Basic designs of high pressure fuel pumps: axial piston pump (left); radial piston pump (center); in-line pump (right)

The following main designs of high pressure fuel pump are possible; **figure 4.7**.

- Axial piston pump
- Radial piston pump:
  - in one or multiple cylinder design as a star or in-line arrangement
  - with displacement ring or direct on cams
- In-line pump

The different designs, with their respective advantages and disadvantages, are set out in the overview below. The most commonly used designs are the radial piston pump and the axial piston pump, in which mostly multiple-cylinder designs are chosen for reducing the pulsations in the fuel supply flow. However, single-cylinder pumps for high pressure gasoline injection are also available [18]. Membrane pumps are also used in isolated cases [23].

The engine needs a defined fuel quantity suitable for the operating point. Since the high pressure pump is directly coupled to the camshaft, the required fuel quantity is proportional to the engine speed. The required quantity also increases with the speed. The dependency of the required fuel quantity on the load status of the engine interferes with the speed.

The main techniques used to control the required quantity are as follows:
- Changing the engine speed or displacement (volume)
- Throttling the volume flow to the high pressure pump
- Throttling a fraction of the volume flow to the high pressure pump
- Throttling the volume flow sucked in
- Electrically actuated valve for controlling the actual compression start.

**Overview of the design types and selection criteria of high pressure fuel pumps**

| Principle | Criteria | | | | | |
|---|---|---|---|---|---|---|
| | Service life | Efficiency | Pulsations | Oil/fuel separation | Size | Costs |
| Axial piston pump | o | o | + | o | – | + |
| 3 Cylinder radial piston pump | + | + | + | o | + | + |
| 1 Cylinder radial piston pump | o | o | – | o | ++ | + |
| In-line pump | o | o | + | o | – – | o |
| Membrane pump | – | – | o | ++ | – | – |

The star shaped coupling of the high pressure pump with the camshaft does not allow any adjustment of the speed independent of the engine speed. Also, mechanical adjustment of the displacement proves to be difficult. Due to the fuel supply characteristics of the piston pumps, throttling the complete fuel supply is not advisable. Throttling leads to increased consumption of the pump power since the throttle position causes a pressure increase that cannot be used for the injection system. One of the easiest alternatives for controlling the fuel quantity is to install the pump in such a way that it can deliver the full load quantity with sufficient reliability at every operating point. The excess quantity delivered is then decompressed to the ambient pressure using the pressure control valve. However, during this switch, the energy and the friction work are also used for the reset quantity, i.e. the system efficiency is not optimum.

Another quantity control method that is easy to realize is the throttling of the volume flow that is sucked in, thereby using up the least energy and friction work. Due to throttling in the suction line, vacuum is generated, which could lead to the formation of a negative cavity.

Valves that are electrically operated enable the flexible use of these techniques. At the beginning of the pressure stroke of the high pressure pump, it is also possible to deliberately generate a short circuit of the combustion chamber with the intake side and thus change the piston stroke, whereby the quantity can be controlled without extra energy and friction work and without the risk of cavity formation.

### Designs

*Three-cylinder radial piston pump (e.g. Bosch)*

The fuel pump shown in **figure 4.8** is a radial piston pump with three cylinders offset by 120°, which features an alternately overlapping fuel supply and, because of this, less pulsation. It is driven from the front at the engine camshaft. The drive shaft with one eccentric and one displacement ring generates the displacement stroke of the pump piston. The intake valve is integrated into the hollow piston, whereby fuel flows through the casing of the eccentric ring. The sliding systems of the piston/cylinder and eccentric/gliding block are lubricated using the medium. The supply quantity is proportionately dependent on the engine speed. Fuel supplied in excess (e.g. part load operation) is decompressed to the low pressure level and fed to the intake side.

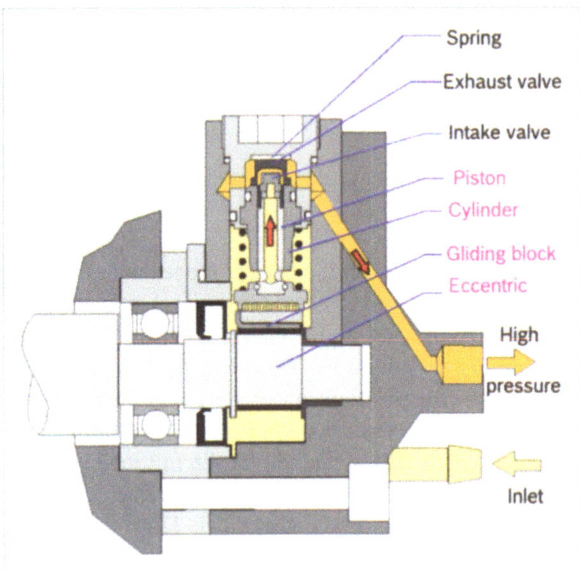


**Figure 4.8:**
Three-cylinder radial piston pump [18]

*Single-cylinder radial piston pump (e.g. Bosch)*

The single-cylinder radial piston pump, figure 4.9, is run as a plug pump with direct drive radial using the camshaft (additional cam lobes). The stroke is transferred directly using the bucket-tappet integrated in the cylinder head. An electrically switchable quantity control valve enables the fuel to be returned to the inlet. Adjusting the fuel supply quantity to that required by the engine reduces the power consumption of the pump and with it the fuel consumption. The pressure pulsations resulting from the supply characteristics of the pump are reduced using a pressure damper. The piston sealing, acting as a separator between the fuel and the engine oil area, is the most important functional element and is decompressed by connecting it to the leakage line that goes to the tank. Lubrication using the media is done only in the sliding system of the piston and cylinder and not between the eccentric and sliding block as in case of the three-cylinder radial piston pump.

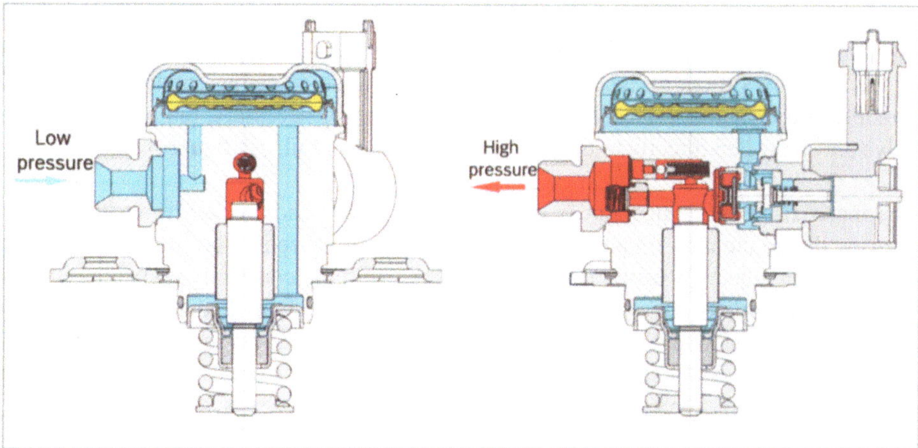


**Figure 4.9:** Single-cylinder radial piston pump, Bosch HDP5 (high pressure pump) [18]

*Three-cylinder axial piston pump (e.g. Siemens)*

**Figure 4.10** shows an axial piston pump manufactured by Siemens [33]. The stroke of the three pistons in this pump is generated using a swash plate that is rotated using the powered shaft. To achieve good efficiency, the fuel supply is controlled using an integrated three-way valve on the inlet side of the pump. There is no need for a return line from the pressure controller to the inlet of the high pressure pump, which simplifies the system. This reduces the fuel temperature at the inlet of the high pressure pump since heated fuel does not flow back. The piston and cylinder gliding system and the swash plate and gliding block are lubricated.

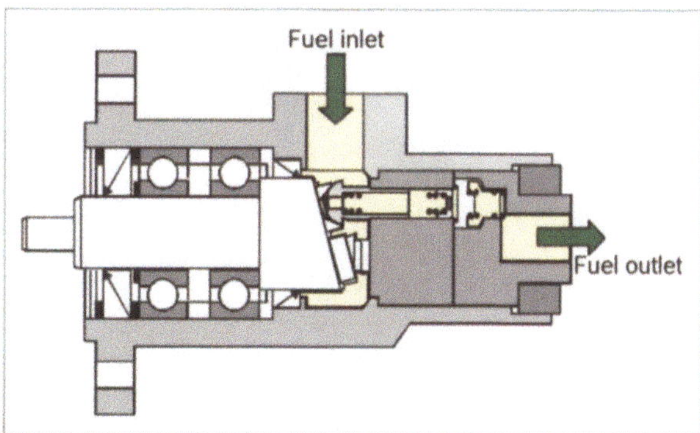


**Figure 4.10:**
Axial piston pump [24]

### 4.2.2 Rail

The common tubular fuel line is known as the rail (fuel distribution rail). The common rail is used to store and dampen the high pressure fuel and supply it to the injection valves of the engine or a cylinder bank. The larger the storage capacity, the better the damping properties. However, the pressure build-up occurs slowly, as does the pressure control (time delay component). This means that the larger the storage capacity is, the slower the response of the injection system to dynamic changes in the engine operating conditions. As far as possible, the injection valves, the pressure sensor and, if necessary, the pressure control valve are connected directly to the fuel storage or the fuel rail; see **figure 4.11.**

The material of the component is subjected to high demands related to endurance and bursting due to the high frequency pressure fluctuations caused in the common rail by the pump and the injection valves (**figure 4.12**) along with the constantly prevailing internal pressure. The earlier rails were made of aluminum. Due to ethanol additives that can lead to alkoxide corrosion and increasing rail pressures, aluminum was replaced by iron based material, i.e. high alloy steel or stainless steel. Tensile stress peaks in the rail bore that could cause fault mechanisms are often impeded or moderated by systematically applying pressure internal stress by plastic deformation (e.g. autofrettage) [35].

The pressure control valve is installed between the rail and the low pressure side of the high pressure pump. The desired fuel pressure in the rail can be set by throttling the fuel supply flow of the high pressure pump using this valve. The excess fuel that is supplied is fed back to the low pressure circuit. The design of the pressure control valve is shown in **figure 4.13** [18]. The flow cross-section of the valve is adjusted by lifting the valve ball (7) more or

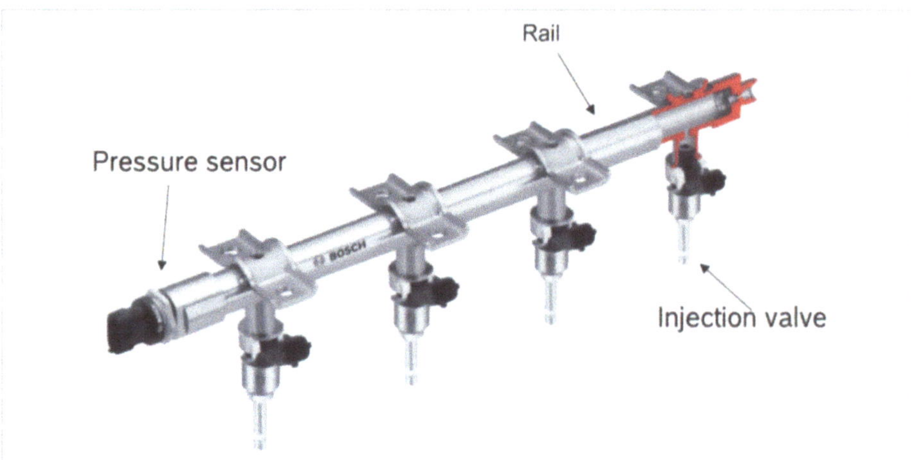


**Figure 4.11:** Common rail with injection valves and pressure sensor [19]

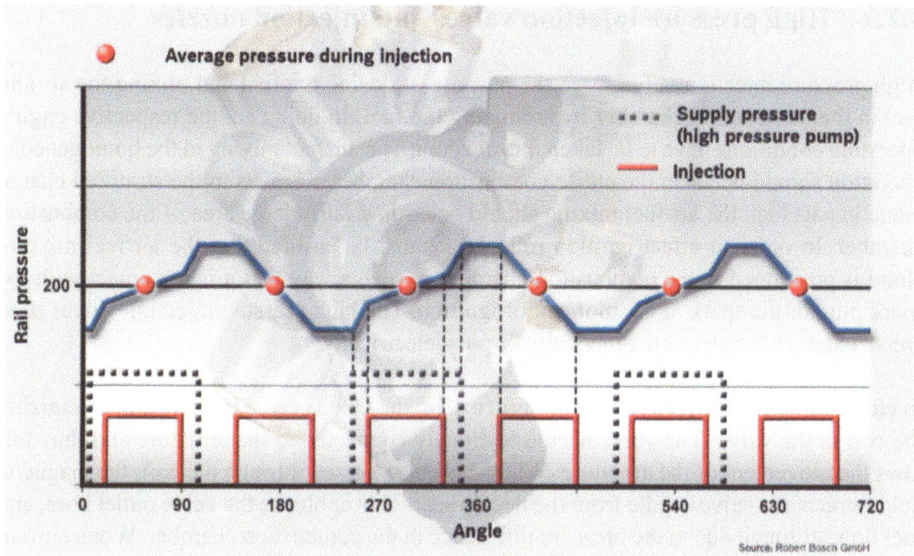


**Figure 4.12:** Rail pressure curve across a working cycle for Mercedes 3.5 l V6 [16]

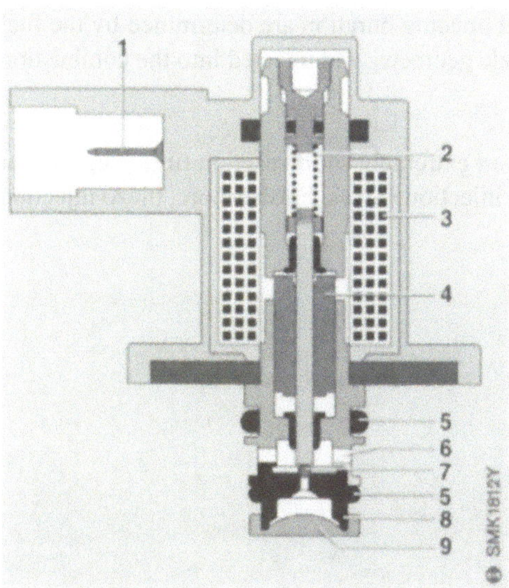


**Figure 4.13:**
Cross-section of pressure control valve [18]
1 Electrical connection
2 Pressure spring
3 Coil
4 Armature
5 Sealing ring (O-Ring)
6 Outlet hole
7 Valve ball
8 Valve seat
9 Inlet with inlet filter

less powerfully from the valve seat (8). The valve ball is lifted from the valve seat using the coil (3), actuated via a modulated signal. The pressure control valve has a zero current in order to ensure the requisite fuel supply even if the electrical actuation fails. A mechanical pressure limiting function is integrated, to protect the components from impermissible high fuel pressure.

### 4.2.3 High pressure injection valves and injection nozzles

High pressure injection valve serve the purpose of dosing the fuel and mixing the air and fuel in the combustion chamber by atomizing the fuel. In doing so, the respective engine operating conditions have to be taken into account. The air/fuel mixing in the homogeneous operation should occur in the entire combustion chamber, whereas in the stratified charge mode in part load the air/fuel mixing should occur in a fairly large area of the combustion chamber. In order to effect ignition reliably, it must be ensured that the air/fuel mixture cloud is positioned in the combustion chamber in such a way that it is in contact with the spark plug or the spark at the moment of ignition. The high pressure injection valves used are actuated either electromagnetically or piezoelectrically.

In **electromagnetic injection valves, figure 4.14**, the valve needle is opened by energizing the coil in the valve. The valve needle is directly connected to the armature and thus follows the movement of the armature. When current is passed through the coil, the magnetic field retracts the valve needle from the needle seat. This unblocks the valve outlet bore, and fuel flows through due to the pressure difference in the combustion chamber. When current does not flow through the coil, the needle closes again due to the spring pressure, and the fuel flow is interrupted. The opening and closing occurs quickly and against the available rail pressure. A constant cross-section is opened during the opening duration. The rail pressure, combustion chamber back pressure and opening duration are determined by the fuel quantity. The fuel is atomized as per the nozzle geometry and injected into the combustion chamber.

**Piezo injection valves** are characterized by an extremely short reaction time – up to four times shorter than in case of electromagnetic injection valves. Furthermore, piezo injection

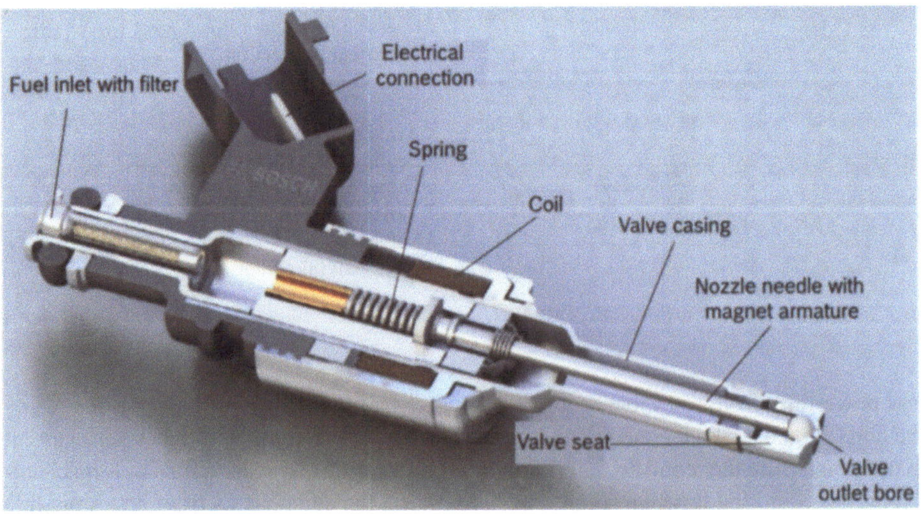


**Figure 4.14:** Electromagnetic injection valve, Bosch HDEV5 (High Pressure Injection Valve)

valves provide the shortest nozzle strokes and precise fuel dosing. By application in gasoline engines with direct injection, extremely good spray atomization and through charge stratification can be achieved.

When applied with an electrical voltage, piezo ceramic materials create very little displacement. Therefore, in one injection valve, several piezo elements are connected one after the other as multi-layered ceramic elements to form what is known as the piezo stack [17]. The distance of 80 μm is covered in less than a thousandth of a second. In this case, the theoretical force with which the injection nozzle needle moves is 6300 N. Since the ceramics cannot withstand any tensile stress, the piezo elements are preloaded using a spring. This avoids the formation of tensile stress in the displaced state thereby protecting the actuator against damage. The maximum displacement that can be achieved with a piezo element depends on its length and the voltage that is applied. Piezo actuators used in injection valves realize a valve needle displacement of 30 μm for a ceramic element of approximately 40 mm at 200 bar fuel pressure.

The design of an injection valve with piezo actuator is shown in **figure 4.15**. The three function groups, injection nozzle, piezo actuator and compensation element (coupler), are arranged linearly. Since the nozzle needle is operated directly by the piezo stack, opening and closing occur without delay. The effect of temperature on the displacement behavior of the piezo element poses a problem. To solve this, a compensation element that offsets the different thermal expansions of the injection casing and piezo stack is installed. The compensation element acts as a hydraulic compensator that should be sufficiently stiff so as not to cause displacement loss in case of longer opening timings. By using the compensation element it is ensured that the needle displacement and with it the flow across the entire temperature range is kept constant during the engine operation.

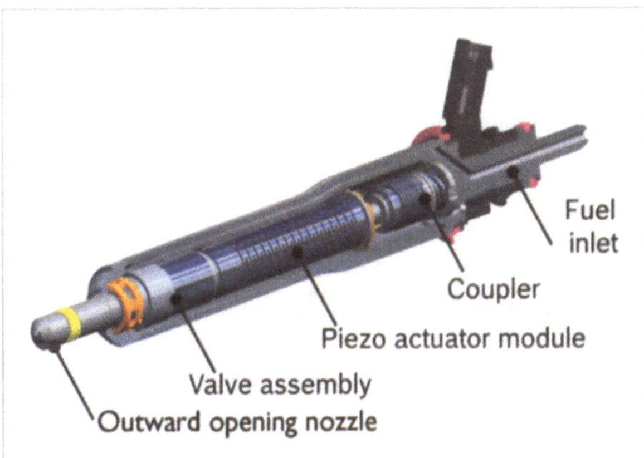


**Figure 4.15:**
Piezo injection valve, Bosch HDEV4

The Siemens injection valve [17] reaches maximum displacement after approximately 200 μs; the closing time is also based on the maximum displacement of approximately 200 μs. In addition to the full displacement, partial displacement can also be set in case of piezo actuators. Transfer from the opening ramp directly into the closing operation is also possible. Alternatively, partial displacement can be maintained for a defined period of time. Due to the ultra small injection timings, reproducibility is very important, particularly when small quantities of injected fuel are involved.

The different nozzle designs used are the swirl nozzle, the multi-hole type nozzle with inward opening nozzle needles, and the A-nozzle that opens outwards via the mushroom valve.

**Swirl nozzle**

Swirl nozzles are inward opening, electromagnetically operated valves that have a swirl generator for atomization of the fuel. The design and functionality is similar to conventional injection valves in intake manifold injection. First the fuel flows axially into the nozzle and is then deflected tangentially, which creates a swirl.

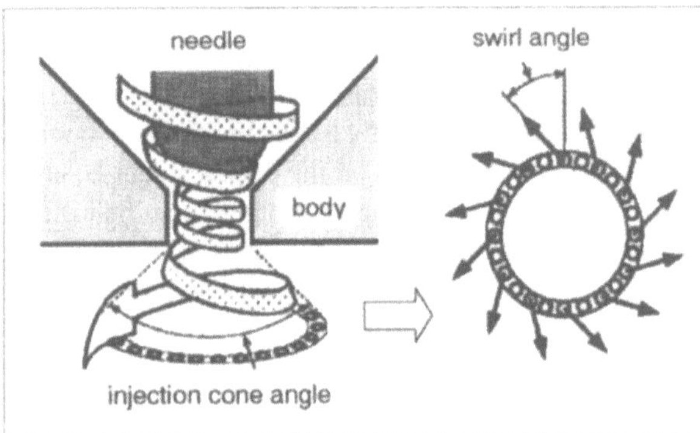


**Figure 4.16:**
Swirl generation in a swirl nozzle [25]

The swirl generation in a swirl nozzle is schematically shown in **figure 4.16** using the commonly used terms. The swirl angle is a measure for the axial movement of the fuel and the injection taper angle is a measure for the expansion of the injection spray.

The axial movement of the fuel is subjected to a tangential movement. This reduces the kinetic energy of the fuel droplets in the axial direction, and it decreases the depth of penetration. A swirl generator is incorporated in the nozzle downstream of the flow. The Swirl Reynolds Number (SRN) is used to designate the fuel swirl. This is defined as:

$$SRN = U\frac{r}{\mu}$$

Where:

| | |
|---|---|
| $U$ | Tangential speed of the swirl movement [m/s] |
| $r$ | Swirl radius [m] |
| $\mu$ | Viscosity [m/s$^2$] |

**Figure 4.17** shows the different ways of generating the required fuel swirl. Where the SRN is identical, the different models show similar spray characteristics [25]. The design in which the fuel swirl is generated through lateral orifices with a tangential slit is generally preferred, for reasons pertaining to manufacturing.

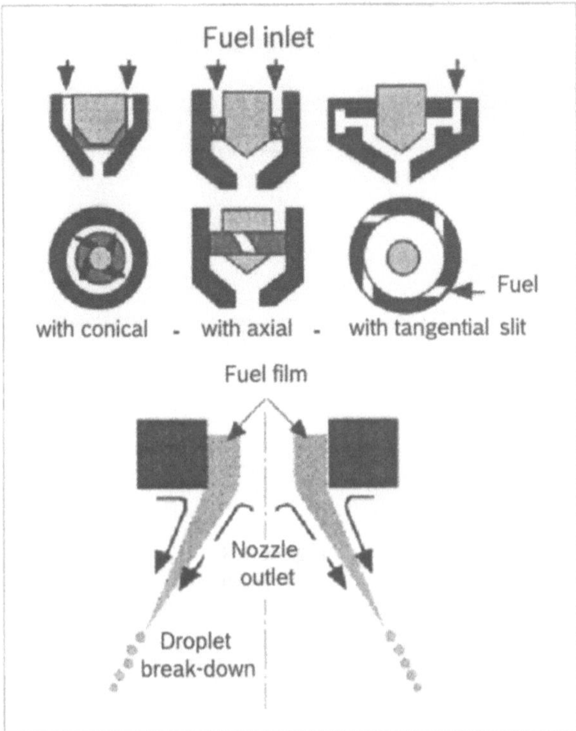


**Figure 4.17:**
Different swirl nozzles and spray formation [25]

**Figure 4.18** shows two examples of models for swirl generation that are used by Bosch [26] and Siemens [27].

In the Bosch model, as seen in the figure above, the spray opening is oblique and not perpendicular. Through this, fuel is sprayed oblique to the injection valve axis. Angles of up to 25° vis-à-vis the injection valve axis can be realized by aligning the injection opening. This can be advantageous for wall-guided and air-guided mixture formation processes as well as for homogeneous mixture formation.

**Figure 4.19** shows the injection process with a swirl nozzle. A hollow taper is clearly visible. The air swirl entering the taper from above diverts the small droplets into the taper and then out of it [28]. The large droplets are initially not diverted. The wall of the taper thickens. However, the continuing injection changes the hollow taper into an increasingly cylindrical structure. At the end of the injection, a cylindrical structure almost completely filled with fuel has been formed.

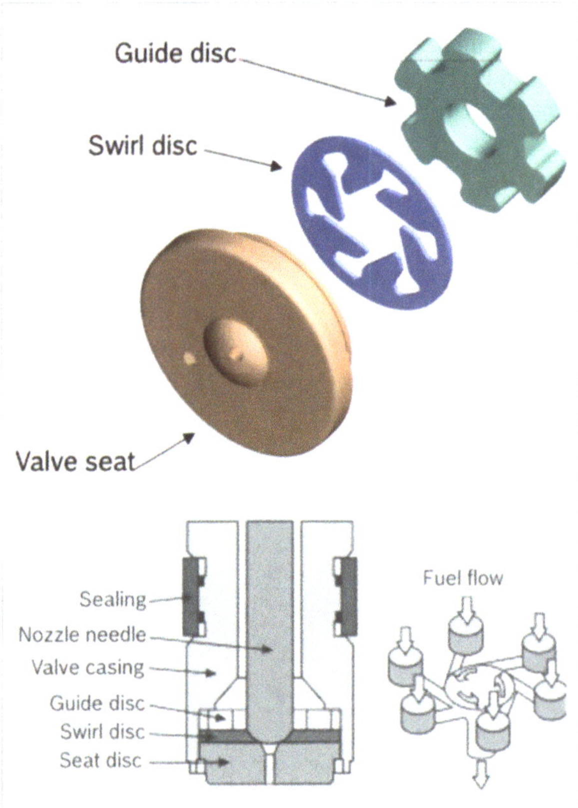


**Figure 4.18:** Swirl generation in Bosch HDEV (above) and Siemens Deka DI (below)

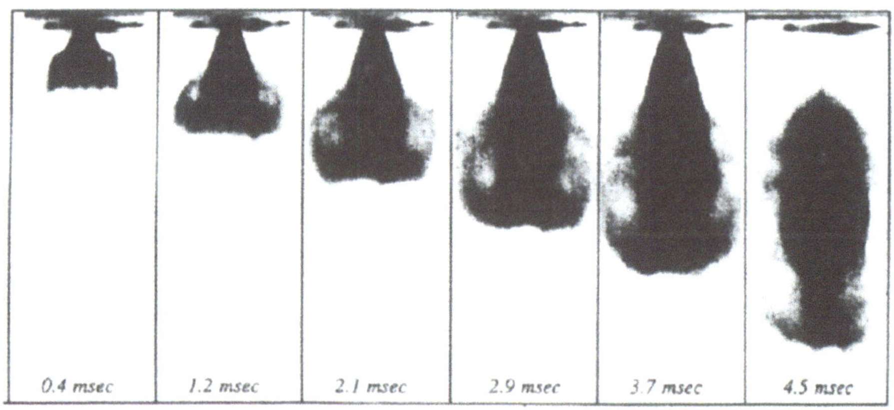


**Figure 4.19:** Injection process in a swirl nozzle [28]

During this process, the fuel pressure (injection pressure) and the spray taper angle have a considerable effect on the diameter of the droplets and their penetration depth. The droplet size increases with the decrease in the injection pressure. Simultaneously, the penetration depth decreases with the decreasing fuel pressure; **figure 4.20.**

All the tests relating to the preparation of the injection spray for high pressure injection show that there is a large number of parameters that influence the injection spray and the mixture formation. For a precisely tuned combustion process with direct injection, it is also important to adjust the injection valve and injection pressure to the engine precisely, also in view of the desired running properties.

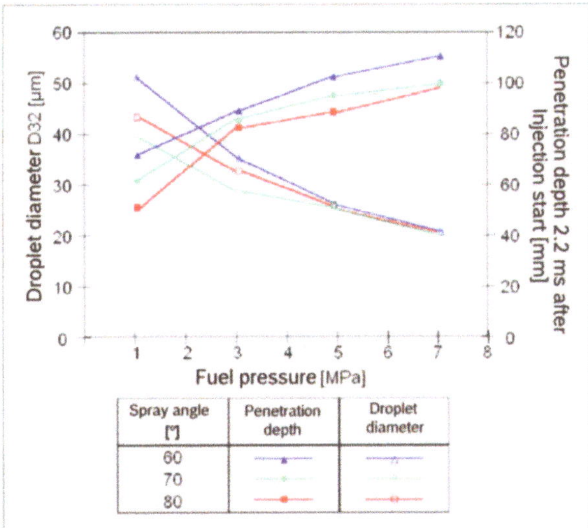


**Figure 4.20:**
Droplet size and penetration depth depending on injection pressure [29, 30]

**Multi-hole nozzle**

The design of multi-hole valves is very similar to that of conventional diesel injection valves. Highly flexible spray patterns are possible using multi-hole nozzles, **figure 4.21**, since the number and configuration of the injection sprays into the combustion chamber can be matched with the demands of the combustion process. Thus, for instance, it is possible to block out areas in the combustion chamber that are critical to emissions, thereby avoiding moistening the open intake valves during injection. Moreover, a spray gap prevents the spark plug becoming moistened with fuel; **figure 4.22**.

As a rule, the variants that have been designed have between 4 and 12 injection holes. It is also possible to incline the injection sprays vis-à-vis the axis of the injection valve, thus increasing the mounting flexibility.

Depending on the design, there is always a dead volume between the valve seat and the injection hole. The fuel that remains in this injection hole after closing of the valve needle vaporizes in the combustion chamber. Fuel residue may cause the formation of carbon

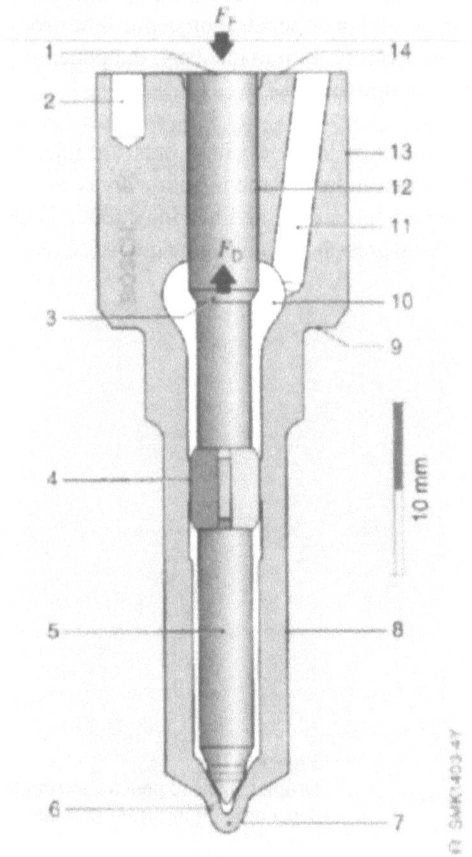


**Figure 4.21:**
Design of the multi-hole nozzle [18]
1 Stroke stop surface
2 Fixing hole
3 Exposed annular area
4 Double needle guide
5 Needle shaft
6 Injection hole
7 Nozzle dome
8 Nozzle body shaft
9 Nozzle body annular area
10 Pressure chamber
11 Inlet hole
12 Needle guide
13 Nozzle body collar
14 Gasket surface
$F_F$ Spring force
$F_D$ Force, resulting from fuel pressure, on the exposed annular area

deposits on the nozzle which may affect how the spray spreads. The geometry of the spray hole may influence the spray spread and the mixture formation.

### Outward opening nozzle (A-nozzle)

The outward opening nozzle creates a ring-shaped gap when opening the nozzle needle. This gap allows the fuel to form a thin film and to flow out at high speed. A uniform hollow cone that features very good atomization properties and that does not require pilot spraying is formed. During this, the spray angle remains almost unaffected due to the back pressure of the combustion chamber. The thickness of the hollow taper cover at the nozzle outlet is determined by the valve stroke. The disintegration of the droplets starts downstream, whereby the droplet spectrum, the spray penetration depth and the spray angle are affected by the needle displacement, the geometry of the injection needle and the valve seat. The size of the droplets can be further reduced using an additional swirl generator in the nozzle; **figure 4.23** c).

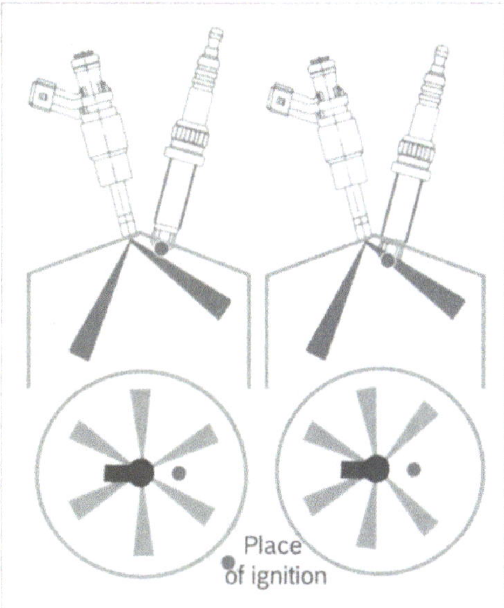


**Figure 4.22:**
Hole arrangement in multi-hole valve [31]

During engine operation, the valve tip is subjected to large temperature variations. The risk of residue deposits and carbonization is lesser in case of swirl nozzles and multi-hole nozzles, since the valve movement has a cleaning effect.

The A-nozzle can be operated either with a twofold electromagnetic set-up or with a piezo actuator. In electromagnetic actuation, the armature plate of the valve needle is positioned between two coils whose distance is defined by needle displacement; **figure 4.23** a) and b).

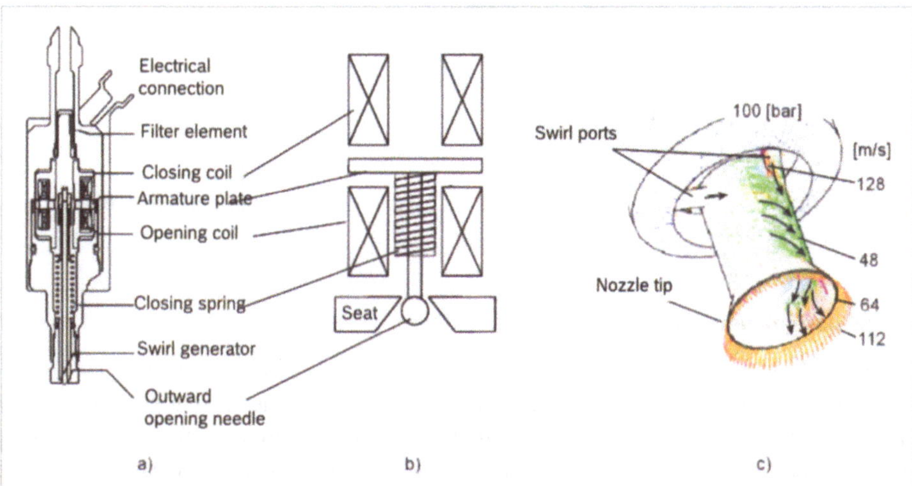


**Figure 4.23:** Outward opening nozzle (A-Nozzle) with electromagnetic operation and swirl generator [32] (Delphi): a) schematic design, b) two-coil operation of the valve needle, c) calculated internal flow

With separate coils for opening and closing, the opening and closing characteristics can be improved. The return spring is useful to keep the valve tightly closed, to support the closing process and to compensate for the thermal expansion of the components.

In piezo injection valves, **figure 4.24**, the switching times for opening and closing are typically less than 0.2 ms, which makes multiple injections easy to realize. Other advantages are the direct actuation of valve needle, high precision and reproducibility of the valve stroke from cycle to cycle. Additional advantages of the piezo actuation are the good mixture preparation and the prevention of post injections towards the end of the injection duration. The injection quantity is dosed with very high precision and at deviations less than 2% [24]. At the same time, the dynamic operating range is large and a higher flow is possible with piezo actuators than with nozzles that open inwards. The large opening and mixture preparation cross-section helps in using the available fuel pressure almost without throttle losses.

Temperature compensation poses a significant challenge, since the piezo stack has a distinctly less thermal expansion than the injection valve casing. The different expansions that occur in the operation and that may well be in the order of magnitude of the valve stroke are compensated for in the closed hydraulic compensation element (coupler). Excellent preparation quality with a Sauter diameter of less than 15 µm, the large linear range resulting from the short switching timings and the high protection against formation of carbon deposits make this injection valve type very suitable for use in the stratified, spray-guided combustion process, despite the significantly high costs.

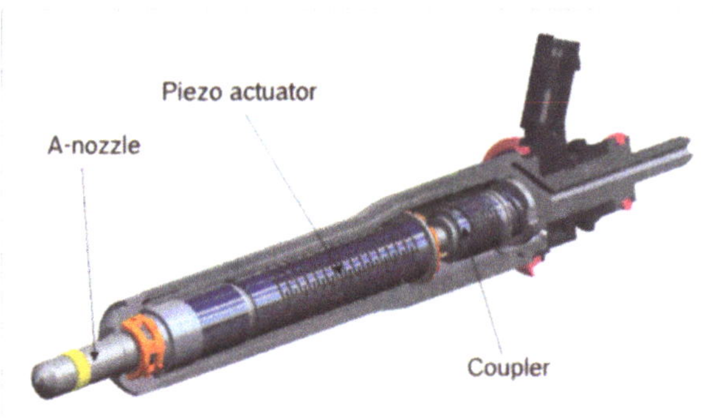


**Figure 4.24:**
Piezo injection valve, Bosch HDEV4

**Nozzle design comparison**

**Figure 4.25** shows the three nozzle types in use today. As mentioned, the nozzles are differentiated primarily according to the opening movement of the closing element [1, 29]. The outward opening nozzle (A-nozzle) releases the nozzle cross-section while opening the nozzle and generates the spray pattern via the taper. This nozzle type can also be used for low pressure injection. The multi-hole nozzle and the swirl nozzle are inward opening (I-nozzles). The multi-hole nozzle is already known for its use in diesel injection technology. The swirl nozzle features a swirl generator, the position of which is upstream.

**Figure 4.25** below shows sprays for the three injection nozzles at the same point of time after the start of injection. The injection pressure during each measurement was 120 bar [33].

The swirl nozzle features excellent atomization quality and high spray flexibility. It is also relatively resistant to dirt and mechanical and thermal impact. The large number of these injection nozzles makes them relatively cheap to manufacture. The spray of these injection valves forms a hollow taper, whereby the taper angle is highly dependent on the pressure of the combustion chamber. This is definitely a disadvantage in a spray-guided combustion process, since it affects the position of the spray with reference to the spark plug. This concept is presently used by most of the developers and manufacturers purely for homogeneous gasoline direct injection and for the stratified charge process with wall-guided mixture formation.

The main advantage of the A-nozzle over the swirl injector is the generation of a uniform hollow taper without a pilot spray. The spray angle is virtually unaffected by the back pressure; **figure 4.25**. If scrutinized, however, the preparation is not as high quality as with the swirl nozzle. This is due to the spray contraction of the taper being too low and to severe limiting of the lubricant spray. Using piezo actuation, it is possible to effect multiple injections at ultra short time intervals with this type of injector. Moreover, this type of actuation enables separate control of the droplet size and the injection quantity, making it superior to conventional magnet actuation [10, 22, 34]. The spark plug for a spray-guided combustion process with this injector type can be positioned in the recirculation area as well as at the surface of the fuel spray. The A-nozzle is a favorable variant for a spray-guided process at significantly high injection pressures in the range of 200 bar and possibly more in the future.

The acutely limited injection sprays are characteristic of the multi-hole nozzle. Despite different designs, the mixture in the combustion chamber homogenizes only partly due to the injection spray pattern, since the atomization quality is inadequate. Rich mixture zones with excess fuel are located directly next to lean mixture zones with insufficient fuel. As a result, after igniting an injected spray – determined by the position of the spark plug – the flame front spreads at a fluctuating speed (acceleration of the flame front in the rich zones of individual injection sprays, reduction in speed of the flame front in the lean zones between the injection sprays) [35] instead of spreading uniformly in the combustion chamber. The number of injection bores thus significantly affects the operating characteristics of the engine. A bore distribution that effects good fuel distribution and stable flame spread between the different injection sprays enables the best engine behavior. Increasing the number of nozzle bores while also reducing the individual flow surfaces can increase the capability of the engine to achieve leaner mixtures.

The three nozzle types are once again presented in a comparative summary assessment in **figure 4.26**.

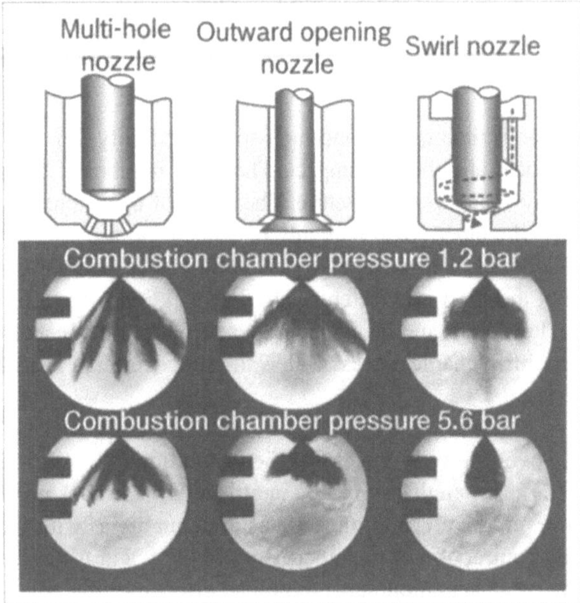


**Figure 4.25:**
Comparison of injection nozzles: multi-hole nozzle (left), outward opening nozzle (center), swirl nozzle (right) [33]

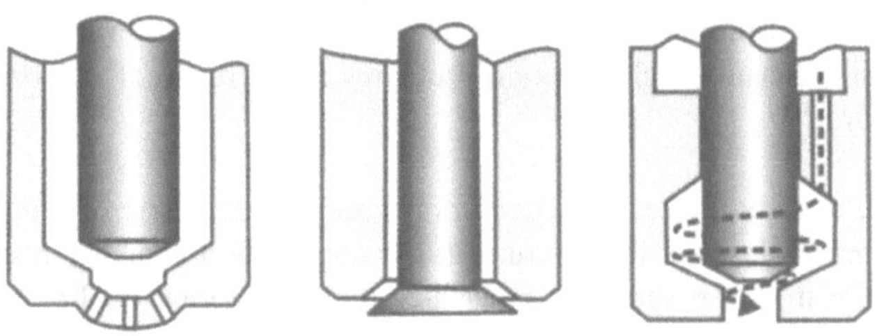

| Criterion | Multi-hole Nozzle | Outward Opening Nozzle | Swirl Nozzle |
|---|---|---|---|
| Flexibility of spray shape | ++ | + | + |
| Possibility of spray inclination | + | – | ++ |
| Quality of preparation at system pressure of 10 Mpa | – | o | ++ |
| Resistance to carbonization | – | ++ | + |

**Figure 4.26:** Nozzle designs and assessment [1]

**Figure 4.27** shows the summary and assessment of combustion processes of gasoline direct injection in combination with the possible injection valves [31]. A gasoline engine with intake manifold injection and running in homogeneous mode is used as the basis for

| Combustion process | Performance | | | | | Quantitative | | Application effort |
|---|---|---|---|---|---|---|---|---|
| | Consumption | Emissions | | | Costs | Consumption | Particulate | |
| | | HC | $NO_x$ | Particulate | | | | |
| Homogeneous, stoichiometric and lean, lateral mounting also with supercharging | + | + | o | - | + | -2 % | +50 % | o |
| Stratified, wall-guided lateral mounting, swirl valve | ++ | - | - | - - - | + | -8 % | +400 % | o / - |
| Stratified, spray-guided central mounting, multi-hole nozzle | +++ | - | - | - - | o | -13 % | +200 % | - |
| Stratified, spray-guided central mounting, A-nozzle, possibly with supercharging | +++ | + | - - | - | - - | -15 % | +50 % | - - / - - - |

+++ very good (little effort), 0 neutral (neutral effort), - - - less good (high effort), as against homogeneously operating gasoline engines with manifold injection

**Figure 4.27:** Assessment of the different combustion processes in terms of specific advantages and disadvantages [31]

the assessment. The quantitative data in the literature varies strongly in parts. Therefore, this summary cannot meet any absolute demands and represents only trends.

With respect to the cost-benefit ratio, the spray-guided variant with electromagnetically actuated multi-hole valve is considered the optimal solution. In view of utilizing the theoretical potentials of the direct injection and notwithstanding the system constants, the spray-guided variants with outward opening piezo actuated injection valve in combination with supercharging are favored [36].

## 4.2.4 Injection strategies

Higher fuel pressure and significantly less time available for the injection and mixture formation in the combustion chamber are the main differences between gasoline direct injection and intake manifold injection. Requirements for the injection time available for gasoline direct injection and intake manifold injection are shown in **figure 4.28**. Up to two rotations of the crankshaft are available for injecting the fuel into the intake manifold

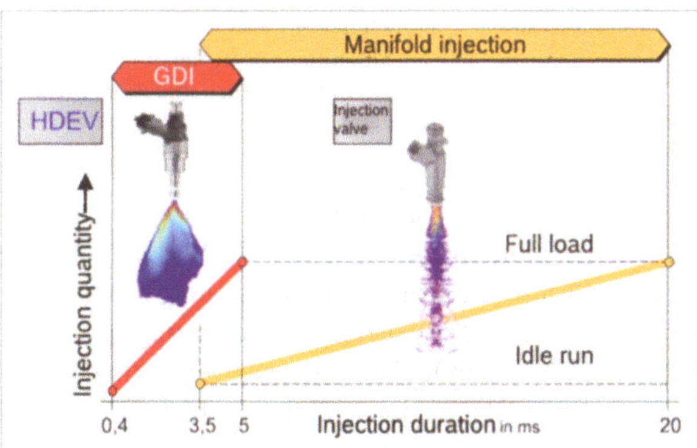


**Figure 4.28:** Injection quantity as a function of the time in the intake manifold and direct injection [18]

(20 ms at 6000 min$^{-1}$) in the case of intake manifold injection, whereas only half a rotation of the crankshaft (5 ms at 6000 min$^{-1}$) is available during injection in the intake stroke (full load operation) during direct injection. In the case of stratified charge mode with injection during compression, the time for fuel injection with subsequent mixture formation until ignition occurs is much shorter. For complete utilization of the mixture stratification in the spray formed mixture formation process, a maximum of 1 to 3 ms is available for injection and mixture formation in an engine approaching idle run. This results in high requirements of the injection system and the respective injection strategy.

The injection valve must be actuated with a complex current path for precise injection; **figure 4.29**. In addition, the pre-magnetization time in which the injection valve is yet to open is also accounted for. At the start of actuation, the valve needle is lifted quickly, due to a high current. As soon as the needle is opened, a low actuation current suffices for constant needle displacement. The injection quantity is then dependent only on the length of time the injection valve is open [18].

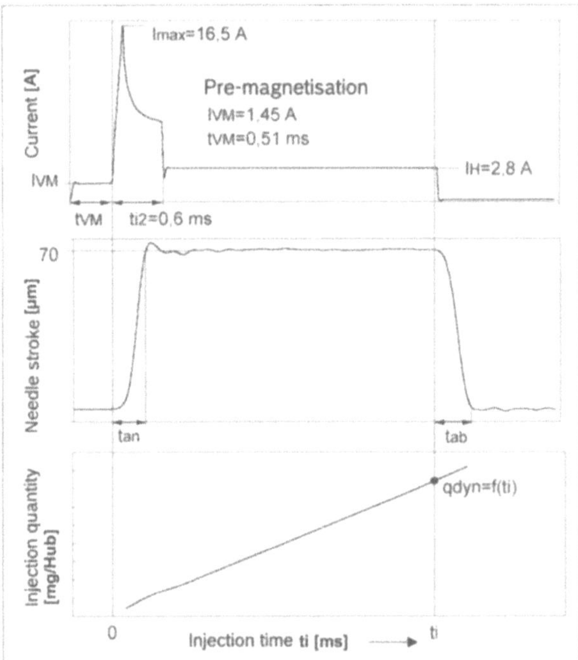


**Figure 4.29:**
Signal curves of the actuation of the high pressure injection valve [18]

The injection point has a significant effect on engine combustion and with it the smooth-running and emissions of an engine with gasoline direct injection [17]. With an advance injection point, very low levels of soot emissions are realized through a long mixture formation time. The hydrocarbon emissions, however, increase due to “overmixing” (too lean, no more burning areas “flame quenching”). The inflammation is delayed and a slow transition to the early combustion phase occurs. Delayed injection leads to low hydrocarbon and NOx emissions but the short mixture formation time increases soot emissions (heterogeneous diffusion combustion). There is also a risk that carbon deposits will form on the spark plug and the injection valve. Inflammation is highly delayed but the transition to the early combustion phase occurs quickly.

**Multiple injection**

Due to the various difficulties in the precise adaptation of the fuel supply and the related mixture formation to the boundary conditions, single injection is often not sufficient to meet the given requirements in the case of second generation direct injection, i.e. the spray-guided mixture formation process. The boundary conditions have to be taken into account in the case of operation in the entire engine map on the one hand and for adhering to the emissions laws on the other. Therefore, there is an increasing transition to multiple injection in the development of the gasoline direct injection process.

Due to their very high dynamics, the common rail injection system and particularly injection valves with piezo actuators enable more than one injection within one working cycle, in fact within one compression stroke. With multiple injections, the mixture stratification in part load operation can be described as largely optimal for combustion. This way, it is certainly possible to avoid overly rich or lean mixture areas that lead either to soot formation or to dying of the flame with a related increase in hydrocarbon and CO emissions. **Figure 4.30** shows a comparison between single and multiple injection for spray preparation, mixture formation and subsequent combustion [37].

By dividing the injection quantity into further rapidly occurring smaller injections, wider transition areas from the rich injection spray to the ambient air can be achieved. Also the area

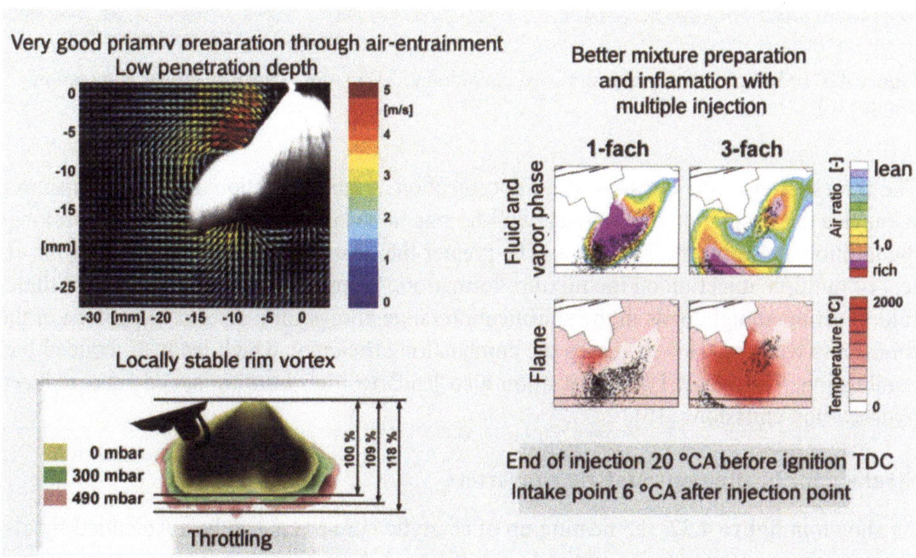


**Figure 4.30:** Potential of multiple injection [37]

of air/fuel mixture with stoichometric ratio increases and has to be kept steady at the spark plug. The ignition window also becomes bigger. Thanks to the improved mixture preparation and distribution, a tapering flame can form and quickly penetrate the mixture cloud. The mixture can also be ignited in case of retarded injection and ignition timings and a thermodynamically suitable center of combustion mass can thus be achieved; **figure 4.31**. Good mixture quality and the use of multiple injection are the prerequisites for extending the stratified operation to higher loads. The stratification area is not restricted in this process by emissions values or soot values, but by the air/fuel ratio that is set and thus by the consumption advantage as against homogeneous operation.

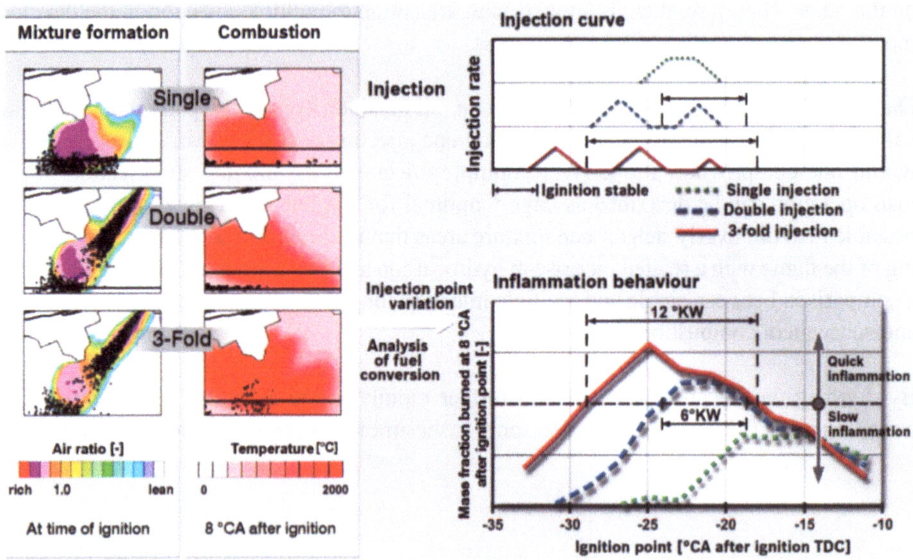


**Figure 4.31:** Multiple injection in stratified operation at 2,000 min$^{-1}$, imep = 3.0 bar: simulation results [16]

The threefold injection consists of a pilot injection, a main injection and a post injection. It enables the generation of compact, well-homogenized mixture clouds and thus prolongs the ignition and robustness window. The greater the quantity of the fuel, the better the effect of multiple injection on the mixture formation. As against the double injection, threefold injection at high loads in the stratification range allows for a distinct expansion of the robustness window. This increases the combustion efficiency, which leads to reduced fuel combustion. The better homogenization also leads to high burning speeds and reduced hydrocarbon emissions [16].

**Strategy for heating up catalytic converters**

As shown in **figure 4.32**, the heating up of catalytic converters can be accelerated by distributing the fuel quantity in multiple injections. A lean homogeneous basic mixture is prepared by early injection during the intake stroke. One or more injections close to the ignition time generate a rich gas mixture in the spark plug area. This guarantees that even in the event of retarded ignition time, definite ignition and inflammation is ensured. The retarded ignition and lean basic mixture increase the temperature of the exhaust gas and cause a quick light-off of the catalytic converter. This can effectively reduce the untreated hydrocarbon emissions [16].

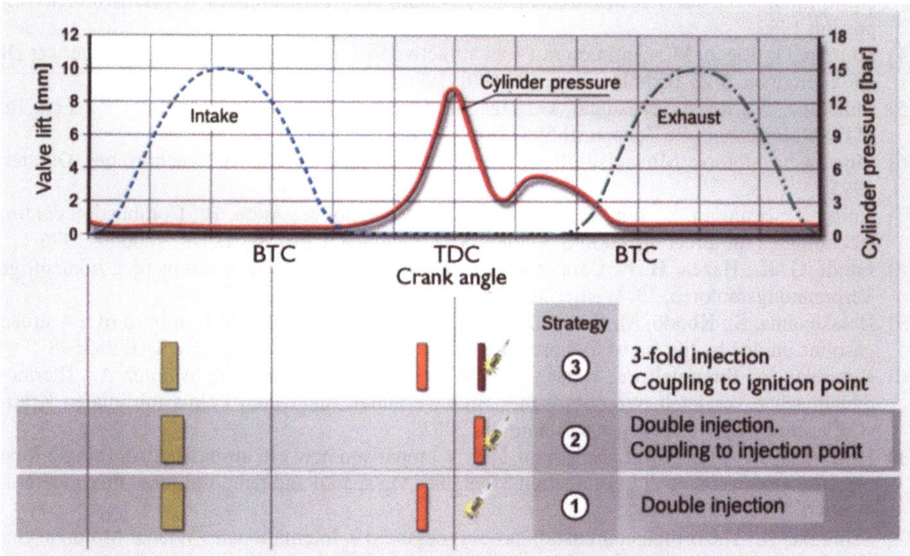


**Figure 4.32:** Injection strategies for heating up exhaust-gas catalytic converters [16]

As mentioned before, multiple injection can also significantly improve the quality of the mixture formation and the emissions behavior in cold start operation. The heat generated by retarded injection, which can no longer be completely used in the process and therefore in the engine power, leads to reduced fuel consumption. Due to multiple injection, there is some deviation from the ideal performance of direct injection with stratified charging. The generated mixture cloud is not as small as it would have to be for ideal stratified charge, as compared to ideal single injection. This can be attributed to the longer time interval of injection and mixture formation. In the case of multiple injection, this may lead to mixture separation in the boundary area of the mixture cloud and with it, to definite leaning. This also has to do with the fact that the potential that is theoretically possible using direct injection is not fully utilized.

## Literature


[1] Moser, W.; Mentgen, D.; Rembold, H.: Benzin-Direkteinspritzung – eine neue Herausforderung fuer zukuenftige Motormanagementsysteme. In: MTZ 58 (1997), Nr. 9, S. 458–464

[2] Stocker, H.; Housten, R.: Luftunterstuetzte Benzin-Direkteinspritzung. Technische Akademie Esslingen, 1998

[3] Stocker, H.; Schuerz, W.; Houston, R. A.; Worth, D.: Spezifische Motorsteuerungsfunktion fuer ein luftunterstuetztes Benzin-Direkteinspritzsystem. 20. Internationales Wiener Motorensymposium 1999

[4] Stumpp, G.; Ricco, M.: Common rail – an attractive fuel injection system for passenger car DI diesel engines. SAE Technical Paper 960870, 1996

[5] Maiorana, G.; Rossi Sebastiano, G.; Ugaglia, C.: Die Common-Rail-Motoren von Fiat. In: MTZ Motortechnische Zeitschrift 59 (1998), 9

[6] Mitsubishi Motors: Mitsubishi Benzinmotor mit Direkteinspritzung. Technisches Dossier, 1995

[7] Kume, T.; Iwamoto, Y.; Iida, K.; Muratami, M.; Akishino, K.; Ando, H.: Combustion control technologies for direct injection SI engine. SAE Technical Paper Series No. 960600, 1996

[8] Fraidl, G. K.; Hazeu, H. P.; Carstensen, H.: DMI – ein Schichtladekonzept fuer zukuenftige Verbrennungsmotoren. 13. Wiener Motorensymposium, 1992

[9] Hatakeyama, S.; Kondo, M.; Sekiya, Y.; Murayama, T.: An attempt at lean burn of a 4 stroke gasoline engine by the aid of low pressure air-assisted in-cylinder injection. SAE 982698

[10] Kemmler, R.; Frommelt, A.; Kaiser, Th.; Schaupp, U.; Schommers, J.; Waltner, A.: Thermodynamischer Vergleich ottomotorischer Brennverfahren unter dem Fokus minimalen Kraftstoffverbrauchs. 11. Aachener Kolloquium, 2002

[11] Hoffmann, J. A.; Eberhardt, E.; Martin, J. K.: Comparison between air-assisted and single-fluid pressure atomizers for direct-injection SI engines via spatial and temporal mass flux measurements. SAE 970630

[12] Schlunke, K.: The orbital combustion process engine. 10. Internationales Wiener Motorensymposium, 1989

[13] Smith, D.; Ahern, S.: The orbital ultra low emissions and fuel economy engine. 14. Internationales Wiener Motorensymposium, 1993

[14] Stocker, H.; Houston, R.: Luftunterstuetzte Benzin-Direkteinspritzung. Technische Akademie Esslingen, 1998

[15] Altenschmidt, F.: Grundsatzuntersuchungen zu einem ottomotorischen Brennverfahren mit Direkteinspritzung und Aufladung. Dissertation, Universitaet Karlsruhe (TH), 2005

[16] Waltner, A.; Lueckert, P.; Schaupp, U.; Rau, E.; Kemmler, R.; Weller, R.: Die Zukunftstechnologie des Ottomotors: strahlgefuehrte Direkteinspritzung mit Piezo-Injektor. 27. Internationales Wiener Motorensymposium, 2006

[17] Warnecke, V.; Achleitner, E.; Baecker, H.: Entwicklungsstand des Siemens VDO Piezo-Einspritzsystems fuer strahlgefuehrte Brennverfahren. 27. Internationales Wiener Motorensymposium, 2006

[18] Robert Bosch GmbH: Ottomotor-Management, 3. Auflage. Wiesbaden: Vieweg Verlag, 2005

[19] Stocker, H.: Benzindirekteinspritzung und Motorsteuerung – Herausforderung und Umsetzung. Motor und Umwelt, AVL 1997

[20] Haentsche, J. P.; Krause, G.; Velji, A.; Spicher, U.: High pressure fuel pump for gasoline direct injection based on ceramic components. SAE Paper 2005-01-2103, 2005

[21] Haentsche, J. P.; Spicher, U.: Ceramic components for high pressure gasoline fuel injection pumps – Keramische Komponenten fuer Hochdruck-Benzindirekteinspritzpumpen fuer Ottokraftstoff. In: Materialwissenschaft und Werkstofftechnik 36 (2005), No. 3/4, 2005

[22] Achleitner, E.; Berger, St.; Frenzel, H.; Klepatsch, M.; Warnecke, V.: Benzin-Direkteinspritzsystem mit Piezo-Injektor fuer strahlgefuehrte Brennverfahren. In: MTZ 65 (2004), Nr. 5, S. 338–349

[23] Schnittger, W.; Koenigstein, A.; Pritze, St.; Poepperl, M.; Rothenberger, P.; Samstag, M.: 2.2 Direct Ecotec – Neuer Ottomotor mit Direkteinspritzung von Opel. In: MTZ 64 (2003), Nr. 12
[24] Achleitner, E.; Amann, R.; Klepatsch, M.; Pasqui, R.; Frenzel, H.; Warnecke, V.; Bauer, Ph.: Die innovative Technologie fuer die Benzindirekteinspritzung mit strahlgefuehrten Brennverfahren. 12. Aachener Kolloquium Fahrzeug- und Motorentechnik, 2003
[25] Yamauchi, T.; Wakasaki, T.; Kaneko, T.; Kaneko, M.; Kato, S.; Hosaka, H.: Numerical analysis of stratified mixture formation in direct-injection gasoline engines. Direkteinspritzung im Ottomotor – Expert Verlag 1998
[26] Robert Bosch GmbH: Kraftfahrtechnisches Taschenbuch, 25. Auflage, Vieweg Verlag, 2003
[27] v. Basshuysen, R., Schaefer, F.: Lexikon Motorentechnik. Vieweg Verlag, 2004
[28] Hoffmann, J. A.; Eberhardt, E.; Martin, J. K.: Comparison between air-assisted and single-fluid pressure atomizers for direct-injection SI engines via spatial and temporal mass flux measurements. SAE 970630
[29] Ohyama, Y.; Ohsuga, M.; Nogi, T.; Fujieda, M.; Shiraishi, T.: Mixture formation in gasoline direct injection engines. Direkteinspritzung im Ottomotor – Expert Verlag, 1998
[30] Nogi, T.; Shiraishi, T.; Nakayama, Y.; Ohsuga, M.; Kurihara, N.: Stability improvement of direct fuel injection engine under lean combustion operation. SAE 982703
[31] Xander, B.: Grundlegende Untersuchungen an einem Ottomotor mit Direkteinspritzung und strahlgefuehrtem Brennverfahren. Dissertation, Universitaet Karlsruhe (TH). Logos-Verlag, 2006
[32] Varble, D.L., Xu, M., VanBrocklin, P.: Entwicklung eines nach aussen oeffnenden Benzin-Direkteinspritzventils fuer Schichtladebrennverfahren. 7. Aachener Kolloquium Fahrzeug- und Motorentechnik, 1998
[33] Ortmann, R.; Wuerfel, G.; Grzeszik, R.; Raimann, J.; Samenfink, W.; Schlerfer, J.: Vergleich von Zerstaeubungskonzepten fuer Benzin-Direkteinspritzung. Direkteinspritzung im Ottomotor III, 2001
[34] Huebner, W.; Witt, A.; Hoss, B.; Albrecht, T.; Durst, B.; Abdelfattah, A.: Methodeneinsatz bei der Entwicklung eines weiterfuehrenden DI-Brennverfahren. Direkteinspritzung im Ottomotor IV, 2003
[35] v. Andrian, S.; Haug, M., Pischinger, F.; Spicher, U.; Spiegel, L.; Knoche, K. F.; Roosen, P.: Abschlussbericht zum FVV-Vorhaben Nr. 486, Direkteinspritzung im Ottomotor II. 1992
[36] Klueting, M.; Melcher, T.: Effiziente Dynamik – ein integraler Ansatz fuer die BMW Antriebe der Zukunft. 14. Aachener Kolloquium Fahrzeug- und Motorentechnik, 2005
[37] Lueckert, P.; Frey, J.; Kemmler , R.; Schaupp, U.; Vent, G.; Waltner, A.: Kunden- und zukunftsorientierte Technologien am Ottomotor – heute und morgen. 26. Internationales Wiener Motorensymposium, 2005

# 5 Power and Torque

One of the most important parameters of an engine is power (work per minute). Power cannot be measured directly at the engine, and is therefore calculated from the torque and engine speed. Torque can be measured directly on an engine test bench using a power brake, whereby the brake takes the power input from the engine and converts it into thermal or electrical output. The effective power of an engine results from the torque and the engine speed in:

$$P_e = M_d \cdot \omega \ , \ P_e \text{ in kW}$$

where

$$\omega = 2 \cdot \pi \cdot n = \frac{\pi \cdot n}{30}$$

$\omega$ angular velocity in $s^{-1}$
$n$ engine speed in $min^{-1}$

Besides the effective power that is delivered at the clutch to the powertrain or transmission, there is also an internal power that results from the total of the process characteristics in the individual cylinders. This is determined by indexing the gas pressure characteristics (charge exchange, compression, combustion and expansion) in the cylinders of an engine. The effective power gained through the work process, and therefore the internal power, are significantly affected by the combustion process.

The internal power $P_i$ (also rated power) derived from the work process is transferred to the piston by the working gas. The following holds true for the change in work done by the gas that affects the piston during the work cycle:

$$dW_{KA} = p \cdot A_K \cdot ds_\alpha$$

$p$ combustion pressure or cylinder pressure
$A_K$ piston surface area
$\alpha$ crankshaft angle
$s_\alpha$ piston travel $= f(\alpha)$
$W_{KA}$ work done by the gas at the piston per work cycle

The integration across a complete work cycle results from the work done at the piston by the gas:

$$W_{KA} = \oint p \cdot dV_\alpha$$

$dV_\alpha$ change in volume $= f(\alpha)$

The internal power $P_{iZ}$ of a cylinder results from:

$$P_{iZ} = n_A \cdot W_{KA}$$

$n_A$ work cycles per time unit

Thus, calculating for cylinder power:

$$P_{iZ} = i \cdot n \cdot W_{KA}$$

$n$ crankshaft rotations per second
$i$ work cycle per rotation;
for four-stroke engines: $i = 0.5$
for two-stroke engines: $i = 1$

The work done by gas $W_{KA}$ with respect to the displacement volume $V_h$ is known as indicated mean effective pressure $p_{mi}$:

$$p_{mi} = \frac{W_{KA}}{V_h} \quad \text{or} \quad p_{mi} \cdot V_h = W_{KA}$$

Thus, cylinder power can also be expressed by the relation

$$P_{iZ} = i \cdot n \cdot p_{mi} \cdot V_h$$

This equation is valid for a single cylinder. The internal power $P_i$ for an engine with multiple cylinders (z = number of cylinders) is:

$$P_i = i \cdot n \cdot \bar{p}_{mi} \cdot V_h \cdot z \quad \text{or} \quad P_i = i \cdot n \cdot \bar{p}_{mi} \cdot V_H \quad , \quad \bar{p}_{mi} = \frac{1}{2} \sum_{j=1}^{z} p_{mij}$$

$\bar{p}_{mi}$ internal mean pressure calculated across all cylinders
$p_{mij}$ j =1, ..., 2 internal mean pressure of the individual cylinder

Due to friction losses, the effective usable power at the crankshaft $P_e$ is less than the internal power $P_i$ generated through work done by gas. Friction losses result, for instance, from piston oscillations and crankshaft rotation as well as the movement of the connecting rod. The total of all individual friction losses (e.g. piston ring friction, piston friction, piston pins friction, connecting rod bearing friction and crankshaft bearing friction) yields the total friction power $P_r$. The effective power is thus calculated as:

$$P_e = P_i - P_r$$

Besides friction losses in the engine, the friction power $P_r$ includes the power required for the ancillaries (e.g. coolant pump, oil pump, cooling fan, servo pumps, etc.).

Similar to internal mean pressure $\bar{p}_{mi}$ the mean effective pressure $p_{me}$ is defined as friction mean pressure $p_{mr}$, thereby deriving the following equations for effective power and friction power:

$$P_e = i \cdot n \cdot p_{me} \cdot V_H$$

$$P_r = i \cdot n \cdot p_{mr} \cdot V_H = i \cdot n \cdot (\bar{p}_{mi} - p_{me}) \cdot V_H$$

$$p_{mr} = \bar{p}_{mi} - p_{me}$$

For calculating effective power using the units usually used for combustion engines, the following quantity equation is derived:

$$P_e = \frac{i}{10} \cdot n \cdot p_{me} \cdot V_H$$

With reference to engine power, the term "horsepower" (hp) is widely known in the automotive industry. This is, however, outdated and is no longer recommended. Rather, the legal term for engine performance, is the SI unit "kilowatt" (kW), which should already be in use. From 2009, power rating in hp will no longer be acceptable. The following applies for conversion:

$$1 \text{ hp} \approx 0.735 \text{ kW}$$

$$1 \text{ kW} \approx 1.358 \text{ PS}$$

If the effective power $P_e$ refers to the displacement $V_H$, it indicates the engine specific power output per liter $P_l$ and with reference to the engine weight $m_M$,the weight-to-power-ratio $m_G$.

$$P_l = \frac{P_e}{V_H} \quad \text{specific power output per liter}$$

$$m_G = \frac{m_M}{P_e} \quad \text{weight-to-power ratio}$$

Specific power output per liter and weight-to-power ratio are variables for engine load and design volume. Higher specific power output per liter means a high engine load but a smaller engine. High weight-to-power ratio means low engine load and a large engine; low weight-to-power ratio corresponds to higher engine load and a small engine. The specific power output per liter of modern passenger car gasoline engines without supercharging is from 30 to 80 kW/l.

In principle, gasoline direct injection offers power and torque advantages over conventional gasoline engines with intake manifold injection. These advantages are the result of injecting fuel directly into the combustion chamber and cooling the cylinder charge that follows vaporization, thereby enhancing the delivery thresholds and power. With internal cooling, the temperature at the start of the compression phase can be reduced. This has a disproportionately positively effect on the temperature curve during the subsequent polytropic compression, reducing susceptibility to knocking. Therefore, with direct injection, it is possible to increase the compression ratio by approximately one to two units. In comparison to the intake manifold system, direct injection increases efficiency and charge, resulting in a 2 to 5% increase in power. **Figure 5.1** illustrates this advantage of direct injection.

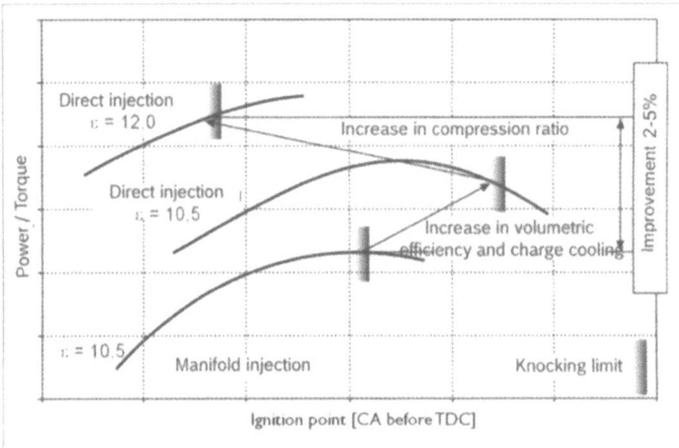


**Figure 5.1:** Effect of direct injection on power and knocking limit [1]

In **figure 5.2**, the torque and power of the Mitsubishi GDI engine with direct injection is compared to the engine variant with intake manifold injection. The combined effect of the previously mentioned positive characteristics helps achieve 10% more power and torque with gasoline direct injection as against the same basic engine with intake manifold injection.

**Figure 5.3** shows the mean pressure and air efficiency $\lambda_a$ across the engine speed, comparing an engine with intake manifold and one with direct injection. Across the entire engine speed range, the engine with direct injection reaches a higher indicated mean pressure with higher air efficiency simultaneously, along with a larger charge [3].

Furthermore, the absolute values for torque and power of a gasoline engine with direct injection are not the only relevant factors. The speed of the torque build-up after the corresponding driver input is also important for the customer. Here, direct injection has a significant advantage over intake manifold injection; the inertia of the fuel wall film that forms in the intake manifold and should first vaporize can be ignored. This leads to distinctly better response behavior. For this reason, gasoline engines with direct injection are much more responsive than intake manifold injection engines with the same power.

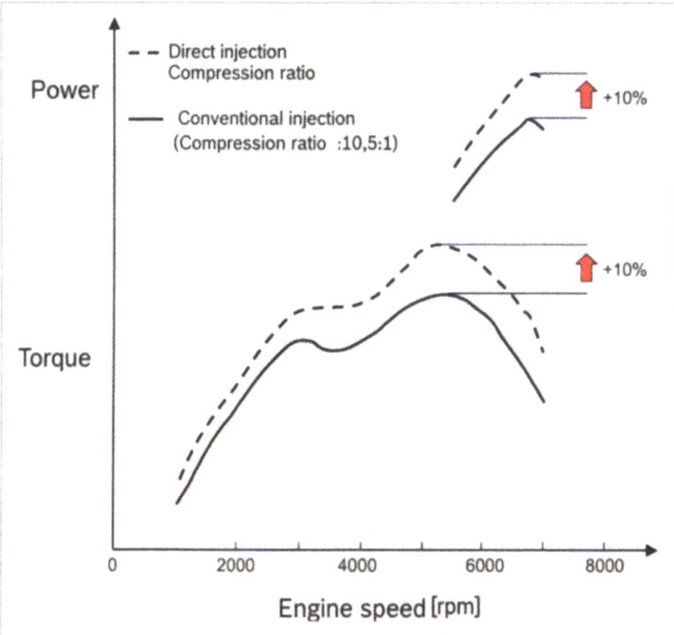


**Figure 5.2:** Power and torque curve of the Mitsubishi GDI engine compared to the engine with intake manifold injection [2]

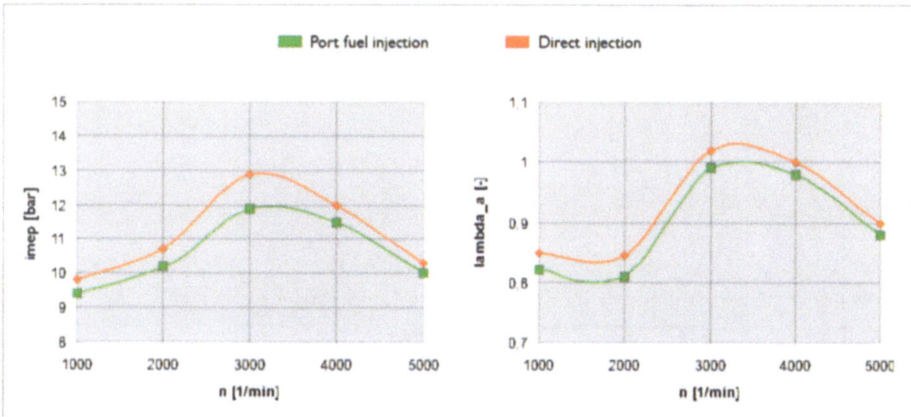


**Figure 5.3:** Comparison of intake manifold and direct injection, full load behavior [3]

**Figure 5.4** shows a comparison of torque build-up over a period of time for a 1.8 l turbo engine with intake manifold injection, with the corresponding torque band scaled up to the combustion chamber of 2.0 l, as well as the torque build-up of a supercharged gasoline engine with 2.0 l and direct injection. It is evident that the torque build-up of the engine with direct injection is more spontaneous as compared to that of the engine with intake manifold injection. Immediately after the throttle valve opens, the engine with direct injection shows an increase in torque of 20 Nm. The engine with direct injection achieves the maximum torque of the intake manifold injection engine in half the time, and reaches a significantly higher rated torque 0.3 s earlier than the engine with intake manifold injection. The curve of the engine with intake manifold injection scaled up to 2.0 l indicates that the benefits of direct injection are two times greater than those that would result by increasing the displacement to 2.0 l.

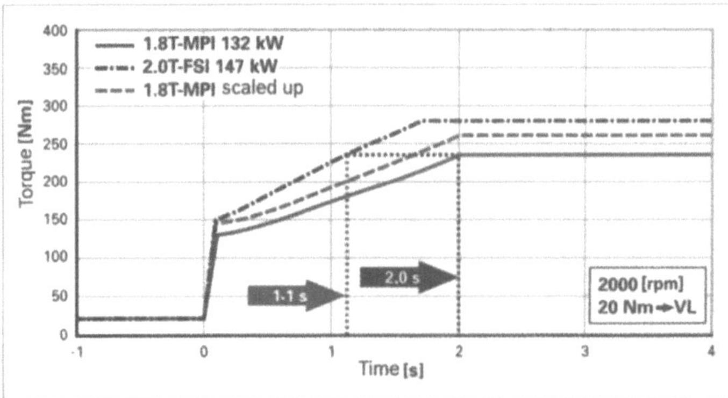


**Figure 5.4:** Torque build-up over time during a load increase at 2000 rpm [4]

**Figure 5.5** shows the dynamic benefit curve of an engine with direct injection in comparison to an engine with intake manifold injection across the engine speed. In this figure, the results of a number of load increases are summarized. The graphic shows the torque values that result one second after a load increase from 20 Nm onwards. Over the entire engine speed range, the gasoline engine with direct injection shows distinct advantages, including in dynamic torque build-up, as against the variants with intake manifold injection. At 2000 rpm and 2500 rpm an engine with direct injection gains a torque of 55 or 65 Nm, which is an improvement of more than 30% [4].

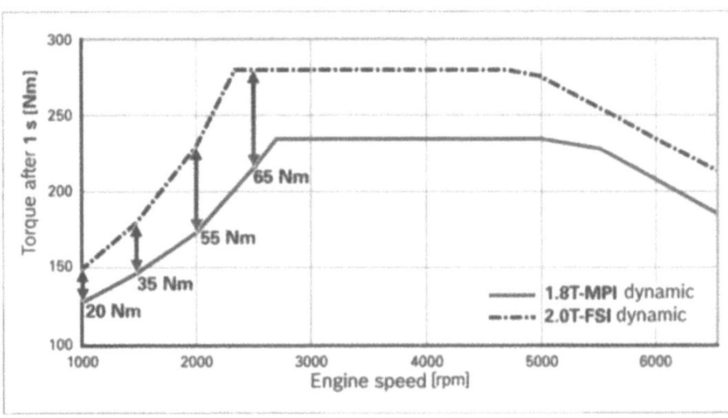


**Figure 5.5:** Comparative dynamic torque build-up [4]

## Possibilities of torque increase

Using the following formula for internal power, the possibilities of increasing the engine power can be considered by assuming constant fuel use (i.e. $H_u$ and $L_{St}$ = constant):

$$P_i = i \cdot n \cdot V_H \cdot \eta_i \cdot \lambda_a \cdot \frac{H_u \cdot p_G}{(L_{St} \cdot \lambda + 1) \cdot R_G \cdot T_G}$$

| | | |
|---|---|---|
| Pi $\uparrow$ : | $i$ | from four-stroke to two-stroke |
| | $n \uparrow$ | increase in engine speed |
| | $V_H \uparrow$ | bore, stroke, number of cylinders |
| | $\lambda \downarrow$ | operation with excess fuel |
| | $\eta_i \uparrow$ | process improvement |
| | $\lambda_a \uparrow$ | flow cross-sections, multi-valve, manifold adjustment, variable valve control timings |
| | $p_G \uparrow T_G \downarrow$ | supercharging, charge air intercooling |

In any given engine (displacement and number of cylinders), an increase in engine speed and supercharging have the greatest potential to increase power. Modern high performance engines achieve their high and specific power output either by supercharging with an exhaust-gas turbo charger or a mechanical compressor (see chapter 6) or use what is known as a high speed engine concept.

## High speed engine concepts

In the case of high speed engine concepts, high specific power output per liter is achieved by increasing the rated engine speed beyond the measure (6000 to 7000 rpm) that is normal for gasoline engines. The rated output is in some cases more than 8000 rpm. These high engine speeds mean a correspondingly high load for the engine mechanics, e.g. average piston speeds at 25 m/s, therefore specific power output per liter is in the range of that achieved by racing engines. In racing engines, however, required running times are less than those of series production engines. High speed engine concepts put into series production need adapted tools and appropriate component design to ensure that a series production engine is able to endure such loads throughout its service life. The disadvantage with these concepts is the friction loss that increases by a power of three above the engine speed.

The powertrain for high engine speed concepts can be designed significantly lighter than torque concepts, where extremely high torque must be transferred to a reinforced heavy powertrain. Together with an adjusted short transmission, there is better acceleration response. The lighter engine weight is a further advantage of the high engine speed concept. High engine speed concepts with direct injection create specific challenges for the injection system, since a large quantity of fuel has to be injected with great precision in the short time that is available, e.g. Audi 4.2 V8 FSI for the RS4 [5]:

| | |
|---|---|
| **Max. Engine Speed:** | 8250 rpm |
| **Displacement:** | 4163 $cm^3$ |
| **Max. Power:** | 309 kW (420 PS) at 7800 rpm |
| **Max. torque :** | 430 Nm at 5500 rpm |
| **Torque:** | 90% of torques available between 2250 and 7600 rpm |

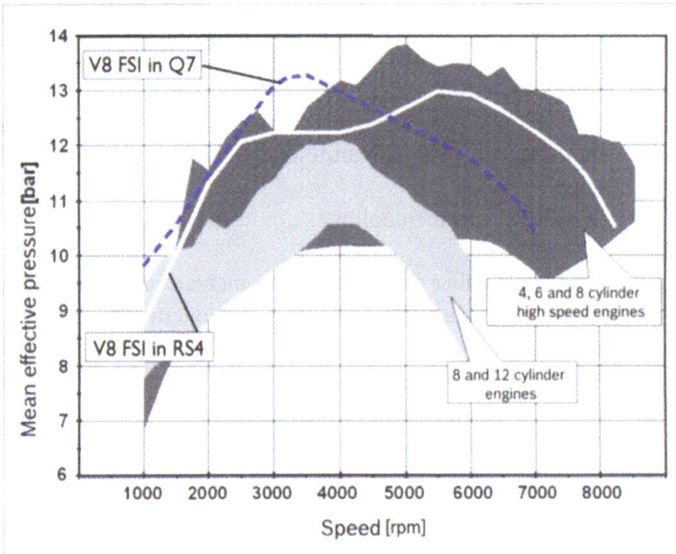


**Figure 5.6:**
FEV scattering range for the mean effective pressure of different engine variants [5]

**Figure 5.6** shows the FEV scattering range of the mean effective pressure for different engine concepts. It is now possible to obtain extremely high maximum mean pressures using direct injection. For the V8 high speed engine concept with direct injection, the engine is in the range of the scattering range which is essentially typical for four- and six-cylinder engines.

In this case as well, direct injection makes it possible to achieve a significant reduction in consumption, shown clearly by the FEV scattering range for specific consumption in **figure 5.7**. As a reference point (2 bar / 2000 rpm), the specific consumption is 360 g/kWh (V8 high engine speed concept) and 355 g/kWh (V8 basic engine) respectively. The significant difference between these two units is the additional control for the coolant temperature in the basic engine. At the best efficiency point, the radically decelerated high engine speed concept achieves a specific combustion of 231 g/kWh. With 235 g/kWh, the basic engine is slightly ahead of it. Thus, both the units are at the lower limit of the scattering range [5].

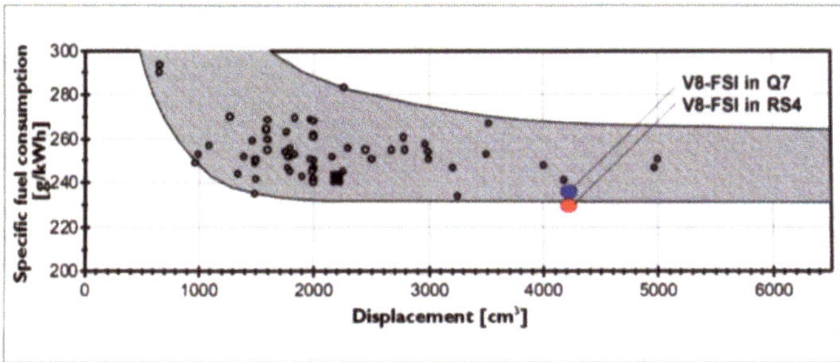


**Figure 5.7:** FEV scattering range for the specific consumption in best efficiency point [5]

# Literature

[1] *Eichlseder, H.; Klueting, M.; Piock, W. F.*: Der Fahrzeugantrieb – Grundlagen und Technologien des Ottomotors. Springer-Verlag, 2007

[2] Mitsubishi Motors: Mitsubishi Benzinmotor mit Direkteinspritzung. Technisches Dossier, 1995

[3] *Roessler, K.; Kraemer, S.*: Einfluss verschiedener Parameter auf das Betriebsverhalten eines Ottomotors mit Benzin-Direkteinspritzung. Direkteinspritzung im Ottomotor, Tagung Haus der Technik, Essen 1997

[4] *Wurms, R.; Kuhn M.; Zeilbeck, A.; Adam, S.; Krebs, R.; Hatz, W.*: Die Audi Turbo FSI Technologie. 13. Aachener Kolloquium Fahrzeug- und Motorentechnik, Aachen 2003

[5] *Koenigstedt, J.; Bach, M.; Brinkmann, C.; Jablonski, J.; Mueller, R.; Hatz, W.*: Die neuen V8-FSI-Motoren von Audi. 14. Aachener Kolloquium Fahrzeug- und Motorentechnik, Aachen, 2005

# 6 Supercharging

Naturally aspirated engines take in the air for combustion from the environment through air filters during the intake cycle. In contrast, supercharged engines use compression units to compress combustion air before it enters the engine. In comparison to naturally aspirated engines, supercharged engines with the same displacement take in the same air volume but due to a higher working medium pressure, larger air masses reach the combustion chamber. From an engine development point of view, this leads to a greater degree of freedom. More fuel can be combusted, resulting in increased engine performance, though the engine speed and volume displacement are still the same. Another option is to increase the amount of air for combustion, so that the engine delivers the same power but the combustion takes place at other air/fuel ratios, which enables lower pollutant emissions when measures for treating exhaust gas are also being implemented. Supercharged engines can be found wherever high performance with good economic efficiency is required. Supercharged engines consume less fuel than equally powerful naturally aspirated engines because they can be built in a smaller size with smaller friction and heat losses. The torque characteristic of supercharged engines can be configured more favorably. Engine supercharging enables smaller power units to achieve the same performance as bigger power units without supercharging. On the one hand, this results in a reduction of friction losses in the engine as smaller engines usually make do with fewer and smaller mounting points. On the other hand, charge exchange losses are reduced due to supercharging. Also, small engines are subject to higher specific loads. As a consequence, in most driving situations the operating point of the engine is shifted to the engine map range that is more favorable in terms of consumption (downsizing principle). A detailed explanation of these relationships is given in chapter 8 Downsizing and Downspeeding.

Pre-compression during supercharging results in charge air heating of by up to 180°C. If the air is cooled in a charge air intercooler before it reaches the combustion chamber, it is possible to further increase the air density and therefore the engine power. In combustion engines, charge air intercooling presents as one of the few measures taken that has a positive effect not only on performance and fuel consumption but also on pollutant emissions. Directly injecting the gasoline engine additionally increases the supercharging potential. As gasoline engines with direct injection present with lower knock sensitivity, it is also possible to select an unusually high compression ratio for supercharged engines. Turbocharged engines also feature a higher air mass flow rate which results from de throttling which counteracts the "turbo lag" in the lower engine speed range. Many manufacturers see a large fuel-saving potential in downsizing concepts in connection with direct injection.

For supercharging, there is a difference between engines with mechanically driven superchargers and engines with exhaust-gas turbochargers. The compressors relevant to the engine can be divided into two groups: displacers and flow compressors. The difference lies in the compression mechanism. Within the displacers' group, a differentiation is made between compressors with or without internal compression. Internal compression is present when the air is subject to increased pressure as it passes through the compressor. Displacers without internal compression transport the air volume up to the outlet of the compressor and it is only here that the air is compressed by ram pressure. With internal compression, the compression to be employed can be reduced and compressor efficiency can be increased [1]. Flow compressors, in contrast, generally work with internal compression.

The basic differences become apparent in the compressor map where the compressor pressure ratio is represented as a function of the flow of air volume that enters the compressor. The compressor map also indicates the curves of constant supercharger speed $n_V$ and the curves of constant isentropic compression efficiency $\eta_{sV}$. **Figure 6.1** shows the basic differences in the maps of displacers and flow compressors. Displacers, **figure 6.1** left, typically feature constant supercharger speed lines that are inclined to the left, according to which the volume flow decreases for increasing charge pressure, are typical of displacers.

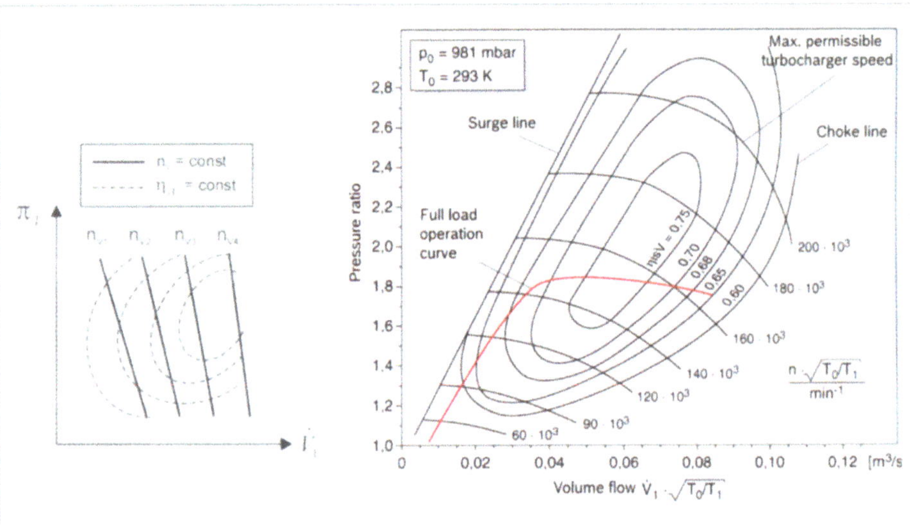


**Figure 6.1:** Maps of displacer (left) [2] and flow compressor (right) (schematic) [3]

The reasons for this are firstly the dead space which, for increasing discharge pressure, leads to a decrease in the volumetric efficiency and therefore in the delivery volume, and secondly, leakage and clearance losses, which increase due to air re-expansion over the gap within the compressor. The achievable charge pressure is thus almost independent of supercharger speed, but depends crucially on design-engineering boundary conditions. The delivery volume, however, is determined directly by supercharger speed and the construction volume of the supercharger. Unlike flow compressors, positive displacement superchargers deliver the air relatively discontinuously. Besides the excitation of pressure vibrations in the charge-air tubes, positive displacement superchargers create higher noise levels and work only up to pressure ratios of approximately $\pi_V = 2$ ($\pi_V$: quotient of total pressure downstream and upstream of the compressor). However, advantages arise from the use of the entire map as there is no unstable operation.

Flow compressors are based on the physical principle of converting kinetic energy into potential energy or pressure energy. The most important flow compressors for use in reciprocating piston engines are the radial compressors (centrifugal compressor) in which the air comes in axially and leaves the compressor wheel radially after compression. The operating behavior of flow compressors is characterized by three parameters: pressure ratio, volume flow and engine speed. In the compressor map, **figure 6.1** right, the pressure ratio is plotted against volume flow with the parameter engine speed, the variables being represented as related via a Mach number similarity for comparability reasons. In the map, the curves of constant efficiency ($\eta_{sV}$) are the diagonal concentric ellipse-like curves. Unlike displacer machines, the pressure ratio of a flow machine depends on engine speed. The useable map range of flow compressors is limited by surge and choke lines and maximum permissible compressor speed.

### Surge line

The surge line limits the left margin of the map; **figure 6.1** right. Strictly speaking, the surge line is not a feature of the compressor but is influenced by the flow conditions upstream and downstream of the compressor. It occurs only in machines based on the flow principle and is thus a feature of this concept. If volume flow is too low while pressure ratios are high, the flow can no longer follow the blade airfoil, which leads to a backflow in the boundary layer with stalling of the airflow as a result. The delivery process is interrupted; the airflow through the compressor is reversed until a stable pressure ratio with positive volume flow is reached again. The pressure builds up again until the flow stalls again at a particular pressure. The process is repeated periodically. The resultant noise is known as "surging". Stable operation at or beyond the surge line is not possible. The surge line can be influenced by taking certain measures at the flow compressor, i.e. its position in the map can be shifted, but it cannot be prevented. Backward bending of the blading at the exit of the running wheel can, for example, shift the surge line to smaller flow volumes, increasing the stable range of the compressor map. The design of the diffuser can also change the position of the surge line in the map.

### Choke line

The maximum volume flow of centrifugal compressors is generally limited by the cross-section at the compressor wheel inlet. When the air at the wheel inlet reaches sonic speed, it is not possible to further increase the flow rate. Steeply falling engine speed curves at the right margin of the map indicate the choke line; **figure 6.1** right. Like the surge line, the choke line can be changed through selective modifications in the geometry of the compressor wheel. Setting back every second blade (splitter blades), reducing the blade thickness or decreasing the number of blades to minimize the blocking effect of the blades shifts the choke line to higher volume flows. Increasing the hub ratio also causes the maximum airflow rate to increase and thus shifts the choke line. During design, it should be taken into consideration that these measures also have an effect on other compressor characteristics and are limited by component strength, required service life and the processes necessary for cost-optimized mass production.

## 6.1 Mechanical Supercharging

During mechanical supercharging, the combustion air is guided to the engine by a positive displacement supercharger, which is driven by the engine itself; **figure 6.2**. Some of the increase in power thus achieved is used up for the driving power required for the positive displacement supercharger (between 10 and 15 kW depending on the engine size in the passenger car), which is the biggest disadvantage of this supercharging method.

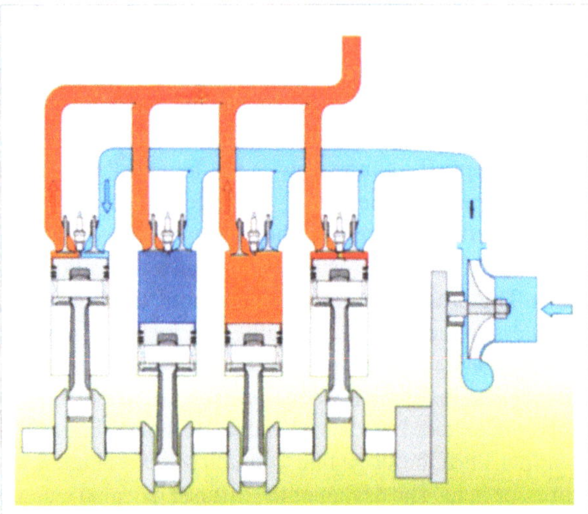

**Figure 6.2:**
Principle of mechanical supercharging [3]

Mechanical supercharging offers advantages compared to exhaust-gas turbocharging, particularly for gasoline engines. Thanks to mechanical coupling, there is the possibility of quickly increasing the charge pressure even at low engine speeds, to improve driving dynamics, increase the torque and improve the vehicle response, especially for a gasoline engine. Furthermore, components that conduct exhaust gas need not be made of high-temperature resistant and cost-intensive materials. The air-conducting components can be mounted at a sufficient distance from the hot components. In the case of exhaust-gas turbocharging, however, there is no mechanical coupling with the engine. As a result, the charge pressure build-up depends on the exhaust-gas mass flow and therefore also depends on the load condition "turbo lag". The turbine is brought to engine speed until the compressor provides the charge pressure by means of a corresponding amount of exhaust gas. In exhaust-gas turbocharging, the exhaust-gas turbine is mounted as closely as possible to the exhaust manifold for efficiency reasons. This causes thermal problems for the remaining adjacent engine parts and imposes high demands on the exhaust-gas sealings. However, the exhaust routing of the compressor engine can be designed favorably in terms of flow and costs, and the catalytic converter can be mounted at the ideal location. The specific advantages and disadvantages of mechanical supercharging are summarized below.

**Advantages**:
- Direct drive by the engine – can be switched off through magnetic coupling
- Rapid charge pressure build-up – high torque at low engine speeds
- Improved vehicle response
- Low complexity of the routing of exhaust gas and fresh air
- Quicker light-off of the catalytic converter.

**Disadvantages**:
- Poor overall efficiency – drive costs about 15 kW of driving power
- For sealing reasons, only charge pressure of up to 1.8 bar is possible
- Large installation space required
- High costs.

The most common terms for mechanical charging units in German-speaking countries are "*Kompressor*" [compressor] or "*Mechanischer Lader*" [mechanical supercharger]. The terms "supercharger" and "blower" are more common in English-speaking countries. Depending on the design, the following types of superchargers are distinguished: screw-type supercharger, positive displacement supercharger, flow compressor supercharger, rotary piston supercharger, and vane-type supercharger.

**Design types of mechanical supercharging (compressor supercharging)**

Rotary piston machines (positive displacement superchargers) with one or two shafts are the most common machines used for mechanical supercharging of combustion engines. As for toothed gearings (external and internal toothing), types with outer axes and inner axes are distinguished for double-shaft machines. In the simplest design, the rotor consists of a circular cylinder. The housing is made of a circular-cylindrical tube and the separator between the suction chamber and the pressure chamber consists of a spring-loaded slide valve. The single-shaft type can be easily and compactly mounted to the combustion engine. The drive can also function without belts or gear wheels when the rotor is directly driven by the crankshaft tip. The central-symmetric line-up of several rotors with one behind the other enables a smooth, almost pulsation-free delivery process. However, the double-shaft, bolted design type with the outer axes based on the Roots design is currently the one most frequently used in series production cars. Screw-type compressors are also increasingly popular. The aftermarket also uses mechanically and electrically driven radial compressors and screw-type compressors. Mechanically driven radial compressor use compact, low-friction, low-noise gears are used, such as . a toothed belt drives. To improve driving dynamics, increase the torque and improve the vehicle response, especially for gasoline engines with downsizing, it is also possible to use electrically driven (auxiliary) compressors which is another way of quickly increasing charge pressure, even at low engine speeds.

### Rotary piston superchargers

These superchargers are also called, named "Roots blowers" named after the inventors of this design principle. They have two counter-rotating rotors with two or three club-shaped, alternately meshing lobes. Similar to a gear pump, the air is drawn in at one side, carried along the inner wall of the oval housing by the lobes, and pressed out on the other side; **figure 6.3**.

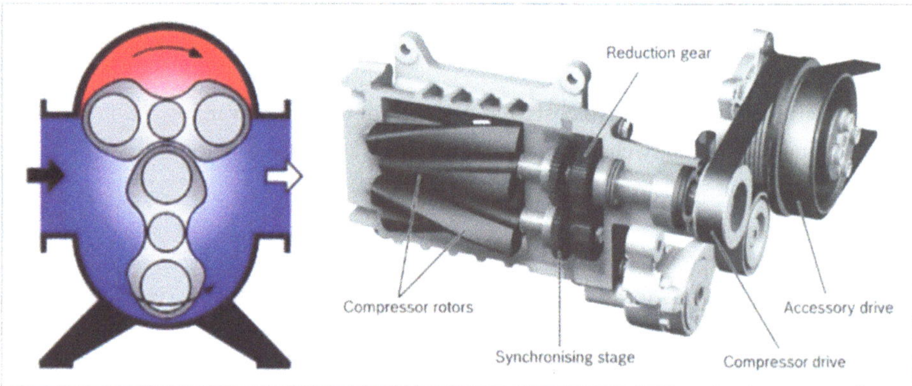


**Figure 6.3:** Principle of the rotary piston supercharger (left, [source: FAG]) and realized charging unit (right, [source: VW])

The shafts of the two rotary pistons are connected by gear wheels outside the housing. The pistons operate without directly contacting each other or the housing. Roots superchargers work without internal compression. Due to their mode of operation, they work effectively only with large amounts of air and are therefore relatively large and heavy. They are durable and economic in production which is due to the lower engine speed, smaller thermal and pressure loads and because they run without making contact. In the past, Roots superchargers were frequently installed in Mercedes racing cars; today, they are mounted only in the Mercedes-Benz compressor models. The rotary supercharger is a further developed model, with one lobe and three internal rotors.

### Screw-type compressors

Screw-type compressors are double-shaft rotary piston machines based on the positive displacement principle with internal compression. During its transport from the intake connector (mounted at the top of the housing) to the pressure connector (mounted at the bottom of the housing), the delivery gas is compressed in steadily diminishing chambers and then delivered. The compression procedure itself is shown in drawings 1 to 4 in **figure 6.4:**

1. The gas to be compressed passes through an intake orifice located in the housing and flows into the tooth space.
2. After reaching the maximum tooth space volume, the tooth points of the rotors sweep over the intake metering edge. This seals the working chamber, and compression begins.
3. The compression process continues until the tooth points pass the exhaust orifices.
4. Further rotation pushes the medium out under pressure.

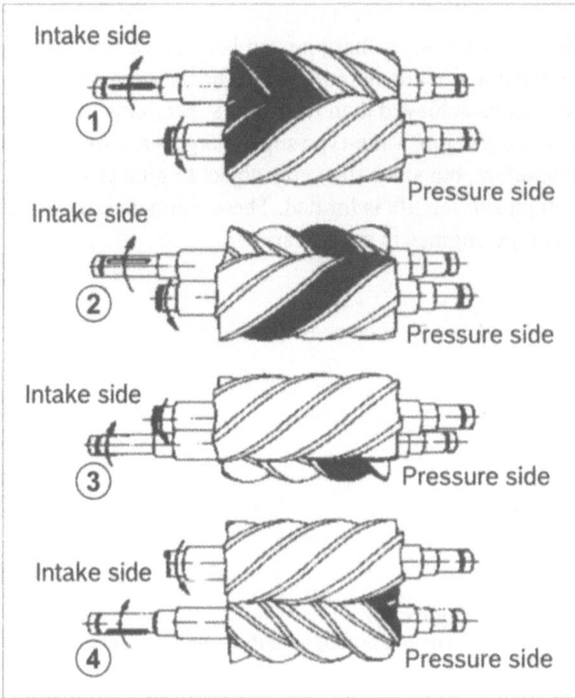


**Figure 6.4:**
Compression process for a screw-type supercharger

Screw-type compressors are characterized by simple design, small dimensions, little mass, even and pulsation-free delivery, quiet running and the lack of oscillating masses and controls.

**Vane-type superchargers**

These are based on the principle of the hydraulic pumps of the same name, or, seen another way, the principle of the "air motors" in many pneumatic tools. An eccentrically mounted rotor runs in a housing with a circular cross-section; **figure 6.5**. One or several guides, mostly radially arranged, are built into the rotor (2). Rotary valves (3) are fitted in these guides. These valves subdivide the space between the stator and the rotor into several chambers. In order to compensate for the changes in distance between the rotor (2) and the stator (1) during a rotation, the rotary valves can move within the guides. They are usually pushed against the inner wall of the stator by a spring (4) mounted at the base of the slot. In the operating state, a spring force or centrifugal force pushes the outer edges of the rotary valves (vanes) against the inner wall of the housing and slides them on the surface of the inner wall. This leads to the formation of enclosed spaces between adjacent vanes, also called cells, in which the air is transported. The housing has an inlet orifice and an outlet orifice. Due to the eccentric mounting, the cells are first increased in size on the intake side during rotation, which creates a light vacuum. Towards the pressure side, they taper until the outlet orifice is reached. This way, the air is pre-compressed and is guided to the intake system of the engine more rapidly. The eccentricity can be changed by means of easily operated adjustment devices, and the supercharging can be adapted without any problems. Vane-type superchargers have a smaller output than turbochargers and Roots superchargers. Higher engine speeds can be achieved than for Roots superchargers, but the speed is limited by centrifugal forces and friction. Vane-type superchargers are small, light and relatively cost-effective to manufacture, but since they are subject to greater wear and tear due to the friction of the vanes, their service life is limited. These characteristics make them suitable primarily for small gasoline engines in sports cars.

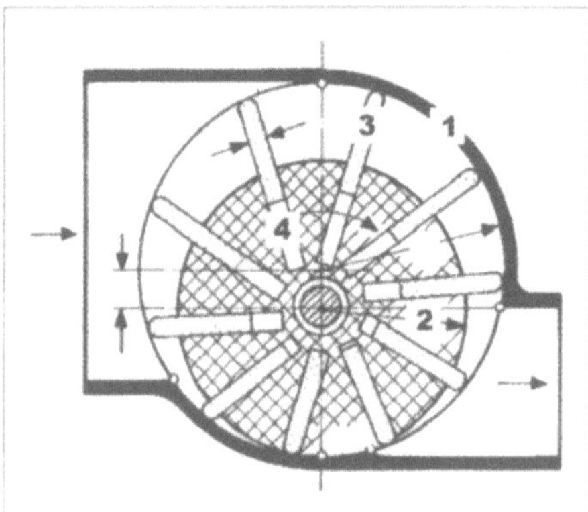


**Figure 6.5:**
Function principle of the vane-type supercharger [4]

## Spiral-type superchargers

The principle of this group of positive displacement superchargers was patented in the USA, at the beginning of the 20th century, but was not widely used, due to the complicated manufacturing and materials requirements. In the 1970's, the idea was taken up again by Volkswagen, and after a test phase with a few hundred samples and many detailed changes, it was used in larger quantities in the 1980's. VW produced two different size models and called it the *G-Lader* [G supercharger] (G40 / Polo and G60 / Golf, Passat and Corrado). The name refers to the size of the spiral-type supercharger; **figure 6.6**.

The housing is round in the cross-sectional view and consists of two halves into which two spiral-shaped bars are cast which, like the spiral-shaped displacers, resemble the capital letter G. The numbers 40 and 60 stand for the height/width of the housing in millimeters. The

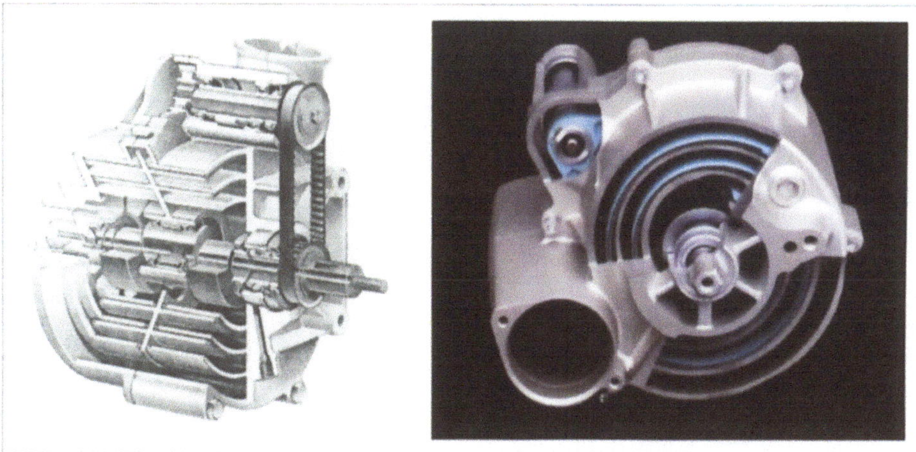

**Figure 6.6:** Spiral type supercharger realized as G supercharger [Source: VW]

displacer is driven by a main shaft which is connected to a crankshaft via a belt drive and is guided by an auxiliary shaft coupled with the main shaft by means of a belt. Both shafts have eccentrics so that the displacer does not rotate but only makes fast oscillatory movements. The air flows tangentially into the housing. It is enclosed between the spiral bars of the housing and the displacer and is transported towards the housing center, from where it reaches the intake system. The oscillatory movement leads to steadily diminishing volumes between the bars. The air is compressed and accelerated. Due to the considerable friction of the complex sealing elements and springs, which are located between the front sides of the displacer and the housing, G superchargers have a short service life if the high-wear components are not checked and replaced regularly. The G supercharger did not catch on for economic reasons as manufacturing costs (a scrap rate of up to 80%) repair and replacement costs were high. VW ceased production in the early 1990's.

## 6.2 Exhaust-gas turbocharging

To drive the charging unit, exhaust-gas turbocharging, **figure 6.7**, uses the hot exhaust gas from the engine, which is otherwise released into the environment with a high content of unused energy.

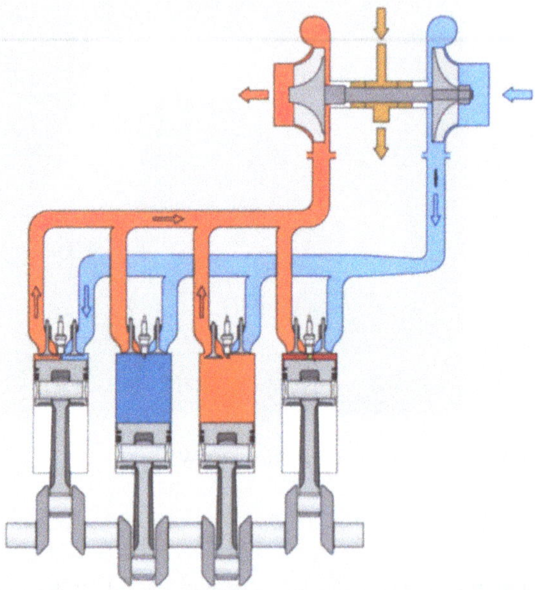

**Figure 6.7:**
Principle of exhaust-gas turbocharging [3]

Turbocharger and combustion engine are, therefore, coupled purely thermodynamically, rendering the following important advantages [1, 3, 5]:

- The fuel consumption of the turbocharged engine is lower when compared to an equally powerful naturally aspirated engine, because some of the otherwise unused exhaust energy contributes to engine performance.
- The turbocharged engine performs much better at high altitudes. The performance of a naturally aspirated engine is considerably reduced at high altitude as a consequence of the decreasing air pressure. In a turbocharged engine, the turbine power increases as there is a larger pressure difference between the almost constant pressure upstream of the turbine and the lower ambient pressure. The lower air density at the compressor inlet is partially compensated for, and the engine loses almost no power.

An exhaust-gas turbocharger consists of a compressor (centrifugal compressor) and a turbine, **figure 6.8**, connected to each other by a common shaft. The turbine converts some of the energy contained in the hot engine exhaust gas into mechanical work; the pressure and temperature of the exhaust gas decrease (polytropic, i.e. dissipative expansion). The mechanical work is then supplied to the compressor via the common shaft. In the running wheel of the compressor, the energy is initially transferred to the intake combustion air in the form of kinetic energy, and some of it is subsequently converted into pressure energy. During this process, the temperature of the intake combustion air increases. Apart from the mechanical and thermal losses, steady-state operating conditions show a power equilibrium between the turbine and the compressor.

Feel good about driving
Less emissions
More driving pleasure

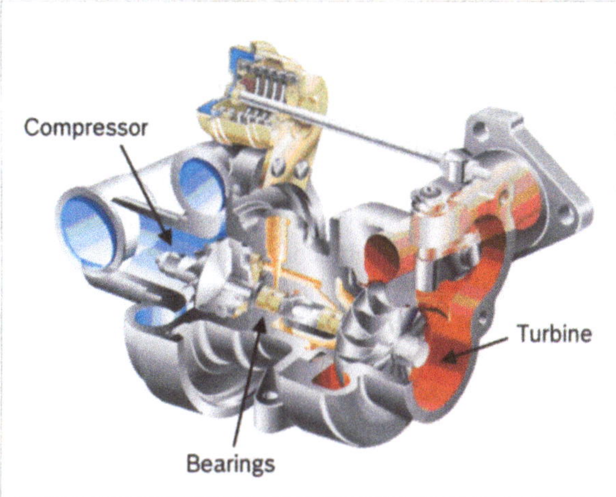


**Figure 6.8:**
Design of the exhaust-gas turbocharger

This power equilibrium results in a relationship which is also called the turbocharger main equation, applying the thermodynamic relationships for the compressor power, the turbine power and the isentropic enthalpy differences,

$$\pi_V = \frac{p_2}{p_1} = \left(1 + \frac{\dot{m}_T}{\dot{m}_V} \cdot \eta_T \cdot \eta_V \cdot \eta_m \cdot \frac{T_3}{T_1} \cdot \frac{c_{p,A}}{c_{p,L}} \cdot \left[1 - \left(\frac{p_4}{p_3}\right)^{\left(\frac{\kappa_A - 1}{\kappa_A}\right)}\right]\right)^{\left(\frac{\kappa_L}{\kappa_L - 1}\right)_L}$$

The main turbocharger equation shows for a particular inlet condition in the turbocharger ($p_1$, $T_1$) that the charge pressure ratio $\pi_V$ is the higher, the greater the isentropic efficiencies for the turbine $\eta_T$ and the compressor $\eta_V$, the smaller the mechanical losses (i.e. the greater the mechanical efficiency $\eta_m$) and the higher the turbine inlet temperature $T_3$ and the pressure $p_3$ upstream of the turbine. By reducing the mechanical losses, at otherwise identical conditions, a specific charge pressure $p_2$ is achieved even at smaller exhaust-gas back pressure $p_3$, which reduces the fuel consumption of the engine and leads to favorable conditions for the scavenging of the exhaust gases from the cylinder. For high overall efficiency of the turbocharger, coordination between the compressor wheel and the turbine wheel diameter is very important. The specification of an operating point in the compressor map results in a specific turbocharger speed. The turbine must be adjusted in such a way that it works in this operating range with the greatest possible efficiency.

It is possible to use the dynamic part ("pulse") of the escaping exhaust gases or to even it out. In constant pressure turbocharging the exhaust-gas pulses from the gas exchange are largely smoothed out by means of collectors with relatively large displacement. As a consequence, the turbine is charged with more or less constant pressure. In pulse turbocharging, however, the exhaust-gas pulses are maintained until the turbine inlet is reached by means of a suitable manifold design (volume as small as possible, separation or a suitable combination of the outlet ports up to the turbine inlet, i.e. firing order separation). Both variants – pulse turbocharging and constant pressure turbocharging – are implemented in the charging technology and have their advantages for specific applications. Constant pressure turbocharging is used in engines that do not need high excess torque for acceleration. Pulse turbocharging, in contrast, improves the torque yield in the lower speed range of the engine and is therefore preferably used to supercharge passenger car engines. The disadvantages of pulse turbocharging for the turbine, as compared to constant pressure turbocharging (poorer efficiency, lower discharge capacity, potential vibrational excitation of the blades), are more than compensated for by utilizing the pulse. For reasons of cost or space, less complex designs are frequently implemented in which the outlets lead to a common tube which is positioned upstream of the turbine and supplies the exhaust gases to the charger. The separation of the firing order, which is not consistent, reduces the potential of pulse turbocharging and has repercussions on the charge exchange. The potential for improvement through pulse turbocharging in the lower engine speed range is obvious; however, it cannot overcome the increasing discrepancy between the steady-state full load curve and transient behavior of turbocharged engines that goes hand in hand with the increasing degrees of supercharging.

In terms of thermodynamics, isothermal compression of the charge air would be the best solution (i.e. the temperature of the medium remains constant during the change of state). Isothermal process control is not possible under real conditions, though. Besides, the actual compression is not free from losses, i.e. it is not a reversible adiabatic process (final compression temperature $T_{2,s}$) but polytropic. As compared to reversible adiabatic process control, more work has to be used from pressure $p_1$ to $p_2$ during dissipative compression, which manifests itself in a higher compressor outlet temperature $T_2$ and is expressed by isentropic efficiency $\eta_{\check{V}}$, i.e. the ratio of isentropic $(h_{2,s} - h_1)$ and actual enthalpy difference $(h_2 - h_1)$ across the compressor.

$$\eta_V = \frac{h_{2,s} - h_1}{h_2 - h_1}$$

The higher final compression temperature $T_2$ leads to the charge air having less density. In order to counteract poorer cylinder charging that results from it, the charge air is carried to a charge air intercooler before it enters the engine so that the charge air can be cooled, to achieve an increase in charge density. This also reduces the thermal load of the engine and has a positive influence on the knock limit. The lower temperature level in the cylinder inhibits the formation of nitrogen oxide. However, the charge air intercooler causes a loss in pressure, which must be taken into consideration when designing the turbocharger. The cooling of the charge air is mainly influenced by the temperature of the coolant and the effectiveness of the cooling system, the heat having to dissipate into the environment through a closed circuit in the case of passenger cars.

### Function and design of the exhaust-gas turbocharger

The centrifugal compressor of a turbocharger consists of a running wheel, a diffuser and a spiral housing. The charge air is drawn in axially and is accelerated to high velocity in the wheel (in the absolute system, i.e. for an observer stationary with regard to the housing). The air generally leaves the compressor wheel in a radial direction. A diffuser is connected right next to the running wheel and is formed from a compressor backplate and part of the spiral housing. The diffuser slows down the velocity of the air, largely without any losses, so that the pressure and the gas temperature both rise. Even for constant diffuser width, the radius increase in flow direction results in an increase of the flow cross-section and therefore a deceleration of the flow results. So far, bladeless diffusers with fixed geometry have generally been used for passenger car applications. They have a wide map range. The air is collected in the spiral housing, where the velocity continues to be reduced before the charge air reaches the compressor outlet, and this leads to another pressure increase. In terms of aerodynamics, the compressor flow is a delayed flow (in the relative system, i.e. for an observer positioned on the wheel). The channels of the compressor wheel are complex diffusers. The flow in the channels is in the direction of a positive pressure gradient. Compressor wheels for passenger car applications have a relatively high hub ratio (ratio between the compressor inlet and compressor outlet diameters), an adjusted number of blades, and thin blades bent backwards. To reduce the component stress during operation, the blades are attached diagonally to the hub.

The turbine of a turbocharger consists of a spiral housing, bladed or bladeless nozzles, and a turbine wheel. The exhaust gas from the engine is retained by the flow resistance of the turbine. The spiral housing is designed in such a way that the exhaust gas is directed to the nozzle as evenly as possible across the circumference. In the spiral housing and in the nozzle, some of the exhaust-gas pressure that has built up is converted into kinetic energy, i.e. the flow is accelerated and directed to the turbine wheel. The turbine wheel converts some of the enthalpy of the hot exhaust gas into mechanical energy that drives the compressor of the turbocharger via the shaft. Due to the mass flow and pressure ratios that arise, only centripetal turbines that receive airflows radially and expel them axially are used for passenger car turbocharging. In regard to their comparatively low moment of inertia, axial turbines would be quite interesting; the small mass flow, together with the Mach and Reynolds numbers necessary for an effective energy conversion, lead to very small flow cross-sections so that the boundary layer effects and clearance losses clearly limit the efficiency. In terms of aerodynamics, turbine flow is an accelerated flow; the channels of the turbine wheel are complex nozzles. The flow in the channels is in the direction of a negative pressure gradient. Certain flow phenomena, as can occur in compressors depending on the operating point (surge), cannot occur in a turbine. The turbine performance increases with increasing pressure gradients across the turbine. At higher engine speed, more exhaust gas is built up and the expansion work increases. Turbine performance also improves with increasing exhaust-gas temperature due to the higher energy content of the exhaust gas. The retaining behavior of the turbine is determined by the free flow cross-section at the crossing point from the inlet port to the spiral, or throat cross-section. If the throat cross-section is reduced, more exhaust gas is retained by the turbine, the turbine performance increases as a result of the higher pressure ratio and a higher charge pressure can be generated in the compressor. The throat cross-section is specifically set during the design of the turbocharger, i.e. when it is matched to the required performance of the engine. In addition to the throat cross-section of the turbine housing, the passage cross-section upstream of the wheel inlet also influences the flow rate characteristic of the turbine. The machining of the wheel contour makes it possible to adjust the cross-section and thus to influence the charge pressure. Contour enlargement results in a larger flow cross-section of the turbine. Turbocharger manufacturers usually offer turbine wheels of the same diameter with different contours within a production series; these can be made from the same blank.

**Variable turbine geometry**

Depending on the operating point, the free flow cross-section or the direction of the flow can be influenced by movable elements located in the spiral housing or the nozzle. While the turbocharger with "variable turbine geometry", or VTG, has established itself as the standard charging device for diesel engines, this technology is still a huge challenge for gasoline engines due to the high exhaust-gas temperatures. Besides the solutions for adjustable elements presented and examined here [6], in particular pivot mounted guide blades (VTG principle), **figure 6.9**, or an axially slidable sleeve (VST principle), **figure 6.10**, are the other main solutions.

The VTG turbocharger has a spiral housing with fixed geometry. Between the spiral housing and turbine wheel inlet, pivot mounted turbine guide blades are arranged so that the flow cross-section and the angle of approach to the turbine wheel can be changed. While straight guide blades were installed in the first VTG series turbochargers, today increasingly complex cambered blade profiles with much greater efficiency and favorable control behavior are used. The individual guide blades are connected with a ring by means of an adjusting lever charged by an actuator via a multi-linkage lever. The exhaust gas in modern

**Figure 6.9:** Turbocharger with pivot mounted guide blades (Variable Turbine Geometry / VTG)

gasoline engines can reach up to 1050 °C. The development goal is to reliably increase the working temperature of turbines with variable geometry to this value.

The constant, reliable movement of the guide blades in the hot exhaust-gas stream imposes high demands on the materials used and the design. Detailed adjustment of the tolerances of the individual components concerned within the turbine is indispensable. Irrespective of the size of the turbocharger, a certain clearance is required in order to guarantee that the blades will remain moveable during the entire service life of the vehicle. As a result, the clearance losses increase as the size decreases.

At low engine speeds, the flow cross-section of the turbine is reduced by closing the guide vanes. The charge pressure and therefore the torque of the engine increase as a result of the higher pressure difference between the turbine inlet and the turbine outlet. At high engine speeds, the guide vanes open. The desired charge pressure is reached for lower exhaust-gas back pressure, i.e. for a smaller turbine pressure ratio. The result is a favorable utilization of the engine's primary energy. During acceleration phases from low speeds, the guide blades close for higher accumulation levels and therefore provide plenty of excess power for the quick acceleration of the rotor. With increasing engine speed, the blades open continuously according to the run-up characteristic of the engine. The first turbocharger with pivot mounted guide vanes for gasoline engines with exhaust-gas temperatures of 980°C was introduced in series by BorgWarner Turbo Systems for the Porsche 997 in 2006 [7, 8].

A special feature of the original VST turbocharger is a twin flow turbine housing and a sliding sleeve, figure 6.10, which can be moved partially over the turbine wheel axially in the direction of the shaft axis and can thus directly change the flow cross-section at the turbine inlet. In the lower engine speed range, one of the two channels of the inlet spiral is completely covered, almost simulating a turbocharger with small turbine throat cross-section. From a certain operating point onwards, with increasing engine power, the second

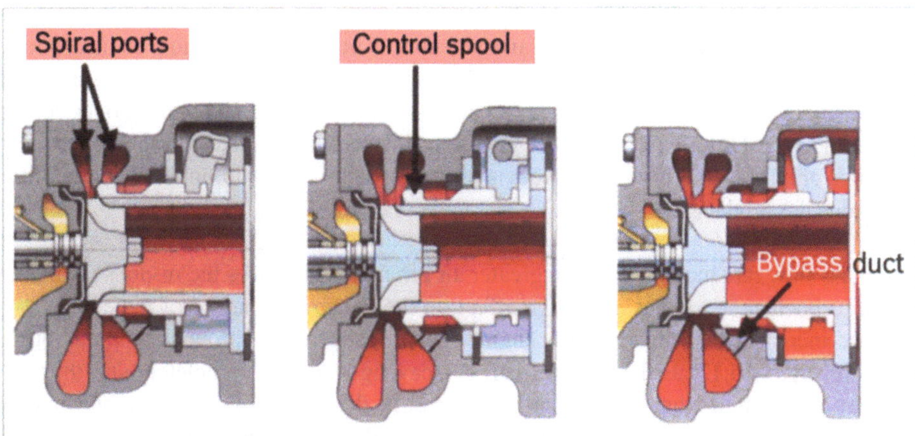


**Figure 6.10:** Axially slidable sleeve (VST) principle

channel is continuously exposed until it is completely open. When the amount of exhaust gas continues to increase, the sliding sleeve axially exposes a ring-shaped bypass duct. The sliding sleeve is directed by an adjusting fork which in turn is moved by an actuator by means of a lever system [3]. Further development of the VST principle in terms of optimized efficiency and control abandons the twin flow housing; a fixed radial guide blade is placed in the passage cross-section directly upstream of the turbine wheel to generate a directed flow.

Depending on the load and engine speed, the axially slidable sleeve increases or decreases the passage cross-section for optimal adjustment to the operating point. As in the original VST turbocharger, a bypass duct can be exposed using the sliding sleeve to decrease the basic charge pressure in the upper operating range. So far the VST principle has not been able to replace the principle of rotating blades.

## 6.3 Synergies between exhaust-gas turbocharging and direct injection in gasoline engines

As for modern diesel engines, the combination of direct injection and turbocharging in gasoline engines has great potential. In addition, gasoline engines have variable valve control as another important component. Therefore, the historical disadvantages of turbocharged gasoline engines – a compression ratio needing to be reduced due to knock tendency, efficiency disadvantages due to the enrichment required for the protection of components in certain operating ranges, and delayed response when accelerating from low engine speeds – can be effectively and sustainably counteracted. The combination of direct injection, turbocharging and variable valve timing opens up new possibilities for improving the engine process. Variable valve timing makes it possible to reduce the residual gas amount of the full load and to realize favorable torque behavior even at low engine speeds. If the scavenging slope is positive, i.e. the charge pressure is higher than the pressure at the turbine inlet, scavenging with fresh air can be achieved by setting a long valve overlapping phase; **figure 6.11**. In gasoline direct injection, the fuel is injected into the combustion chamber only after the exhaust valve has closed, which is why no unburnt hydrocarbon emissions are produced.

The mass flow rate increased by the greater air efficiency improves the response behavior of the turbocharger and has a positive effect on the dynamic behavior of the engine during load increase. Scavenging at low engine speeds leads to an improvement in the torque behavior, **figure 6.12**, as the cylinder can be scavenged almost completely and thus can be kept almost free from residual gas. As a result, the torque is built up more quickly during a load increase and dynamics is increased. BMW [9] additionally sets an air/fuel ratio of $\lambda = 0.9$. This leads to an optimal combustion speed (knocking and pre-ignition are

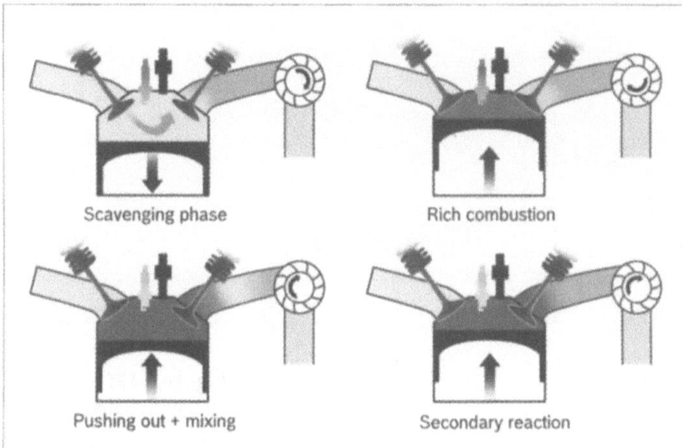


**Figure 6.11:** Principle of scavenging in the low-end torque range [9]

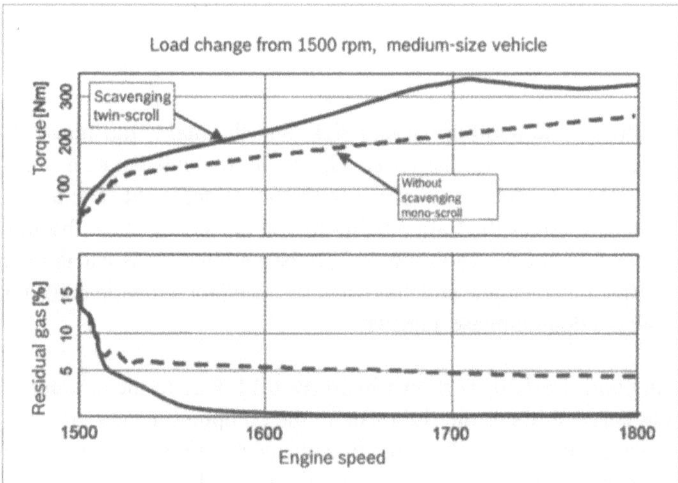


**Figure 6.12:** Torque build-up during scavenging [10]

avoided). In addition, the unburnt fuel of a cylinder reacts with the scavenged fresh air of another cylinder present in the manifold. This secondary reaction increases the enthalpy upstream of the turbine and leads to a stoichiometric air ratio upstream of the three-way catalytic converter, depending on the quantity of the scavenged fresh air.

**Figure 6.13** shows the transient torque build-up at a constant engine speed of 1500 $min^{-1}$, starting from $p_{me}$ = 2 bar up to full load without phase adjuster at the camshaft, with intake phase adjuster and with double phase adjuster.

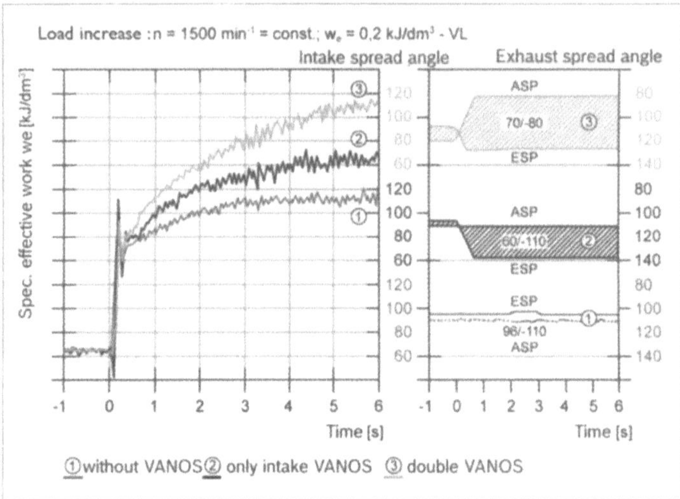


**Figure 6.13:** Dynamic behavior: variation of intake and exhaust spread angle [11]

Without the phase adjuster, the torque build-up is very slow and the maximum mean effective pressure remains on a low level. With the phase adjuster at the intake side, a much faster torque build-up and a higher absolute value of $p_{me} \approx 18$ bar are achieved. The use of a double-phase adjuster further improves the torque build-up and the steady-state absolute level reaches values of $p_{me}$ over 23 bar for 1500 min$^{-1}$. The torque build-up is carried out almost as fast as in a naturally aspirated engine and thus suggests excellent dynamic behavior of the turbocharged engine with a double-phase adjuster.

Another effect of scavenging with fresh air is shown in **figure 6.14**. Due to the amount of scavenged air, the operating curve shifts to the right side in the compressor map, towards higher mass flows. The increased distance to the surge line opens up new freedoms for the selection of the turbocharger. For example, a larger compressor that has more favorable efficiencies in the nominal power range or offers more potential for power increase can be used.

The interaction of the technologies mentioned enables high cylinder charging and lower knock tendency as compared to intake manifold injection. The result is gasoline engines that have high torques even at low engine speeds while maintaining a high speed spread angle. From today's point of view, steady-state performance characteristic values are possible in the order of magnitude of 200 Nm/l and 100 kW/l with extremely responsive dynamic behavior; i.e. with regard to specific torque, the gasoline engine with direct injection can match the ambitious values of the diesel engine; with regard to specific performance, it clearly exceeds the values [11, 12, 13].

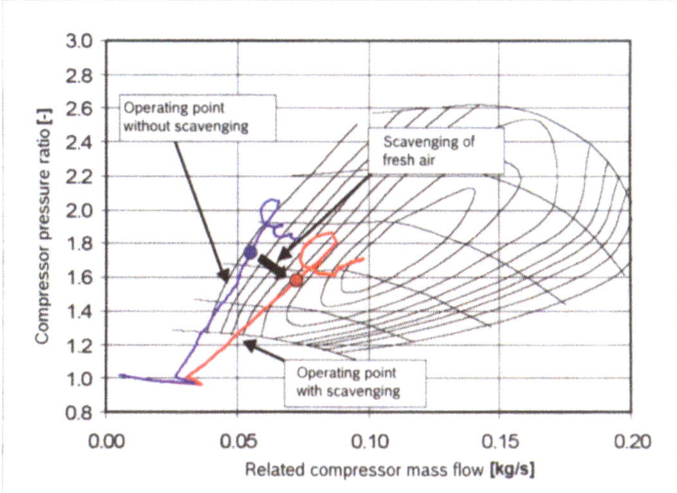


**Figure 6.14:** Shift of the operating points in the compressor map due to scavenging [10]

## 6.4 High pressure supercharging with an electrically driven supercharger

It is in particular the realization of attractive small-displacement gasoline engines that is inseparably connected with a high degree of boosting, among other measures. For such applications, the supercharging system must have a noticeable effect, also in the lowest possible engine speed range. The discrepancy between steady-state and transient behavior increases with increasing supercharged engine power output per liter, even if engine tuning and turbocharger design are optimal, i.e. during the acceleration of the vehicle, the steady-state full load curve is reached at higher and higher engine speeds. This increasing discrepancy can be covered by additional supercharging, whose operating power should be practically independent of or at least be decoupled from the operating state of the engine. Electrical energy is a good choice for this purpose. The development of an electrically assisted supercharging system, however, must always be seen in the context of the possibilities and limitations of the vehicle electrical system. Since the introduction of a 42 V vehicle electrical system with increased capacity has been delayed for various reasons, efforts are focusing on a modified 12 V vehicle electrical system with increased power. This development is currently driven by the increasing number of electrical high power loads, i.e. there is no direct relationship between this development and an electrically driven system. However, the supercharging system can benefit from innovation in the field of vehicle electrical systems. In this context, it should be pointed out that ultimately, electrically assisted supercharging can be classified as mechanical supercharging.

or controlled flap that reduces the pressure losses in the intake port when the eBooster is not in action [14, 15]. To be effective, the eBooster must be able to produce a certain pressure ratio in a very short period of time. On the other hand, the electrical power consumption of the system should be as low as possible in order to remain within the capacity limits of the vehicle electrical systems currently available. This target conflict leads to a compromise in the design of the eBooster power unit with regard to operating speed and moment of inertia of the shaft or the wheel diameter of the compressor. Since the booster power unit is spatially separated from the exhaust-gas system, water or oil cooling is not necessary; the power unit can be realized with air cooling and the shaft can be guided via roller bearings with long-life lubrication. Besides, the power electronics of the electromotor can be integrated onto the eBooster power unit, which leads to very short electrical supply lines. The conceptual design of a suitable drive motor for the e-Booster depends on both the requirements with regard to the torque demand of the compressor wheel and the extremely short run-up time, which creates a quite significant additional torque demand for the acceleration of the rotating masses. The demand for a design that is both robust and cost-effective is another key focus [16]. The effectiveness of the eBooster turbocharging system especially for small displacement gasoline engines with high pressure supercharging has been proved in advanced development projects. It remains to be seen whether the system can establish itself in the market, including with regard to cost. The increasing electrification of vehicles that started some time ago may create the appropriate technical conditions.

## 6.5 Complex supercharging

In engine development, the following goals are paramount: starting torque as high as possible, good propulsion power, and high performance that complies with the legal emissions standards for pollutants and carbon dioxide. Good acceleration behavior requires a rather small turbocharger. This, however, leads to a high back pressure in the nominal power point, with high fuel consumption and lower performance as a result. High performance in the nominal power point along with fuel consumption being as low as possible requires the exhaust-gas back pressure to be as low as possible, i.e. a rather large turbocharger, which, would not have a very satisfactory acceleration behavior due to the shaft's relatively large mass moment of inertia, though. This target conflict is more and more solved by means of a complex, sometimes multi-stage interconnection of charging units. Additional supercharging virtually independent of the operating condition of the engine is considered as one option; see section 6.4.

### Multi-stage supercharging

In multi-stage supercharging, **figure 6.16**, the supercharging of the engine is done via two smaller turbochargers connected in parallel instead of by one large turbocharger. In addition to the turbochargers, a multi-stage supercharging system includes several switching devices and sensors that facilitate a smooth change from a single turbocharger operation to a multi turbocharger operation. Only one turbocharger is active in the lower engine speed range. From a certain engine speed onwards, both turbochargers operate in parallel. Due to the fact that the air supply is distributed between two turbochargers, torque characteristics are responsive and nominal power is high. The significantly lower mass moment of inertia of the turbocharger shaft, as compared to a correspondingly designed single turbocharger unit, furthermore results in a very favorable response behavior by the engine. The system combines high performance with relatively low fuel consumption and impressive acceleration. In the lower engine speed range, all the exhaust gas and all of the charge air flow over only one turbocharger until the corresponding setpoint charge pressure is reached.

The turbine sequence valve and compressor sequence valve for the second turbocharger are closed; the second turbocharger is not active. If the desired charge pressure is built up and if more exhaust-gas enthalpy is available than is required to reproduce charge pressure, it is directed to the turbine of the second turbocharger which is connected in parallel and subject to acceleration through this step.

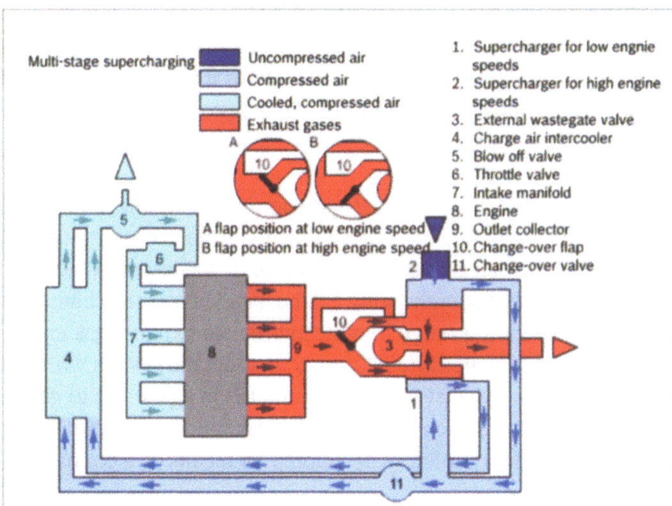


**Figure 6.16:** Principle of multi-stage supercharging

The compressor of the second turbocharger begins to deliver but the compressor sequence valve is still closed; the air mass flow being that is being delivered is slowed down via a relief valve and is introduced upstream of the compressor of the first turbocharger. The engine now works in controlled single turbocharger operation; the bypass valve is closed, the turbine sequence valve is operational. From a certain engine speed onwards, sufficient exhaust-gas enthalpy is available for both turbochargers and the second compressor to be connected. For this purpose, the turbine sequence valve is fully opened and the relief valve is closed. The pressure downstream of the second compressor increases rapidly and opens the compressor sequence valve. Now both turbochargers are operational and each handles approximately half of the engine's air supply. Excess exhaust-gas energy is expelled via a waste gate; the charge pressure is controlled by means of the bypass valve.

**Two-stage supercharging**

In contrast to multi-stage supercharging, two-stage supercharging connects the turbochargers in such a way that serial operation is carried out, at least in certain operating phases. Different connection variants have been examined for diesel engines [17] and it seems reasonable to apply this forward-looking approach to gasoline engines as well. In this extremely efficient system, turbo-chargers of different sizes are used, i.e. the flow rate behavior of the two turbochargers is different. As a result, the enthalpy gradient can be better utilized for identical mass flow. The advantage of this supercharging system as compared to the single-stage technique is the increase of the nominal power with a simultaneous improvement of the steady-state torque at low engine speeds and the acceleration behavior of the engine due to the rapid build-up in charge pressure. There is no "transition lag" when coordination and control are carried out in this way.

**Combined supercharging: mechanical supercharger plus turbocharger**

To improve the starting torque of turbocharged combustion engines, additional mechanical supercharging is possible. One way to reduce "turbo lag" is to use a combination of exhaust-gas turbocharging and mechanical supercharging. This system was used for the first time in 1985, in the Lancia Delta.

In the lower engine speed ranges, in which little exhaust-gas energy is available to drive the turbocharger turbine, the charge air is supplied by a mechanically driven compressor. The mechanically driven compressor can be placed either upstream or downstream of the compressor of the turbocharger. The complete supercharging can be handled by the exhaust-gas turbocharger only starting from an operating range in which sufficient exhaust-gas energy is available to generate the required charge pressure. During transition, the system works as two-stage supercharging. As soon as the exhaust-gas turbocharger is in a position to generate the charge pressure on its own, the mechanically driven supercharger can either be completely bypassed using a bypass valve or completely decoupled by means of a coupling.

Decoupling, e.g. via a magnetic coupling, has energetic advantages, as the drive energy of the mechanical supercharger is no longer needed due to this, but requires greater effort in terms of control engineering, mechanical durability and costs. In order to have sufficient charge pressure even at low engine speeds, it is necessary to make a correspondingly high transmission of the mechanically driven supercharger. This results in high engine speed gradients when switching the mechanical supercharger on and off, which in turn leads to a high mechanical load on the drive. With regard to the thermodynamic design of the exhaust-gas turbocharger in combination with the mechanically driven supercharger, advantages arise due to the fact that the turbocharger can be designed mainly for the medium and higher engine speed range. This leads to the use of relatively large turbochargers, which have good efficiency, especially for high mass flow rates, and thus enable a high maximum engine performance with favorable fuel consumption. In spite of the energetically unfavorable direct coupling of the mechanical supercharger to the crankshaft of the engine, very favorable fuel consumptions arise in the normal driving cycle, due to the optimal design of the turbocharger. Moreover, due to the mechanically driven supercharger, full advantage can be taken of downsizing as there is no longer any perceptible “turbo lag”, even for high degrees of supercharging.

Despite the undeniably favorable characteristics of this supercharging system, the related manufacturing costs are regarded critically. Taking the measures required to improve the critical acoustics behavior of a mechanically driven positive displacement supercharger into consideration, significant technical costs arise, and these are reflected in the system price. Volkswagen’s recently introduced VW 1.4 l TSI engine with combined supercharging impressively succeeded in demonstrating the technical potential; **figure 6.17**.

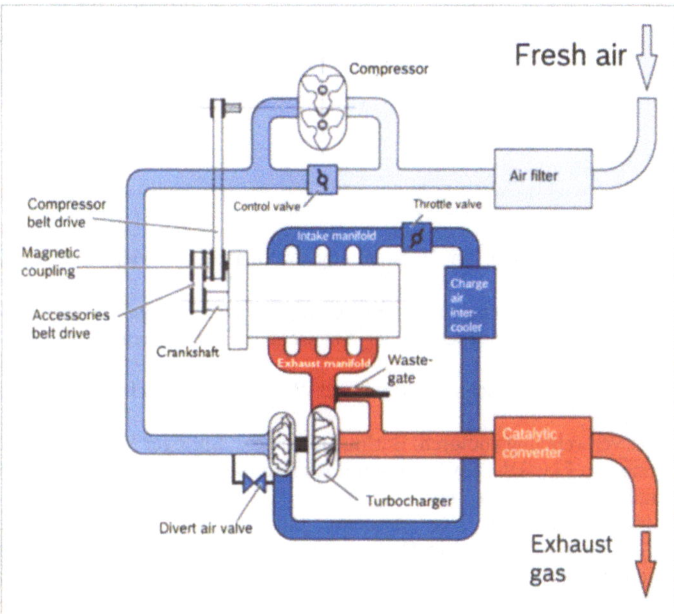


**Figure 6.17:** Principle of additional mechanical supercharging [18]

A mechanically driven Roots blower is connected upstream of the exhaust-gas turbocharger, the latter being linked to the engine via a clutch and a gearbox. The combination is coordinated so that the exhaust-gas turbocharger is always active and the mechanical supercharger is activated in the lower load and engine speed ranges; **figure 6.18**. Additional mechanical supercharging makes it possible to achieve both a high specific power (via turbocharging) and good response behavior at low engine speeds (via mechanical supercharging). However, the additional supercharging system clearly increases the complexity of the power unit and the coordination. The future will show how this system holds up against the other possible supercharging technologies, including with regard to system costs [15].

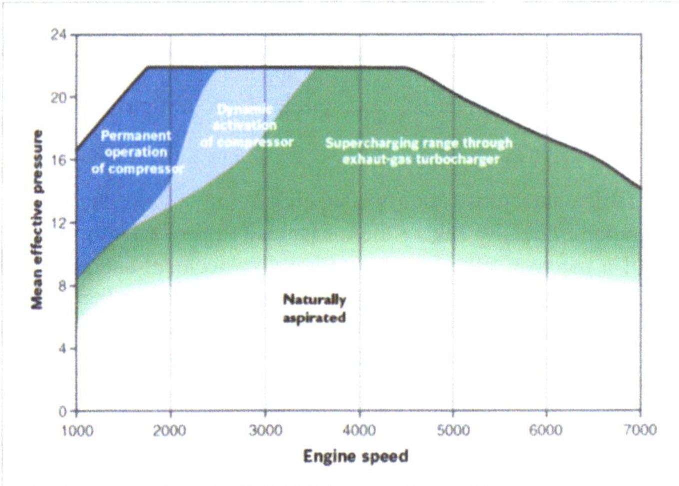


**Figure 6.18:** Distribution for additional mechanical supercharging [18]

# Literature


[1] *Hiereth, H.; Prenninger, P.*: Aufladung der Verbrennungskraftmaschine. Wien, NewYork: Springer Verlag, 2003

[2] *Golloch, R.*: Downsizing bei Verbrennungsmotoren. Berlin, Heidelberg: Springer Verlag, 2005

[3] *Mayer, M.*: Abgasturbolader : Sinnvolle Nutzung der Abgasenergie. Landsberg: Verlag Moderne Industrie, 2001

[4] *Pischinger, S.*: Skript Kolbenarbeitsmaschinen. Rheinisch-Westfaelische Technische Hochschule Aachen, 2006

[5] *Zinner, K.*: Aufladung von Verbrennungsmotoren. Berlin, Heidelberg: Springer Verlag, 1985

[6] *Schmalzl, H. P.*: Aufladung von Pkw DI-Ottomotoren mit Abgasturboladern mit variabler Turbinengeometrie. Dissertation, TU Dresden, 2006

[7] *Gabriel, H.; Lingenauber, R.; Ramb, T.*: Der Turbolader mit variabler Turbinengeometrie (VTG) fuer den neuen Porsche 911 Turbo – Ein Meilenstein in der Ottomotorenaufladung. 11. Aufladetechnische Konferenz, Dresden, 2006

[8] *Sterner, A.; Hofstetter, M.; Kerkau, M.; Beer, M.; Ronneburger, R.; Knirsch, S.*: Die variable Turbinengeometrie fuer die ottomotorische Anwendung beim neuen 3,6 l Biturbo-Motor des Porsche 997 Turbo. 11. Aufladetechnische Konferenz, Dresden, 2006

[9] *Klueting, M.; Missy, St.; Schwarz, Ch.*: Potenziale des strahlgefuehrten Benzin-DI-Brennverfahrens in Verbindung mit Aufladung. 26. Internationales Wiener Motorensymposium, 2005

[10] *Reulein, C.; Kellerer, H.; Schwarz, Ch.*: Methodeneinsatz bei der Potenzialbeurteilung aufgeladener Verbrennungsmotoren. Motorprozesssimulation und Aufladung, Haus der Technik Tagung, Berlin, 2005

[11] *Ardey, N.; Klueting, M; Schmitz, K.*: Aufladung von direkteinspritzenden Ottomotoren – ein Ansatz zur Steigerung der effizienten Dynamik. 10. Aufladetechnische Konferenz, Dresden, 2005

[12] *Borrmann, D.; Brinkmann, F.; Walder, K.; Pingen, B.; Wojahn, J.; Behrends, P.*: Benzineindirektspritzung mit Turboaufladung – ein interessantes Downsizingkonzept. 11. Aachener Kolloquium Fahrzeug- und Motorentechnik, Aachen, 2002

[13] *Fraidl, G. K.; Kapus, P.; Piock, W.*: Otto-Direkteinspritzung mit Aufladung – Die Konkurenz zu dieselmotorischen Antrieben. 26. Internationales Motorensymposium, Wien, 2005

[14] *Hoecker, P.; Jaisle, J. W.; Muenz, S.*: Der eBooster – Schluesselkomponente eines neuen Aufladesystems von BorgWarner Turbo Systems fuer Personenkraftwagen. 22. Internationales Motorensymposium, Wien, 2001

[15] *Lang, O.; Habermann, K.; Wolf, K.; Pischinger, S.*: Anwendung der Zusatzaufladung bei abgasturboaufgeladenen Ottomotoren. 9. Aufladetechnische Konferenz, Dresden, 2004

[16] *Muenz, S.; Schier, M.; Schmalzl, H. P.; Bertolini, T.*: Der eBooster – Konzeption und Leistungsvermoegen eines fortgeschrittenen elektrisch unterstuetzten Aufladesystems. 8. Aufladetechnische Konferenz, Dresden, 2002

[17] *Guilain, S.; Lefebvre, A.; Doleac, L.; Schreiber, G.; Muenz, S.*: Optimization of a small two-staged turbocharged diesel engine. 11. Aufladetechnische Konferenz, Dresden, 2006

[18] *Middendorf, H.; Krebs, R.; Szengel, R.; Pott, E.; Fleiss, M.; Hagelstein, D.*: Der weltweit erste doppelaufgeladene Otto-Direkt-Einspritzmotor von Volkswagen. 14. Aachener Kolloquium, Aachen, Oktober 2005

# 7 Fuel Consumption

The practical parameter of greatest interest to drivers is fuel consumption. This is true in western Europe, and increasingly the case in other countries, such as the USA. Fuel economy has become topical because supplies of crude oil, from which fuel is made, are limited, and because the cost of fossil fuel has increased drastically over the last few years.

Fuel consumption is directly connected to the efficiency with which an engine is operated in the entire load range. Furthermore, fuel consumption directly affects the emission of carbon dioxide ($CO_2$) which is also one of the factors held responsible for the possible greenhouse effect, i.e. global warming. Therefore, reducing fuel consumption and taking measures to improve engine efficiency are becoming more important than ever before.

While fuel consumption is defined in liters per 100 km (l/100 km) for the customer, i.e. as absolute consumption, or in the USA as driven miles per gallon of fuel (miles/gallon), specific fuel consumption $b$ is used in the development of combustion engines. The fuel quantity consumed in a given time (the fuel mass flow rate $\dot{m}_K$) is expressed in terms of power for reference analyses:

$$b = \frac{\dot{m}_K}{P} = \frac{1}{\eta \cdot H_u}, \; b \text{ in g/kWh or kg/kWh}$$

If only the internal process of the engine is considered, the internal power is calculated. The effective power output of an engine comes into play as soon as mechanical losses that result from friction in the engine and from driving ancillaries are calculated. It can, therefore, be defined as the power output that is delivered from an engine to the power train or from a clutch to the transmission. Specific fuel consumption is therefore expressed in terms of both the internal power and the effective power:

$$b_i = \frac{\dot{m}_K}{P_i} = \frac{1}{\eta_i \cdot H_u} \qquad \text{internal specific fuel consumption}$$

$$b_e = \frac{\dot{m}_K}{P_e} = \frac{1}{\eta_e \cdot H_u} \qquad \text{effective specific fuel consumption}$$

**Figure 7.1** shows the effective specific fuel consumption of today's gasoline engines for a part load operating point with a mean effective pressure of 2 bar and an engine speed of 2000 $\text{min}^{-1}$. This operating point corresponds to a mean load point in the New European Driving Cycle (NEDC) and is often used for comparisons, where appropriate.

Hyperbolic functions can be used to represent the relationship between efficiency and specific fuel consumption for different fuels with specified calorific values; **figure 7.2**. The curve directly indicates the effective efficiency of an engine for a specific calculated fuel

Open comparative processes can be used to estimate the potential of different engines and combustion processes. For gasoline engines, it is assumed that combustion theoretically takes place very fast at constant combustion chamber volume (constant volume cycle). Here, the piston is in its highest position, i.e. in the top dead center. For gasoline engines with direct injection, the fuel is injected directly into the combustion chamber, either during intake or after the inlet valves are closed during or towards the end of the compression phase. This way, as is the case with diesel engines, air is taken in, rather than fuel/air mixture.

If a theoretical estimation is carried out envisioning consistent implementation of gasoline direct injection for the gasoline engine of the future, thermal efficiency can be taken into consideration and analyzed for the perfect constant volume cycle engine process. **Figure 7.5** shows the theoretically achievable efficiencies of the perfect engine in the case of air intake depending on the compression ratio $\varepsilon$ and the air/fuel ratio $\lambda$.

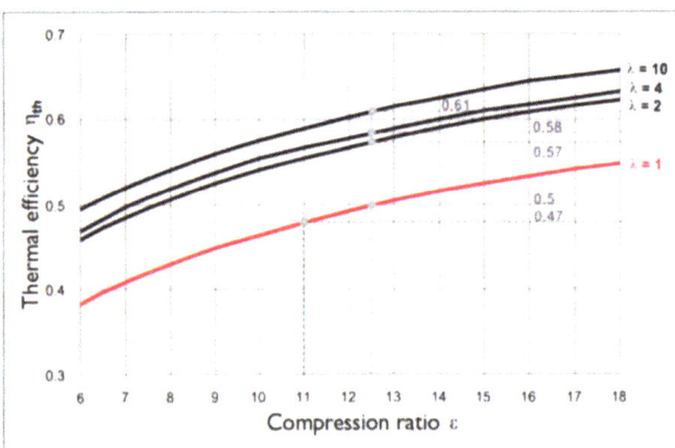


**Figure 7.5**: Efficiency of the perfect engine with constant volume combustion [10]

By considering the possibility that the compression ratio can be increased by approximately 1.5 through direct injection due to internal cooling as a result of fuel vaporization and the related improved knocking behavior, it should be possible, in the load point $p_{me}$ = 2 bar and $n$ = 2000 rpm, to achieve an increase in thermal efficiency from approximately 47% for the compression ratio $\varepsilon = 11$ and intake manifold injection with the air/fuel ratio $\lambda = 1$ to approximately 58% for the compression ratio $\varepsilon = 12.5$ and direct injection with the air ratio $\lambda = 4.0$ with ideal stratified charge. This corresponds to an efficiency increase of approximately 22% with respect to the initial value of 47%. This increase in efficiency, however, is attributed only to the improvement of the high pressure process as a result of the enleanment of the mixture due to stratified charge. A further increase in efficiency can be achieved through unthrottled operation with stratified charge, which should be in the range of approximately 12 to 13% for the operating point under consideration. Thus a total efficiency increase of approximately 35% can be estimated, which will be implemented in the future as well.

If the fuel consumption of the conventional gasoline engine with intake manifold injection is set to 100% at the operating point $p_{me}$ = 2 bar and $n$ = 2000 rpm, the fuel consumption with optimum gasoline direct injection is reduced to approximately 65%; **figure 7.6**. This results in a saving of approximately 35%, which is distributed as follows:

- 19% can be attributed to the lean, stratified mixture,
- 4% to the increase of the compression ratio, and
- 12% to the reduction in throttle losses.

In low idle operation, fuel savings of approximately 60% are possible with gasoline direct injection. The distribution is as follows:

- 23% can be attributed to the lean, stratified mixture,
- 7% to the increase of the compression ratio and
- 30% to the reduction in throttle losses.

Fuel savings of 6% at full load is due exclusively to the increase in the compression ratio.

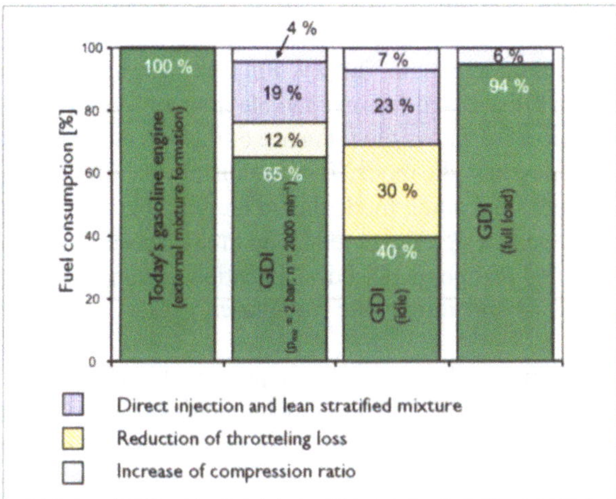


**Figure 7.6:**
Theoretical fuel savings with gasoline direct injection [11]

**Fuel consumption of commercial engines**

The engine of the Mitsubishi Carisma GDI (Gasoline Direct Injection) presented at the International Motor Show in Frankfurt in 1995 was the first modern example of a gasoline engine with gasoline direct injection. It worked on the basis of a wall-guided combustion process. Mitsubishi stated a reduction in fuel consumption of approximately 40% during the low idle mode as a consequence of lean combustion and idle speed reduction [12]. For low to medium-high part load, consumption advantages of 15 to 25% are specified in comparison with the conventional Mitsubishi engine with intake manifold injection (15% at 100 km/h, 20% at 60 km/h, approximately 18% at 45 km/h on average and 25% at 25 km/h on average). These are typical speeds in the Japanese test cycle; **figure 7.7**. Significantly lower fuel savings were achieved in normal driving operation with the European version of

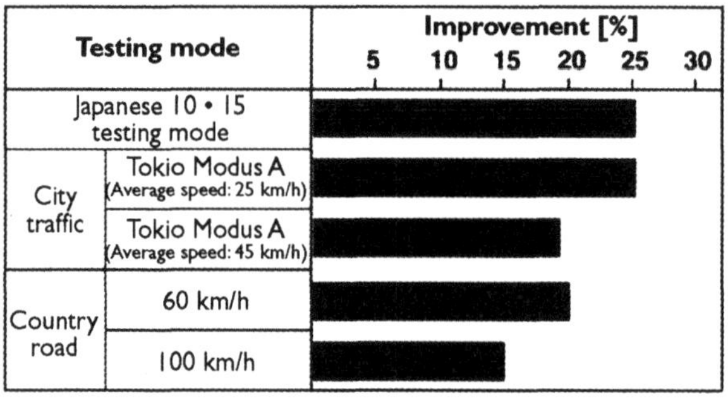


**Figure 7.7:**
Consumption reductions [12]

the engine. At that time, state-of-the-art engines of German manufacturers demonstrated a consumption similar to that of the Mitsubishi Carisma with GDI engine.

Volkswagen presented the *Fuel Stratified Injection* process (FSI) in 1999. Volkswagen indicated fuel savings of 15% at lean stratified charge for this engine as compared to the equivalent basic power unit with intake manifold injection. The VW FSI basic power unit uses a combined wall/air-guided combustion process that features tumble flow and an injection nozzle which is located at the inlet. **Figure 7.8** shows the influence of the operating modes on fuel consumption at part load and during low idle mode. The combustion stability which was already increased in the homogeneous operating mode due to tumble flow and direct injection in the combustion chamber permits a significant reduction in idle speed. The low idle mode is even steadier in the stratified charge mode. For application reasons, though, the same idle speed as in the homogeneous operating mode was chosen [13].

The *Injection Directe Essence* (IDE) engine with direct injection presented in the Renault Mégane in 1999 was operated only homogeneously with the stoichiometric air ratio ($\lambda = 1$) due to its exhaust-gas treatment. The spark plug and injector were arranged very close to each other, as is required for a spray-guided combustion process. This led to the conclusion that Renault would favor the spray-guided combustion process for the stratified operating mode. For the homogeneously operated engine, Renault showed a consumption advantage of 15% as compared to predecessor engines. Seven percent of it was attributed to the high precision optimization of the engine and the remaining 8% was achieved by the internal vaporization of fuel on account of direct injection, the higher compression ratio and the increased exhaust-gas recirculation as compared to the engine with intake manifold injection [14].

The 2.0 l Turbo FSI engine by Audi combines homogeneous direct injection with exhaust gas turbocharging. In this engine, the charge movement was systematically adapted to the requirements of a supercharged engine with direct injection. The combination of gasoline direct injection and supercharging leads to an efficient combustion process characterized by high knock resistance and fast and efficient combustion [15]. As compared to the predecessor engine, both the torque in the lower and medium engine speed range and the response of the engine were considerably improved through this combination. The consump-

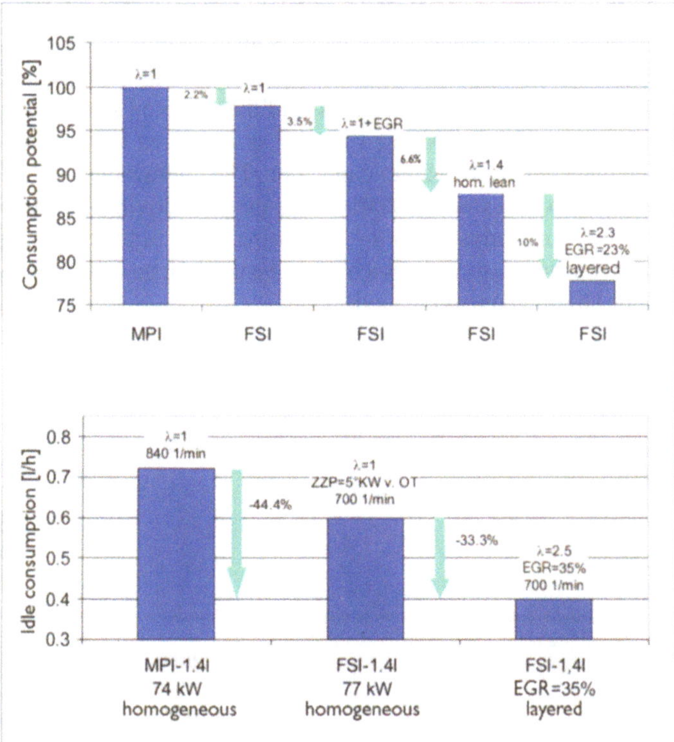


**Figure 7.8:** Comparison of VW 1.4 l FSI (77 kW) with VW 1.4 l MPI (74 kW) in the operating point $n = 2000$ rpm, $p_{me} = 2$ bar (top) and at low idle (bottom) [13]

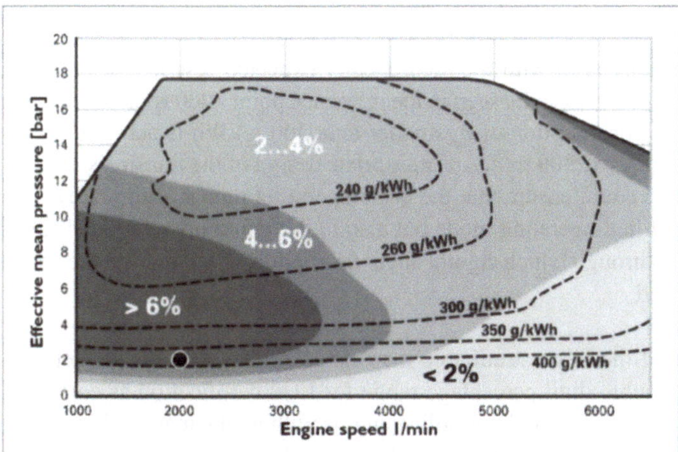


**Figure 7.9:** Specific fuel consumption of the 2.0 T FSI engine in comparison to the 1.8 T [15]

tion advantage of the 2.0 T FSI with 147 kW as compared to the 1.8 T engine with 132 kW is shown in the map in **figure 7.9**. In spite of the higher power, the advantages are greater than 6% in a large map range.

The first engines with for what is known as the second generation of spray-guided direct injection, which allow a stratified charge mode, show much higher savings in fuel consumption than engines with homogeneous direct injection or those with wall-guided direct injection. **Figure 7.10** shows the fuel consumption advantage of the new 3.5 l six-cylinder engine with spray-guided direct injection by DaimlerChrysler as compared to the corresponding 3.5 l six-cylinder engine with intake manifold injection. The improvement in fuel consumption can be achieved due to the extended stratification range in driving operation over a large speed range [3]. Additional advantages for this engine result from combining it with a seven-gear automatic transmission, with which it can be used for speeds of up to approximately 150 km/h in the stratified operating mode.

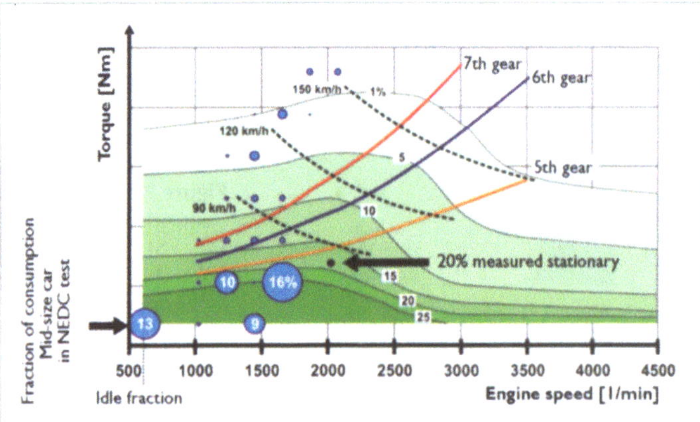


**Figure 7.10:** Consumption advantages of second-generation gasoline direct injection [3]

The measurement results tally well with the theoretically expected results. An improvement in consumption of up to 20% can be seen in the reference point 2000 rpm, $p_{me} = 2$ bar. For $p_{me} = 2$ bar, a specific consumption of $b_e$ smaller than 300 g/kWh is achieved in the engine speed range from 1000 to 2300 rpm. An appropriate design of the transmission stepping and the rear axle ratio could ensure that the basic operating points of the NEDC are driven not only in the stratified operating mode but also in the upper operating and speed ranges outside the NEDC, through which significant consumption advantages in the actual driving operation are gained.

Good thermodynamics and the large stratifiable map range lead to savings in fuel consumption that are in the double-digit percentage range for EU4 certification as compared to the current fuel-saving 3.5 l six-cylinder gasoline engine with intake manifold injection; **figure 7.11**. Such fuel savings can also be achieved in the field driving cycle used by the magazine *auto, motor und sport*.

The consumption map shows the very favorable values for the effective specific consumption $b_e$ achieved by this engine; **figure 7.12**. The minimum is at 235 g/kWh. A consumption of only 290 g/kWh is achieved at the reference point 2000 rpm and $p_{me}$ 2 bar. This value is below the consumption values achievable by passenger car diesel engines [16].

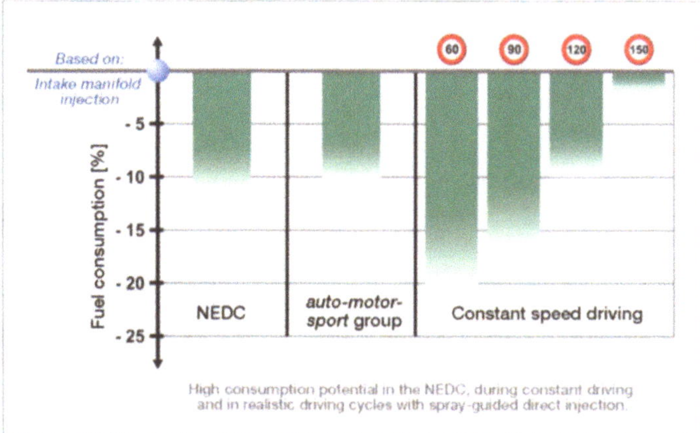


**Figure 7.11:** Consumption potential in the NEDC, the *auto-motor-sport* group and during constant driving [3]

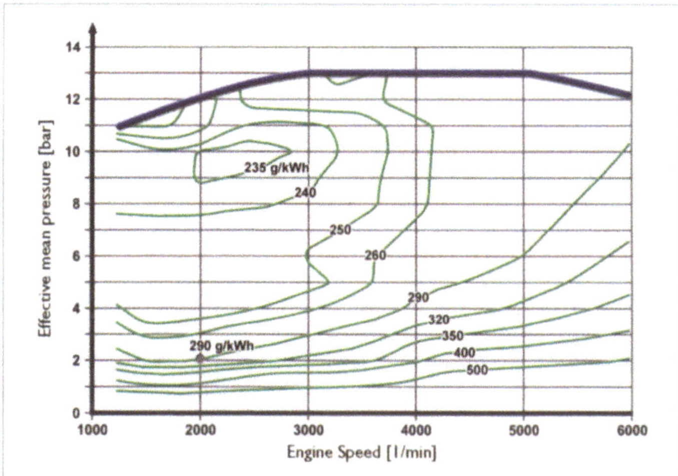


**Figure 7.12:** Consumption map of the Mercedes Benz 3.5 l six-cylinder engine with spray-guided direct injection [16]

Modern engines with fully variable valve-trains reach an effective specific consumption of 340 g/kWh at 2000 rpm and $p_{me} = 2$ bar; **figure 7.13** (FEV scattering range). BMW's spray-guided gasoline direct injection process beats the magic mark of 300 g/kWh in specific fuel consumption with a value of 295 g/kWh [17]. Here, too, diesel engines achieve optimum values for this load point.

In addition to the favorable consumption in the reference point, the large map range, which can be operated in stable stratification mode, (**figure 7.14**) is vital to achieving the consumption potential. Moreover, a range of hybrid injection control, which also uses the advantages of the large volume of excess air, is added to the classical stratified operating range. As a whole, these operating ranges cover such wide limits with regard to engine speed (limit at approximately 4500 rpm) and load that they practically need not be violated due to the combustion process in the European Driving Cycle and in realistic field cycles in city traffic and on country roads at all times [17].

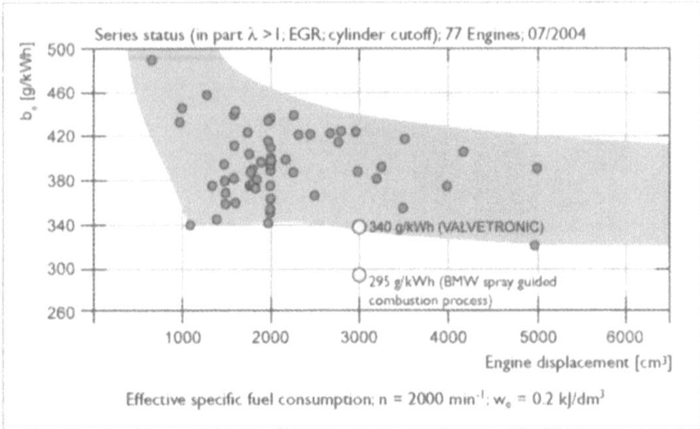


**Figure 7.13:** FEV scattering range and fuel consumption for gasoline direct injection in the BMW 3 l six-cylinder engine [17]

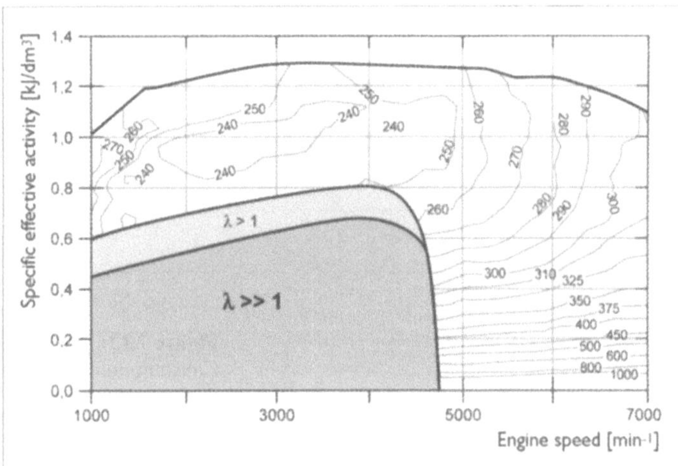


**Figure 7.14:** Stratification range of the BMW combustion process [17]

In addition to the consumption potential, it is also important to take the requirements of exhaust-gas treatment into account to meet the exhaust emissions standards. This is possible with an emissions optimization strategy adapted to the combustion process and the overall system. In NEDC, the engine start is followed by a short phase in which the catalytic converter is quickly heated, followed by a warm-up phase in the homogeneous operating mode. In the further test process, the stratified fuel-saving operation is interrupted only by regeneration phases of the nitrogen oxide storage catalyst. This way, it is possible to adhere to the EU4 limit values and to limit the influence of the emissions measures, causing an increase in fuel consumption to approximately 4% in the EU test. With the BMW combustion process, it is possible to show a consumption improvement of 20% through a significantly extended stratification range and a design typical of gasoline engines [17]; c.f. **figure 7.15**.

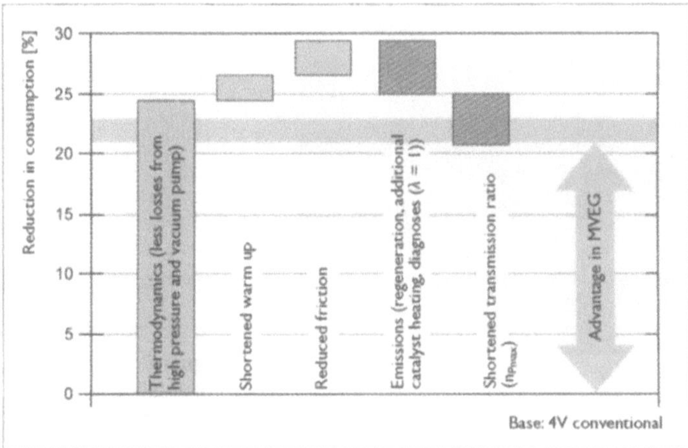


**Figure 7.15:** BMW combustion process, consumption in the New European Driving Cycle [17]

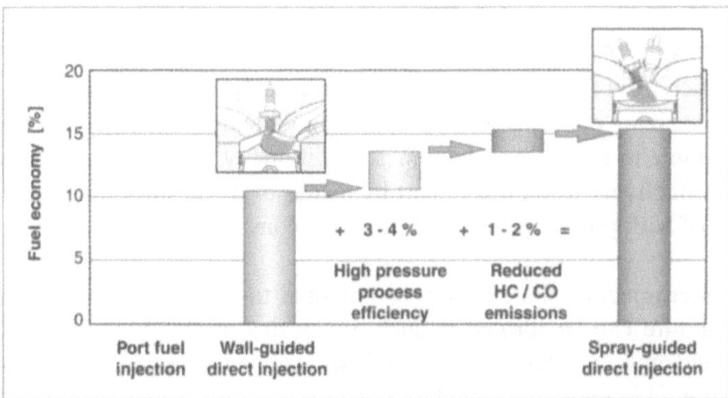


**Figure 7.16:** Consumption improvement of a spray-guided DI combustion process (mean value of the NEDC-relevant range) [18]

**Figure 7.16** shows the potential of an engine with spray-guided direct injection as compared to intake manifold injection and a wall-guided combustion process. As opposed to manifold injection, the wall-guided combustion process achieves fuel savings of approximately 11%. Due to the improved high pressure efficiency and the lower hydrocarbon and CO emissions, the spray-guided combustion process can reduce fuel consumption by an additional 4 to 6%.

Combining gasoline direct injection with supercharging makes it possible to exploit new potential with regard to consumption and make vehicles fun to drive [3]. Combined with exhaust-gas turbocharging, it is even possible to further extend the stratification range and to improve fuel consumption in a spray-guided combustion process due to the larger volume of excess air as a consequence of supercharging; **figure 7.17**. Large volumes of excess air result in more favorable thermophysical properties and better calorific values, through which the thermodynamic advantages are obtained.

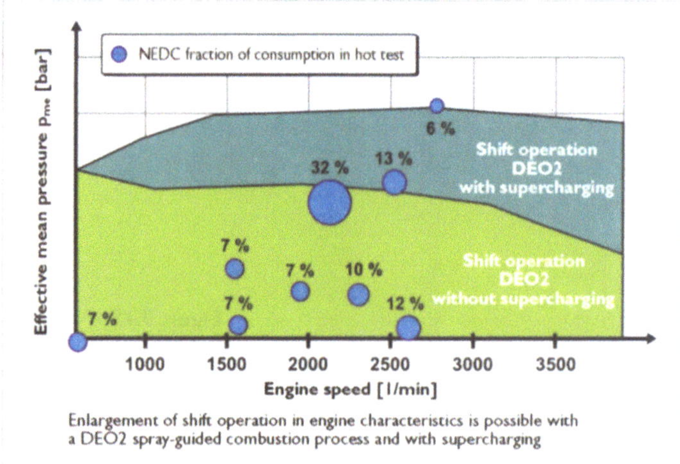


**Figure 7.17:** Potential of spray-guided gasoline direct injection with supercharging [3]

The combination of stratified direct injection with supercharging also enables the implementation of attractive downsizing concepts with good drivability and favorable consumption at a $CO_2$ level comparable to that of diesel engines. Additionally, start/stop operation can be implemented cost-effectively using direct start by means of direct injection, which is also beneficial in terms of both pollutant emissions and fuel consumption [3].

Gasoline engines with direct injection have enormous potential to reduce fuel consumption, provided that the mixture can be formed in such a way that an ideal stratification between air/fuel mixture and ambient charge of air or exhaust can be implemented reliably for exhaust-gas stratification [19] under all boundary conditions during which a stratified charge mode is desirable. It is therefore of special significance to the injection system that fuel be introduced in the vicinity of the spark plug.

If the idea of gasoline direct injection with optimum stratified charge and unthrottled operation is successfully implemented in the future, it could mean a significant advance for the gasoline engine with regard to low fuel consumption with simultaneously low emissions at little additional cost as compared to today. If, for example, it is taken into consideration that reduction in consumption of approximately 30 to 35% appears to be possible with this combustion process, provided that it is consistently developed and implemented (reduction of present $CO_2$ emissions of approximately 165 g/km to approximately 105 to 110 g/km), and if the consumption reduction of this combustion process is compared with that in the hybrid powertrain, e.g. Toyota Prius, the same consumption values can be achieved at significantly lower cost.

# Literature

[1] *Voss, E.; Schmittger, W.; Königstein, A.; Scholten, I.; Pöpperl, M.; Pritze, St.; Rothenberger, P.; Samstag, M.*: 2,2 l ECOTEC DIRECT – Der neue Vollaluminiummotor mit Benzindirekteinspritzung für den Opel Signum. 24. Internationales Winer Motorensymposium, 2003

[2] Robert Bosch GmbH: Ottomotor-Management, 3. Auflage. Vieweg Verlag, 2005

[3] *Lückert, P.; Frey, J.; Kemmler , R.; Schaupp, U.; Vent, G.; Waltner, A.*: Kunden- und zukunftsorientierte Technologien am Ottomotor – heute und morgen. 26. Internationales Wiener Motorensymposium, 2005

[4] *N. N.*: Forum der Meinungen. In: MTZ 63 (2002), Nr. 12

[5] *Klaus, B.; Drexler, G.; Eder, T.; Eisenkölbl, M.; Luttermann, Ch.; Schleusener, M.*: Weiterentwicklung der vollvariablen Ventilsteuerung BMW-Valvetronic. In: MTZ 66 (2005), Nr. 9

[6] *Ademes, N.; Lang, O.; Lauer, S.; Salber, W.; Jené, H.*: Ventiltriebsvariabilitäten für moderne Ottomotoren. In: MTZ 66 (2005), Nr. 12

[7] *Schnittger, W.; Königstein, A.; Pritze, S.; Pöpperl, M.; Rothenberger, P.; Samstag, M.*: 2.2 Direct Ecotec – Neuer Ottomotor mit Direkteinspritzung von Opel. In: MTZ 64 (2003), Nr. 12

[8] *Königstedt, J.; Müller, R.; Jablonski, J.; Dirschnabel, T.; Uhrich, J.; Hatz, W.*: Der neue V-10-FSI-Motor von Audi. 27. Internationales Wiener Motorensymposium, Wien, April 2006

[9] *Middendorf, H.; Krebs, R.; Szengel, R.; Pott, E.; Fleiß, M.; Hagelstein, D.*: Der weltweit erste doppelaufgeladene Otto-Direkt-Einspritzmotor von Volkswagen. 14. Aachener Kolloquium, Aachen, Oktober 2005

[10] *Spicher, U.; Weimar, H. J.*: Direkteinspritzung bei Ottomotoren: Hintergrund und theoretische Grundlagen. Tagung: Direkteinspritzung im Ottomotor, Haus der Technik, Essen, März 1997

[11] *Spicher, U.; Kubach, H.; Häntsche, J. P.*: Die strahlgeführte Direkteinspritzung als Zukunftskonzept für Ottomotoren. MTZ-Konferenz Motor 2006, Stuttgart, 2006

[12] Mitsubishi Motors: Mitsubishi Benzinmotor mit Direkteinspritzung. Technisches Dossier, 1995

[13] *Krebs, R.; Spiegel, L.; Stiebels, B.*: Ottomotoren mit Direkteinspritzung von Volkswagen. 8. Aachener Kolloquium Fahrzeug- und Motorentechnik 1999

[14] *Goppelt, G.*: Der neue Ottomotor mit Benzin-Direkteinspritzung von Renault. In: MTZ 60 (1999), S. 530–534

[15] *Wurms, R.; Kuhn, M.; Zeilbeck, A.; Adam, S.; Krebs, R.; Hatz, W.*: Die Audi Turbo FSI Technologie. 13. Aachener Kolloquium Fahrzeug- und Motorentechnik, 2003

[16] *Lückert, P.; Waltner, A.; Rau, E.; Vent, G.; Schaupp, U.*: Der neue V6-Ottomotor mit Direkteinspritzung von Mercedes-Benz. In: MTZ 67 (2006), Nr. 11, S. 830–840

[17] *Klüting, M.; Missy, St.; Schwarz, Ch.*: Potenziale des Strahlgeführten Benzin-DI-Brennverfahrens in Verbindung mit Aufladung. 26. Internationales Wiener Motorensymposium, 2005

[18] *Wirth, M.; Zimmermann, D.; Friedfeldt, R.; Caine, J.; Schamel, A.; Storch, A.; Ries-Müller, K.; Gansert, K. P.; Pilgram, G.; Ortmann, R.; Würfel, G. ; Gerhardt, J.*: Die nächste Generation der Benzin Direkteinspritzung – gesteigertes Verbrauchspotenzial bei optimierten Systemkosten. 12. Aachener Kolloquium Fahrzeug und Motorentechnik, 2003

[19] *Sarikoc, F.; Kettner, M.; Velji, A.; Spicher, U.; Krause, A.; Elsaesser, A.*: Potential of reducing the NOx emissions in a spray guided DI gasoline engine by stratified exhaust gas recirculation (EGR). SAE Technical Paper, 2006-01-1261, 2006

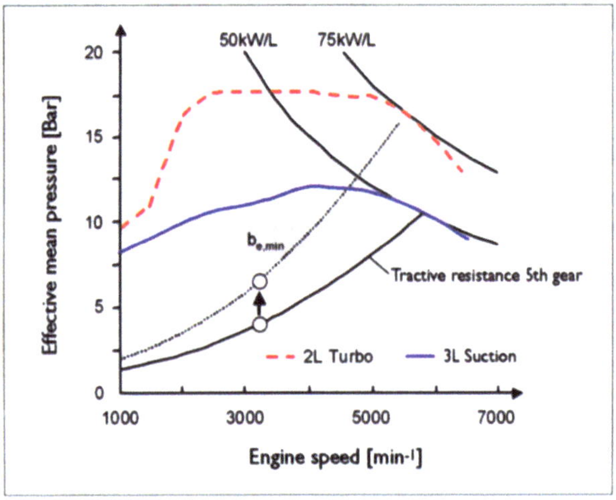


**Figure 8.1:** Increase of specific power by downsizing

Shifting the operating points reduces losses due to heat transfer through walls, charge transfer losses and mechanical losses. Losses caused by actual combustion can increase, however, due to a longer period of combustion or retarded ignition angles, but on the whole, there is a distinct reduction in fuel consumption through downsizing, especially in the consumption cycle [2].

**Figure 8.2** and **8.3** show two downsizing concepts. The first concept is based on reduction of displacement, keeping the number of cylinders the same [3]. Though the ratio of surfaces to volume increases, fuel savings of 4% are possible through supercharging, in order to achieve the specified power of 10 kW. The second concept is based on an increase in displacement of each cylinder while reducing the number of cylinders. With supercharging, this concept enables fuel savings of 9% to achieve the specified power of 10 kW, the engine speed for both concepts remaining unchanged at 2000 rpm.

Due to the precompression during supercharging, the charge air heats up to a temperature of 180°C. The air density and performance can be further increased if this air is cooled in an intercooler. Intercooling is one of the few measures in a combustion engine that not only has a positive effect on the performance and fuel consumption but also on emissions. The higher air density reduces the tendency of knocking in gasoline engines. Tests conducted by BMW on a downsizing concept with displacement reduction, direct injection, exhaust-gas turbocharger with increased turbine inlet temperature and geometrical variation of the compression show a reduction in fuel consumption of up to 25% in the New European Driving Cycle, while also improving the performance behavior [4]. According to BMW, the additional costs are still less than those for a pure concept with direct injection.

The potential of turbocharging is increased by direct injection in gasoline engines. As direct injection gasoline engines present with lower knock sensitivity it is also possible to select exceptionally high compression ratios for turbocharged engines. Moreover, in turbo-

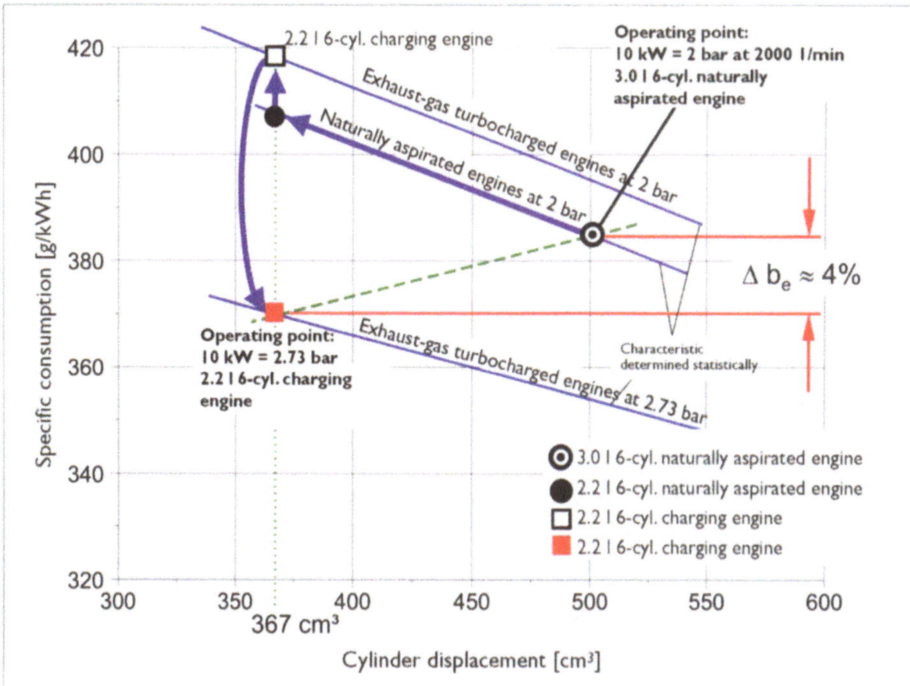


**Figure 8.2:** Downsizing concept with a constant number of cylinders but reduced cylinder displacement [3]

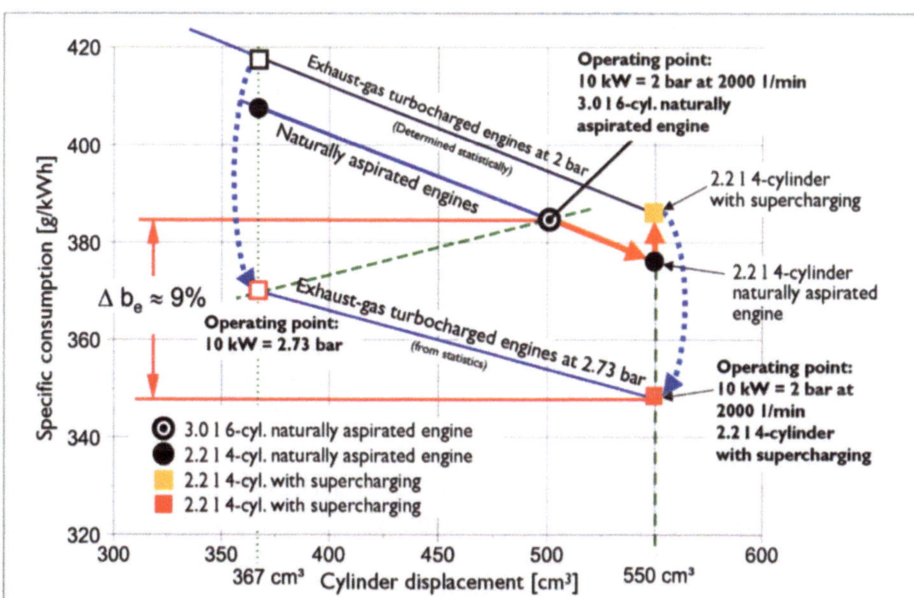


**Figure 8.3:** Downsizing concept with a reduced number of cylinders [3]

charged engines, the higher air mass flow rate on dethrottling counteracts the "turbo lag" in the lower engine speed range. Direct injection also enables larger valve overlapping, which can improve residual gas scavenging and higher volumetric efficiency. Furthermore, systematic variable charge movement can improve mixture homogenization and residual gas compatibility.

Many manufacturers see a great fuel-saving potential in downsizing concepts when combined with direct injection. In the medium term, stoichiometric operation is a convenient option for downsizing concepts, as it allows both the use of a three-way catalytic converter and the potential of direct injection. The full potential of the combination of downsizing concepts and direct injection can be fully exploited only through spray-guided combustion processes with stratified charge. However, the production of charge stratification for supercharged engines is much more difficult to implement than for naturally aspirated engines [2].

**Figure 8.4** shows the dynamic behavior of an engine whose size has been reduced by exhaust-gas turbocharging, in relation to the stationary torque of an uncharged basic engine [5]. To compare the two, the torque was measured after 1 s or 0.5 s following the load increase. Although there is a clear excess of stationary torque in the entire speed range that was subject to consideration, there is evidence of considerable losses as compared to the basic naturally aspirated engine, particularly in the lowest speed range. From approximately 1400 $min^{-1}$ onwards, however, there is a constant advantage, even in the dynamic operation of the exhaust-gas turbocharged engine.

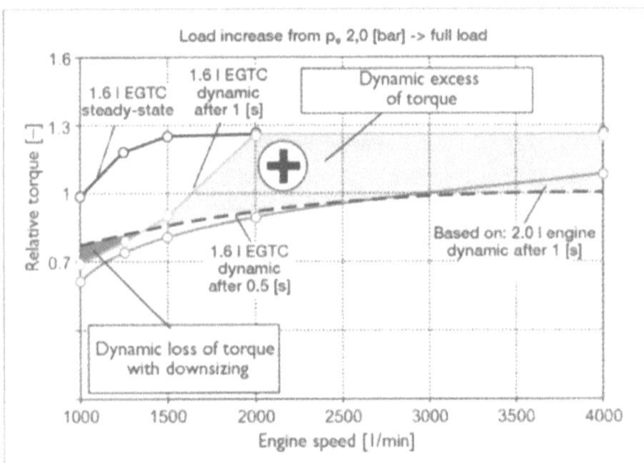


**Figure 8.4:**
Dynamic torque build-up in a turbocharged downsized engine [5]

The dynamics and drivability in the lower engine speed range can be further improved with two-stage supercharging units or through electrically assisted, combined supercharging systems. By combining two exhaust-gas turbochargers, using electrically assisted supercharging or combining exhaust-gas turbocharging and mechanical supercharging, charge pressure may build up faster, thus enabling direct torque generation comparable to that of a naturally aspirated engine with large displacement.

Downsizing is an important technology for gasoline engines in terms of fuel savings. However, it is important to take the entire powertrain into consideration in order to meet customer demand for general vehicle characteristics such as drivability, vehicle response, comfort and acoustics, thereby gaining appropriate customer acceptance.

Examples of downsizing engines with gasoline direct injection in series production are the VW 1.4 TSI and the Mercedes C180 CGI. Both engines reach performance values comparable to those of engines with significantly larger displacement. Mechanical supercharging and exhaust-gas turbocharging are combined in VW engines. In spite of the small displacement, fast charge pressure build-up at low engine speeds can be achieved using the mechanical supercharger. The performance data of Mercedes' four-cylinder engine is approximately equal to that of a six-cylinder engine.

## 8.2 Downspeeding

Downspeeding generally refers to reducing the engine speed by modifying the overall transmission ratio. A fuel saving drive transmission enables fuel consumption improvements similar to those achieved through reduction in displacement [6]. Due to the higher performance level and mean effective pressure level of supercharged engines, the operating point can be shifted towards lower speeds and higher loads. This means that, in terms of fuel consumption, the engine can be operated in a more favorable map range. Downspeeding concepts are generally combined with a concept for downsizing. This combination is especially suitable for gasoline engines with direct injection and supercharging.

**Figure 8.5** shows the relationship between displacement reduction, engine speed decrease and consumption potential. Fuel savings of 15% can be achieved, for example, by reducing

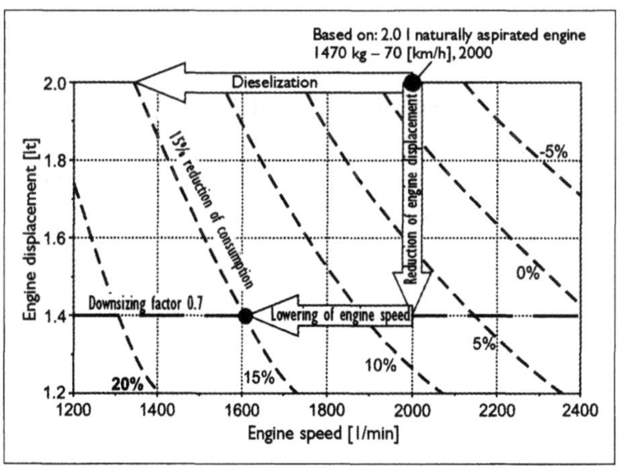


**Figure 8.5:** Downsizing and downspeeding, displacement reduction and engine speed decrease [5]

the displacement from 2.0 to 1.4 l and by decreasing the engine speed by approximately 400 min$^{-1}$. The same reduction in consumption can be achieved by lowering only the engine speed by approximately 650 min$^{-1}$. If the reduction in fuel consumption is achieved mainly via a decrease in the engine speed, this process is often referred to as "dieselization".

**Figure 8.6** shows a comparison between a gasoline engine with direct injection and downsizing concept (high load turbocharging), a conventional engine with intake manifold injection and an engine with a high speed concept and intake manifold injection. It clearly shows that the downsizing concept enables considerable fuel savings in the entire map when shifting the load point towards higher engine torques and lower engine speeds.

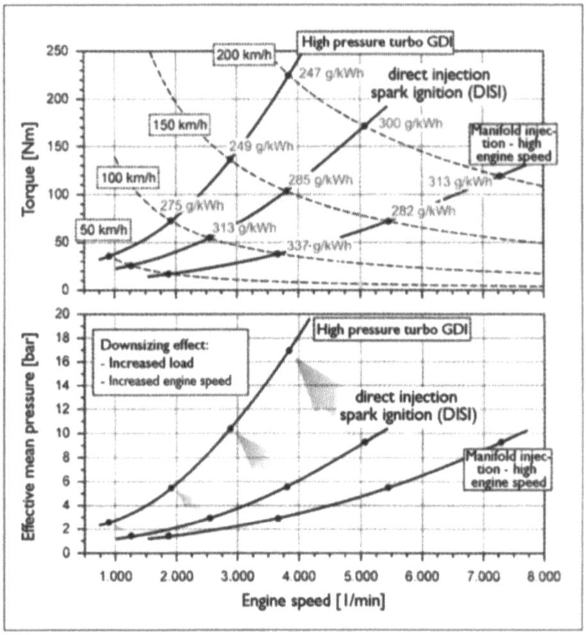


**Figure 8.6:**
Comparison between downsized engine, intake manifold injection and high engine speed concept [1]

Achieving good response behavior is important to ensure the acceptance of such concepts among customers. The turbocharger should therefore be designed keeping the start-up and dynamic behavior in mind. The design of such concepts should also take the minimum speeds relevant to comfort into consideration. An optimum engine design with downspeeding can achieve fuel savings of at least 10%.

# Literature


[1] *Golloch, R.*: Downsizing bei Verbrennungsmotoren. Springer Verlag, 2005

[2] *Golloch, R.; Merker, G. P.*: Downsizing bei Verbrennungsmotoren – Grundlagen, Stand der Technik und zukünftige Konzepte. In: MTZ 66 (2005), Nr. 2, S. 126–131

[3] *Heil, B.; Weining, H. K.; Karl, G.; Panten, D.; Wunderlich, K.*: Verbrauch und Emissionen – Reduzierungskonzepte beim Ottomotor. In: MTZ 62 (2001), Nr. 11, S. 900–915

[4] *Fröhlich, K.; Meyer, R.; Miersch, J.; Reulein, C.; Schwarz, Ch.; Witt, A.*; Downsizing: $CO_2$-Potentiale – Technische Realisierbarkeit. 9. Aachener Kolloquium Fahrzeug- und Motorentechnik, 2000

[5] *Prevedel, K.; Piock, W. F.*: Aufladung beim Direkteinspritz-Ottomotor. VDI-Tagung Innovative Fahrzeugantriebe, Dresden, 2004

[6] *Bandel, W.; Fraidl, G. K.; Friedl, H.; Kapus, P. E.*: Mehrwertorientierte Techniken für ottomotorische Antriebssysteme. In: MTZ 67 (2006), Nr. 2, S. 110–117

# 9 Exhaust Emissions and Pollutant Reduction

All combustion processes, including those occurring in engines, involve oxidation by oxygen. Fuels used in engines are generally composed of hydrocarbons; gasoline consists of a mixture of different hydrocarbons. When a hydrocarbon molecule is completely burnt, theoretically only carbon dioxide and water are formed. The following equation is a rough representation of the chemical reactions that would take place in the case of ideal, complete combustion of fuel containing carbon and hydrogen:

$$C_xH_y + \left(x + \frac{y}{4}\right)O_2 \rightarrow xCO_2 + \frac{y}{2}H_2O$$

## 9.1 Formation of pollutants during engine operation

Due to non-ideal conditions during combustion, the exhaust emissions of gasoline engines mainly comprise the following pollutants, in addition to carbon dioxide ($CO_2$) and water ($H_2O$):

- Carbon monoxide (CO)
- Unburnt hydrocarbons (HC)
- Nitrogen oxides ($NO_x$).

Exhaust emissions also contain nitrogen from the atmosphere, and possibly oxygen when the engine is operated with excess air. **Figure 9.1** shows the exhaust-gas composition of gasoline engines run with a stoichiometric air/fuel ratio ($\lambda = 1$). The measurement of exhaust emissions generally involves determining the quantity of pollutants in ppm (parts per million) or their volume share as a percentage. Almost no sulfur dioxide ($SO_2$) is formed during the combustion process in gasoline engines, as current and future fuels contain almost no sulfur. Like diesel engines, gasoline engines with direct injection can produce particulate emissions when fuel injection is not adjusted and mixture formation is therefore incomplete during combustion. However, these emissions are usually lower than those of diesel engines

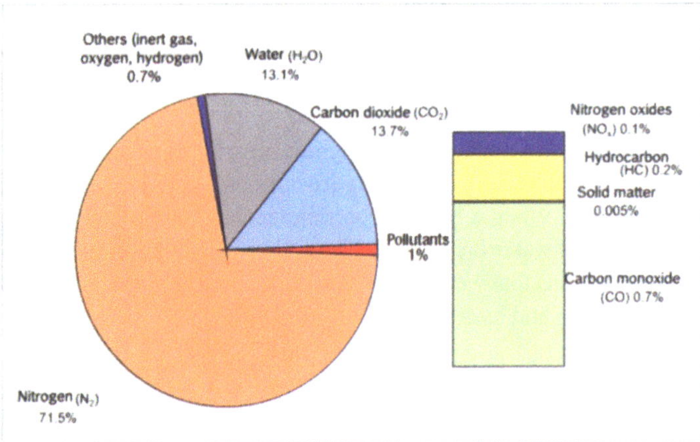


**Figure 9.1:** Exhaust-gas composition in gasoline engines with $\lambda = 1$ [1]

The main pollutants contained in engine emissions have the following effects:

- **Carbon monoxide** is a colorless, odorless gas. It bonds more strongly (factor 240) with hemoglobin, the $O_2$ carrier of blood, than oxygen. Therefore, there is a danger of suffocation even at low CO concentrations.
- **Unburnt hydrocarbons (HC)**, depending on their composition, have a more or less narcotic effect and induce irritation of the mucous membranes. Certain components have a carcinogenic effect (aromatics, e.g. 3, 4 benzopyrene, benzene). Pre-oxidized hydrocarbons (aldehydes) have an effect similar to that of aromatics.
- **Nitrogen monoxide** is a colorless gas and causes a change in lung function. It induces irritation of the mucous membranes and oxidizes with $O_2$ to form $NO_2$. There is a risk of nitric acid formation. Nitrogen monoxide is unstable under environmental conditions and converts to nitrogen dioxide ($NO_2$).
- **Nitrogen dioxide** is a reddish-brown gas with a pungent odor. It irritates the lungs even in small concentrations, causes tissue damage and induces irritation of the mucous membranes. As with nitrogen monoxide, there is a risk of acid formation.
- Criticism of diesel engines and gasoline engines with direct injection is growing, due to **particulate emissions** (carcinogenic potential). Particulates may be inhaled and are dangerous because they adsorb harmful substances. They also contain soot, which can be pure carbon or contain adsorbed hydrocarbons. New findings indicate that the soot core can also have a carcinogenic effect.
- **Carbon dioxide** does not affect human beings directly in the concentration present when an engine is running, but contributes indirectly to long-term damage to the environment through changes in the atmosphere (greenhouse effect). When the concentration is too high, there is a risk of suffocation.

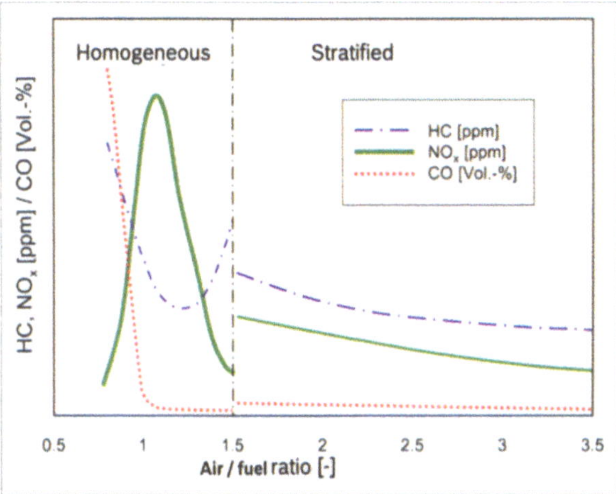


**Figure 9.2:**
Pollutant concentration in gasoline engines with direct injection in the homogeneous and stratified operating mode

**Figure 9.2** shows the basic dependence of pollutant concentrations on the air/fuel ratio for a gasoline engine with direct injection. For an air/fuel ratio between $\lambda = 0.8$ and 1.5, measurements were taken when the engine was running homogeneously, i.e. with intake stroke injection. The result is a pollutant concentration curve similar to that of a conventional engine with intake manifold injection. The curve shows emissions of unburnt hydrocarbons, nitrogen oxides and carbon monoxide when the stratified charge mode for air/fuel ratios is greater than 1.5. Carbon monoxide is formed as the product of incomplete combustion in the area deficient in air. The concentration of CO in the cylinder is at its maximum during combustion. During the subsequent expansion, some of it oxidizes to form $CO_2$.

As in the case of carbon monoxide, air deficiency leads to incomplete combustion and therefore results in the emissions of unburnt and incompletely burnt hydrocarbons. Other hydrocarbon emissions originate from the combustion chamber zones which are not covered, or only partially covered, by the combustion process, such as boundary layers close to the walls in which the flame extinguishes due to cooling (*squish effects*), as well as gaps into which the flame cannot penetrate; **figure 9.3**. Besides hydrocarbons from squish effects and from the ring-gap area between the piston and the cylinder (top land area), hydrocarbons are also formed from deposits and due to desorption (a process in which foreign atoms or molecules leave the surface of a solid body) from the oil film. During expansion and exhaust, the hydrocarbons are mixed with the hot exhaust gas and oxidized by secondary reactions, which require high temperatures and available oxygen. This is one reason for the minimal hydrocarbon emissions with little excess air. When the amount of excess air is extremely high, the number of hydrocarbons increases drastically due to *flame quenching* [2]. The combustion temperatures are not sufficient to maintain flame propagation in cooler parts of the combustion chamber [2, 3].

This leads to increased emissions of unburnt hydrocarbons in the stratified charge mode. Wall-guided combustion processes, in particular, can produce 80% more hydrocarbon emissions than conventional engines with intake manifold injection, due to fuel coating on the

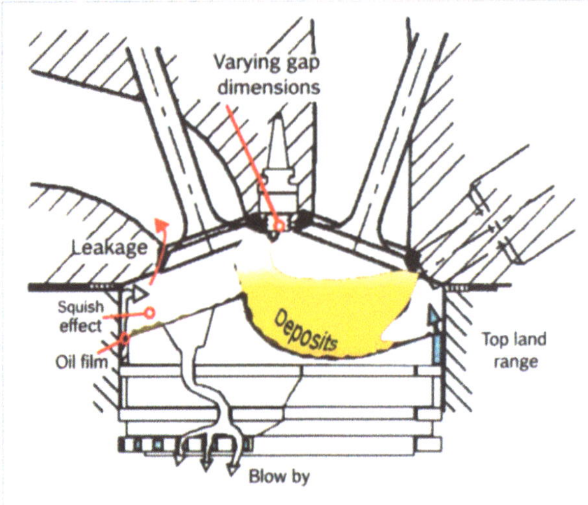


**Figure 9.3:**
Hydrocarbon sources in direct injection

piston [4]. In general, hydrocarbon emissions are also slightly higher in homogenous direct injection than in intake manifold injection.

Nitrogen oxide emissions ($NO_x$) are a combination of nitric oxide NO (approximately 90%), nitrogen dioxide $NO_2$ (approximately 9%) and nitrous oxide $N_2O$ (approximately 1%) concentrations. The maximum combustion temperature and the duration of its effect have a decisive influence on the $NO_x$ concentration, and especially on the NO concentration. The maximum combustion temperature in the engine is reached when the air/fuel ratio is $\lambda \approx 0.95$. The maximum NOx concentration is in the lean range, however, at $\lambda \approx 1.05$ to 1.1.

The reason for this is that in addition to the high temperatures, a high enough concentration of oxygen in the burnt part is required. Compared to homogeneous mixture formation, when operating with direct injection and in stratified charge mode, the $NO_x$ emissions are higher due to the high combustion temperatures at certain locations.

The reactions that form nitrogen oxides are substantially slower than the reactions that form CO. The NO is basically formed as per the following elementary reactions [5]:

- $O_2 + M \rightleftarrows O + O + M$ (M = collision partner)
- $N_2 + O \rightleftarrows NO + N$ (Zeldovic chain reaction)
- $O_2 + N \rightleftarrows NO + O$ (Zeldovic chain reaction)
- $OH + N \rightleftarrows NO + H$

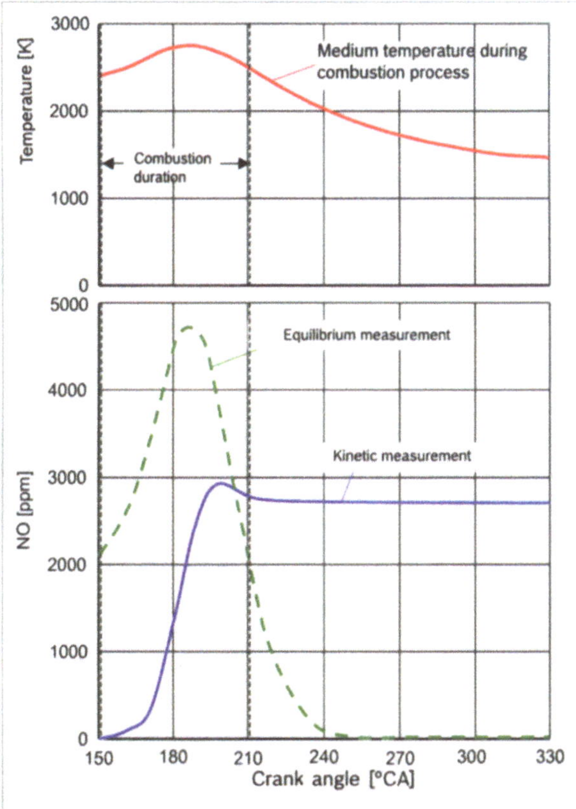


**Figure 9.4:**
NO concentration and mean temperature in the gasoline engine [5]

The sequence of these reactions in the flame front (which is relatively thin) is negligible, due to the short the retention time. NO is formed only behind the flame in the burnt part. In spite of the high temperatures, the reactions take place so slowly that a state of equilibrium cannot be reached; **figure 9.4**.

As **figure 9.4** shows, immediately after combustion, the formation of nitrogen oxides falls considerably behind the equilibrium concentration. At the start of expansion, after reaching a maximum value, slight NO back formation occurs which ceases immediately, however, because the reactions freeze below approximately 2200 K. Therefore, the formation of NO cannot be described using equilibrium analyses, but rather by taking the time behavior of the reactions into consideration. The formation of nitrogen oxides can be influenced considerably through the combustion process sequence and the non-homogeneities of mixture formation.

## 9.2 Statutory regulations

The problem of air pollution first became known at the beginning of the 1950's, starting in Los Angeles. Since then, regulations to control air pollution and reduce the pollutant emissions of automobiles have been introduced progressively throughout the world. In 1959, "*Air Quality Standards*" were defined in California and emissions limits for automobiles, which were to come into effect in California from 1966, were derived from them. In 1968, legislative limitation of pollutant emissions was stipulated for the USA. Stricter regulations have since gradually come into effect.

Legal limits on the exhaust emissions of passenger cars with gasoline engines have existed in the EU and thus in Germany since 1971, while legal limits with regard to noise emissions have existed in Germany since as early as 1958. Stricter regulations can be expected in the future for both exhaust emissions and noise emissions and this will be a key factor in the further development of combustion engines.

In almost all highly motorized countries, vehicle emissions are limited by statutory regulations. To maintain reproducibility and comparability, the pollutants are measured according to specified test procedures (test cycles). These are more or less accurately derived from typical driving behavior in road traffic. The oldest drive program was introduced in California in 1966 with what is known as the "*California test cycle*". The disadvantage is that many countries have now determined different test procedures, evaluations and limit values, which makes comparison impossible or only partially possible. Furthermore, the variety of test procedures makes development unnecessarily expensive.

The emissions limits valid in individual countries are not interchangeable, due to specific measurement methods and different test cycles. A number of test procedures are used to measure emissions and determine the fuel consumption of brand new vehicles within the scope of type approval. The most important procedures for passenger cars are:

- US procedure FTP 75 and US Highway with the supplements SC03 (air conditioning) and US06 (aggressive driving)
- New European Driving Cycle (NEDC)
- Japanese 10/15 mode hot test and 11 mode cold test.

The different test cycles can be attributed to the demand to simulate behavior as representatively as possible for the driving conditions in critical traffic zones. The US FTP 75 test cycle involves a higher speed level with fewer idling phases than the European and Japanese test cycles and imposes high demands on transient driving behavior. In the European and Japanese test cycles, however, the catalytic converter must respond much more quickly after cold start.

**Emissions legislation in Europe**

The EU Commission specified European emissions limits for CO and hydrocarbons for the first time in Directive ECE-R 15. A modification came in the form of directive ECE-R 15/04, in which the limit values for CO were further reduced and the limit values for hydrocarbons and $NO_x$ were grouped together. From 1985, tax benefits were granted for gasoline engines with catalytic converters. Further tightening of the limit values, first with the Euro 1 standards (from July 1, 1992) and Euro 2 standards (from January 1, 1996), and the tax incentives led to almost all vehicles being equipped with three-way closed-loop catalytic converters from about 1990. The introduction of the Euro 3 standard (from January 1, 2000) implied the mandatory introduction of *On Board Diagnostics* (OBD), similar to the US standard. The Euro 4 standard has been in effect since January 1, 2005. The limit values for the future Euro 5 standard will apply to new vehicles from September 1, 2009. The Euro 6 level will come into force from September 1, 2014. **Figure 9.5** gives an overview of the emissions limit values.

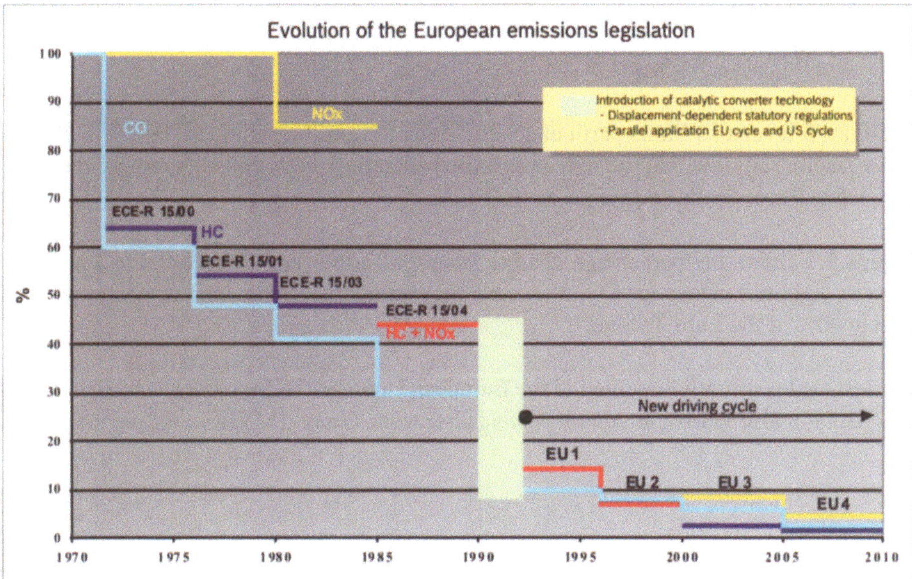


**Figure 9.5:** Development of European emissions limit values

**Figure 9.6** shows the values for the previous levels Euro 1 to Euro 3, the present level Euro 4 and the future Euro 5 and 6 for gasoline and diesel engines. Euro level 3, which has been in effect since 2000, and Euro level 4, in effect since 2005, both distinguish between hydrocarbon and $NO_x$ emissions.

The limit values have been continuously tightened over the different levels of emissions legislation. For gasoline engines, for example, during the transition from Euro 3 to Euro 4, the limit for hydrocarbon emissions was halved and the CO emissions limits were reduced by more than half.

| | | Date of introduction | CO [g/km] | HC [g/km] | HC+NOx [g/km] | NOx [g/km] | PM [g/km] |
|---|---|---|---|---|---|---|---|
| Diesel engine | | | | | | | |
| | **Euro 1** | 01.07.1992 | 2.72 | - | 0.97 | - | 0.14 |
| | **Euro 2, IDI** | 01.01.1996 | 1.0 | - | 0.7 | - | 0.08 |
| | **Euro 2, DI** | 01.01.1996 a) | 1.0 | - | 0.9 | - | 0.10 |
| | **Euro 3** | 01.01.2000 | 0.64 | - | 0.56 | 0.50 | 0.05 |
| | **Euro 4** | 01.01.2005 | 0.50 | - | 0.30 | 0.25 | 0.025 |
| | **Euro 5** | 01.09.2009 | 0.50 | - | 0.23 | 0.18 | 0.005 |
| | **Euro 6** | 01.09.2014 | 0.50 | | 0.17 | 0.07 | 0.005 |
| | | | | | | | |
| Gasoline engine | | | | | | | |
| | **Euro 1** | 01.07.1992 | 2.72 | - | 0.97 | - | - |
| | **Euro 2** | 01.01.1996 | 2.2 | - | 0.5 | - | - |
| | **Euro 3** | 01.01.2000 | 2.30 | 0.20 | - | 0.15 | - |
| | **Euro 4** | 01.01.2005 | 1.0 | 0.10 | - | 0.08 | - |
| | **Euro 5** | 01.09.2009 | 1.0 | 0.10 | - | 0.07 | 0.005 b) |
| | **Euro 6** | 01.09.2014 | 1.0 | 0.10 | - | 0.07 | 0.005 b) |

**Figure 9.6:** EU emissions limit values: IDI: diesel engines with indirect injection DI: diesel engines with direct injection – a) until September 30, 1999 (subsequently, the limit values for IDI will apply), b) only for engines with gasoline direct injection that work either partially or exclusively in lean mixture operation

The Euro 5 level also limits particulate emissions for gasoline engines. This will also address gasoline engines that run in lean stratified operating mode and may require the use of particulate filters for these engines as well.

**Figure 9.7** shows the percentage change from the Euro 3 emissions level to Euro 5 via Euro 4. The limit values for CO, NOx and unburnt hydrocarbons for Euro 5 are approximately 40% of the Euro 3 value.

The test cycle currently specified in the European Union is the *New European Driving Cycle* (NEDC), also known as *Motor Vehicle Emissions Group* (MVEG); **figure 9.8**. This is

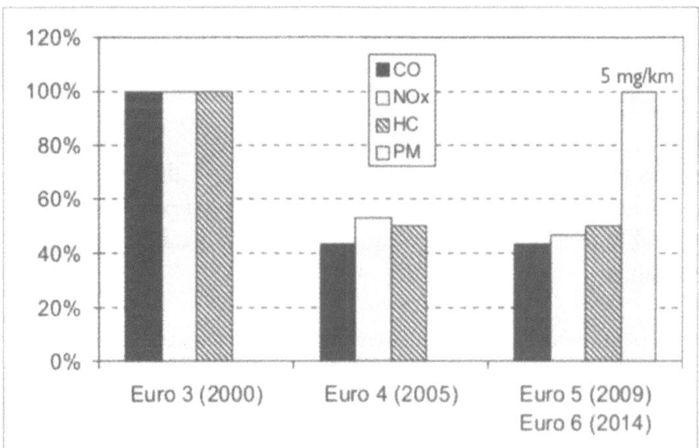


**Figure 9.7:** Percentage change from Euro 3 to Euro 5 for gasoline engines (for limit values, see figure 9.6)

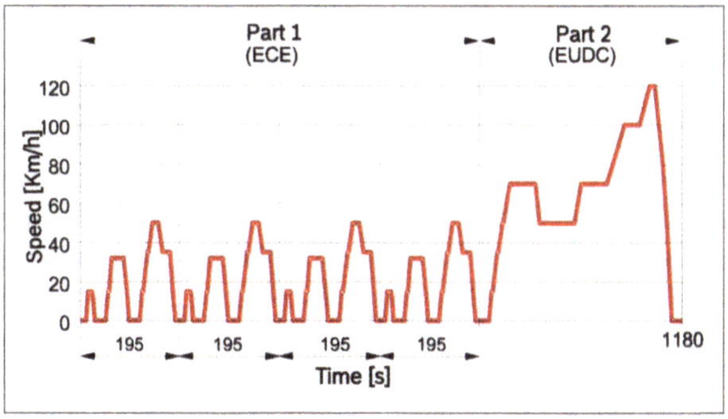


**Figure 9.8:**
European test cycle NEDC
Test length: 11.007 km
Test duration: 1180 s
Average cycle speed: 33.6 km/h
Maximum speed: 120 km/h
Sampling commences upon engine start

divided into an urban driving cycle (part 1) and an extra-urban driving cycle (part 2). Part 1, starting from a cold start, is carried out four times. The city cycle contains extremely high percentages of idling time (31%) and trailing throttle. The average driving speed is only 19 km/h. Since the Euro 3 level was introduced in 2000, the first 40 seconds of idling have been dropped from the test cycle, which was not previously the case; that is, measurement of emissions commences directly after the start. Since 1992, Part 2, the *Extra Urban Driving Cycle* (EUDC), which includes speeds of up to 120 km/h, has also been in effect. This second part takes country road and highway driving into consideration and is carried out only once.

**Emissions legislation in the USA / California**

Due to its special climatic conditions, California has always been a pioneer in emissions legislation and has substantially influenced US emissions legislation. After the "*Clean Air Act*" of 1968, new limit values determined in 1977 stipulated a 90% reduction and led to the introduction of three-way catalytic converter technology. Since that regulation, exhaust mass emissions have been measured according to the FTP 75 test cycle (*Federal Test Procedure*).

The years 1994 and 1998 saw a further gradual tightening, including the introduction of "*On Board Diagnostics*", OBD, an auto-diagnostic system for monitoring all emissions-relevant components for malfunction. Based on planning by CARB (*California Air Resources Board*), the *Low Emissions Vehicle* (LEV) program has become the basis of emissions standards in the USA. New categories of vehicles and emissions – TLEV (*Transitional Low Emissions Vehicle*), LEV and ULEV (*Ultra Low Emissions Vehicle*) – were defined in California. The US EPA (*Environmental Protection Agency*) has enacted separate legislation at state level.

Furthermore, CARB passed the ZEV (*Zero Emissions Vehicle*) mandate. This stipulates that from 2005, increasing percentages of new vehicles sold should either cause even lower emissions or no emissions. Medium-size manufacturers can fulfill the ZEV mandate with a PZEV (*Partial Zero Emissions Vehicle*). PZEV refers to a vehicle whose emissions are

**Emissions legislation in Japan**

At approximately the same time as in the USA, the first emissions limit values for CO were specified in Japan on the basis of the four mode test cycle. In 1973, hydrocarbons and NOx were also limited and determined on the basis of the 10 mode test cycle. From 1976, the introduction of catalytic converter technology brought about a reduction of the $NO_x$ emissions by 90%. In Japan, a distinction is made by different criteria (drive concept, engine concept, fuel type, origin of the vehicles). The current emissions standards are determined from a combination of the 11 mode and 10.15 mode test cycles which replace the 10 mode test cycle; **figure 9.13**. The Japanese emissions limit values are shown in **figure 9.12**. More levels of the *New Long Term Emissions Standards* are planned, including the replacement of previous cycles with the new JC08 cycle by 2008 or 2011 (the limit values remain the same). The new driving cycle is more clearly based on actual driving habits in Japan.

In the European test and in the FTP 75 test, the pollutants in exhaust gas are currently evaluated using the CVS (*Constant Volume Sampling*) measuring method on a dynamic

| CO | $NO_x$ | HC |
|---|---|---|
| 1.15 g/km | 0.05 g/km | 0.05 g/km |

**Figure 9.12:**
Japanese emissions limit values

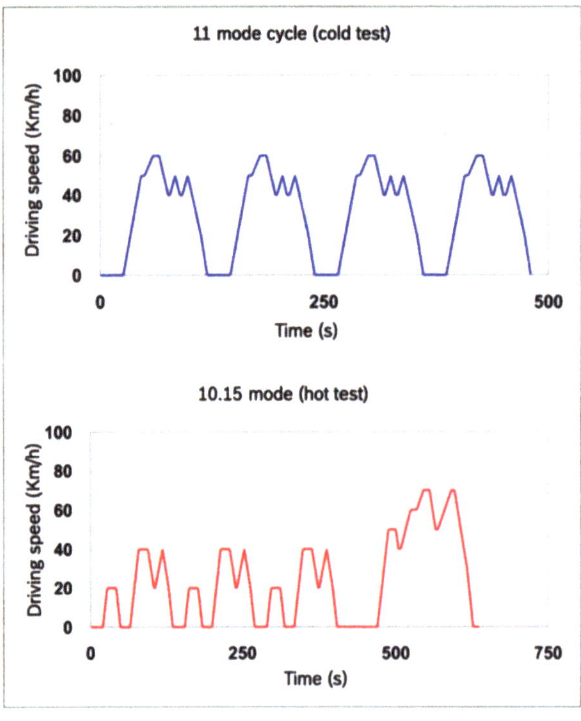


**Figure 9.13:**
Japanese 11 and 10.15 mode cycle

power analyzer. During the test, the exhaust gases of the vehicle are diluted and a sample is collected in a pouch; **figure 9.14**.

The dynamic power analyzer is equipped with a power brake with which the operating data can be set. In line with the temporal course of speed (test cycles), the engine is activated in particular load-speed ranges (driver aid system). Road resistances, such as air resistance and rolling resistance, are determined by tests and are made available and taken into consideration using appropriate curves.

The exhaust gases are added to the ambient air entering through the filter. The diluted air is drawn in by the blower through a heat exchanger and a heater which is controlled by a temperature controller. The constant volume flow delivered by the blower is calculated based on the tachometer reading, and a temperature display is required to determine the mass flow rate. A metering pump feeds samples into a pouch. To control the air quality, the supplied air that has been filtered and diluted is also collected in the pouch using the metering pump. The contents of the pouch are analyzed after the test. The pollutant masses emitted in each test can be calculated from the total amount of gas delivered with the pollutant concentrations in the pouches. They are determined in relation to the distance covered in the cycle, i.e. are given in the unit [g/km].

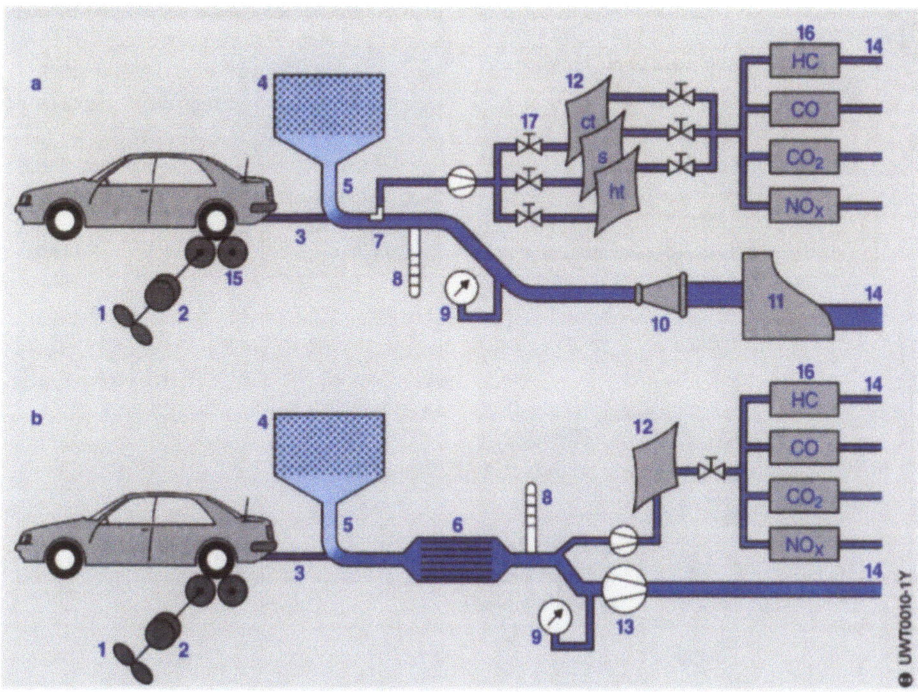


**Figure 9.14:** Dynamic power analyzer with CVS: a) US federal test (here with Venturi equipment) and b) European test equipment (here with rotary piston blower) [1]

**Regulations on fuel consumption reduction**

The regulations on fuel consumption reduction have different rationales, including the fact that reserves of fossil fuels are finite and therefore limited, and that a vehicle's $CO_2$ emissions are a direct measure of its fuel consumption and contribute to the greenhouse effect.

In 1985, the first fuel consumption regulations were introduced in the USA with the *Corporate Average Fuel Economy* (CAFE) and the *Gas Guzzler Tax* (GGT), an additional tax on vehicles with high fuel consumption. The main motivation has been to save resources. Consumption is determined by the weighted average of the FTP 75 and the US highway (= US Combined). While a fleet value is defined within the scope of the CAFE regulations,

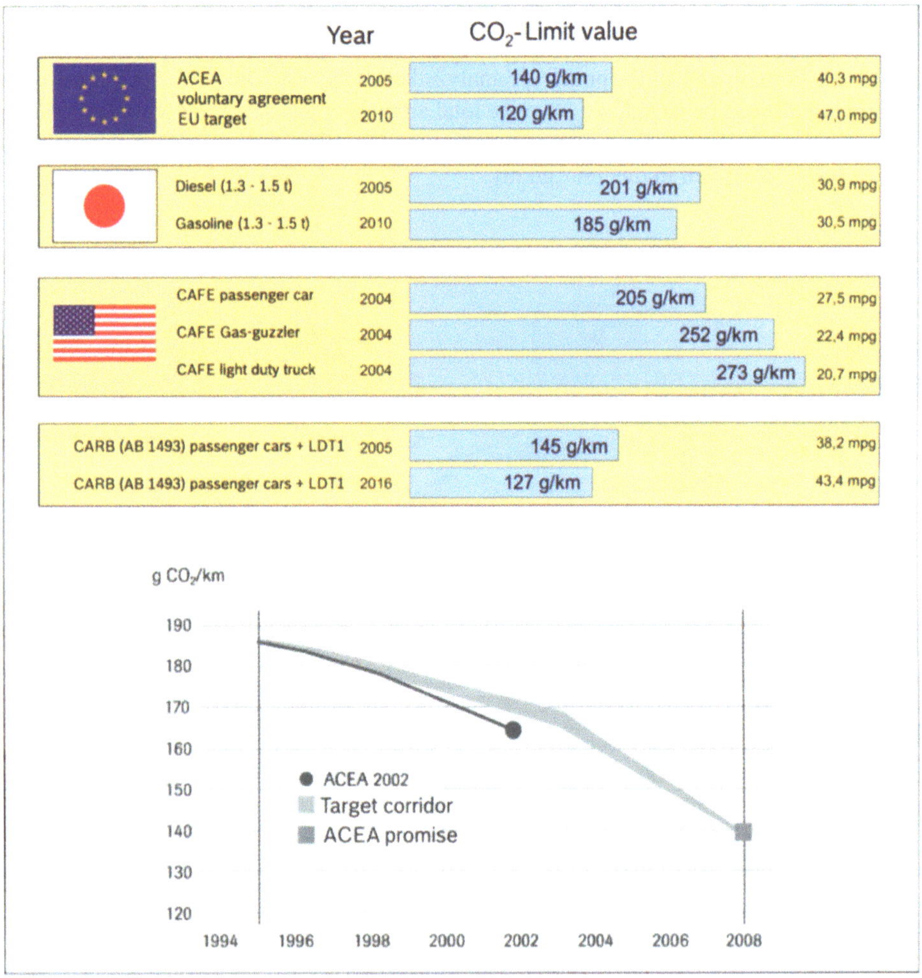


**Figure 9.15:** $CO_2$ emissions and fuel consumption targets in USA (CAFE) and Europe (ACEA voluntary agreement [Source: ACEA])

GGT refers to the value of an individual vehicle. If the limit values are exceeded, a fine must be paid.

The European Automobile Manufacturers Association (ACEA) has entered into a voluntary agreement with the European Commission that stipulates an average reduction in the fleet emissions of ACEA member companies of approximately 25%, with a target value of 140 g $CO_2$ per km, in the period from 1995 to 2008. A more far-reaching recommendation by the European Commission calls for another reduction of the target value to 120 g $CO_2$ per km by 2012. Unlike consumption specifications in the USA, the primary motivation in Europe is climate protection; **figure 9.15**.

Many of the countries that have signed the Kyoto Protocol are demanding an appropriate contribution by the automobile industry to reduce $CO_2$ emissions. One of these countries is Japan where a 20% reduction of $CO_2$ emissions is specified for the period from 1995 to 2010. Some countries, such as Korea and Taiwan, are trying to establish a maximum consumption value relevant to type approval that may not be exceeded. China has introduced such a regulation from 2005.

## 9.3 Pollutant reduction including particulates

The amount of different pollutants present in the exhaust gas of gasoline engines can be influenced by a variety of measures, which can be divided into three main categories:

- Design parameters, e.g. design of the cylinder head and combustion chamber, displacement, stroke to bore ratio, compression ratio, spark plug position, number of valves.
- Operational parameters, e.g. ignition point, injection point, control timings, mixture preparation, internal and/or external exhaust-gas recirculation, overrun fuel cut-off.
- Exhaust-gas aftertreatment, e.g. catalytic converters, secondary air injection.

In practice, a combination of different measures is usually used. A selection of the most frequently used measures is explained below by way of example.

### Engine modifications

The compression ratio can be influenced by the design of the combustion chamber. For fuel consumption reasons, the compression ratio should be as high as possible (high thermal efficiency). With increasing compression, however, the $NO_x$ and hydrocarbon emissions rise. The nitrogen oxide emissions increase due to the increased combustion temperatures in the combustion chamber while the hydrocarbon emissions increase due to the crevices of the combustion chamber (relatively larger number of gaps). For smaller cylinder units, the hydrocarbon emissions rise with an unfavorable surface/volume ratio. The combustion chambers should therefore be as compact as possible. **Figure 9.16** shows the relationship

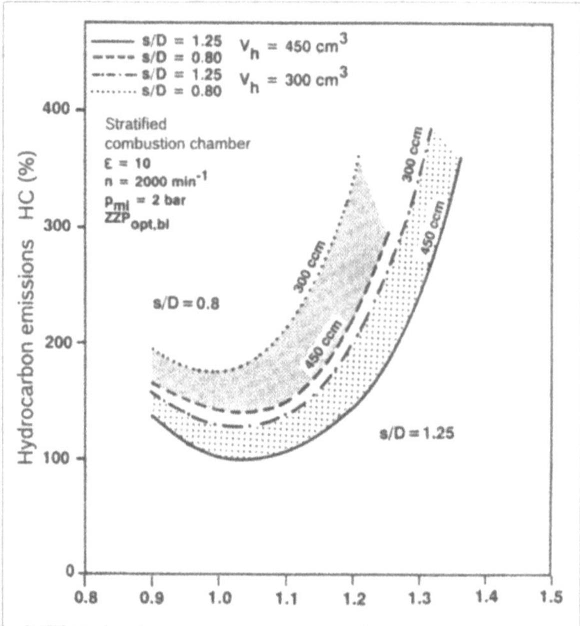


**Figure 9.16:** Influence of the surface/volume ratio and air/fuel ratio on hydrocarbon emissions [6]

between hydrocarbon emissions depending on the air/fuel ratio for different combustion chambers [6].

With increasing compression, the exhaust-gas temperature decreases due to increased efficiency and the secondary reactions of unburnt hydrocarbons and carbon monoxide deteriorate in the exhaust.

The ignition point influences fuel consumption and exhaust emissions (hydrocarbon, $NO_x$) in different ways. Starting from the optimum fuel consumption, a retardation of the ignition point leads to increasing exhaust-gas temperatures and decreasing combustion temperature (peak values). As a result, it is the $NO_x$ and hydrocarbon concentrations in particular that decrease. At the same time, however, efficiency decreases. Therefore, retarded ignition is a useful measure only for low part load and for idling. At full load, retarded ignition is used to prevent knocking combustion.

**Exhaust-gas recirculation**

Exhaust-gas recirculation means the supply of exhaust gas to the fresh mixture. There are two types:

- External exhaust-gas recirculation
- Internal exhaust-gas retention.

In external exhaust-gas recirculation, the exhaust gas is recirculated from the exhaust-gas system to the intake system via external supply; **figure 9.17**.

Internal exhaust-gas retention is carried out by means of a reduction of the valve overlap, i.e. by corresponding valve timing and therefore poor cylinder scavenging. As a result, the amount of residual gas, i.e. the amount of exhaust gas in the cylinder, increases.

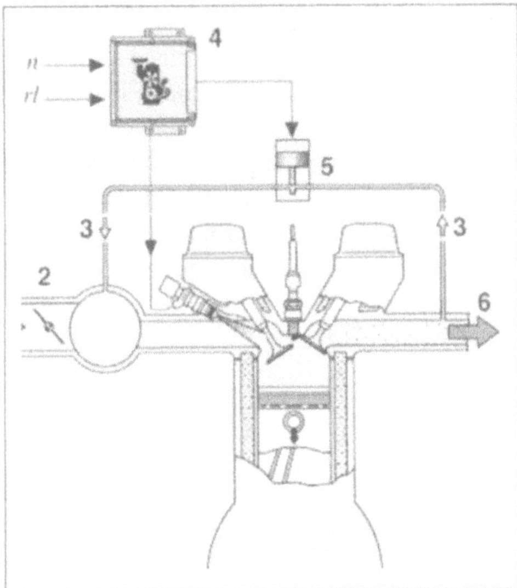


**Figure 9.17:**
Schematic diagram of an external exhaust-gas recirculation system [1];
1 Fresh intake air
2 Throttle valve
3 Recirculated exhaust gas
4 Engine control unit
5 Exhaust-gas recirculation valve (EGR valve)
6 Exhaust gas
*n* Engine speed, *rl* relative ventilation

As shown in **figure 9.18**, the aim of exhaust-gas recirculation is primarily to reduce $NO_x$ emissions. Lower recirculation rates usually have only limited influence on fuel consumption at $\lambda \approx 1$. At high recirculation rates, efficiency deteriorates significantly. Generally, the exhaust-gas recirculation rate is between 5 and 10% per volume. The hydrocarbon emissions usually increase slightly when exhaust-gas recirculation is used.

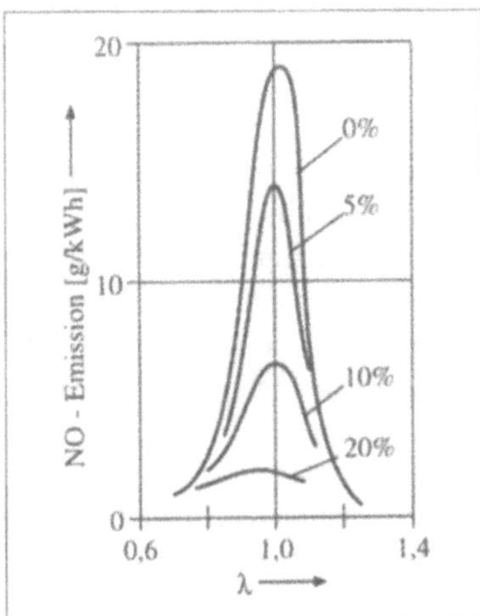


**Figure 9.18:**
Influence of EGR on $NO_x$ formation (gasoline engine) [1]

**Exhaust-gas treatment during stoichiometric operation**

For exhaust-gas treatment in gasoline engines, the three-way catalytic converter remains the most effective measure for achieving a significant reduction in exhaust-gas components, the concentration of which is limited. In catalytic converter technology, the conversion of the exhaust-gas components depends, besides temperature, mainly on the catalytic effect of the materials used and on the mixture setting (air/fuel ratio). To initiate a chemical reaction that results in the conversion of a particular exhaust-gas component, it is necessary to cross a certain energy threshold (activation energy). A catalytic converter works on the principle of lowering the energy threshold for initiating the processes of oxidation (hydrocarbon, CO) and reduction ($NO_x$). The three-way catalytic converter has the characteristic of simultaneously reducing all three components, i.e. hydrocarbons, CO and $NO_x$, to a large extent. However, a complex control system ($\lambda$-control) is required to achieve optimum results for exhaust-gas control; **figure 9.19**.

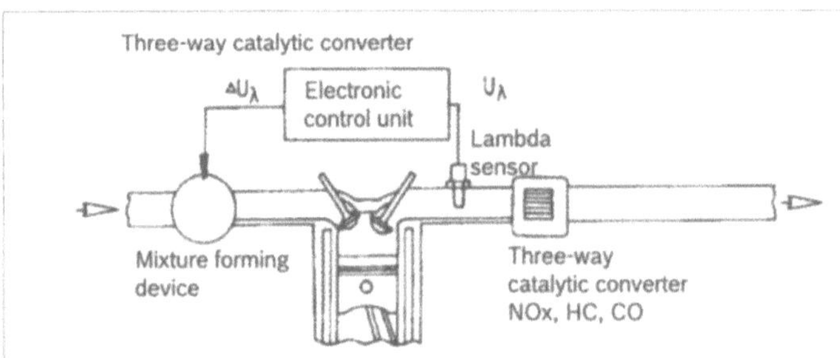


**Figure 9.19:** Schematic diagram of a three-way catalytic converter [1]

$\lambda$-control is carried out with a $\lambda$-sensor located in the exhaust tract. The sensor detects the air/fuel ratio of the previous working cycles and transmits a corresponding signal to the ECU. Here the actual air/fuel ratio is compared with the setpoint value and, if necessary, the fuel quantity is corrected through an actuator. **Figure 9.20** shows a $\lambda$-sensor which is currently used in gasoline engines with three-way catalytic converters in series production.

An oxygen sensor made of zirconium dioxide ($ZrO_2$) is mounted in the exhaust pipe in such a way that the outer platinum layer on the $ZrO_2$ is surrounded by exhaust gas while the inner platinum layer stays in contact with the ambient air. From approximately 300°C, the $ZrO_2$ ceramic becomes conductive for oxygen ions. This causes an electrical voltage between the two interfaces when there are different amounts of oxygen. The oxygen concentration present in the exhaust gas strongly depends on the air/fuel ratio, resulting in an erratic characteristic of the sensor voltage in the range $\lambda = 1$.

The control in the range $\lambda = 1$ makes it possible to allow both oxidation reactions (CO, HC, $H_2$ to $CO_2$, $H_2O$) and reduction reactions ($NO_x$ to $N_2$, $H_2O$, $CO_2$) in the catalytic converter. In case of excess air, oxidation takes place and in case of air deficiency, reduction. How-

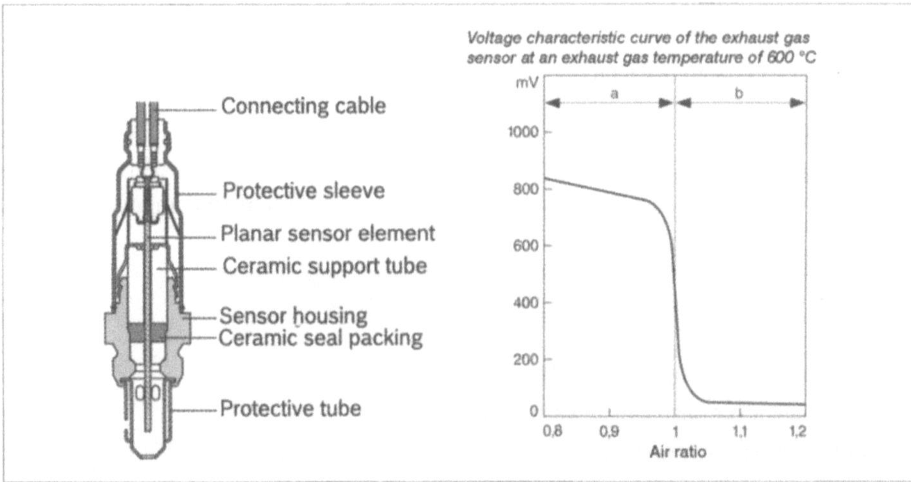


**Figure 9.20:** $\lambda$ sensor; a: rich mixture (air deficiency), b: lean mixture, (excess air) [1]

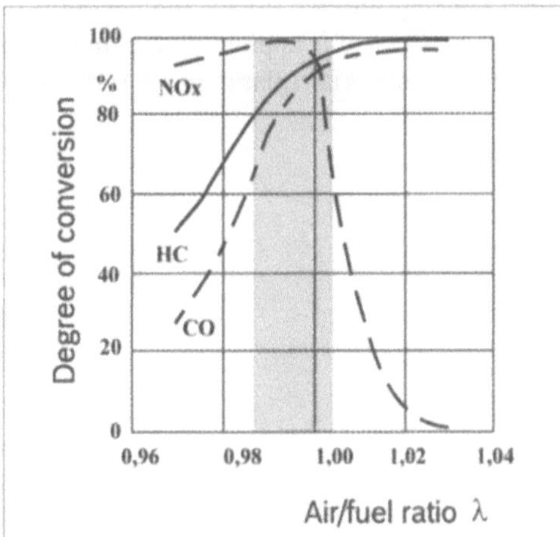


**Figure 9.21:**
Control range of the $\lambda$ sensor [1]

ever, the conversion behavior of a catalytic converter is sufficient only in a narrow range around $\lambda = 1$; **figure 9.21**.

Oxidation equations:

$$2\,CO + O_2 \rightarrow 2\,CO_2$$

$$C_xH_y + \left(x + \frac{y}{4}\right)O_2 \rightarrow xCO_2 + \frac{y}{2}H_2O$$

Reduction equations:

$$2\,NO + 2\,CO \rightarrow N_2 + 2\,CO_2$$

Outside the range $\lambda = 1$, very small conversion rates are observed, especially of CO and hydrocarbons for $\lambda < 1$ and of $NO_x$ for $\lambda > 1$. Therefore, the $\lambda$-control is active mainly in the range between 0.99 and 1.003, to ensure that at least a 90% of all components will be converted.

A catalytic converter basically consists of four main parts:

- Housing
- Carrier material (ceramic or metal)
- Intermediate layer (also called wash coat)
- Catalytically active layer.

**Figure 9.22** shows the structure of a catalytic converter. A honeycomb carrier body (monolith) with fine channels is tightly fixed in a steel housing through honeycomb support or knitted steel mesh. A porous intermediate layer, the wash coat, is applied on the thin walls of the carrier body. The layer is coated with catalytic material (platinum (Pt), rhodium (Rh), or palladium (Pd))

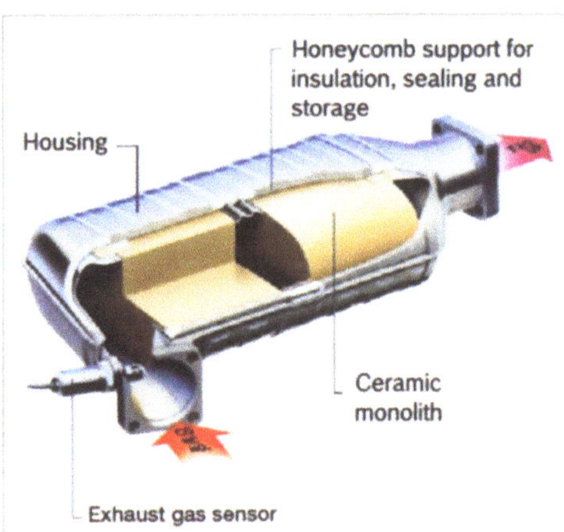


**Figure 9.22:**
Underfloor catalytic converter
[Source: Eberspaecher]

Exhaust-gas treatment will also prevent hydrocarbon emissions peaks during cold start, to comply with future limit values, although the catalytic converters are not at operating state temperature (i.e. below light-off temperature) at the start of the test cycle. The measures used to reach this goal include the improvement of mixture formation (e.g. prevaporization of the fuel, air assistance) and combustion behavior, mixture enrichment and retarded ignition, external heating of the catalytic converter, and secondary air injection.

**Figure 9.23** shows the schematic design of secondary air injection (thermal treatment of the exhaust gas). The injection of additional air directly after the combustion chamber leads to afterburning of the hot exhaust gas. This exothermic reaction reduces the exhaust gas components hydrocarbon and carbon monoxide and warms up the catalytic converter. As a result, the conversion rate of the catalytic converter is significantly increased in the warm-up phase:

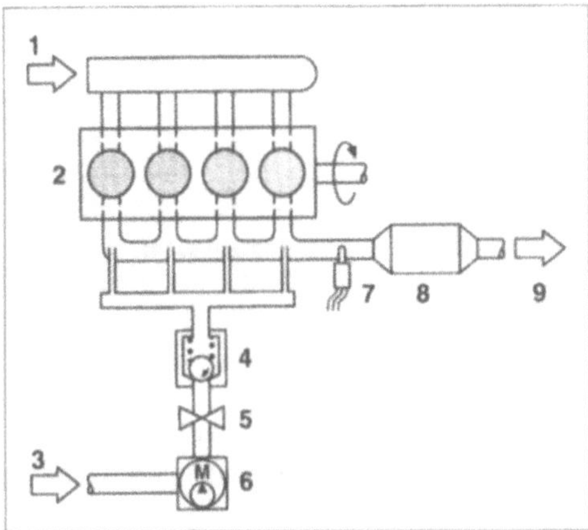


**Figure 9.23:**
Secondary air injection
1 Intake air
2 Engine
3 Secondary air
4 Back pressure valve
5 Secondary air valve
6 Electrical secondary air pump
7 $\lambda$ sensor
8 Catalytic converter
9 Exhaust gas

$$C_xH_y + \left(x + \frac{y}{4}\right)O_2 \rightarrow xCO_2 + \frac{y}{2}H_2O$$

$$CO + \frac{1}{2}O_2 \rightarrow CO_2$$

$$CO + H_2O \rightarrow CO_2 + H_2$$

$$H_2 + \frac{1}{2}O_2 \rightarrow H_2O$$

In addition to three-way catalytic converters, oxidation catalytic converters (e.g. for diesel engines) allow only oxidation reactions (conversion of CO and HC) and what are known as DENOX catalytic converters (reduction catalytic converters which reduce NOx also during excess air). Heating strategies for exhaust-gas catalytic converters in connection with direct injection are explained from page 238.

**Exhaust-gas treatment for lean mixtures (direct injection)**

The variation range of the air/fuel ratio in gasoline engines with direct injection – from extremely rich mixtures ($\lambda > 1$) in unthrottled or partially throttled part load operation to stoichiometric and even lean mixture operation in full load – imposes high demands on exhaust-gas treatment. Low exhaust-gas temperatures, especially in low part load operation, complicate exhaust-gas treatment.

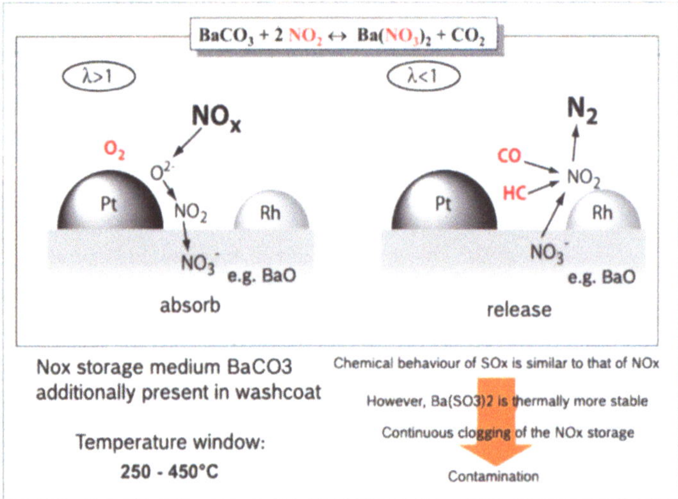


**Figure 9.24:** Reduction mechanism of $NO_x$ adsorption

Due to the large amount of oxygen during lean operating phases, unburnt hydrocarbons and carbon monoxide can be oxidized easily, whereas the nitric oxides are almost impossible to reduce. Development activities currently focus on the $NO_x$ absorption catalyst, also called storage catalyst, whose operating principle is shown in **figure 9.24**. In the lean operating mode, NO oxidizes to form $NO_2$ at the platinum contact and is absorbed in the form of nitrate ($NO_3$) by means of alkaline storage components (barium in this case) included in the wash coat of the catalytic system. A regeneration phase initiated by a brief period of operation at $\lambda < 1$ should take place at the latest when the $NO_x$ storage capacity is exhausted. The stored $NO_2$ is released and reduced to nitrogen $N_2$ in a reduction reaction with the exhaust gas components hydrocarbon and C present due to the fact that the engine is temporarily running in rich cycle. The engine management triggers the enrichment of the mixture at specific intervals.

**Figure 9.25** shows the exhaust-gas treatment system of a Volkswagen FSI engine, operated in stratified charge mode. High hydrocarbon conversion and exhaust-gas control in stoichiometric operation is achieved through a three-way catalytic converter (start catalyst) which is mounted close to the engine. A $NO_x$ storage catalyst with corresponding $NO_x$ sensor is integrated downstream. **Figure 9.26** shows the operational mode of the engine during the use of a $NO_x$ storage catalyst.

**Figure 9.27** shows another exhaust-gas system for gasoline direct injection. To reduce $NO_x$ raw emissions, the engine is operated with exhaust-gas recirculation. The exhaust-gas system also consists of a three-way catalytic converter (start cat) in order to achieve high hydrocarbon conversion and for exhaust-gas control in stoichiometric operation. A downstream main catalytic converter consists of both a $NO_x$ catalyst and a three-way catalytic converter.

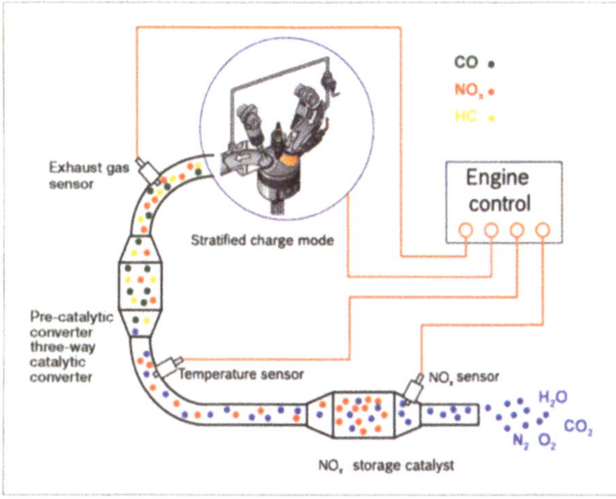


**Figure 9.25:** Exhaust-gas concept for FSI engine [Source: VW]

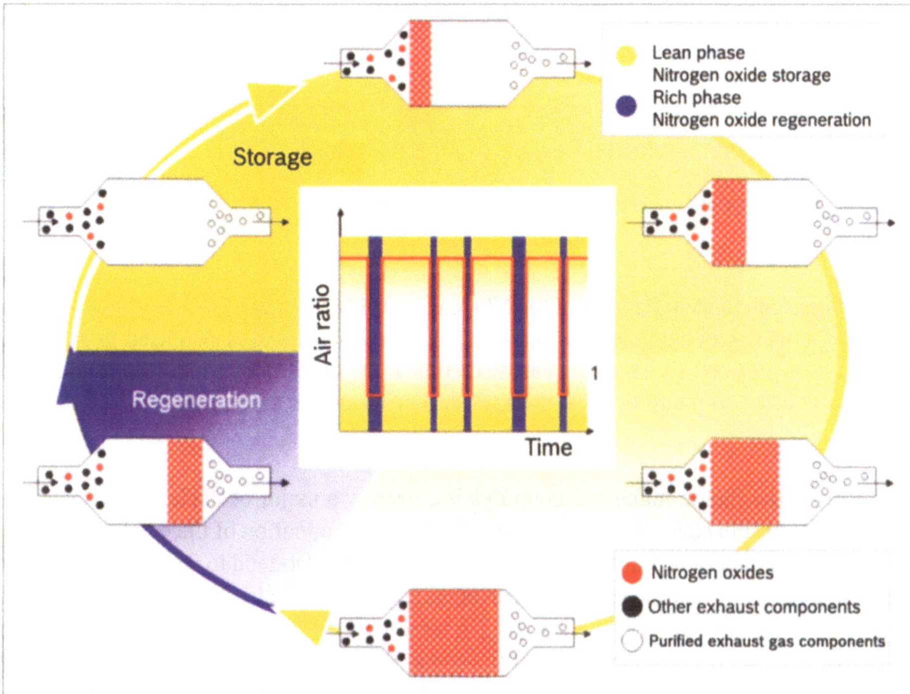


**Figure 9.26:** Operational mode of a $NO_x$ storage catalyst [Source: VW]

In this concept, the two-point exhaust-gas sensor LSF is used as reference sensor during stoichiometric operation to control the $NO_x$ catalyst and diagnose the $NO_x$ and $O_2$ storage capacity of the main catalytic converter. The temperature sensor makes it possible to diagnose the preliminary catalytic converter and, by means of an exhaust-gas temperature

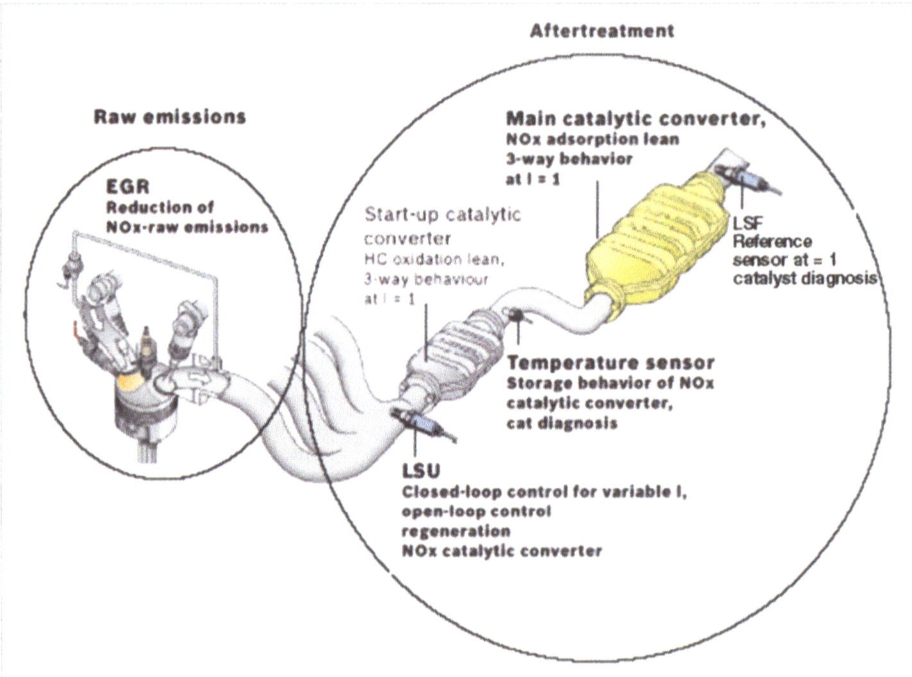


**Figure 9.27:** Exhaust-gas concept for gasoline direct injection [8]

model, determine the temperature of the $NO_x$ catalyst. The $\lambda$ required during lean operation is set by the wide band exhaust-gas sensor LSU. The LSU signal is also used to diagnose the $NO_x$ storage capacity. The LSU can also control the air/fuel ratio $\lambda$ during rich engine operation in order to purge the $NO_x$ catalyst according to the specifications of the storage catalyst control.

The sulfur in the fuel in numerous countries is currently a major obstacle to the operation of a $NO_x$ adsorption catalyst. The problem of sulfur contamination of the catalyst is due to the fact that all materials that are suitable for $NO_x$ storage also tend to store sulfur dioxide by forming respective sulfates. During lean operation, the $NO_x$ adsorption catalyst oxidizes the $SO_2$ in the exhaust gas to $SO_3$. The $SO_3$, in turn, reacts with the stored oxide to form the respective sulfate. The problem with sulfur contamination is that the sulfate has a higher thermal stability than the nitrate and therefore a separate sulfate regeneration process is required when the temperature of the catalyst increases.

**Figure 9.28** illustrates the influence of the sulfur content of the fuel on the $NO_2$ conversion rate without regeneration. With a sulfur content of 50 ppm in the fuel, the conversion rate of the catalyst decreases to only 50% after only approximately 5000 km.

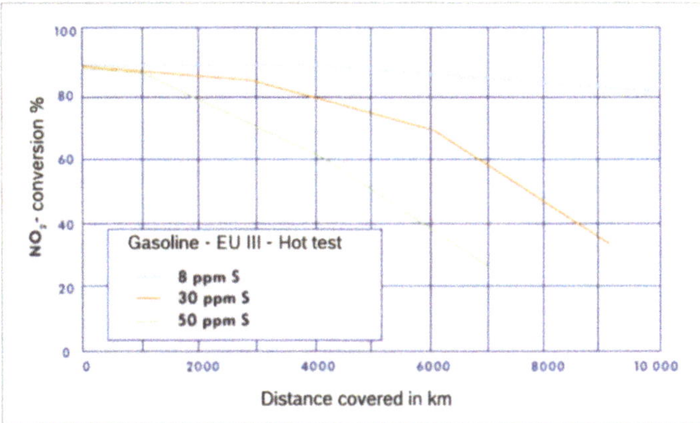


**Figure 9.28:** Influence of sulfur content in fuel on catalyst effectiveness [Source: VW]

In summary, recent tests have shown that sufficient regeneration of $NO_x$ storage catalysts contaminated with sulfur can be achieved. However, the inevitable associated extra consumption of fuel reduces potential fuel savings facilitated by the direct injection technology. For this reason, the lowest possible sulfur content in fuel is desirable, especially since the frequency of the desulfurization cycles decreases progressively with decreasing sulfur content in the fuel and, as estimated by the automobile industry, desulfurization of the fuel in a refinery is more economic for the end consumer than the extra expenses associated with increased fuel consumption.

**Heating strategy for exhaust-gas catalytic converters**

Double injection with retarded ignition (Strategy 1 in **figure 9.29**) is a well-known heating strategy for exhaust-gas catalytic converters in gasoline engines with direct injection. Some of the fuel quantity is injected during the intake stroke, the rest during the compression stroke [9]. Considerably improved mixture inflammation is achieved by coupling the second injection directly to the ignition point (Strategy 2). As a result, ignition can be retarded in order to increase the exhaust gas temperature.

Dividing the injection quantity into three injections (Strategy 3) offers further advantages. The first injection occurs during the intake stroke, the second during the compression stroke and the third, which is only a very small fuel quantity, is retarded and takes place directly on the spark plug. This enables an extremely retarded point of ignition with further improvement of engine smoothness and increased exhaust-gas temperatures for heating the catalytic converter. This strategy makes it possible to achieve higher temperatures in the manifold and catalytic converter with less fuel. It also reduces hydrocarbon raw emissions. Thus, an injection strategy is available for heating catalytic converters during the cold start of gasoline engines with spray-guided combustion process and piezo injection valve in central position, which meets all current emissions regulations and promises further potential if the standards are tightened in the future [9].

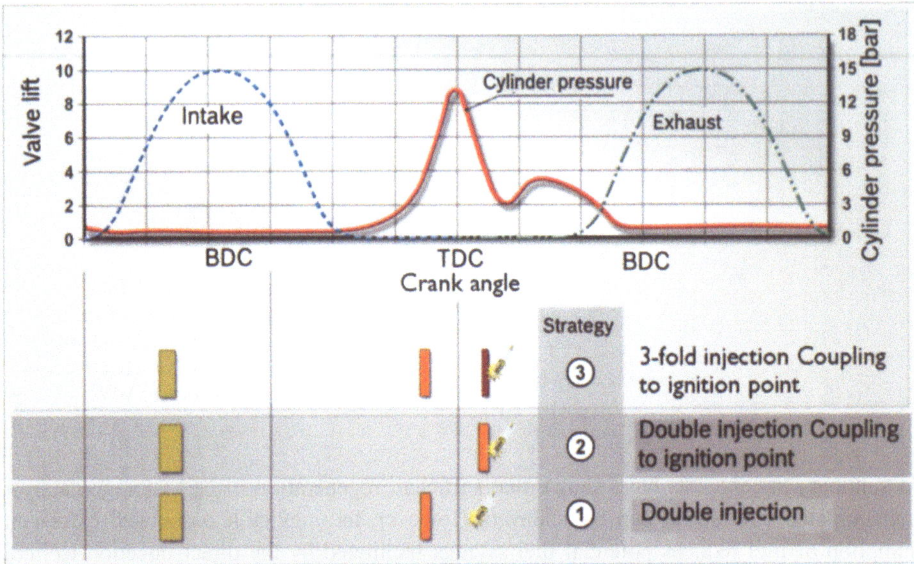


**Figure 9.29:** Injection strategies for heating exhaust-gas catalytic converters [9]

**Figure 9.30** shows the reduction of hydrocarbon raw emissions by means of double injection. The first injection during the intake phase creates a homogenous lean basic mixture. The second injection during the compression stroke creates a rich zone around the point of ignition, enabling reliable inflammation even for very retarded ignitions. The retarded combustion and the excess oxygen in the remaining combustion chamber increase the exhaust gas temperatures and therefore reduce the hydrocarbon raw emissions. The ignition adjustment to retarded is limited by an increase in $NO_x$ emissions.

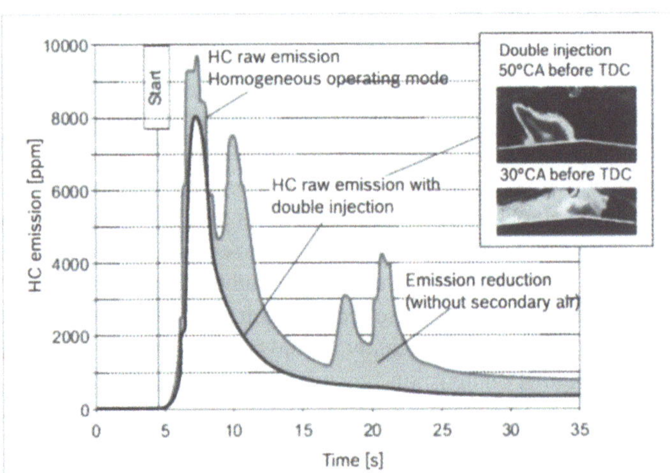


**Figure 9.30:** Advantages for hydrocarbon emissions through double injection for catalytic heating [10]

Multi-point injection significantly improves both mixture formation quality and emissions behavior, also during cold start operation. However, the heating that occurs due to retarded injection, which can no longer be fully converted into process work and thus into engine power, leads to extra fuel consumption. Multi-point injection means a more or less significant deviation from the ideal concept of direct injection with stratified charge, as the mixture produced is not as dense as is required for ideal stratified charge, due to the longer duration of injection and mixture formation as compared to an ideal single injection. In multi-point injection, this leads partly to a de-mixing in the periphery of the mixture vapor and therefore to greater lean adjustment in that area. As a result, the potential theoretically offered by direct injection cannot be completely utilized.

Except for very low temperatures, Audi [11] uses what is known as the high pressure stratified charge start. In this method, the fuel is injected exclusively during compression stroke. This way, the compression heat is used for mixture formation. Homogenization is better than at low pressure starts and the emissions of unburnt hydrocarbons are reduced. After a successful start, the homogeneous split operating mode (HSOP) with secondary air is used. A double injection strategy ensures that the fuel quantity is divided between the intake and compression strokes. This, together with adjusted tumble flaps, makes it possible to retard the ignition point considerably. The reactions due to secondary air and the retarded combustion phasing lead to a large heat flow which heats the catalytic converter to operating temperature quickly.

**Particulate filters**

Stricter emissions standards can make it necessary to use particulate filters for gasoline engines with direct injection, too, as engine modifications alone are no longer sufficient to fulfill the statutory regulations. There are different designs and working principles of particulate filters for diesel engines. The most common designs are presented below.

**Ceramic monolithic filters or honeycomb filter**

These particulate filters are manufactured from cordierite or silicon carbide (SiC) by means of extrusion. The design of the filters is similar to that of a cell catalytic converter, except that the cells are closed alternately. The exhaust gas flow is guided through the filter wall (*wall flow filter*). The soot particles are filtered and with increasing saturation of the filter and a layer of soot forms in the intake ports of the filter. This is represented schematically in **figure 9.31**.

For diesel particulate filters, the diameters of the pores of a ceramic monolithic filter average between 9 and 35 μm. They are therefore much larger than the particulates. Filtering first occurs because of the diffusion and deposition of particulates. Only then are layers, which increase with the duration of filtering, formed on the ceramic surface. Filter efficiency varies depending on the pore diameter. In the case of large pore diameters of 35 μm, 40 to 50% of the particulates are retained only after particulate deposition in the pores at the beginning of filter saturation. Filters with pore diameters of 12 μm, however, retain more than 90% of the particulates even at the start of the filtering process. If pores are large,

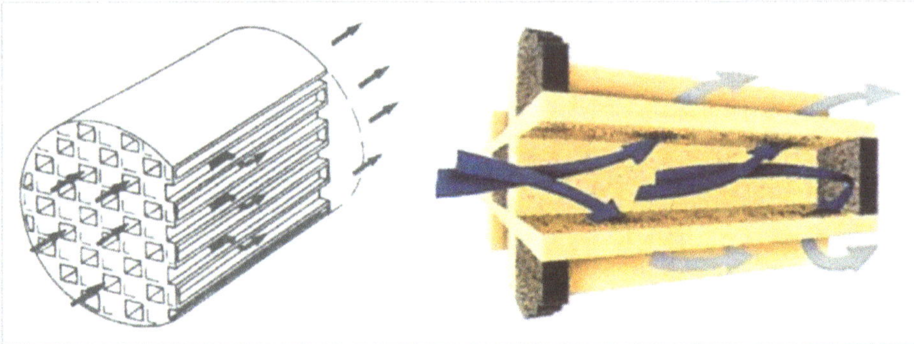

**Figure 9.31:** Design and mode of operation of a ceramic monolithic filter

particulate deposition first takes place directly in the pores. This makes the filter more efficient and increases flow resistance. If pores are smaller, there is very little deposition in the pores.

### Sintered metal filters

Sintered metal filters are developed on the basis of metallic materials. Their structure is similar to that of the ceramic monolith, i.e. the sintered metal plates divide the filter into inlet and outlet areas. When flowing through the filter plate, the soot particulates are deposited on its surface almost from the start. This process is called surface filter process.

In comparison to ceramic filters, sintered metal filters are relatively heavy but very robust. Recent designs feature lighter weight bellow type structures and are made of filter plates. In these filters, the metallic fibers are sintered and form filter plates using metal powder. Different filter designs are possible with these plates. **Figure 9.32** shows the design of a sintered metal filter, also called a jet filter, as developed by HJS for commercial vehicles and by Bosch for passenger cars. Due to factors including the heavy weight and extra cost as compared to ceramic filters, development has been abandoned.

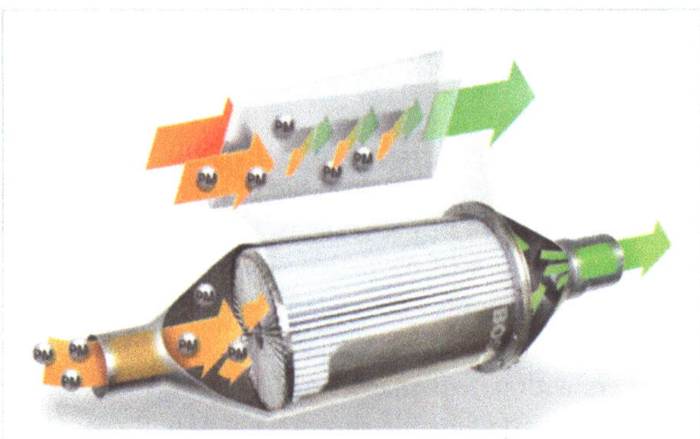

**Figure 9.32:**
Sintered metal jet filter by Bosch

**Open filter structures**

A completely different principle is used in open filter structures. In these filters, some of the exhaust gas flow is directed through a metallic fiber fleece by means of deflecting blades. The particulates present in this partial flow are deposited in the fleece and oxidize continuously with $NO_2$ [12]. **Figure 9.33** shows the design and function of the filter.
Unlike the closed filter structures of monolithic filters, open filter structures have neither the disadvantage of increasing exhaust-gas back pressure nor the danger of filter clogging due to incomplete regeneration. In this design, the pressure loss is independent of saturation and is lower than with closed filter types. However, the particulate deposition only amounts to approximately 30 to 70% [13]. This filter structure is especially suitable for retrofitting purposes.

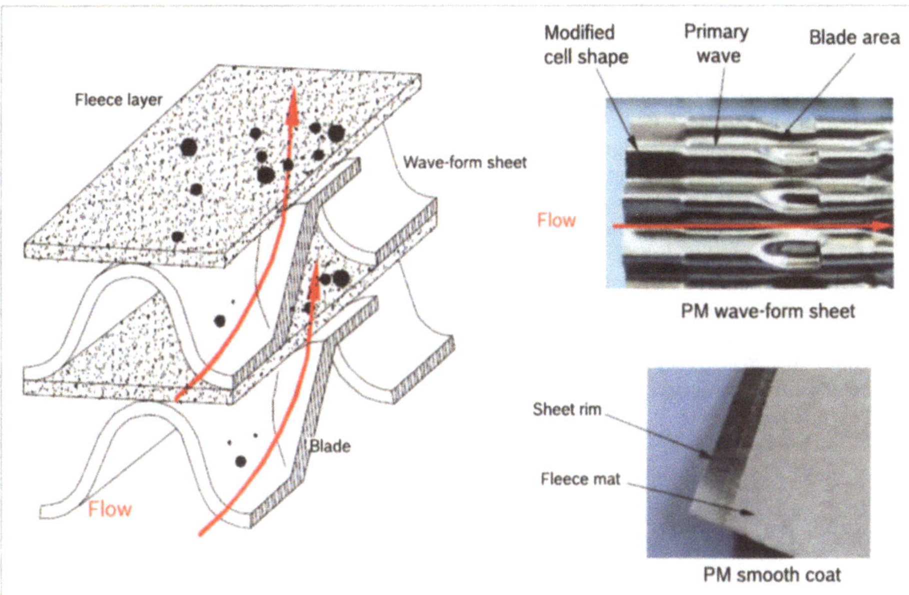


**Figure 9.33:** Design of a PM KAT™ from open filter structures [12]

# Literature


[1] Robert Bosch GmbH: Ottomotor-Management, 3. Auflage. Vieweg Verlag, 2005

[2] *Spicher, U.; Kubach, H.; Haentsche, J. P.*: Die strahlgefuehrte Direkteinspritzung als Zukunftskonzept fuer Ottomotoren. MTZ-Konferenz Motor 2006. Stuttgart, 2006

[3] *Warnatz, J.; Maas, U.; Dibble, R. W.*: Verbrennung – Physikalisch-Chemische Grundlagen, Modellierung und Simulation, Experimente, Schadstoffentstehung, 3. Auflage. Springer-Verlag, 2001

[4] *Li, J.; Matthews, R. D.; Stanglmaier, R. H.; Roberts, C. E.; Anderson, R. W.*: Further experiments on the effects of in-cylinder wall wetting on HC emissions from direct injection gasoline engines. SAE Technical Paper 1999-01-3661, 1999

[5] *Kleinschmidt, W.*: Untersuchung des Arbeitsprozesses und der NO-, $NO_2$- und CO-Bildung in Ottomotoren. Dissertation RWTH Aachen, 1974

[6] *Bick, W.*: Einfluesse geometrischer Grunddaten auf den Arbeitsprozess des Ottomotors bei verschiedenen Hub-Bohrungs-Verhaeltnissen. Dissertation RWTH Aachen, 1990

[7] *Krebs, R.; Spiegel, L.; Stiebels, B.*: Ottomotoren mit Direkteinspritzung von Volkswagen. 8. Aachener Kolloquium Fahrzeug- und Motorentechnik, 1999

[8] *Moser, W.; Kuessell, M.; Mentgen, D.*: Bosch Motronic MED7 – Motorsteuerung fuer Benzin-Direkteinspritzung. 19. Internationales Wiener Motorensymposium, 1998

[9] *Waltner, A.; Lueckert, P.; Schaupp, U.; Rau, E.; Kemmler, R.; Weller, R.*: Die Zukunftstechnologie des Ottomotors: strahlgefuehrte Direkteinspritzung mit Piezo-Injektor. 27. Internationales Wiener Motorensymposium, 2006

[10] *Prevedel, K.; Piock, W. F.*: Aufladung beim Direkteinspritz-Ottomotor. VDI-Tagung, Innovative Fahrzeugantriebe, Dresden, 2004

[11] *Koenigstedt, J.; Mueller, R.; Jablonski, J.; Dirschnabel, T.; Uhrich, J.; Hatz, W.*: Der neue V10-FSI-Motor von Audi. 27. Internationales Wiener Motorensymposium, 2006

[12] *Brueck, R.; Diringer, J.*: Der PM-Filterkatalysator: Ein Diesel-Oxidationskatalysator mit Partikelminderung der naechsten Generation. 2. Emission Control, Dresden, 2004

[13] *Maus, W.; Brueck, R.; Kaiser, F. W.*: Fortschrittliche PKW Dieselabgasnachbehandlung: Potential fuer niedrigste Emissionsgrenzwerte? 23. Internationales Wiener Motorensymposium, Wien, 2002

# 10 Noise Emission

The noise emitted by a combustion engine can be divided into four different individual noise sources:

- Direct combustion noise
- Indirect combustion noise
- Mechanical noise
- Flow noise.

While direct combustion noise is caused by combustion, indirect combustion noise is caused by the response of the engine structure to combustion excitation. Mechanical engine noise can generally be detected by applying a thrust while the throttle valve is closed. The proportion of flow noise can generally also be parameterized and determined in the first approximation, with different throttle valve positions and by taking into consideration the respective valve timings of the engine under motored operation. Different types of procedures [1] have been established in order to predict the engine noise even in the early development stage of an engine or in the case of a new type of combustion system in an existing engine structure: procedures based on combustion excitation and unified structure transfer function and procedures that go further as they predict both individual noise ratios through gas and torsional force percentages, and flow noise with the help of correlation analysis. For some time now, noise analysis methods have been investigated taking the load dependent structure transfer function into consideration [2]. However, tools are now available that can predict the engine noise in the time domain while using load dependent transfer functions, finally allowing the prediction of relevant sound quality parameters, too [3]. Furthermore, research is being carried out on prediction methods to quantify the noise quality, and the use of binaural transfer path analysis for the auralization of the new engine in the virtual vehicle is encouraged Generally, the earlier the changes in engine noise emission can be detected and quantified in the development process, the lower the resulting change costs will be for the acoustic measures that need to be implemented in the engine, the entire powertrain or the vehicle.

## 10.1 Legal requirements

During homologation of a motor vehicle, the drive-by noise and the noise of the exhaust orifice are measured and checked for compliance with the laws of the respective country. The permitted levels have been reduced in the last few years and vehicle production is being constantly monitored. New limits and measuring methods are under discussion; **figure 10.1**.

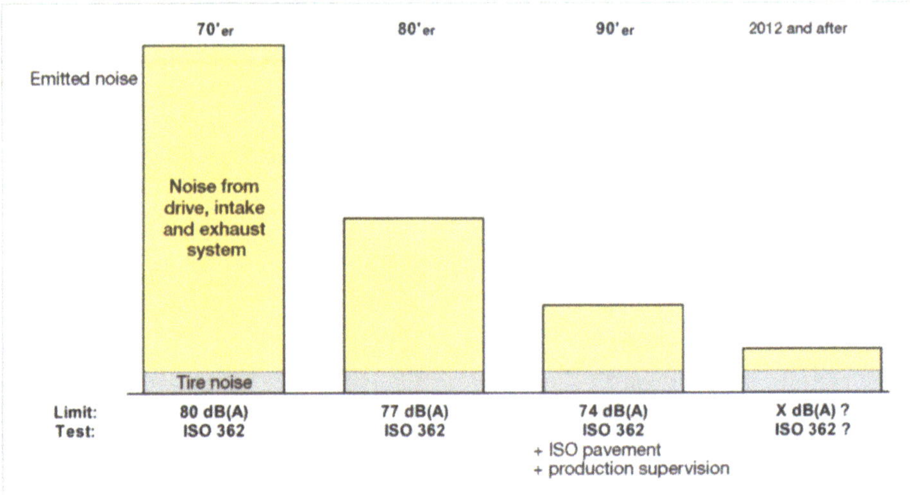


**Figure 10.1:** Periodic development of permitted levels for the drive-by noise of passenger cars

In the early days, the combustion engine was the major contributor to drive-by noise, whereas nowadays tire noise, as well as the exhaust and intake orifice noise, are the major contributors [4].

The power density of diesel and gasoline engines has increased in the course of current downsizing strategies[1] Due to significantly increasing peak cylinder pressures, this can, in the medium term, make it necessary to concentrate on the internal combustion engine as one of the noise sources during the homologation process. In gasoline engines with direct injection and turbocharging (GDI turbo, downsizing), increased noise emission must be expected due to the high specific output, high torque and high cylinder peak pressures; shown in **figure 10.2** as a direct comparison with an engine without turbocharging, having equal power and a correspondingly large displacement in the engine speed range of 1500 to 2700 rpm [5, 6]. Also interesting to note is the comparison between the basic characteristics of the curve of the noise emitted during full load in the middle engine speed band (approximately 3250 rpm onwards) of the downsizing concept shown and the engine with equal power but larger displacement; the downsizing engine, however, tends towards a lower sound pressure level due to lower excitation by the inertia forces at high engine speed (approximately 6000 rpm onwards). As already indicated in section 3.5.5, while introducing a new combustion process or new combustion strategy such as stratified mode or Atkinson cycle, it is taken into account that higher noise levels can be generated as a result of increased peak pressures and pressure gradients.

[1] Downsizing: substituting a highly supercharged engine with smaller displacement for a large displacement engine, thereby reducing the partial loads and thus the total fuel consumption.

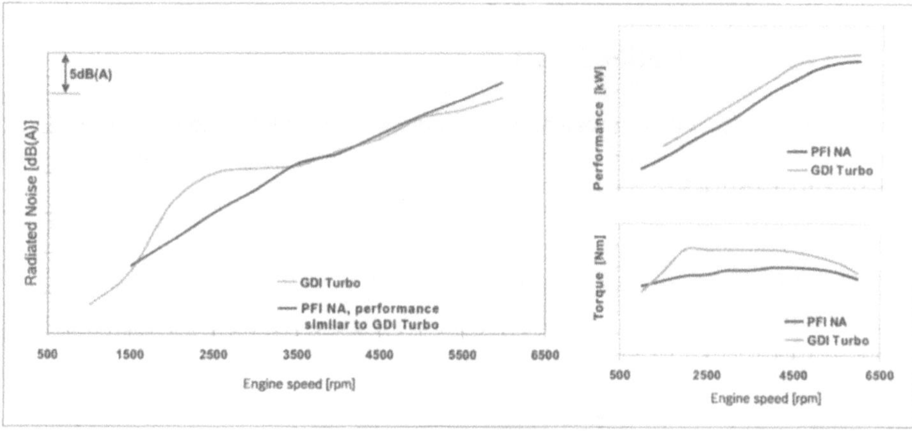


**Figure 10.2**: Noise and power data – GDI turbo (downsizing) as compared to an engine having the same power with intake manifold injection without supercharging (PFI NA)
Left: comparison of noise emitted; Right: comparison of power and torque

Timing and dethrottled part load operation adjusted for the new combustion process, as can be the case for controlled auto-ignition (CAI) for instance, can significantly increase the noise of the intake orifice or the general flow noise. To comply with the homologation criteria, which are expected to become increasingly strict, it is advisable to take the relevant aspects into account well ahead of time while developing the targets for upcoming engines.

## 10.2 Combustion noise

The direct combustion noise of an engine is generated due to the increase in pressure in the combustion chamber during combustion. The peak pressure in the combustion chamber can be considered a major factor affecting the noise level. Depending on the engine speed, peak pressure affects the excitation spectrum in the low frequency range (< 250 Hz). The rate of increase in the pressure decisively affects the excitation of the engine structure in the 0.5 to 5 kHz frequency range. Together with the respective structure transfer function, it also has a decisive effect on the noise level. This is generally specified in the requirements specifications, in order to meet the durability targets for the crankshaft drive components and sealing and the acoustic target values. However, the frequency ranges and levels of combustion noise depend on the engine structure and its attenuation. The time structure of the combustion excitation also plays a very important role with regard to noise quality (roughness, knocking, etc.).

### 10.2.1 Comparison between gasoline direct injection (homogeneous GDI) and stratified GDI

A comparison of the cylinder pressure for the homogeneous and stratified modes of a spray-guided combustion system [6] and the result of a short-term frequency analysis (wavelet) is shown in **figure 10.3**.

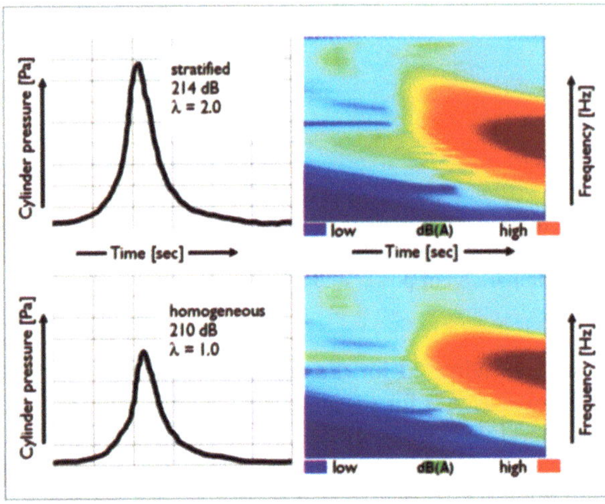


**Figure 10.3:**
Combustion excitation for $n = 2000$ rpm, $p_{me} = 2$ bar
Stratified combustion with supercharging compared with homogeneous combustion (right half: wavelet analysis of the combustion excitation)

The analysis shows that increasing cylinder peak pressure and pressure gradient may occur during stratified charged mode operation. This results in an increase in the excitation level of combustion, especially in the 0.3 to 5 kHz frequency range. This can be seen from the short-term frequency analysis of the cylinder pressure in the right half of **figure 10.3**. However, a projection of this behavior on other engine speed points and load points in the engine map range with stratified mode cannot be linearly linked to the acoustic behavior. Considering the influence of stratified combustion on the noise emitted, no significant change is observed in the higher engine speed and load ranges as shown in **figure 10.4**. At

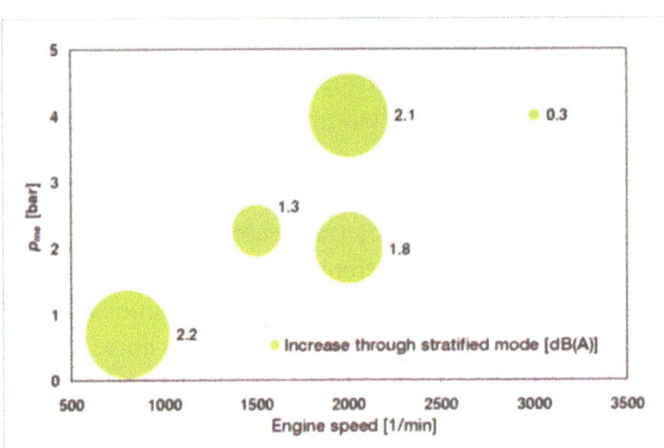


**Figure 10.4:**
Increase in the noise level emitted during stratified mode for selected part load operation points

low engine speeds and loads, the change in engine acoustics is more distinct because the air efficiency increases, which in turn results in a higher peak pressure and pressure gradient.

A further consequence of the stratified mode is the increase in torsional vibrations associated with it, as shown in **figure 10.5** [7]. It may require countermeasures for the affected engine such as special dampers for the auxiliary drive, or acoustic measures such as the application of a dual-mass flywheel on the powertrain side [5, 7].

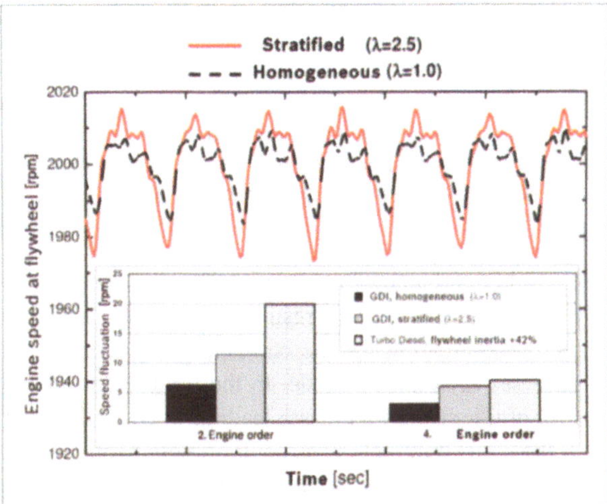


**Figure 10.5:**
Effect of stratified combustion on rotational vibration excitation
($n = 2000$ rpm, $p_{me} = 2.6$ bar)

## 10.2.2 Effect of the swirl and tumble level on combustion excitation

In order to achieve efficient combustion, it can be advantageous to generate particular swirl or tumble ratios in the combustion chamber during the intake phase. This is usually managed using controllable throttle valves in the intake channels in front of the valve. The resulting combustion, which occurs in a somewhat different manner, leads not only to an increase in the excitation level but can also result in a change of the noise signature – the engine's radiated noise. **Figure 10.6** shows a comparison of the combustion excitation of a four-valve engine during spray-guided operation alternating between homogeneous and stratified mode operation, with open intake port valve (tumble[2] dominated) and a valve closed at one port (swirl[3] dominated).

[2] Tumble: the dominating flow movement rotates in the cylindrical pipe at 90° inclined to the cylindrical axis, i.e. generally concentric or parallel to the crankshaft axis.

[3] Swirl: the flow conditions in the cylinder are dominated by a rotational motion concentric to the cylindrical pipe.

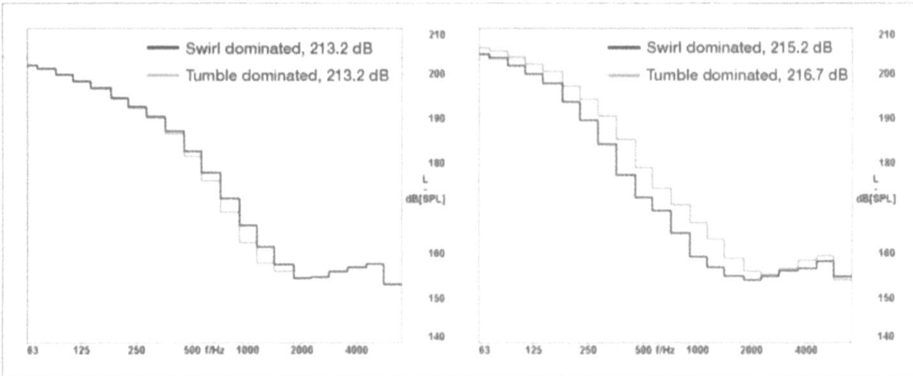


**Figure 10.6:** Effect of swirl on combustion excitation (grey: tumble dominated, black: swirl dominated), third octave spectral analysis of cylinder pressure characteristics: Left: combustion excitation during homogeneous mode; Right: combustion excitation during stratified mode

Because of the slightly lower pressure gradients that arise at open intake ports (tumble) in homogeneous mode, correspondingly lower excitation levels result that are in the frequency range 0.5 to 1.5 kHz for the subjective evaluation for the corresponding combustion (grey line). The excitation behavior reverses in stratified mode. In the tumble dominated case, the peak pressure and the pressure gradient both increase, leading to overall higher combustion excitation as compared to the swirl dominated case and homogeneous mode; **figure 10.6**. Conclusion: with regard to the overall noise level, the characteristics of the intake flow in the homogeneous mode are less important with respect to the sound pressure level than they are in the stratified mode. However, further analysis of sound quality aspects corresponding to the cycle-to-cycle combustion characteristics, such as the roughness of the noise, can be beneficial for both modes of operation.

## 10.2.3 Effect of the exhaust-gas recirculation rate (EGR) on the combustion noise

Exhaust-gas recirculation is used in GDI operation to slow down the combustion in order to meet the corresponding emissions and fuel consumption requirements [8]. As a higher exhaust-gas recirculation rate slows down the combustion, i.e. heat release, the peak pressure and the pressure gradient are also reduced (process dependent). This results in a lower excitation level and, and, as shown in the example, in a reduced noise level in the frequency band of 250 to 700 Hz; **figure 10.7**.

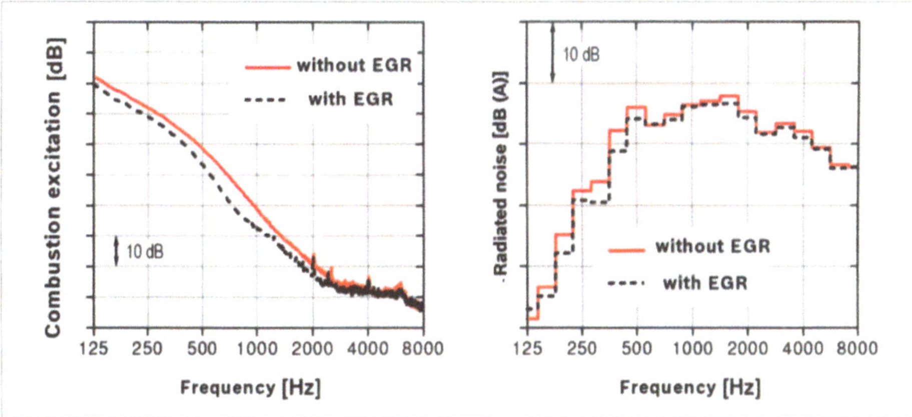


**Figure 10.7:** Effect of exhaust-gas recirculation on combustion excitation [9]

### 10.2.4 Effect of changes in operating mode

Due to the thermodynamic boundary conditions, operating a GDI engine in stratified charge mode is generally possible only in the lower to middle part load range. Thus it is absolutely essential to switch the engine repeatedly between stratified and homogeneous modes during the transient engine operation. This results in the following tasks:

- The combustion noise profile must be similar to those of the previously operated mode
- The peak pressure profile must be adjusted in such a way that no disturbing acceleration excitations are triggered in the vehicle structure
- The engine torque that is delivered should be generated smoothly
- No change should be detected in the noise of the intake and exhaust orifices.

By suitably controlling the mixture composition, the valve timing, the injection, and the throttle valve position, it is possible to switch between operating modes so that is not noticeable in the vehicle from the acoustic point of view. **Figure 10.8** shows an applied strategy. The internal noise level in stratified mode as compared to that in homogeneous mode is just 0.8 dB (A). The frequency-time signature of the internal noise during the switchover from "stratified" to "homogeneous" shows an increase which is subjectively imperceptible, except in the 500 to 600 Hz frequency range in the stratified mode. This is due to increased combustion excitation in this frequency range.

During the acoustical tuning of such an engine, this frequency range and the time structure of the sound signal, which can clearly influence the noise quality, also have to be taken into consideration.

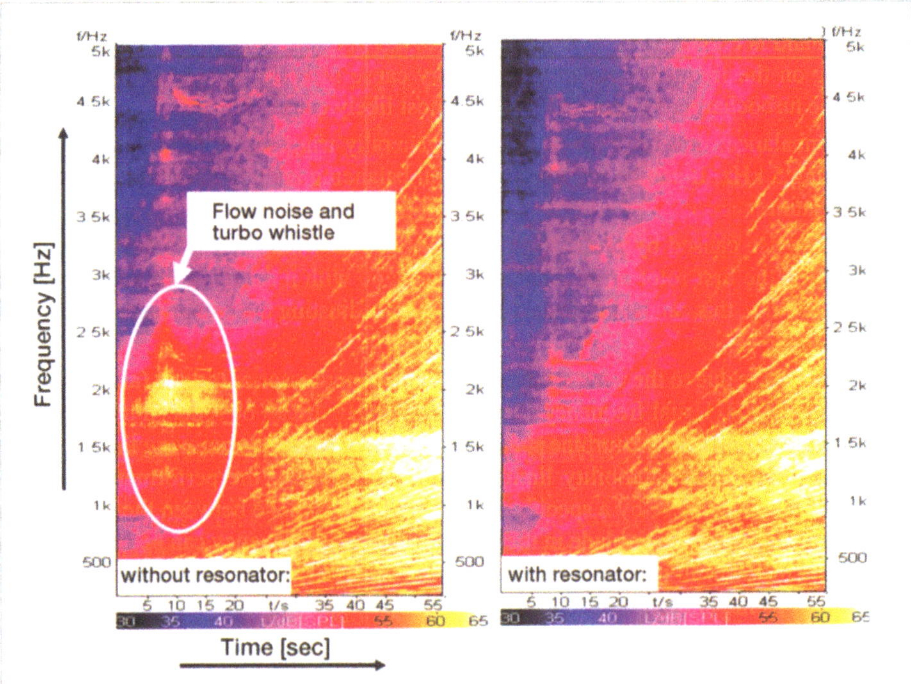


**Figure 10.9:** Reduction of flow and whistling noises in turbo engines by using resonators with band stop characteristics in the air intake duct

direct injection. Resonators that work based on the Helmholtz principle have been used in this case for the last few years. Usually installed in the intake system in front of the throttle valve and the airflow meter, these amplify selected frequency ranges using specifically tuned membranes and systematically direct them into the passenger compartment.

## 10.2.7 Effect of fuel quality on acoustics

Modern downsizing concepts with direct injection and active knock control allow the engine to operate with varying fuel qualities, without any change in the delivered performance. This is indispensable due to the country-specific variation in the quality of fuel even within the European Union. Particularly the engine concepts with high power density, as represented by the downsizing concepts, show, with respect to the standard fuel quality 95 RON and 98 RON, a different behavior regarding the dependence of the emitted noise on the load. Here, the influence in the lower engine speed range becomes more significant with increasing load as shown in **figure 10.10**. This can be explained by means of the higher peak pressures and pressure rise rates in case of operation with 98 RON fuel. To attain the same performance at full load with 95 RON as with 98 RON, the charge pressure should be correspondingly increased in order to reach the required maximum torque. However, this can bear the risk of an increased flow noise on the intake side [6].

If the generation of noise level at a constant engine speed above the load is taken into account, the level of noise emitted for the two qualities of fuel does not differ until ~ 80% of the maximum torque is reached. Only with the further increase in load does the noise level for 98 RON fuel increase, whereas in the case of 95 RON the ignition timing should be shifted to "retarded" because the knock limit is reached earlier. Retarded ignition timing in turn effects both lower peak pressures and lower pressure rise rates and results in a converging behavior of the emitted noise level as is illustrated in **figure 10.10**. An analysis of the roughness of the emitted noise can also be useful. With increasing load, the roughness of the radiated noise for the 98 RON operation increases; this can be directly attributed to higher pressure rates; **figure 10.11** and cylinder to cylinder combustion characteristics.

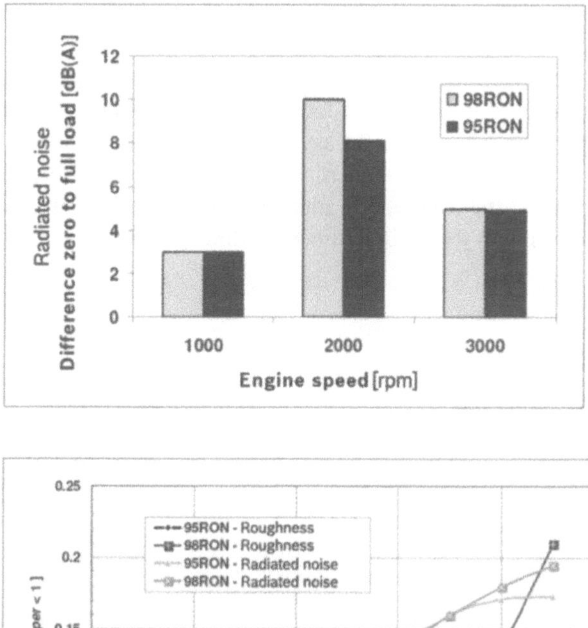


**Figure 10.10:**
Load dependency of the emitted noise for different fuel qualities

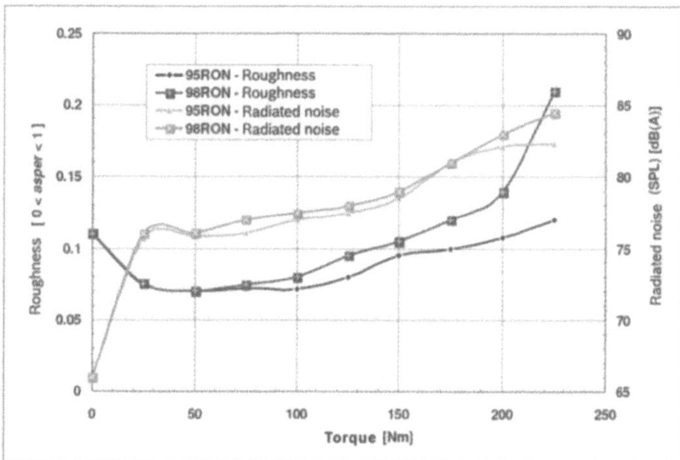


**Figure 10.11:**
Effect of the fuel quality on the load dependent noise emitted at $n = 2000$ rpm [6]

For both qualities of fuel, the slight increase in the roughness close to idle load is due to the increased mechanical noise, justified by the movements of the engine components which are not guided in a sufficiently defined manner by the lower gas and mass forces occurring at this particular load and speed.

## 10.3 Mechanical noise

In gasoline direct injection, the piston and sometimes the valvegear are modified. This may affect the mechanical noise of the engine. While calibrating CAI[4] with the help of negative valve overlapping, the order structure of the valvegear noise can also be changed too.

The piston can be a possible source of unwanted noise, especially in case of wall-guided and air-guided processes. This is due to the relatively complicated and mostly asymmetrically designed piston heads, which are necessary for such combustion control, especially during air-guided operation [10]. The resulting higher piston mass leads to an increase in the inertial forces and thus to an increase in the engine mount vibration; as depicted in **figure 10.12**. Stratified charge operation can require an asymmetric piston head design, which results in higher deviational components in the inertial tensor of the piston. If there is secondary movements of the pistons, this can lead to altered tipping behavior, which in turn results in higher contact forces with the cylinder liner and finally results in an increased level of piston noise.

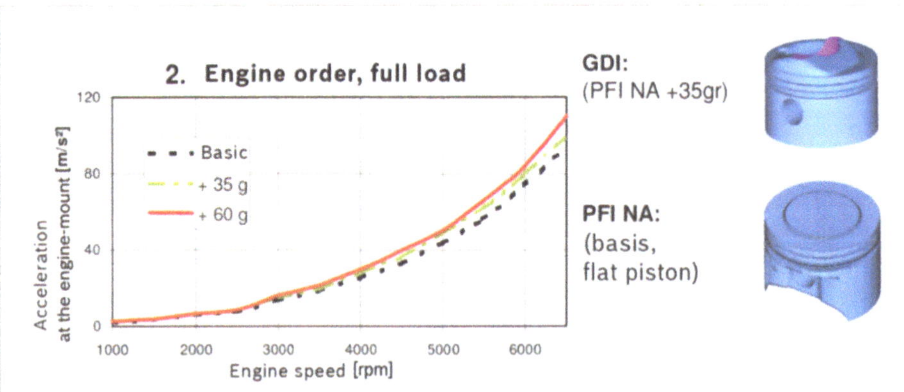


**Figure 10.12:** Effects of piston weight on vertical engine bearing acceleration and the internal noise level of the vehicle of the second engine array [9]

4 CAI: Controlled Auto-Ignition in the gasoline engine; see also section 3.5.5

If the combustion process with direct injection leads to high pressure rise rates and high peak pressures, then an investigation of the tertiary movements of pistons and the oscillatory behavior of the connecting rods can also be useful. In CAI mode with negative valve overlap, because of precompression the piston "free flight phase" can be changed into a more definite piston movement inside the cylinder during scavenging TDC.

### 10.3.1 Optimization of the knock control system

Engines with gasoline direct injection are generally equipped with an electronic knock and charge pressure control system. Engine concepts with fewer cylinders or an odd number of cylinders, such as the in-line engine with three cylinders, can generate particularly interesting tasks related to this. An analysis, which was required to optimize the knock control system in a three-cylinder turbo engine with gasoline direct injection, is described Below. Initially, only one knock sensor was installed between the second and the third cylinder. In one of the test engines examined, a disturbance in the knock control signal resulted due to the ignition angle or injection timing, which is 240° in the case of three cylinders instead of 180° in the case of a four-cylinder in-line engine . Moreover, the position of the knock sensor was not centralized, for feasibility reasons. Measurements at the components integrated into the system showed that the closing impulses of an injector of a cylinder, which at that time was not involved in the combustion process, occurred precisely in the knock window of that cylinder in which the combustion was taking place, i.e. at the time of potentially knocking combustion, the knock sensor acquires interference due to the closing impulse, which subsequently makes it difficult to detect knocking. The signals measured during a complete engine cycle are shown in **figure 10.13**.

The results of a modal and transfer path analysis of the engine structure showed that among other reasons, the different lengths of the structure-borne noise paths to the knock sensor contributed to different intensities of the structure-borne noise signals (impulses) in the knock window. Based on transfer path analysis, improvements were made in the structural transfer at the block and an additional knock sensor was installed between the first and second cylinders, in order to ensure equal transmission paths from every combustion chamber to the sensor. To suppress the unwanted noise even further, washers were installed between the injector and cylinder head. In this way, the transfer of structure-borne noise in the engine structure due to the closing impulse of the injector can be effectively reduced in order to achieve robust system behavior under all conditions. It may be worthwhile to mention here that, in combination with stratified mode or CAI, the combustion-induced eigenmodes of the combustion chamber also drop to lower frequencies, due to the lower combustion temperatures. [11].

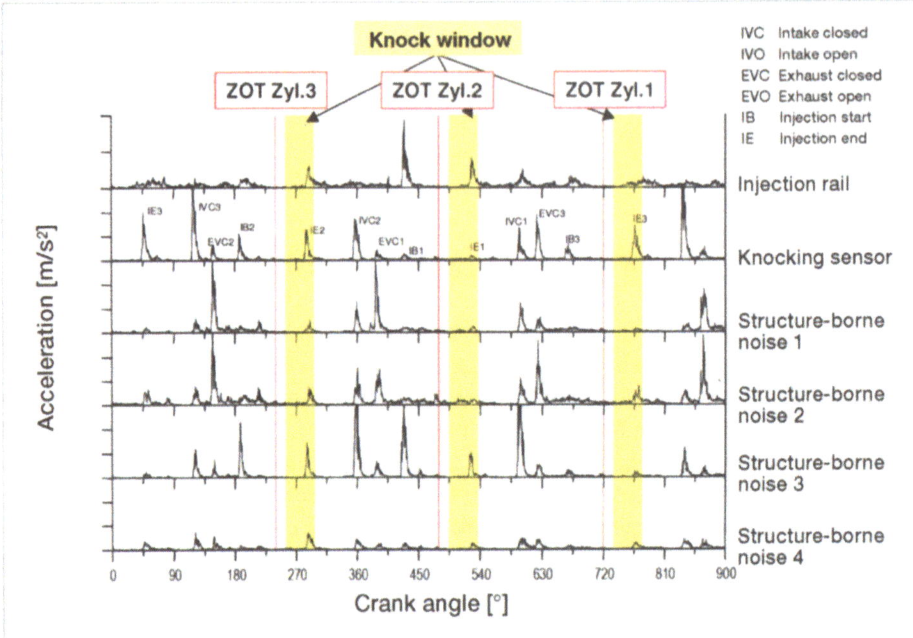


**Figure 10.13:** Acceleration measurements at the knock control system components, the engine structure and the injection system, at 3000 rpm, full load

## 10.3.2 Effect of high pressure fuel distribution system

Modern concepts with gasoline direct injection can be effectively implemented only by using higher injection pressure. Single or multiple-cylinder piston pumps can provide engine-side fuel pressures of up to 200 bar in a common rail arrangement of fuel rails and injectors. This can lead to "ticker noise"; caused either due to pressure fluctuations in the fuel system (which in turn are caused due to the closing impulse of the injectors), or due to pulsations generated by the high pressure pumps. The time frequency structure of such a noise phenomenon, analyzed from a radiated noise signal during low idle operation, is shown as an example in **figure 10.14**.

In this example the "ticker noises" are mainly caused by the pressure pulsations created due to the opening and closing of the injector nozzles, which coincide with the second flexural resonant frequency of the fuel rail. Ticker noise thus gives rise to a classic machine-acoustics task with the aim of separating the excitation from the operating modes. As an example, **figure 10.15** shows an optimization process for such an injection rail. In this case, as a consequence of the calculation results, among others, the fuel rail is reinforced by means of optimal positioning of the bolting points connecting it to the intake system at the cylinder head,. With that the second flexural resonant frequency increases from 2114

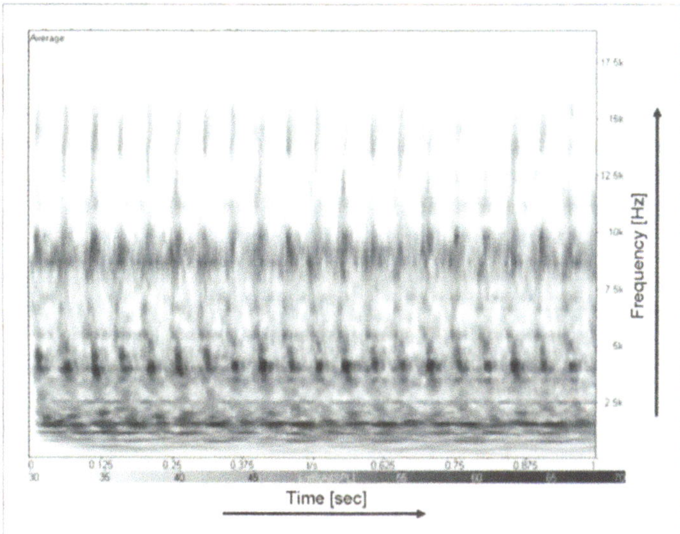


**Figure 10.14:** Time-frequency structure of ticker noise (radiated noise) in engine low idle

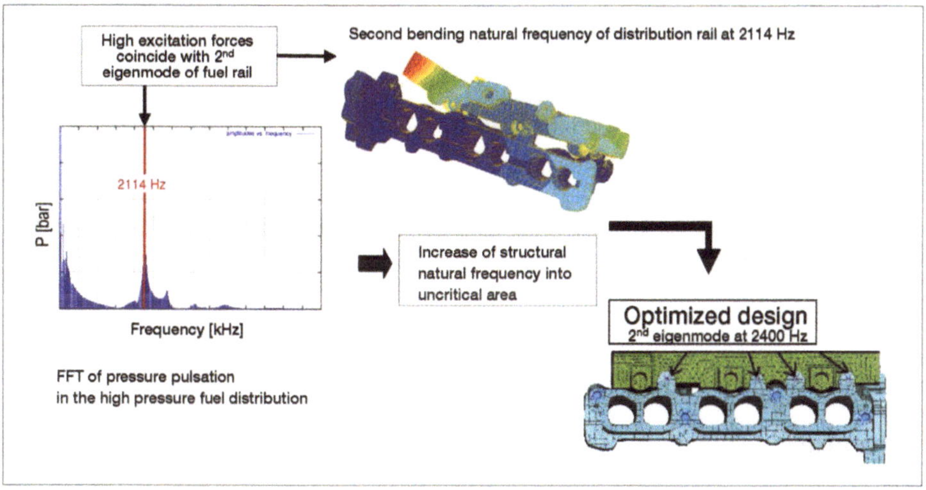


**Figure 10.15:** CAE based structure optimization for a high pressure fuel rail

Hz to 2400 Hz, as seen in **figure 10.12**. By implementing these measures, the system can now work non-critically without creating noise disturbance.

Other airborne portions of ticker noise can be caused by standing pressure waves within the fuel rail. This fraction of the noise can be reduced by means of a pressure pulsation damper in the form of an orifice plate installed between the fuel rail and the injector nozzle as depicted **figure 10.16**.

# 11 Effects of Gasoline Direct Injection on Other Engine Systems

## 11.1 Ignition systems

Despite consistent efforts to develop alternative ignition systems like plasma, laser and other ignitions for series production, traditional coil ignition (CI) has been widely accepted due to its cost-benefit ratio. In exceptional cases, capacitor discharge ignition (CDI) is used with integrated ionic current measurement [1]. Coil ignition was significantly improved by replacing the mechanical contact breaker with a semiconductor and removing the high tension distributor. Later, the size was further reduced. An advance ignition angle, which depends on the operating point, can be realized by improving the electronic control without mechanical adjustment devices like a centrifugal advance mechanism or vacuum unit. The requirements with respect to ignition characteristics are not so high for combustion processes with intake manifold injection and stoichiometric mixture composition, since stoichiometric mixtures are highly ignitable. Chemically inactive mixtures, as available in high exhaust-gas recirculation rates and lean air/fuel ratios, show fluctuations in retarded ignition timings or even misfire when using conventional ignition systems.

The optimization of the ignition system – required for the modern combustion processes of gasoline engines with direct injection – with respect to the inflammation characteristics and adjustment to the combustion chamber requirements, has been examined and carried out only to a limited extent. So far, the spark plug has been a wearing part that requires complex adjustments and creates maintenance costs [2].

### 11.1.1 Spark ignition systems

Below is an introduction to the different spark ignition systems which describes the possible high voltage sources, the various spark plug types and the different ignition systems and operating strategies.

**Coil ignition (CI)**

The inductive ignition system in the form of a coil ignition system, as shown in **figure 11.1**, has been widely accepted for use in the automobile industry. Each power stroke of the combustion engine generates the high voltage required for the sparkover [3]. The energy required for this is taken through the ignition switch (1) from the battery (2) and stored temporarily in the ignition coil (3). Using the appropriate ignition output stages (4), the engine control unit logically assigns the ignition pulses to the ignition coils, producing an ignition spark at the spark plugs (5). In the case of the single-spark ignition coils, every cylinder is assigned an ignition output stage and an ignition coil. The engine control unit actuates the ignition output stages according to the firing order of the engine cylinders.

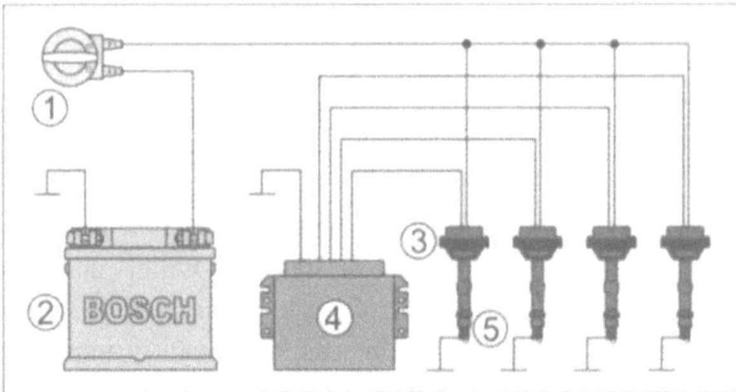


**Figure 11.1:**
Schematic design of an inductive coil ignition system for a multi-cylinder engine with single-spark ignition coils [3]

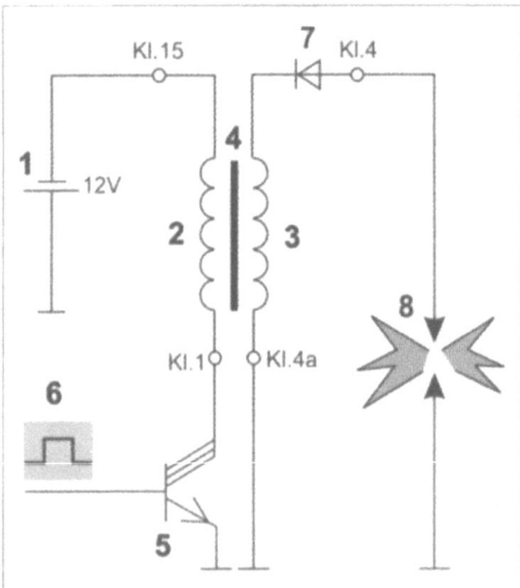


**Figure 11.2:**
Basic design of high voltage generation with an ignition coil [3]

The high voltage required for igniting the air/fuel mixture is generated from the ignition energy saved prior to the sparkover. This is the function of the ignition coil, the basic design of which is shown in **figure 11.2**. State-of-the art ignition coils consist of two magnetically coupled copper windings – the primary winding (2) and the secondary winding (3) – as well as an iron core made of individual metal sheets in a cast plastic casing, whereby the two windings fit on the entire iron core. The primary winding consists of thick wire with fewer windings. One end of the winding is connected to the positive terminal of the battery (1) by using the ignition switch. The other end is connected to the ignition output stage (5), which can switch the primary current using that end. The secondary winding consists of thin wire with several windings. The transformation ratio, which is the numerical ratio of the windings to each other, is usually between 1:50 and 1:150.

The primary function of the ignition coil is to save the energy required for generating the high voltage. The high voltage generation system, which should deliver up to at least 30,000 volts in individual cases such as turbocharger engines, follows the principle of magnetic induction. In the ignition coil of a gasoline engine, the high voltage is generated as follows: The engine control unit (6), as shown in **figure 11.2**, enables the ignition output stage while calculating the closing time (time taken to energize the primary coil). The primary current in the ignition coil rises during this time, thereby setting up a magnetic field in the coil. If the electric circuit remains closed, then the primary current increases until it is restricted by the resistors in the primary circuit or by an active limitation via the output stage. Depending on the ferromagnetic material used, magnetic saturation occurs in the ferromagnetic circuit from a specific current intensity onwards. Thus, the inductivity decreases and the losses within the ignition coil increase heavily. Therefore, the operating point of the coil is set below the magnetic saturation. The magnitude of the primary current and that of the primary inductance of the ignition coil determine the amount of energy that is stored in the magnetic field. The ignition output stage cuts off the flow of current at the time of ignition and secondary voltage is induced in the secondary winding of the ignition coil by changing the magnetic field. The maximum possible secondary voltage depends on the following factors [3]:

- Turns ratio of primary and secondary winding in connection with the primary voltage limitation of the ignition output stage
- Winding capacitor and secondary load capacitors (e.g. spark plug, spark plug slot, ignition cable)
- Stored energy
- Shunt (e.g. spark plug wetted with fuel or covered with soot).

The available secondary voltage (available ignition voltage) should always be more than the voltage required for the spark flashover at the spark plug (required ignition voltage) (8); **figure 11.2**. It must be ensured that in all conditions, the ignition system provides the required ignition voltage. The required ignition voltage depends on [3] the:

- Density of air/fuel mixture in the combustion chamber and thus also on the timing of the ignition
- Composition of the air/fuel mixture
- Electrode geometry of the spark plug
- Electrode material
- Electrode gap.

After the primary current is cut off at the time of ignition, the voltage in the secondary winding of the ignition coil increases within a short time (maximum approximately 30μs [3]) up to the ignition voltage. If the required voltage is exceeded, the spark gap of the spark plug becomes conductive between the center electrode and the ground electrode and an electrical gas discharge occurs [4].

The current and voltage curve of an ignition process, shown in **figure 11.3**, has already been explained in detail in section 3.4. The discharge is subdivided into three phases: the breakdown phase (I), the arc phase (II) and the glow phase (III) [5, 6, 7, 8]. The capacitors that are charged to the ignition voltage in the secondary circuit of the ignition coil discharge suddenly in a spark head (K in **figure 11.3**, also known as flashover). This electric breakdown takes place in the nanosecond range and is marked by voltage of several kV [4]. The intensity of the static breakdown voltage depends, among others, on the following physical variables [9, 10]:

- The electrode gap
- The gas composition
- The combustion chamber pressure
- The roughness of the electrodes.

After the electrical breakdown, an electrically conductive plasma channel is formed between the electrodes of the spark plug, in which only the energy stored in the spark plug capacitor is initially converted. During the short period of the breakdown phase (approximately 5 ns), currents of up to 100 A flow into this plasma channel. The temperature reaches approximately 6000 K, thereby building pressures of up to 300 bar [4]. Subsequently, during a spark duration of typically 1 to 2 ms (tF in **figure 11.3**), the stored energy of the ignition coil is converted to a discharge equaling the spark voltage (B in **figure 11.3**, also known as spark tail S). The combustion duration of the spark can be subdivided in two phases: arc discharge and glow discharge. During the arc discharge phase (II), the capacitively-stored energy in the input lead for the spark plug and the ignition system is converted. Connected to the arc phase is the glow discharge phase, during which the electrical energy stored in the ignition system is discharged into the gas [3, 7, 11]. The magnitude of the spark voltage is basically affected by the length of the plasma channel [4]. If, for example, the spark is deflected due to an airflow, then an increase in the spark voltage is noticed. In the post-oscillation phase that follows, the remaining energy of the ignition coil is released. The energy stored in the ignition coil (ignition energy) is divided, as shown in **figure 11.4**, in the phases of the voltage curve illustrated in **figure 11.3**:

- **Spark head:** the energy stored in the secondary capacitance of the ignition circuit (ignition cable, etc.), increases in quadratic proportions along with the available voltage.
- **Spark tail:** the remaining energy stored in the ignition coil is subsequently released. It results from the difference between the total energy stored in the ignition coil and the energy released through the capacitor discharge; **figure 11.4**. Thus, with the increasing ignition voltage requirement, the energy component in the spark head also increases. If high ignition voltage is required, the energy available in the spark tail is, under certain circumstances, inadequate to supply the required spark duration for inflaming the mixture or reigniting the extinguished flame by a follow-up spark; as is usual for inflaming lean mixtures. If the ignition voltage requirement increases further (dotted arrow in **figure 11.4**), the misfire limit is achieved very quickly because the available energy is not sufficient to generate a sparkover and instead dies out in a dampened vibration. The ignition coil, therefore, must be carefully tuned to the other ignition components, as well as to the requirements of the respective engine concept.

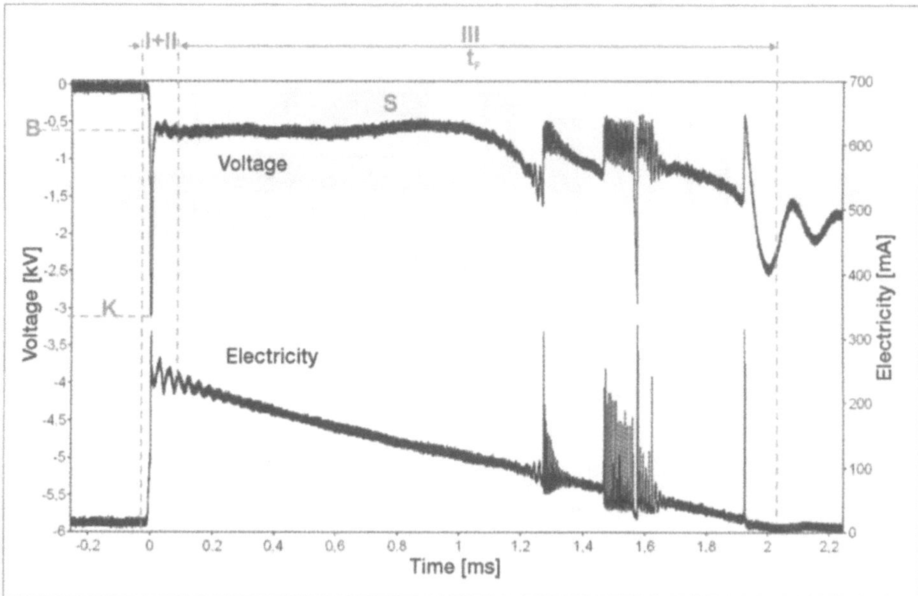


**Figure 11.3:** Time curve of voltage and current measured at the electrodes of a spark plug [5]

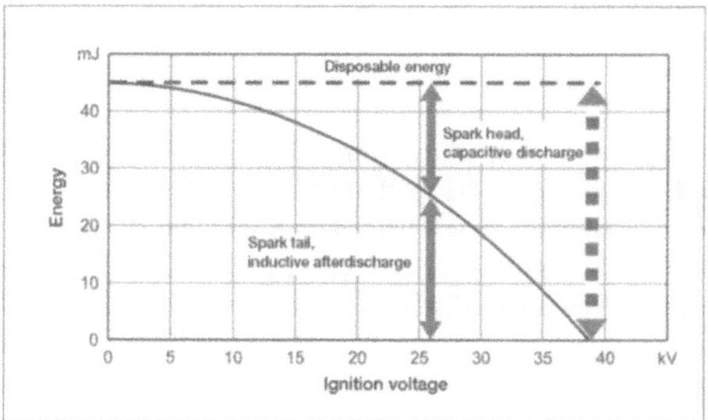


**Figure 11.4:**
Energy balance of an ignition system [3]

The function of the spark plug is to create ignition energy in the combustion chamber and start the combustion of the air/fuel mixture by generating an electrical spark between the electrodes. The spark plug has to be installed in such a way that the cylinder head is always well-insulated against the high voltage to be transferred and the combustion chamber is sealed to the outside. **Figure 11.5** shows the basic design of a spark plug.

A spark plug is a ceramic insulated, gas-proof high voltage bushing in the combustion chamber, with a central electrode (1), **figure 11.5**, and one or multiple ground electrodes (2); **figure 11.5**. The electrode gap is marked EA. It is the smallest distance between the central electrode and the ground electrode and determines, among others, the length of the

**Figure 11.5:** Basic design of a spark plug as per [3]

ignition spark. The smaller the electrode gap, the lower the voltage required to form an ignition spark. If the electrode gap is small, only a small flame kernel forms in the electrode area. Using the contact surfaces with the electrodes, it is again de-energized. Due to this quenching effect, the flame kernel spreads very slowly. As the quenching losses minimize with the increasing electrode gap, the inflammation conditions improve; the required ignition voltage, however, increases sharply. Thus, the ignition voltage reserve reduces for a specified ignition voltage of the ignition coil [3]. Study of the effect of lean operating limit [12] showed that the electrode gap can positively affect the inflammation up to a certain value.

The position of the spark gap, i.e. the point of the sparkover through the electrode gap relative to the combustion chamber wall, is defined by the spark plug discharge gap, marked f in the **figure 11.5**. In modern engines, especially in gasoline engines with direct injection, the effect of the spark discharge gap on the combustion is obvious [3, 12, 13, 14]. Unlike gasoline engines with external mixture formation, distinct charge movements are set to support the mixture formation in direct injection. Thus the flow at the spark plug changes in magnitude and direction at different operating points. A spark discharge gap that projects deep into the combustion chamber can thus be of advantage for the inflammation.

While operating the spark plugs with electrical ignition systems, voltages of over 30,000 V may result. This must not lead to disruptive discharges along the insulator under any circumstances. Residues such as soot, carbon deposits and ashes deposited from the fuel combustion process, and also from oil additives in small quantities, are electrically conductive under certain thermal conditions. However, in this case too, sparkover must not occur through the insulator. The types of loads that a spark plug is subjected to during a working cycle are shown in **figure 11.6**.

The spark type and thus the spark plug type determine the position of a ground electrode(s). This is shown in **figure 11.7**. If the ground electrode is placed opposite the central electrode, then this spark plug type is known as the air-gap spark plug (a). If ground electrodes are laterally engaged, the spark plug is purely of the surface-gap type (b) or a combination

| Clock phase | Compression | Combustion and operation | Exhaust | Intake |
|---|---|---|---|---|
| Gas temperature | 300 to 600 °C | 2000 to 3000 °C | 1300 to 1600 °C | Up to 120 °C |
| Gas pressure | 8 to 30 bar | 30 to 90 bar | 1 to 5 bar | 0.9 bar |

Piston position

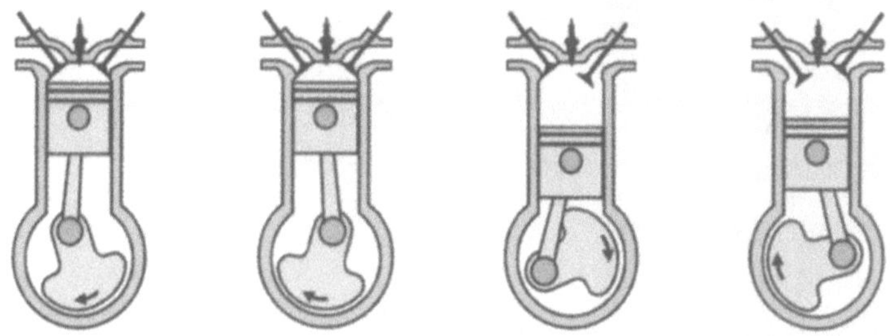

| Crankshaft angle | 0° TDC | 180° BDC | 270° TDC | 540° BDC | 720° TDC |
|---|---|---|---|---|---|

**Figure 11.6:** Loads on a spark plug during a working cycle [3]

of air- and surface-gap spark plugs (c). The above-mentioned spark plug concepts have the following characteristics:

- **Air-gap concept:** In air-gap spark concepts, the ground electrode is positioned in relation to the central electrode in such a way that the ignition spark jumps directly between the electrodes and the air/fuel mixture between the electrodes ignites.
- **Surface-gap concept:** With a defined position of the ground electrode to the ceramic, the spark first jumps from the central electrode over to the surface of the insulator tip and then jumps over to the ground electrode through a gas gap. Since a low ignition voltage is required for discharge over the surface as compared to discharge through an equally large air gap, the surface-gap spark can bridge larger electrode gaps than the air spark for an equal ignition voltage requirement. Thus, a bigger flame kernel develops from the subsequent inflammation, and the inflammation characteristics improve distinctly [14]. At the same time, in the cold restart, overrun and long idle run phases, the surface-gap spark has a cleansing effect or hampers the deposit of combustion residues on the tip of the insulator [3]. In [14] it has been shown that particularly in case of advance ignition angles in stratified charge operation with this spark plug concept, combustion fluctuations and number of misfires can be reduced. However, the surface spark gap should be designed in such a way that it reaches a high temperature for self-cleaning while also preventing overheating of the insulator surface.
- **Semi-surface gap concept:** In these spark plug concepts, the ground electrodes are installed at a fixed distance from the central electrode and the ceramic end face. This results in two alternative spark gaps, which facilitate both previously mentioned discharge patterns and require different ignition voltage values. Depending on the operating conditions and spark plug condition, ignition occurs as an air spark or semi-surface gap spark

[3, 15]. Normally, ignition is effected by air sparks. If the insulator is dirty, the spark still flashes across the dirty surface because it is less resistant than the surface-gap spark, thereby cleaning the insulator.

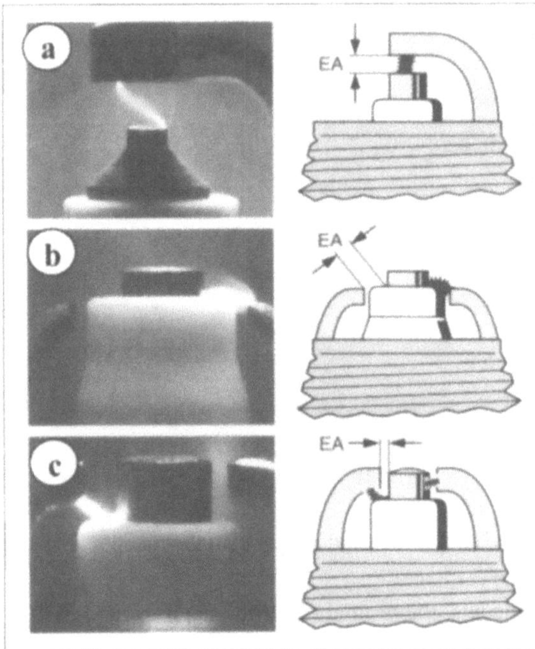


**Figure 11.7:**
Spark behavior of the different spark plug types according to [3]

Additionally, the electrode design affects the formation of the ignition spark and the inflammation [3, 14, 16] to a large extent. During the flame kernel formation and flame expansion, the ignitability can be improved by minimizing the thermal loss using the electrodes [14, 17, 18]. This can be achieved by designing thin electrode points. Furthermore, the required ignition voltage decreases due to thin electrode points, since a heavy electrical field is generated. The high electrode temperatures that result due to the exposed electrode geometry are a disadvantage in this case, as they lead to a very heavy wear and tear. The problem of the endurance limit can be countered by selecting a suitable material, however [3, 19].

**Non-purged prechamber spark plug**

By using prechamber spark plugs, both the inflammation and the main combustion can be improved; mainly in lean mixtures; **figure 11.8** [20]. During the compression stroke, the pressure gradient between the main combustion chamber and the prechamber causes a flow through the overflow bores, leading to a good air/fuel mixture because of better mixing and thereby to improved mixture homogeneity. After the inflammation of the mixture at the prechamber spark plug, torch flames emerge from it into the main combustion chamber. The torch flame impulse is determined by the pressure difference between the chamber and the main combustion chamber [21] and thus depends on the rate of the initial combustion and the throttling effect of the overflow bores. The prechamber ignition generates several flame kernels at different points in the combustion chamber by using the torch flames, which facilitate faster main combustion.

**Figure 11.9** shows fuel mass conversion that is achieved significantly faster with the prechamber spark plug as compared to the spark plug, which is characterized mainly by a distinctly lower advance ignition requirement. According to [21], in the case of the prechamber spark plug, the lean operating limit of the spark plug is higher because of the high residual gas content at the point of ignition.

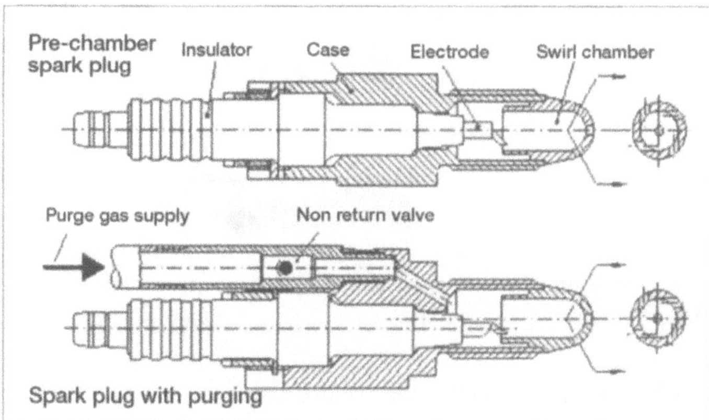


**Figure 11.8:** Prechamber spark plugs with and without purging [21]

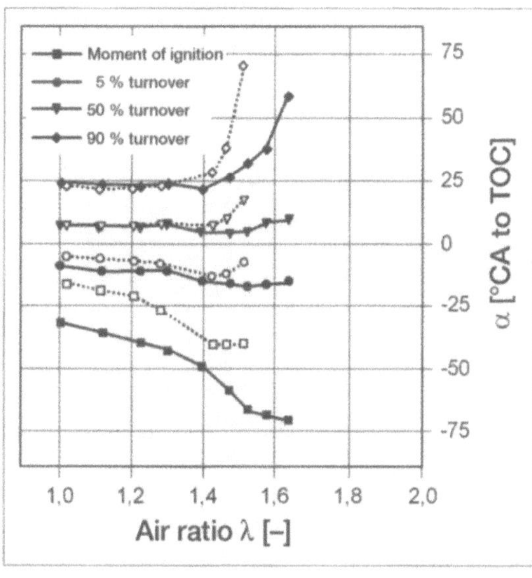


**Figure 11.9:** Homogeneous lean operation during part load with spark plug and prechamber spark plug; continuous line: standard spark plug; dotted line: basic swirl chamber spark plug [21]

**Figure 11.10** shows the inflammation achieved with a prechamber spark plug through the wide-ranging effect of the torch flames (right), as compared to inflammation with a standard spark plug (left). High performance large engines operated in homogeneous mode with a lean mixture even during full load are often equipped with prechamber spark plugs. According to [22], with the TCG 2020 V20 gas engine with 2000 kW power, an efficiency of over 44% is achieved in combination with a modern supercharging technique and prechamber spark plugs.

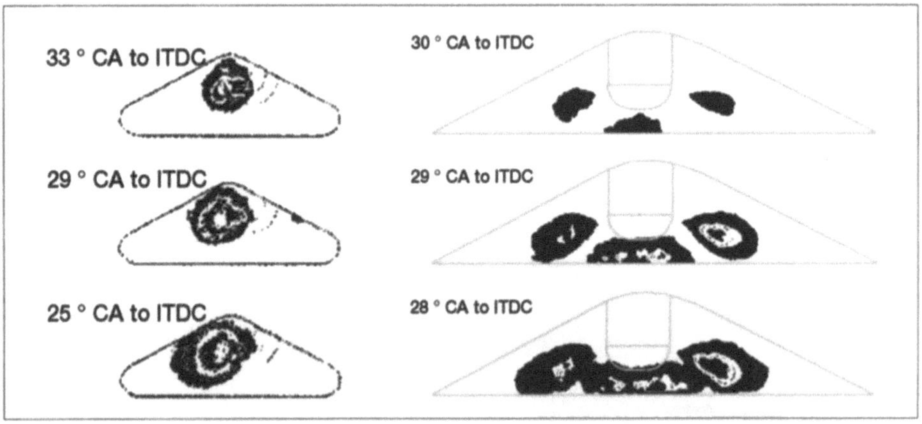


**Figure 11.10:** Visualization of the inflammation in a spark plug [21] (left) and in a prechamber spark plug with six bores (right)

According to [23], prechamber spark plugs with a service life of approximately 6000 h are very durable. Due to improved inflammation and increased turbulence in the combustion chamber by means of the torch flames from the prechamber spark plug, the charge swirl generated because of the intake port shape does not significantly affect the combustion. In **figure 11.11**, a Multitorch [24] prechamber spark plug for use in large gas engines is shown.

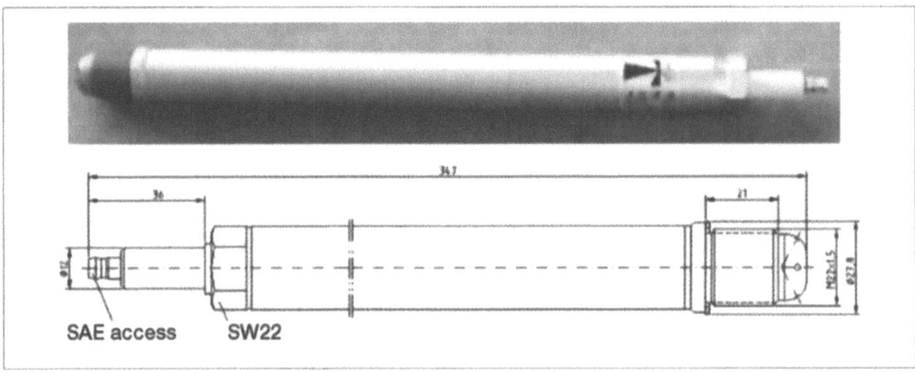


**Figure 11.11:** Prechamber spark plug for large gas engines [24]

**Purged prechamber spark plug**

Regardless of the mixture composition in the main combustion chamber, purged prechamber spark plugs, figure 11.8, can be used to reduce the residual gas content and/or achieve an ideal ignitable mixture composition at the prechamber spark plug. By purging the prechamber spark plug with air or combustion gas, the homogeneous lean operating limit increases significantly as compared to the standard spark plug and the non-purged prechamber spark plug; **figure 11.12**.

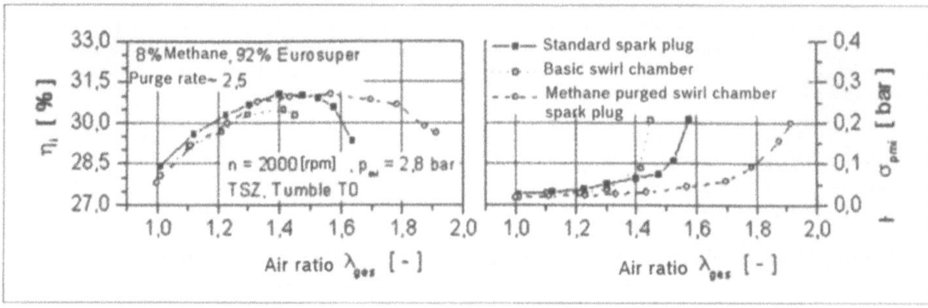


**Figure 11.12:** Lean operating behavior with purged prechamber spark plug [21]

## 11.1.2 Laser ignition

The use of laser ignition, which seems very promising, particularly for inflaming lean mixtures, is currently being tested. In laser ignition, a laser beam coupling takes place in the combustion chamber through an optical access. By focusing with a lens, the energy density at the focal point is increased to such an extent that the air/fuel mixture ignites. According to [25], the spark duration of 25 to 30 ns is extremely short as compared to the conventional spark ignition of 1000 ns. Thus, the risk of the ignition plasma drifting during the turbulent flow in the ignition area is low. The power density required for the ignition is in the range of 10 to 100 GW/cm$^2$. The ignition energy required is in approximately 10 mJ. It is also possible, therefore, to effect a power consumption similar to conventional ignition systems by using a diode laser with a low efficiency of 5%. A bright, hot plasma is generated by focusing the short laser impulse in air or in the mixture. This behavior is known as "optical breakdown" and is attributed to the multiple quantum ionization. The laser beam is absorbed by the plasma generated by ionization. This results in quick heating, thereby spreading the shock wave quickly [25]. The laser light forms an electromagnetic wave. In this case, the minimum energy depends on the gas pressure: $E_{Z,\text{min}} = 1/p^n$. Unlike conventional spark ignition, high back pressure leads to an early optical breakdown [26].

The radius of the focal point $w_F$ depends on the focal length $F$ of the optics used, the wavelength $\lambda$ and the diameter $D$ of the laser spray. Additionally, the beam quality characteristic number K results from the discrepancy between the actual and ideal beam profiles.

$$w_f = \frac{1}{K} \cdot \frac{2}{\pi} \cdot \frac{\lambda F}{D}$$

According to [26], requirements for a laser system are as follows:

| | |
|---|---|
| Pulse duration | 10 ns |
| Pulse energy | Some 10 mJ |
| Pulse frequency | ≥ 100 Hz |

In addition to high reliability and efficiency, the option of miniaturizing the laser system becomes necessary for application in passenger cars. Hence only pulsed lasers are used, of which the following three types meet the requirements:

- TEA $CO_2$ laser
- Excimer laser
- Nd:YAG laser.

For the necessary miniaturization, the Nd:YAG laser, having solid laser medium instead of gas, is an appropriate system. Moreover, the emitted wavelength of 1064 nm allows coupling by means of the light guide. Here, either a flash lamp or a diode laser is used as a pump source. A flash lamp as a pump source is cheaper, but the service life is only in the range of 1000 h. Large amounts of heat are generated as a result of the wide emission spectrum of the lamp along with the narrow absorption band of the Nd:YAG laser, whereby a low efficiency of approximately 2 to 5% is reached. Since the emitted wavelength can be adjusted, significantly high efficiencies of more than 50% are reached by using diode lasers as a pump source. Diode lasers have a sufficiently long service life of several tens of thousands of hours. The pulse duration of the laser beam is reduced to values below 10 ns using a Q-switch. For example, this can be achieved by means of baffles as well as through electromodulators or acousto-optic modulators. The aim is to generate a small, high-energy impulse. Light guides are suitable for the beam guide, as they are easier to handle and flexible to use, unlike deviating mirrors, which have to be adjusted. The coupling in the combustion chamber takes place through a window in the combustion chamber wall that usually functions as a "laser spark plug". According to [25], a Nd:YAG solid state laser (pump source: flash lamp) was used for a single-cylinder engine research unit with beam-guided combustion process. The wavelength can be 1064 nm or 532 nm. The maximum impulse energy is 140 mJ, the impulse duration 6 ns, the power 10 W, the power consumption 1 kW and the beam diameter 6 mm.

Due to separate assembly of the combustion chamber window and the focusing lens, separated optics make it easy to adjust the focal point along the beam axis. While configuring in [25], the focal point can be up to 20 mm and corresponds to almost the entire combustion chamber height. Synthetic quartz glass was used for the lens and the window material. The main advantage of laser ignition is the arbitrary selection of the ignition point in the combustion chamber. In the case of beam-guided combustion process, it is also possible to ignite in the injection spray. This will not result in any disadvantages for inflammation or flame expansion by heat dissipation via the electrodes or by fuel wetting ignition-relevant components. According to the tests in [25], contamination formed on the combustion chamber side of the window due to combustion chamber deposits are safely burnt off by the laser beam.

The laser energy reduces due to reflection at the boundary surfaces and absorption in the lens, reflection on the combustion chamber window, and at the deposits. The transmission of the combustion chamber window is 93%. The losses due to reflection at the boundary surfaces are approximately 4%. A total optical loss of approximately 15% is realistic. The intensity required for the optical breakdown is $10^{11}$ W/cm$^2$. Deposits formed on the window surfaces while the engine is running are safely removed [25] by the cleaning effect of the laser beam.

**Figure 11.13** is a diagram of the focusing of a laser beam in the combustion chamber of an engine [25]. In the case of separated optics, the focusing lens and the combustion chamber window are designed separately, whereas for combined optics, the combustion chamber window has a lens-shaped design. One advantage of laser ignition over spark ignition is that the ignition point can be selected randomly by adjusting the focal length of the lens system.

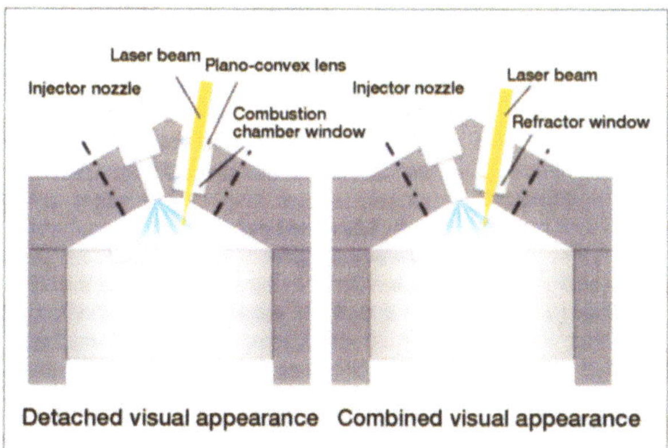


**Figure 11.13:**
Optical combustion chamber access for laser ignition [25]

**Figure 11.14** shows the effect of the focal point on the running of an engine in stoichiometric operation. With respect to the conversion point and the specific fuel consumption, an optimum ISFC is observed for the 12 mm focal point. According to [25], the air gulp limit can be shifted, to significantly higher air/fuel ratios by using laser ignition rather than spark ignition.

This makes the inflammation much less sensitive to variation in the air/fuel ratio. Compared to spark ignition with an iridium spark plug and a standard spark plug (FLR8LDCU), as shown in **figure 11.15** for homogeneous operation, laser induced ignition has a very low advance ignition requirement.

Tests carried out by [25] on laser ignition in a gasoline engine with direct injection, beam-guided mixture formation process and a multi-hole valve show that consumption and emissions can be improved simply by using the option of having the ignition point in the injec-

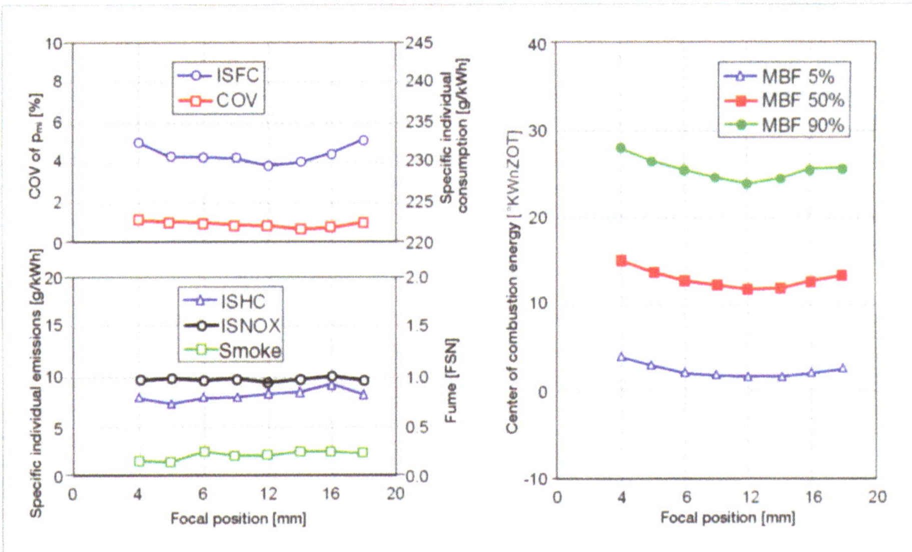


**Figure 11.14:** Effect of the focal point on the process cycle and emission in stoichiometric operation ($p_{mi}$ = 8 bar, $n$ = 2000 min$^{-1}$) [25]

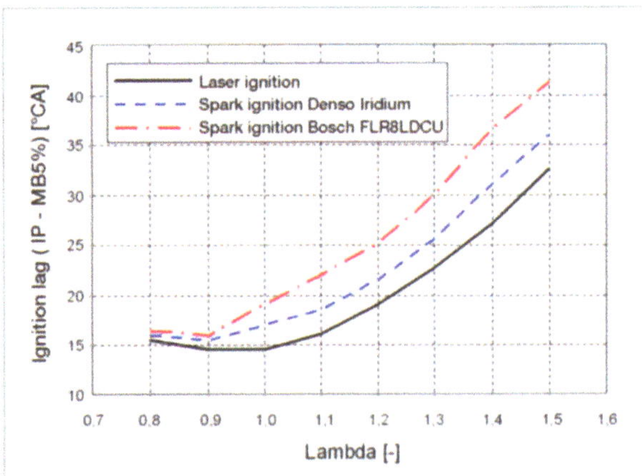


**Figure 11.15:**
Retarded ignition using the air/fuel ratio $\lambda$ with laser ignition and spark ignition at $n$ = 2000 min$^{-1}$ and $p_{mi}$ = 3 bar in homogeneous operation mode [26]

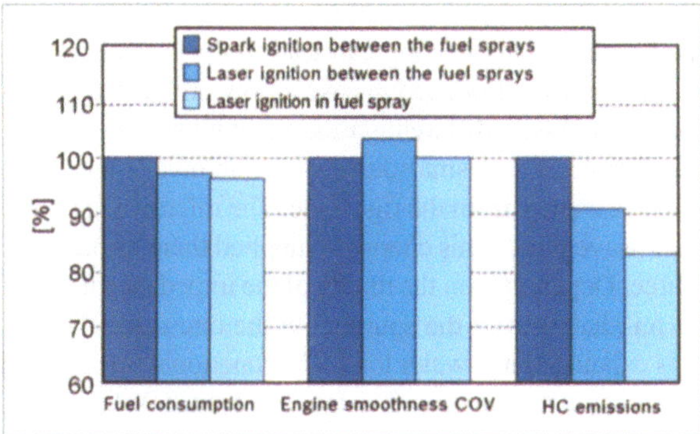


**Figure 11.16:**
Comparison between spark ignition and laser ignition at $n = 2000\ \text{min}^{-1}$ and $p_{mi} = 3$ bar in stratified engine operation mode with AGR in the beam-guided process [25]

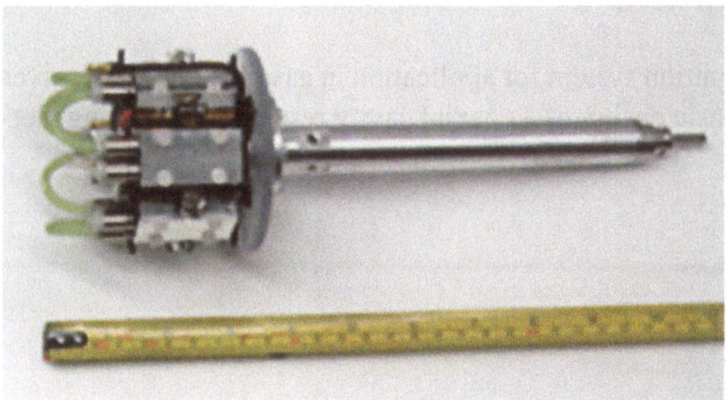

**Figure 11.17:**
CTR AG miniature laser [27]

tion spray. **Figure 11.16** shows that fuel consumption and hydrocarbon emissions can be improved by means of the flexible spark gap with the option of executing a wear-and-tear-resistant ignition directly into the fuel spray.

According to [26], the main objective in the development of laser ignition systems is to design a laser system that corresponds to the requirements of each specific engine. A laser prototype developed by CTR (Carinthian Tech Research AG) has been realized in the form of a miniature laser with sufficient power for igniting the air/fuel mixture in the free space of a conventional ignition system; **figure 11.17**. This ignition system can therefore be calibrated relatively easily to series production engines or engines close to series production. The electronics of the laser ignition can be operated as per the manufacturer's specification with the conventional 12-V vehicle electrical system [27].

### 11.1.3 Microwave ignition

Microwaves are electromagnetic waves with a wavelength between 1 mm and 30 cm. The frequency is 2.45 GHz in most of the applications. Through microwaves, the varying electrical field causes the available dipole molecules to begin to vibrate. This leads to heating of the dipolar media. Under certain circumstances, the heating can lead to the formation of a plasma. Generally, industrial plasma sources are introduced in case of low pressure up to atmospheric pressure. Plasmas can be generated with different gases, e.g. with air, $O_2$, $H_2$, $N_2$, $C_xH_y$ or Ar. **Figure 11.18** shows the principle of a plasma source with microwave feed. The plasma source is provided with microwave energy from the right side. The microwave power coming from a magnetron is guided to a waveguide. This energy is supplied there through the circulator and the E-H tuner of the source. Depending on the tuning of the impedance through the E-H tuner, the reflected power is reradiated from the source into the circulator. The reflected energy is dissipated there and is exhausted in a water load. The functional principle of the plasma generator shown here is based on a resonator with rotating antennas. While using a power of 1 to 2 kW, industrial plasma sources generate a plasma diameter of approximately 70 mm at a low operating pressure ranging from 0 to 1000 mbar [28].

An introduction to a microwave ignition system for application in gasoline engines is given in [29]. **Figure 11.19** shows the basic design of a coaxial cavity resonator. A varying elec-

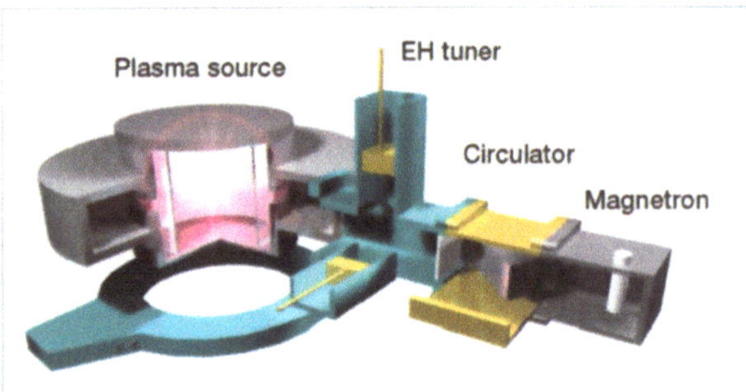


**Figure 11.18:**
Basic design of an industrial plasma generator [28]

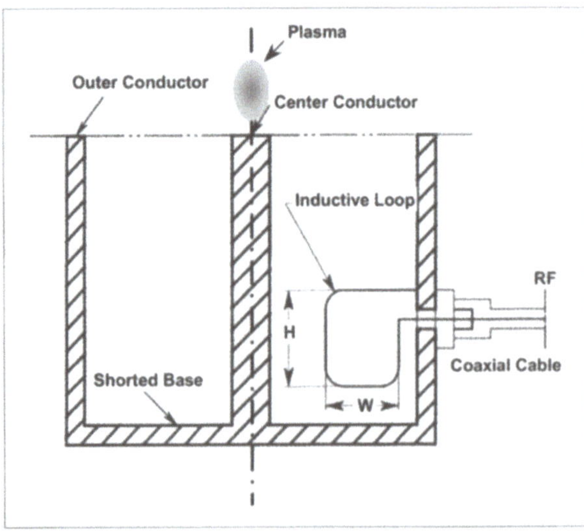


**Figure 11.19:**
Basic design of a coaxial cavity resonator for microwave ignition [29]

trical field is generated by coupling a high frequency alternative voltage with 900 to 1260 MHz. The length of the resonator in this case is one quarter of the wave length. Thus, the intensity of the electrical field is highest at the open end of the resonator. Due to the varying electrical field, ions and polar molecules in the combustion chamber are accelerated and generate a plasma at the end of the central conductor. Here, the ignition does not take place through a high current arc, as in conventional spark ignition, but through a gas discharge. Any desired ignition duration can be set using the microwave ignition. In addition to optimizing the resonator with respect to an ignition plasma that can be used in the gasoline engine, generating the high frequency AC voltage is a challenge under automotive boundary conditions.

## 11.2 Pistons

The piston has various functions in the internal combustion engine: besides being a part of the combustion chamber, it has to transmit the gas pressure that results from the combustion as force to the connecting rod via the piston pin. The connecting rod, in turn, is eccentrically linked to the crankshaft for converting the oscillatory motion into rotary motion. Together with the piston rings, the piston should also form a seal between the combustion chamber and the crankcase so that sufficient pressure can build up during the compression stroke that the combustion pressure does not leak into the crankcase in an uncontrolled way during the combustion and that the exhaust of the combusted mixture and intake of the required air volume occurs most efficiently. The engine's oil consumption, and the blow-by gases that enter the crankcase through the leakages in the ring set, should be restricted to permissible limits. The acting forces (combustion forces, inertia forces resulting from acceleration, lateral and guiding forces) should neither lead to an unacceptable noise level nor to component failure due to overload. The heat flow resulting from the fuel combustion affects the piston along with the said mechanical loads. Moreover, tribological loads act on different areas of the piston (skirt, piston ring grooves, pin bore). Last but not least, the combustion chamber surface may be subjected to erosion due to the effect of injected fuel spray in gasoline direct injection engines.

Current trends in gasoline engine development are gasoline direct injection, downsizing with increased power density, supercharging, high speed concepts (engine running at higher speed), and alternative fuels. These developments often lead to higher specific outputs, associated with increased heat input into the pistons and higher peak cylinder pressures. Depending on the type of mixture formation for stratified operation (wall-guided), gasoline direct injection sometimes requires complex combustion chamber geometries on the piston crown that can affect the piston weight and manufacturing processes. The following demands on the piston result from the operational requirements mentioned above:

- Temperature resistant
- Durable and fatigue resistant
- Deformation resistant
- Wear resistant
- Weight optimized
- Noise optimized
- Friction optimized
- Compliance with functional values (oil consumption, blow-by)
- Operationally reliable under extreme conditions (cold start, seizure resistance, etc.)
- Surface quality
- Optimized fine geometry (skirt, pin bore).

These requirements can be attributed to three main areas: material and material selection, manufacturing and processing, and design. A detailed description of these aspects is given in the sections that follow.

## 11.2.1 Materials

Currently, pistons for passenger car gasoline engines are produced only from aluminum alloys that are essentially alloyed with silicon, copper, nickel and magnesium. The silicon content is typically close to or above the eutectic point so that small primary silicon crystals that help to improve the wear behavior can be formed in the structure. Copper and nickel increase the high-temperature resistance, creep resistance and dimensional stability of the alloy. Examples of the alloys generally used are given in **figure 11.20**. Various structural phases and intermetallic phases that significantly affect the characteristics of the alloys can be seen in the micrograph in **11.21**.

The selected alloy compositions represent a compromise between requirements of good castability, machinability, wear resistance and thermal stability in the case of increased temperatures. The formation of local "hot spots" is avoided by good heat conductivity. However, it can be necessary to spray oil from the lubricating oil system circuit of the engine onto the crown of the piston for cooling. Cooling galleries, as frequently used in diesel pistons, are only used in very rare cases to further improve the cooling efficiency of gasoline pistons.

The durability of the alloys reduces considerably as component temperature increases, **figure 11.20**. Thus accurate knowledge of the local piston temperatures while the engine is running is important for defining the proper shape and design of the piston. Various methods can be used for determining the actual temperatures of engine operation, such as: fusible plugs (inserted plugs made of various alloys with precisely defined melting points indicate what temperatures have or have not yet been exceeded), templugs (inserted plugs

| | | Mahle 124<br>AlSi12CuMgNi | Mahle 138<br>AlSi18CuMgNi | Mahle 142<br>AlSi12Cu3Ni2Mg |
|---|---|---|---|---|
| Si | Wt.-% | 11.0 to 13.0 | 17.0 to 19.0 | 11.0 to 13.0 |
| Cu | Wt.-% | 0.8 to 1.5 | 0.8 to 1.5 | 2.5 to 4.0 |
| Mg | Wt.-% | 0.8 to 1.3 | 0.8 to 1.3 | 0.5 to 1.2 |
| Ni | Wt.-% | 0.8 to 1.3 | 0.8 to 1.3 | 1.75 to 3.0 |
| Fe | Wt.-% | max. 0.7 | max. 0.7 | max. 0.7 |
| Ultimate strength $R_m$ in MPa | 20 °C | 200 to 250 | 180 to 220 | 200 to 280 |
| | 250 °C | 90 to 110 | 80 to 110 | 100 to 120 |
| Yield strength $R_{p0,2}$ in MPa | 20 °C | 190 to 230 | 170 to 200 | 190 to 260 |
| | 250 °C | 70 to 100 | 70 to 100 | 80 to 110 |
| Elongation of fracture $A_5$ in % | 20 °C | <1 | 1 | <1 |
| | 250 °C | 3 | 1.5 | 1.5 to 2 |
| Rotation bending strength $\sigma_{bw}$ in MPa | 20 °C | 90 to 110 | 80 to 100 | 100 to 110 |
| | 250 °C | 45 to 50 | 40 to 50 | 50 to 55 |
| Young's modulus $E$ in MPa | 20 °C | 80000 | 84000 | 84000 to 85000 |
| | 250 °C | 72000 | 75000 | 75000 to 76000 |
| Thermal conductivity $\lambda$ in W/mK | 20 °C | 155 | 143 | 125 to 130 |
| | 250 °C | 159 | 150 | 135 to 140 |
| Thermal expansion $\alpha$ in $10^{-6}$ 1/K | 20 to 100 °C | 19.6 | 18.6 | 18.5 to 19.5 |
| | 20 to 300 °C | 21.4 | 20.2 | 20.5 to 21.2 |
| Density $\varrho$ in g/cm$^3$ | 20 °C | 2.68 | 2.67 | 2.75 to 2.79 |

**Figure 11.20:** Composition and typical characteristics of some of the aluminum alloys used for gasoline pistons

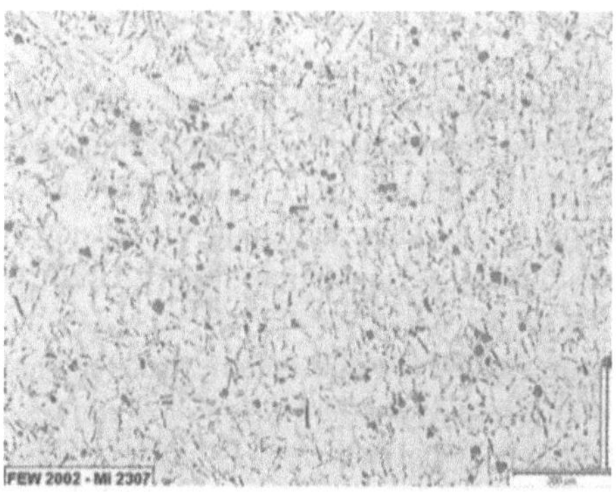

**Figure 11.21:** Micrograph of a typical eutectic AlSi piston alloy

made of special steel are evaluated for their change in hardness after the test run), NTC measurement (temperature measurement during engine operation by continuously measuring the resistance of NTC semiconductor sensors), or thermocouple measurements (using a grasshopper linkage system to guide the connecting leads, or using telemetry), in addition to estimating average thermal load based on the change in piston hardness after an engine test. These measurement techniques differ considerably in terms of effort and accuracy.

As in the case of diesel engine pistons, a cast-in, intermetallically bonded Ni-Resist ring carrier made of lamellar austenitic cast iron can be used if the wear resistance in the first piston ring groove is insufficient to withstand the mechanical and thermal load.

### 11.2.2 Manufacturing process

The above-mentioned materials can be used for cast pistons or for producing billets from continuous casting to manufacture forged pistons. Piston forging produces a uniform, homogenous microstructure with improved material ductility, but it is more expensive. Especially for pistons of direct injection gasoline engines with complex combustion chamber geometry, the objective is to create the largest possible portion of the crown without additional machining. To achieve this, an appropriate mold design and mold coating are required in the case of cast pistons. In critical cases, the casting conditions and if necessary the location of cooling ducts in the tool and their controlled activation, as well as the position of the gate and the riser, must be optimized using solidification simulation in order to minimize the risk of shrinkage porosity within the material. **Figure 11.22** shows an example of a non-optimized tool design.

After casting, the piston blank is subjected to thermal ageing until it reaches an overaged material state, because volume changes that may occur as a result of segregation processes due to thermal load when the engine is running must be suppressed as much as possible; otherwise skirt seizures could occur. Forged pistons are subjected to complete heat treatment (solution annealing and thermal ageing) so that a uniform state of material with minimized residual stresses and within the specified hardness tolerance can be achieved.

Sometimes, a considerable amount of material accumulates above the pin bores due to the large wall thickness as a result of the desired combustion chamber geometry of the piston crown. To reduce the piston weight as much as possible in such cases, relief contours can be created on the casting blank by means of swiveling mold inserts shaping the outside of the pin bores. This reduces the formation of such material accumulations, **figure 11.23**.

Optimized processing procedures and machines are applied for high-volume production of pistons. As a main component of the internal combustion engine, the piston incorporates a number of areas to be machined within tightest tolerances and to extremely high precision, such as the piston ring grooves (tolerance range typically 20 μm), the pin bore (different

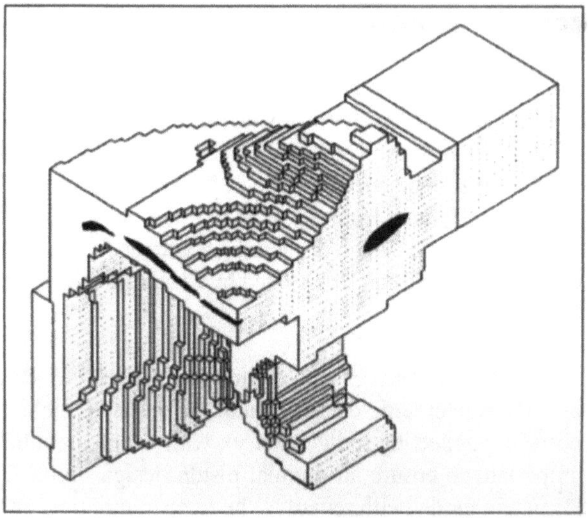

**Figure 11.22:**
Result of solidification simulation for a non-optimized tool (dark surfaces = risk of shrinkage porosity)

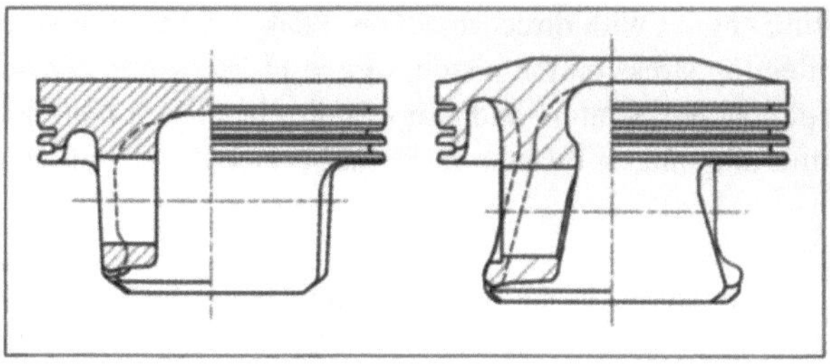

**Figure 11.23:**
Weight optimization by "relief" on the outside of the piston bores: (left) die directly pulled; (right) formed with a swiveling mold insert

microgeometries, tolerance range approximately 6 μm) and the complex skirt geometry (barrel-shaped, varying ovalities, tolerance range at the nominal diameter, e.g. 15 μm).

Many different functional platings have been developed for gasoline engine pistons and introduced into series production. Tin plating of pistons for gasoline engines to enable some dry-running operation has been almost completely replaced by screen printed polymer coatings filled with graphite or iron particles, or a galvanic iron layer (to be used for engines with uncoated hyper-eutectic aluminum crankcases). Narrow installation clearances in the order of 10 to 30 μm can be achieved with high reliability. In many cases, a hard anodized oxide coating ($Al_2O_3$, produced directly from the piston material by a galvanic process)is used as an additional protective layer against the adhesive wear between the piston ring and the ring groove.

### 11.2.3 Design and component strength

Various zones of the pistons for gasoline engines are subjected to high loads and, therefore, must be carefully designed and dimensioned by taking into consideration the expected local component temperature. This includes the piston ring zone (land height, ring groove corner radius), the pin bores, the pin bore support area (transition of the pin bore areas into the piston undercrown) and the transition area from the skirt into the box-type wall between the skirt and the pin bore. The geometric design of these piston areas follows design guidelines which are based on experience and can be checked by FEA and verified by dedicated component tests in the laboratory.

Reliable information about the load (peak cylinder pressure, engine data, specific engine power, incidental heat flux, etc.) as well as geometric design data of the engine (combustion chamber shape, compression height, connecting rod geometry), with which an initial piston design can be created, is important to ensure an optimal piston design draft. FE analysis can be used to analyze this piston model with regard to the temperature field and the stresses and deformations. **Figure 11.24** shows an example for a temperature field calculation in a piston for a gasoline engine with direct injection. Peak cylinder pressure results in medium voltage and deflection stress on the piston, **figure 11.25**, which can be evaluated by considering the component temperature and temperature-dependent material fatigue properties in the fatigue-limit diagram, or by examining the service life for a given engine load profile.

It is possible to change the piston temperatures at the piston by modifying the boundary conditions on the combustion side, modifying the combustion parameters or spray cooling the undercrown of the piston using an oil spray nozzle installed in the crankcase.

Information about the true peak cylinder pressures and the local (assumed, calculated or measured) component temperatures is required for proper design of the areas with high loads mentioned above. In the case of piston land dimensioning, it has to be taken into consideration that although the reduced ring and piston land heights may make it possible to reduce the piston compression height and the piston weight, piston and piston ring problems related to the ring land strength or even to the oil consumption and blow-by behavior of the system may occur. Therefore, the design must be carefully optimized. Simulation programs are available to predict the piston ring behavior (axial movement, sealing behavior). These also require an input of precisely defined boundary conditions to obtain useful results.

The piston pin is most often designed to be fitted to the piston in a "floating pin" design; i.e. the pin moves freely between the connecting rod and the pin bore and is held in position axially by the circlip rings assembled in appropriate grooves in the pin bore. The pin bores of the pistons for gasoline engines are predominantly designed with an oval bore for better strength. A shaped pin bore (an expansion of a few μm) is introduced towards the connecting rod, in order to reduce the edge load and to avoid excessive leaking of the lubricating film. In cases where lubrication problems occur in the pin bore, machined cavities (known

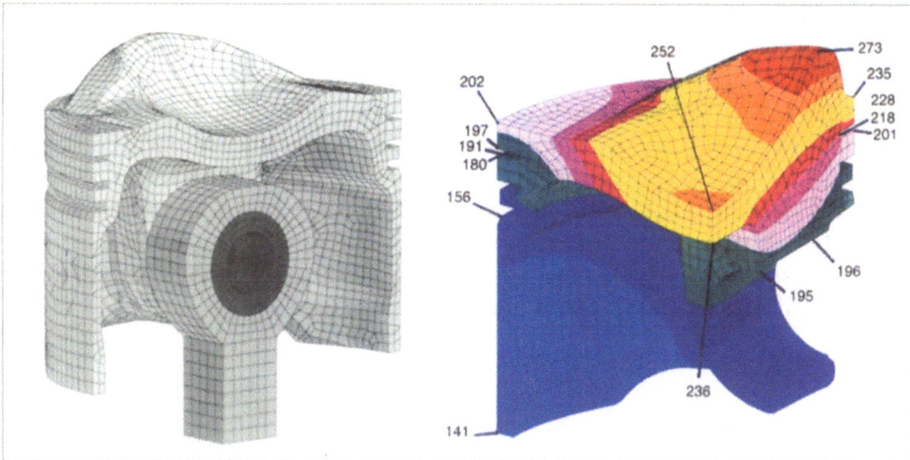


**Figure 11.24:** Temperature distribution within a piston for a gasoline engine with direct injection at rated power (50 kW/l)

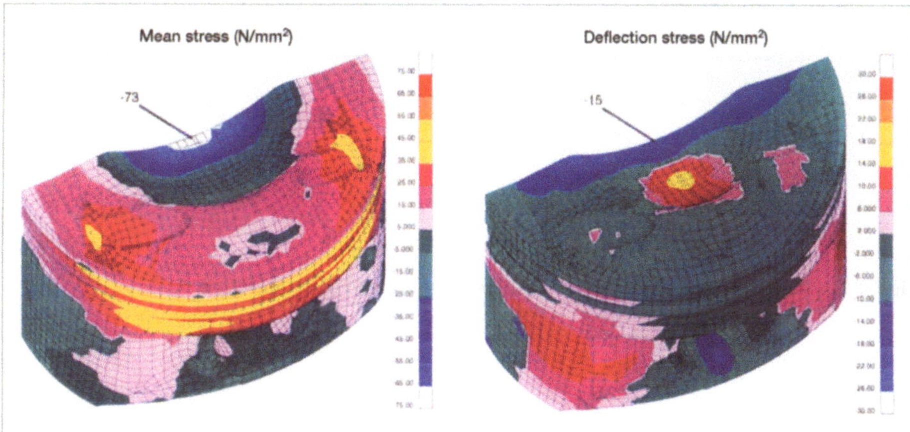


**Figure 11.25:** Mean stress and stress amplitude on the warm piston under the effect of peak cylinder pressure

as slots or side reliefs) can be introduced along the pin bore axis. Circumferential grooves in the pin bore can also be useful to ensure sufficient lubrication in the pin/bore contact.

The piston design is significantly affected by the combustion bowl shape on the piston crown. In the early days, wall- or air-guided combustion process designs were preferred for gasoline engines with direct injection. These designs are characterized by deep, asymmetric bowls, some with undercuts; **figure 11.26**. Nowadays, spray-guided combustion is preferred as it has greater potential for reducing fuel consumption. As a result, piston geometries are again being designed flat and with less distinct ridges; **figure 11.27**. Such geometries are easier to manufacture and assist in reducing piston weight. However, they often demand partial machining to maintain the required tolerances (combustion chamber volume, roughness).

**Figure 11.28** shows another example of a piston for gasoline engines with direct injection. This piston has a ready-cast crown, with the exception of a machined spherical bowl manufactured by turning. Under extreme conditions at the sharp edge of the recess, as seen in the **figures 11.26** and **11.28**, cracks may develop at the piston crown due to thermal cyclic loading. In such cases, local hard anodization (as for the piston in **figure 11.26**, right) can improve the service life of the piston. For durability in this area, a neatly rounded, smooth transition radius should be introduced instead of a sharp edge.

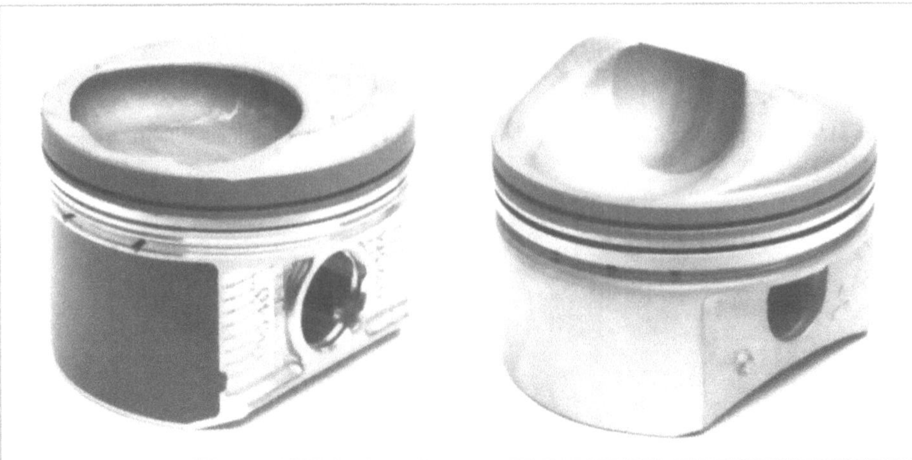

**Figure 11.26:** Earlier piston designs for gasoline engines with direct injection

**Figure 11.27:** Combustion chamber geometries for the spray-guided combustion process in gasoline engines with direct injection

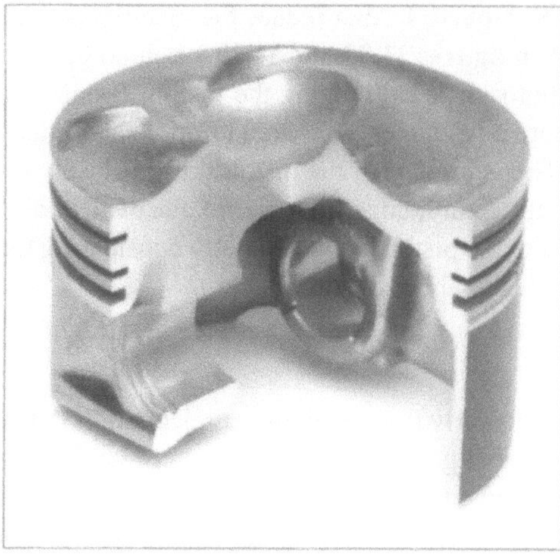

**Figure 11.28:**:
Piston with spherical bowl for a gasoline engine with direct injection

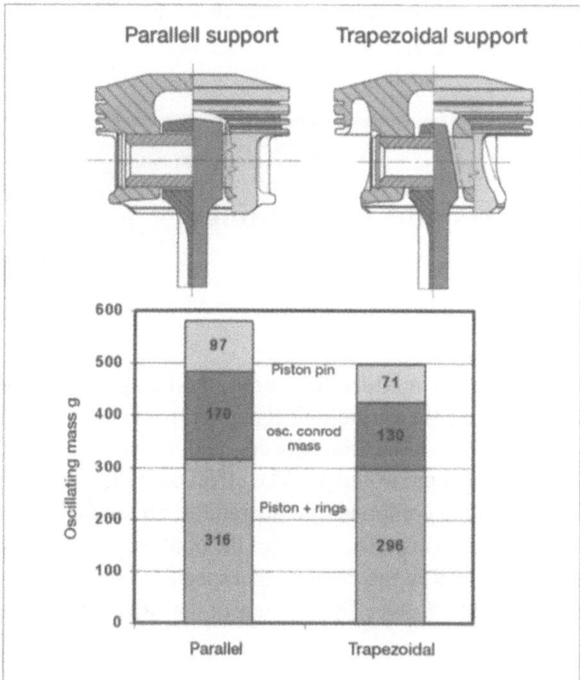


**Figure 11.29:**
Weight optimization of the power cell assembly (piston, pin, connecting rod)

In order to reduce the total oscillating masses as much as possible, despite the often unavoidable material clusters on the piston head in the case of gasoline engines with direct injection, connecting rods with a trapezoidal small end are being used more frequently. In this way, along with the reduced oscillating mass of the connecting rod in combination with acceptable specific loading in the piston pin bore and the small end connecting rod bore, the piston-pin length may be reduced, thereby further reducing the weight of the moving masses; **figure 11.29**.

Moreover, by using a shorter piston pin it is possible to further reduce the material outside of the piston bore by a cast relief – as shown in **figures 11.23** and **11.29** – behind the ring zone in the pin direction. Thus, a system approach to optimizing the piston, connecting rod and pin system can yield significantly more advantages compared to implementing an individual measure on a single component. The geometry of the piston skirt is closely related to the running properties (susceptibility to seizure, noise). Therefore, the oval, barrel-shaped design of the skirt is carefully adjusted to the piston blank geometry considering the engine-specific characteristic features (block design, bore distortion, etc.).

**Conclusion**

The design of pistons for gasoline engines, particularly those with direct injection, is determined to a large extent by the combustion chamber geometry at the piston crown. A large number of individual measures and detailed optimizations help to achieve the best possible balance between manufacturing and processing as well as stability and functional characteristics. Due to the increasing peak cylinder pressure, future piston design may become more similar to diesel engine piston design which may even mean that it becomes necessary to use a ring carrier or even a cooling gallery.

## 11.3 Valves, valve seat inserts and valve guides

### 11.3.1 Valves and valve seat inserts

During the transition from external to internal mixture formation, the charge exchange components (the intake and exhaust valves and their tribological partners), valve guide and valve seat inserts are directly exposed to the combustion process, and undergo complex changes in their load spectrum and tribological environment. The effect of the combustion on these components can bring a change in the component temperatures, component temperature distributions and operating pressures. This effect changes the boundary conditions of the design of valves, valve guides and valve seat inserts such that specific characteristics have to be taken into consideration. **Figure 11.30** gives an overview of the valves and valve seat inserts.

All the points mentioned in the following section relate essential to a reliable and wear-resistant valve design, the thermo-mechanical load on the valve and full load operation relevant to the tribological system comprising valve, valve seat insert and valve guide as shown in **figure 11.31.**

Since better internal cooling and higher knock resistance enable gasoline direct injection to achieve high compression ratios, high combustion peak pressures also usually occur. These are important with respect to a durable design of the valve. Valves are designed increas-

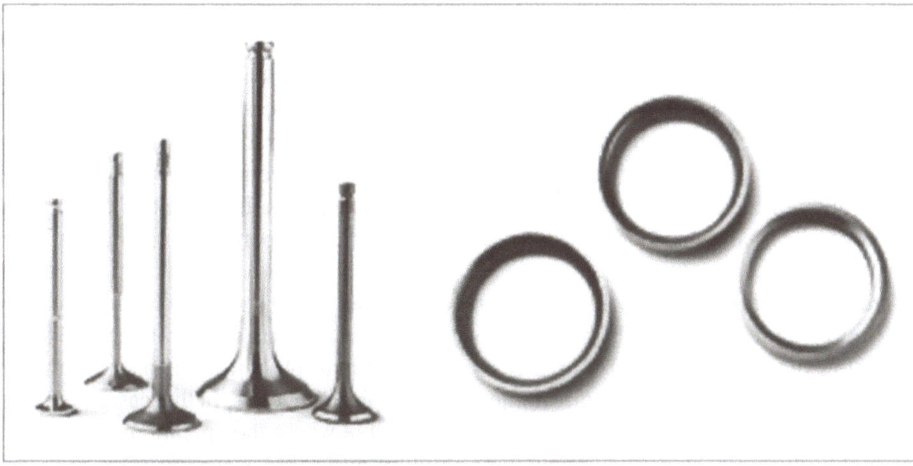

**Figure 11.30:** Valves and valve seat inserts [Source: MAHLE]

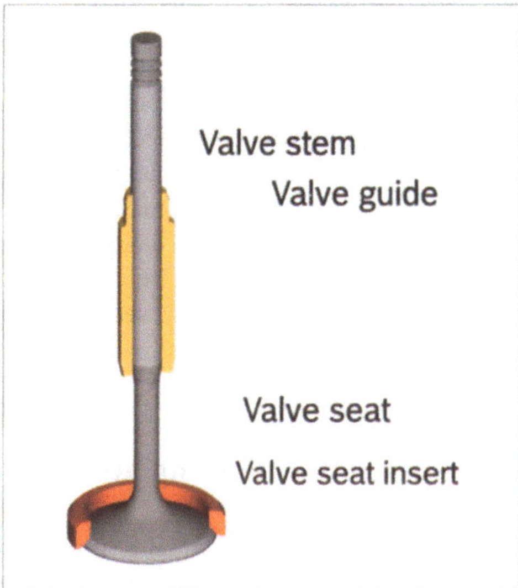


**Figure 11.31:**
Tribological system
Valve stem/value guide/valve seat insert/
valve guide

ingly slimmer in order to reduce the moving mass, e.g. by inserting a spherical calotte at the bottom of the disc combined with a flow-optimized fillet and cone design. Such slim, weight-optimized valve designs can lead to increased elastic deformations under combustion pressure due to their reduced rigidity. Apart from the potential reduction in service life due to the possible larger strain values, the resulting relative movements between the valve seat and valve insert lead to increased wear [33]. Furthermore, the relative movement between the valve stem and the valve guide that results from the valve disc deformation under combustion chamber pressure can affect the safe functioning of the valve-train. Contradictory requirements such as mass reduction and component rigidity complicate the design solution.

This especially applies to gasoline engines with direct injection and supercharging, which are increasingly penetrating the market., As the downsizing concepts, due to good internal cooling, work with high basic compression and relatively high charge pressures, they can also overload the common exhaust valve materials' high temperature stability and resistance to hot gas corrosion because of high exhaust gas temperatures with resulting high component temperatures that are required for optimizing the combustion process. In this case, hollow drilled sodium-filled valve stems that increase the load limit of the exhaust valves can be used, as is done in heavily loaded, supercharged port-injected engines.

In order to achieve extremely low emission values after the cold start by having the catalytic converter respond quickly, high exhaust gas temperatures are rapidly induced (10 seconds after the start above 700°C) partly by means of multiple injections in the warm-up phase for spray-guided gasoline direct injection with piezo injectors. In this case, a reaction to the thermo-mechanical load of the valves is possible where the operational conditions are highly dynamic, since the temperature build-up during the warm-up after cold start can turn out to be higher as compared to the temperature build-up of the valves in the port-injected engines. However, no test results on the consequences of the nominal thermo-mechanical load are available yet.

It also has to be taken into account that, due to the internal mixture formation, there is no lubricating fuel film between the valve seat insert, as is the case with port-injected engines. Tests show that, in contrast to comparable port-injected engines with intake manifold injection, higher alloyed valve seat insert materials usually have to be selected, if necessary with a valve seat hard facing. During exhaust, the above-mentioned turbocharged engines with gasoline direct injection and having higher operational temperatures require cobalt hard facing on the valve insert along with highly specified materials for the valve seat insert.

## 11.3.2 Intake valve deposits on gasoline direct injection engines

In part load operation in engines with gasoline direct injection, there is often a strong tendency to form carbon deposits in the combustion chamber and on the intake valves; **figure 11.32**. This can lead to a loss of power due to the progressing cross-sectional reduction at the valve opening gap and have a negative effect on the consumption and emission behavior. In individual cases, dynamic problems, where the intake valve is not able to follow the specified curve, can arise due to the continuous increase in mass of the valve. In the worst case, the engine function may be disturbed by the detachment of accumulated deposit fragments, since these can hamper the sealing function at the valve seat under certain circumstances.

Carbon deposits may form on the intake valves due to the oil content in the blow-by gases fed to the intake tract, the high exhaust recirculation rate strategy for reducing emissions, the oxidized engine oil components and the valve surface temperatures that support the formation of such deposits [34]. Insufficient fuel, and therefore lack of additives to prevent the formation of deposits on the intake valve cone surface, is identified as the main reason for these deposits. Also, new additive compositions to counteract the formation of deposits are being researched by the mineral oil industry in view of the increasing market share of gasoline engines with direct injection.

**Figure 11.32:**
Intake valves with deposits (example)

Improving the oil separation of the blow-by gases is one of the conventional measures taken against deposits on the intake valves. Additionally, a strategy to prevent the temperature range between 170 and 280 °C (critical to the formation of deposits on the component surfaces) is available. The long-chain hydrocarbon molecules do not disintegrate below the mentioned range; beyond it, the residue burns off. Since continuous operation below the temperature limit mentioned above does not seem possible, a limited attempt is made to raise the surface temperature of the cone area of the intake valve. This is done by increasing the intensity of heat transfer from the lower side of the valve disc to the upper side using, for instance, a contoured increase of the spherical callote surface area. An attempt is also made to use catalytic layers to shift the range relevant to the formation of the carbon deposits to lower temperatures. Trials using a vanadium pentoxide layer ($V_2O_5$) are common. Alternatively, anti-adhesive layers such as polytetrafluoroethylene, or Teflon©, are used to hamper the formation of deposits. Trials of the various strategies to counter the problem of carbon deposits have not been able to provide a consistent solution as to which process can ensure lifelong reliable operation with acceptable formation of deposits in the part load range.

### 11.3.3 Valve guides

The valve guides and the tribological system of the valve and valve guide are least affected by the internal mixture formation; **figure 11.33**. Here, operating conditions differ only slightly from those for port injection. Theoretically, the lack of fuel wall film in the intake port can ensure improved lubricating conditions in the valve guide gap on the intake side. It is currently not possible to assess to what extent the high maximum exhaust gas tem-

**Figure 11.33:**
Valve guides made of sintered steel
[Source: MAHLE]

peratures in turbocharged gasoline engines with direct injection may lead to problems with commonly used non-ferrous metals or sintered materials.

### 11.3.4 Lightweight valves

A new valve technology, such as the one being developed for series production by MAHLE, can reduce the severity of problems related to most of the areas mentioned above. The lightweight valves already introduced are made from many individual parts [35, 36]. These valves are up to 50% lighter than conventionally designed valves; **figure 11.34**. The materials used have considerably better high-temperature stability and corrosion resistance than the exhaust valve materials currently used in series production. Individual parts manufactured by means of stamping and deep-drawing processes are connected using various joining methods and the finishing of the raw part thus manufactured follows the same procedure as for conventional valves.

With a flow-optimized outer contour that keeps the throttle losses low, the lightweight valves have a rigid dome-shaped valve head design. Measurements under operating conditions have shown that this high structural rigidity considerably reduction the elastic deformations of the valve disc center. In addition to the seating impulse being reduced due to the lesser mass, the resulting reduced relative movements between the valve seat and the valve seat insert are behind the very good wear behavior of this type of valve. Trial results show that a very good wear pattern with low wear rates can be achieved in the intake with seat insert materials that have relatively simple specifications. The high-temperature materials also used on the exhaust valve seat area form wear-resistant aluminum and chrome oxide layers on the hot side of the exhaust. In combination with suitable valve seat insert materials, this ensures very good wear behavior with about half the wear rates of conventional valves.

**Figure 11.34:**
Lightweight valve with sodium filling

High compression ratios are desired for improving efficiency through reduced knocking sensitivity in gasoline engines with direct injection. In the lightweight valve application, this is supported by the convex bottom of the valve disc. However, the use of conventional valves with large calottes made for weight optimization complicates the design of such compact combustion chamber geometries and gives the combustion chamber an undesirable rugged contour resulting in a disadvantageous surface/volume ratio.

## 11.4 Air intake systems in modern gasoline engines

Air intake systems in gasoline engines have drastically changed over the past 20 years. These changes are based partly on innovations in the field of technical plastics and, more importantly, on innovations in fuel mixture generation. With the combination of direct injection and supercharging, gasoline engine intake manifolds are currently undergoing a further change. In future, the intake systems of such gasoline systems will be similar to those of the modern diesel engines. This chapter deals with the changes mentioned above and with the current status, before looking ahead.

The term "air intake system" describes the entire air inlet system of the combustion engine, from the atmospheric air intake opening to the cylinder head. Thus the intake system is one of the components in the engine compartment that requires the greatest amount of space. As a result, a number of different functions were linked to the intake system, e.g. design applications or fixing of components. The intake system for naturally aspirated engines consists of four main elements as shown in **figure 11.35** [37]: the atmospheric (or fresh) air duct, the air filter, the clean side air duct and the intake manifold. **Figure 11.35** shows the airflow of a four-cylinder engine with the most important functions and some of the important fixed parts.

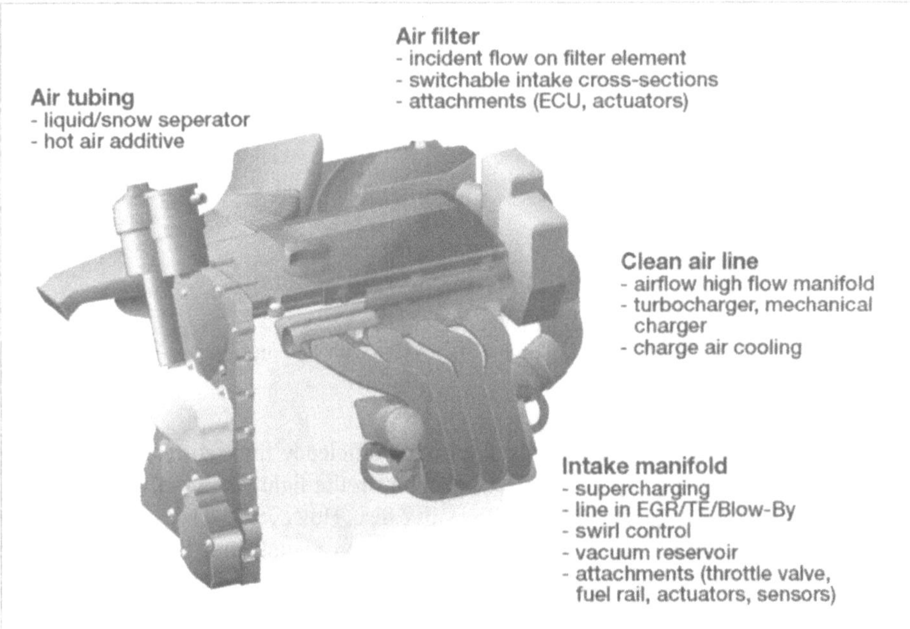


**Figure 11.35:** Airflow system of a combustion engine (schematic)

## 11.4.1 Thermodynamics of the air intake system

The thermodynamics of the airflow system depends on the type of combustion process (gasoline or diesel) and the principle of supercharging. The non-filtered air line and air filter are similar in all the variants. However, the systems differ considerably downstream of the air filter depending on the supercharging principle [37], **figure 11.36**.

Turbocharged engines have an extensive pure air area with a compressor and an intercooler, whereas the intake manifold has a simple structure as it is used only to supply air to the cylinders. On the other hand, naturally aspirated engines have a simple clean air duct but a highly complex (switchable) intake system for improving the cylinder charge.

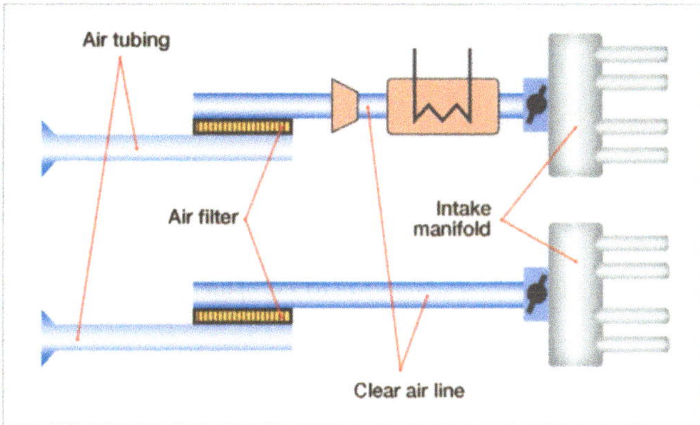


**Figure 11.36:** Airflow in naturally aspirated engines (below) and supercharged engines (above)

## Non-filtered air line

The non-filtered air line, i.e. the part of the intake system between the air inlet and the air filter, conducts the flow, mixes the hot air and separates the coarse particles. Hot air mixing affects the operational behavior of the engine, especially in the cold start phase. Constantly improving fuel mixture generation means that this function is now rarely required; it is used predominantly in cold regions.

Furthermore, an appropriate non-filtered air line can separate the coarse particles (raindrops, snow, dirt) using deflection and with a less pressure loss. This preseparation makes the job of the air filter easier and protects the air filter element from moisture. Today, particle separation and pressure loss at deflection can be reliably predetermined by flow simulation.

## Air filter

Generally, an air filter has an air filter element inside the housing. In addition to the acoustic effect, the air filter performs the function of inducting air for optimal flow to the filter element. Optimal in this case means an inflow that is as uniform as possible, with the speed through the filter element across the entire filter surface being as equal as possible. If the incoming flow is not uniform, the pressure loss in the filter element increases and the engine performance deteriorates. A homogenous inflow makes the best used of the dirt absorbing capacity (dust capacity) of the filter material. Three-dimensional flow simulation (CFD) is used when designing the flow in air filters, which makes it possible to determine an optimum geometry at a very early stage, with minimum testing efforts.

As shown in the example in **figure 11.37**, it was possible to reduce the size by 30% with the same pressure loss and dust capacity. Here, the incoming flow was approximated to within 3% of the ideal test bench value parameters. Reliable pressure loss calculations are state-of-the-art for intake system designs, especially of air filters. In the meantime, with

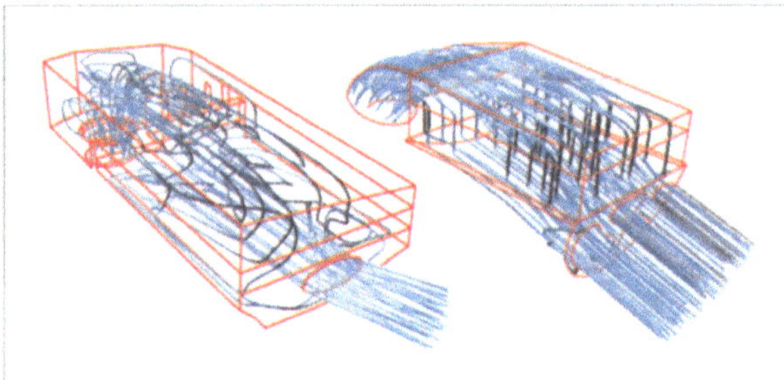

**Figure 11.37:**
Inflow into filter elements: non-uniform with high pressure drop (left) and optimized (right)

innovative calculation methods it is also possible to determine the dust charging of the filter element as well as the pressure loss across the lifetime; **figure 11.38**.

There are two objectives when precalculating the charge profile: making the best use of the filter element during the entire service life, and predicting the flow characteristics in the air filter across this time period. The background of the above-mentioned point is e.g. an adequate constant signal at the air mass meter (AMM) during the service life of the element. Maintaining the flow conditions at the hot film mass meter (AMM) is becoming more and more important due to increasingly tight emissions and consumption regulations.

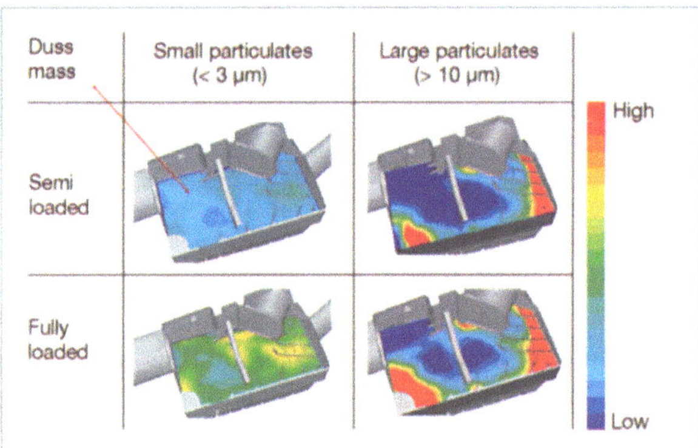


**Figure 11.38:**
Dynamic calculation of charging of the filter element across the service life for two different particle sizes

## Clean air duct

As shown earlier, the incoming flow at the AMM for the intake air on the clean air side is examined by a CFD simulation in case of the new intake system in order to ensure a uniform flow. In addition to a possible change in the flow profile due to charging effects at the filter element, further inspection of the situation at the AMM is necessary with regard to the tighter emissions thresholds. Gradual degradation of the sensor due to deposits of the oil dripping from the crankcase or from the exhaust gas recirculation (EGR) can also be reduced drastically by CFD simulation of the flow induction.

The gas pulsations produced by the combustion engine are more intensive downstream of the air filter. The thermodynamics and acoustics must be observed at the same time at least in the clean air duct, as both the fields have an effect on the airflow. Acoustic components (shunt resonators, $\lambda/4$ pipes) that affect the gas exchange work are present in the clean air duct. 1D simulation tools that provide volumetric efficiency and orifice noise at a very early stage in the designing phase are used for this. The modeling effort can be considerably reduced because one calculation model yields both the results.

Turbocharged engines have a longer intake system than naturally aspirated engines. In engines with exhaust-gas turbocharging (ATL), the intake air from the front module is supplied to the ATL via the air filter near the exhaust manifold. The compressed air is then returned to the front module charge air cooler. Finally, the clean air duct again leads to the intake manifold at the engine.

## 11.4.2 Intake manifolds for naturally aspirated engines

The pressure pulsations induced by the piston are used by naturally aspirated engines to increase the volumetric efficiency in the combustion chamber. The process of "resonance supercharging" can be described with the help of **figure 11.39**. After the intake valve opens, the piston generates a negative pressure wave during its downward stroke. This pressure wave moves against the flow direction away from the combustion chamber and along the tuned intake runner. The negative pressure wave is reflected at the plenum due to a sudden change in the cross-section. The returning high pressure wave can be used to improve the cylinder charging if it reaches the combustion chamber before the intake valves close.

At a constant sound velocity, the optimum pipe length is indirectly proportional to the engine speed $n$. At present, switchable intake manifolds that can switch between the short and long tuned intake runners are used in all vehicle categories in order to achieve good cylin-

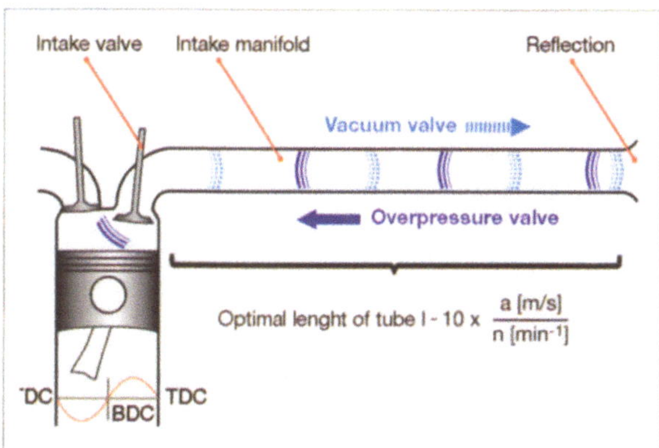


**Figure 11.39:**
Principle of resonance supercharging

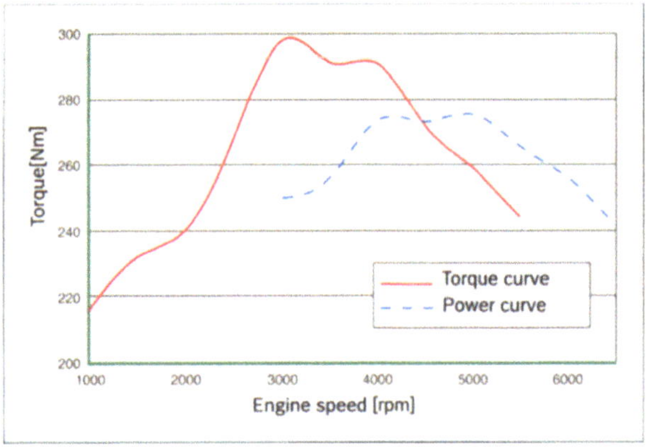


**Figure 11.40:**
Torque curve of a six-cylinder engine with switchover intake manifold

der charging across a wide range of engine speeds. A typical torque curve for a switchable intake manifold with two tuned intake runner lengths is shown in **figure 11.40** [38].

With the increasing complexity of intake systems, increasing the volumetric efficiency depends more and more on improving the quality of the manufacturing and materials. Taking sealing quality as an example, **figure 11.41** shows qualitatively the how volumetric efficiency depends on leakage. Sealing that enables an increase in the volumetric efficiency in a two-stage switchable intake manifold, can on the other hand, lead to complete deterioration of the air efficiency in three-stage or multi-stage intake manifolds.

One more series of variables that affect the gas exchange work in the intake manifold are available in addition to the leakage through the switching elements. **Figure 11.42** gives an overview of the possible sources of loss [37].

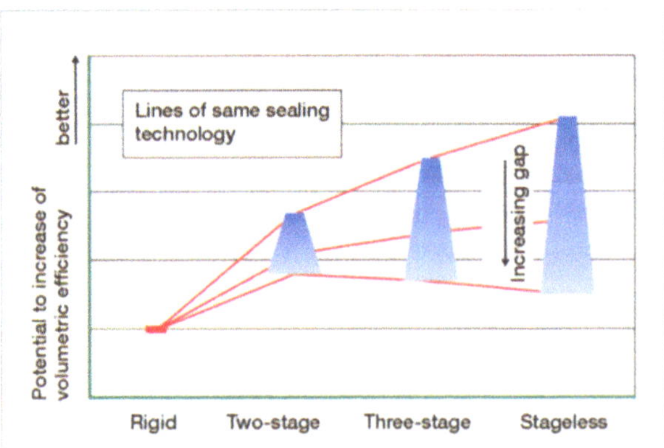


**Figure 11.41:**
Effect of sealing technology on volumetric efficiency in the case of switching flaps

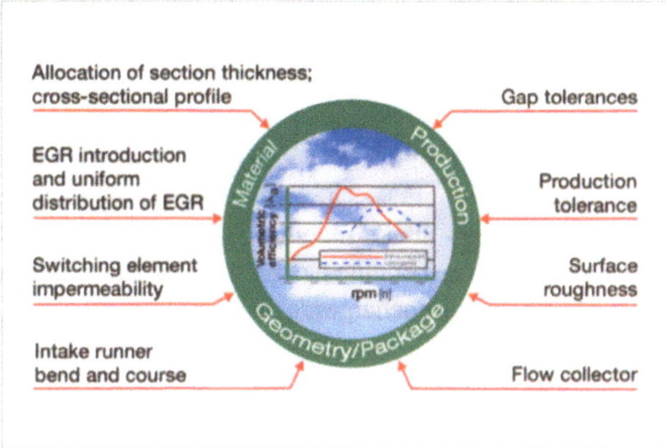


**Figure 11.42:**
Factors affecting volumetric efficiency in switch-over intake manifolds for naturally aspirated engines

As a result, the suppliers of modern intake systems must define the entire intake system thermodynamically and mechanically early in the development. This requires the integration and linking of all the CAE tools (Computer Aided Engineering) at the very beginning of the development [37].

## 11.4.3 Intake manifolds for supercharged gasoline engines

Engines with mechanical or exhaust-gas turbocharging require intake systems for distributing air charge to the cylinders. Short intake manifolds with less pressure loss and equal distribution to the cylinders are desirable for this. Though the pressure pulsations effects act as post-charging in these engines, they play a secondary role and are used partially for improving selected map points.

**Figure 11.43** shows a comparison of two intake manifolds for gasoline engines with comparable performance in the range of 147 to 165 kW. Shown here is a switchable intake manifold

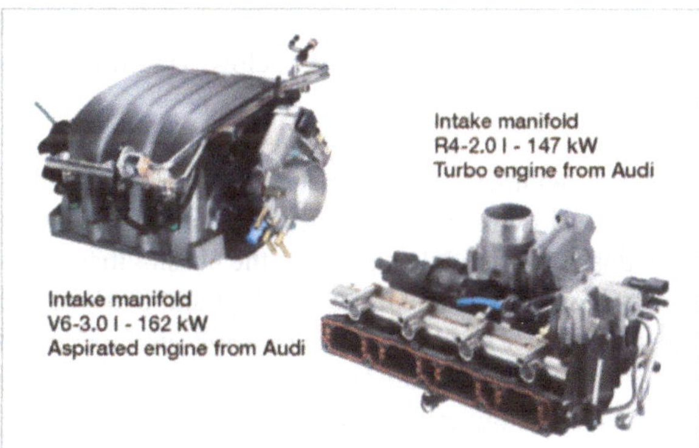


**Figure 11.43:**
Comparison of intake manifolds for Audi gasoline engines: V6-3.0 l-162 kW naturally aspirated engine (left) and R4-2.0 l-147 kW turbocharged engine (right)

by gases is the main aspect that forms the background for legal specifications [41, 42] that stipulate recirculation of the crankcase ventilation gases into the intake system in gasoline engines. Lubricating oil droplets of various sizes, which are carried along with the crankcase ventilation gas flow, should be separated as much as possible by means of the oil mist separator located in the ventilation tract, especially in order to avoid both direct and indirect increases in exhaust gas emissions caused by contamination or deposits in the intake system. **Figure 11.46** shows, with an illustration of the blow-by gas path, the structure of a crankcase ventilation system for a supercharged gasoline engine with direct injection. The coarse oil (droplets > 10 μm, spray, fuel wall film) is separated in the coarse oil separators (e.g. flow dependent areas analogous to the volume separator, flapper, spiral or coarse oil cyclone separators); in most of the current ventilation concepts, fine oil (particle < 10 μm) is separated in fine oil cyclones. Effective coarse and fine oil separation in the mentioned separators, which operate without any auxiliary power supply, requires a minimum differential pressure in the range of 5 to 10 hPa between the crankcase and the intake system of the engine. Because fuel is injected directly into the cylinder, there are considerable deviations related to the system design of the crankcase ventilation as compared to the design in conventional gasoline engines with intake manifold injection. These deviations are described in detail below. Since turbocharging of gasoline engines with direct injection has good potential for reducing the weight to power ratio and fuel consumption – a potential that will significantly enhance these concepts in the future – they will be the focus of the following analyses.

An important consequence of the transition from intake manifold injection to gasoline direct injection is the lack of the cleaning effect of fuel in the intake system. As a result, there is a greater risk of deposits (carbon deposits), especially in the intake ports, turbocharger blades and intercooler, which generally has a negative effect on engine power, fuel consumption and exhaust gas emissions.

Consequently, the lubricating oil components must be kept as separate as possible from the ventilation gas. The impeding factor is the fact that the distribution spectra of oil particle sizes formed in the ventilation gas due to the relatively high mean effective pressures of gasoline engines with direct injection cannot be separated as easily as in the case of the same pressure differences available for the oil separation. The extensive dethrottling of the intake systems of high performance gasoline engines reduces the pressure differences between the intake systems and the crankcase, which can be used for the oil mist separation. Some gasoline engines with direct injection are operated with lean air fuel ratio in stratified charge mode in part load, to optimize fuel consumption. This creates particularly unfavorable conditions for fine oil separation due to the low intake manifold vacuum at this operating point. The changing pressure levels in the intake system of running supercharged gasoline engines generally require two induction points – or even three for some supercharging concepts – and an appropriate number of check valves for the recirculation of ventilation gases into the intake system. The assembly of check valves increases the pressure losses in the ventilation tracks and the complexity levels of the ventilation systems.

In gasoline engines with quality control, considerable quantities of blow-by gases mainly consisting of fuel components near the final boiling point and of water formed during

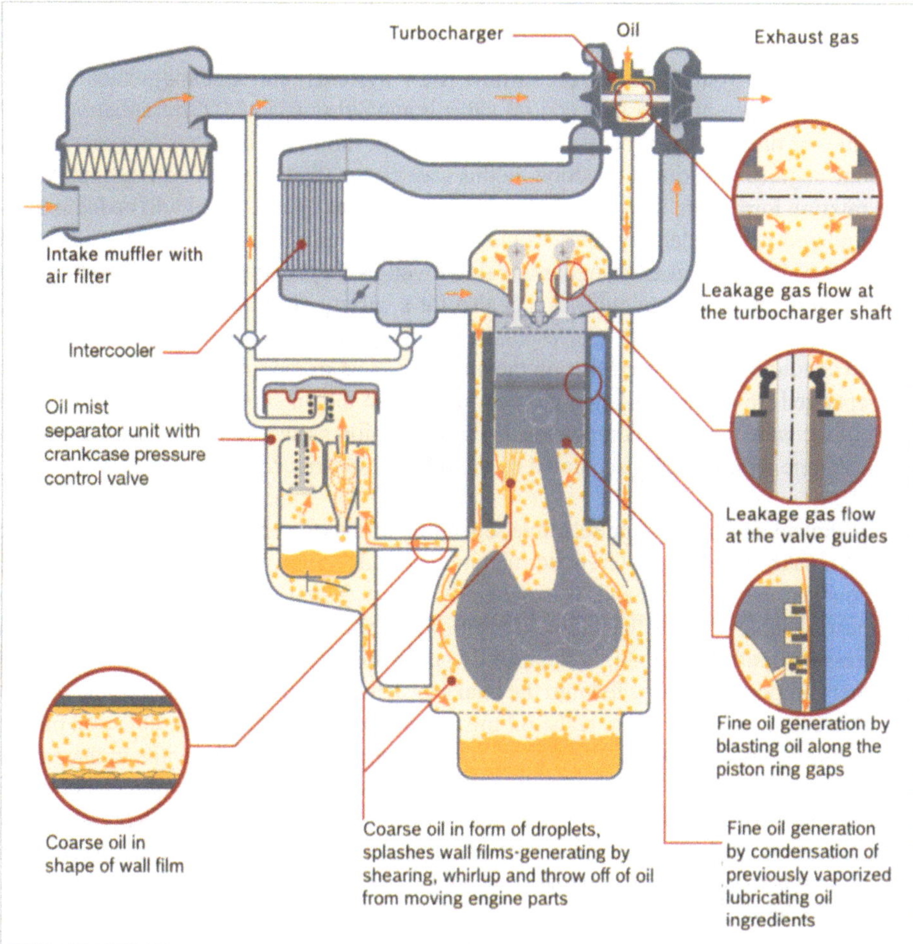


**Figure 11.46:** Structure of the crankcase ventilation system of a supercharged gasoline engine with direct injection and illustration of the blow-by gas path

combustion get into the crankcase. This is in contrast to diesel engines. which operate with excess air in wide map areas. Especially in cold engines, this condensate can accumulate in the crankcase during low load operation and short distance operation. Apart from the resulting reduced lubricating oil quality (decrease in the viscosity), there is a risk that the water and the water-fuel condensate mixture will freeze in winter and lead to functional disturbances in the crankcase ventilation. Additionally, frozen condensate can block the lubricating oil circuit, causing serious damage to the engine. Therefore, PCV systems (*Positive Crankcase Ventilation*) are often used in gasoline engines, especially in those with direct injection, in which the water and fuel condensate are removed more or less completely by scavenging the crankcase with fresh air.

**Figure 11.47** gives an overview on the various crankcase ventilation concepts with their respective advantages and disadvantages. Crankcase ventilation systems can be basically differentiated as systems with fresh airflow (PCV systems) and systems without airflow. The use or absence of a pressure control valve is another criterion of differentiation. Large quantities of fuel must be processed for combustion in short time intervals for supercharged gasoline engines with direct injection. In this way, especially in case of non-optimally realized mixture formation, there is a risk that large quantities of liquid fuel will be fed into the oil sump within considerably shorter time intervals during full load operation. The general tendency of delayed oil exchange intervals as well as the trend towards engines that can be used in many different countries, i.e. also for a variety of oil and fuel qualities, have a critical effect in this context. Corresponding to the given facts, the crankcase ventilation system should be so designed that on the one hand most of the lubricating oil components are separated from the ventilation gas and fed back to the oil sump and on the other hand, most of the water and fuel components in the ventilation gas are fed into the intake system of the engine without being separated by the oil mist separator.

One technical approach to the selective separation of the oil components in the ventilation gas is to make use of the different boiling points of oil on the one hand and water and fuel components on the other hand. It is generally best to install the oil mist separator in the warmest possible areas of the engine-transmission unit, in the direction of the flow behind the pressure control valve i.e. in the negative pressure area. With this arrangement, water and fuel components can be fed into the intake system through the separator to the greatest possible extent in vapor form. If the oil mist separator is installed in the direction of flow behind the pressure control valve, a more or less negative pressure prevails in the separator during throttled part load operation of the engine. However the condition is that the oil separated from the blow-by gas must be fed back into the crank against high pressure. This requires special technical measures such as pumps or locks. An alternative is to store the oil that is separated when the engine is running in a sufficiently dimensioned accumulator and then to feed it back to the crankcase after the engine shut-off. Back-pressure valves that switch open safely in a zero pressure state are required for this.

**Figure 11.47 (right):** Overview of different crankcase ventilation systems for supercharged gasoline engines with direct injection, with their respective advantages and disadvantages: a) controlled PL-PCV system with return in direction of flow; b) controlled PL/FL-PCV system without return in direction of flow; c) controlled FL-PCV system without return in direction of flow; d) controlled PL-PCV system with safety valve; e) controlled PL-PCV system with return in direction of flow; f) controlled ventilation system; g) controlled PL-PCV system;
PL = part load; FL = full load; PCV = positive crankcase ventilation

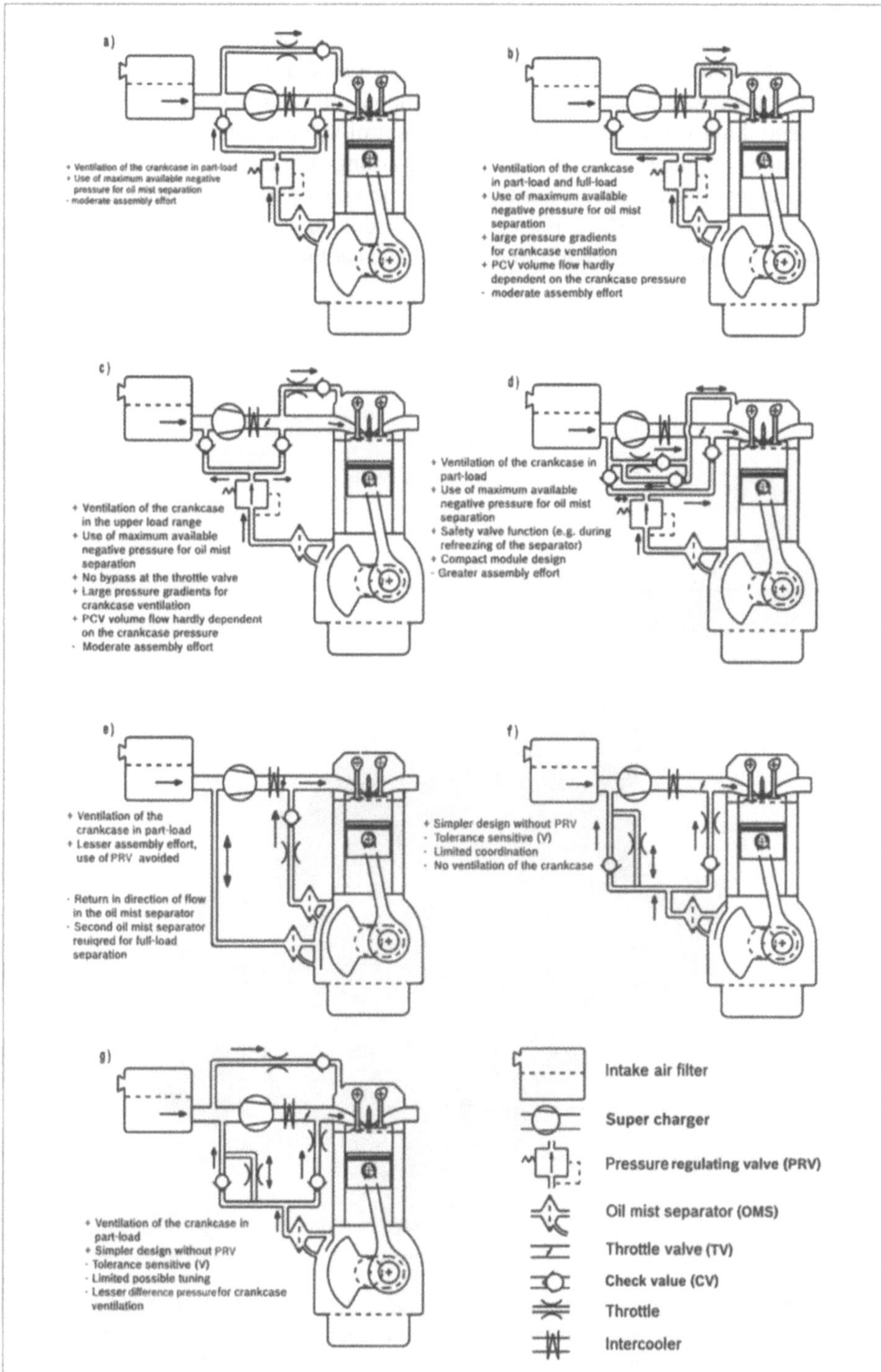
a)
+ Ventilation of the crankcase in part-load
+ Use of maximum available negative pressure for oil mist separation
- moderate assembly effort
b)
+ Ventilation of the crankcase in part-load and full-load
+ Use of maximum available negative pressure for oil mist separation
+ large pressure gradients for crankcase ventilation
+ PCV volume flow hardly dependent on the crankcase pressure
- moderate assembly effort
c)
+ Ventilation of the crankcase in the upper load range
+ Use of maximum available negative pressure for oil mist separation
+ No bypass at the throttle valve
+ Large pressure gradients for crankcase ventilation
+ PCV volume flow hardly dependent on the crankcase pressure
- Moderate assembly effort
d)
+ Ventilation of the crankcase in part-load
+ Use of maximum available negative pressure for oil mist separation
+ Safety valve function (e.g. during refreezing of the separator)
+ Compact module design
- Greater assembly effort
e)
+ Ventilation of the crankcase in part-load
+ Lesser assembly effort, use of PRV avoided
- Return in direction of flow in the oil mist separator
- Second oil mist separator reuiqred for full-load separation
f)
+ Simpler design without PRV
- Tolerance sensitive (V)
- Limited coordination
- No ventilation of the crankcase
g)
+ Ventilation of the crankcase in part-load
+ Simpler design without PRV
- Tolerance sensitive (V)
- Limited possible tuning
- Lesser difference pressure for crankcase ventilation
Intake air filter
Super charger
Pressure regulating valve (PRV)
Oil mist separator (OMS)
Throttle valve (TV)
Check value (CV)
Throttle
Intercooler

**Figure 11.48** shows an example of a crankcase ventilation module for a turbocharged V6 gasoline engine with direct injection. The ventilation concept implemented with this module corresponds to the function diagram in **figure 11.47a**. The functional components: coarse oil separator, fine oil separator, discontinuous oil recirculation via the collection chamber and drain valves, crankcase pressure control valve and the check valves on the clean gas side, are integrated in a single module arranged in the v-shaped space between the two cylinder banks. Moreover, integrated in the upper part of the casing are coreless molded coolant channels, which assist the entire module to warm up more rapidly, thereby preventing the ventilation system from freezing, even at low ambient temperatures.

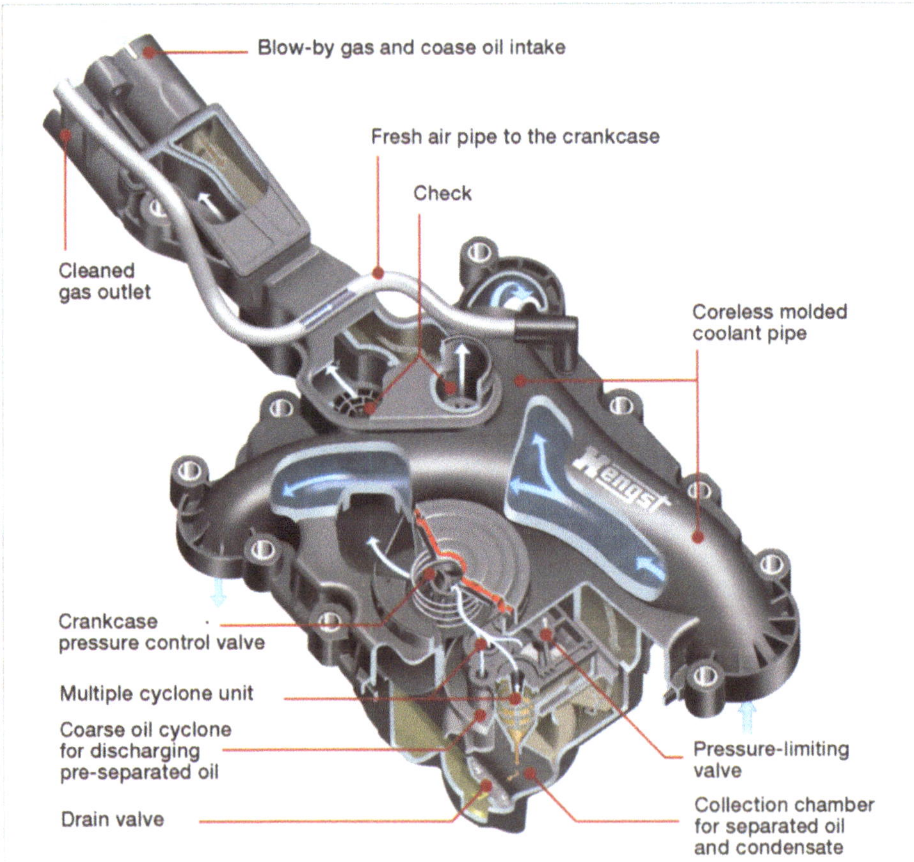


**Figure 11.48:** Sectional view of a crankcase ventilation module for a turbocharged DI gasoline engine with a coarse oil separator, fine oil separator, discontinuous oil recirculation via the collection chamber and drain valves, crankcase pressure control valve and check valves on the clean-gas side [Source: Hengst]

A high integration level like the one implemented in this module not only has advantages related to the cost of installation space and components, assembly and logistics, but also reduces the number of design interfaces to the rest of the engine. This significantly increases the operational reliability of the crankcase ventilation.

## 11.6 Air cycling valves

In the coming years, gasoline engines will still be the dominant drive concept in motoring in various markets. The tighter emissions standards for new engines must be met, along with greatly increased demands for reduced fuel consumption, due to increasing oil prices. Customer demands with regard to engine performance and vehicle dynamics are also increasing continuously. To meet these partly contradictory demands, it is absolutely essential to adopt new technologies for developing charge exchange and combustion processes. The use of air cycling valves for load control is a promising method to meet the above-mentioned demands, which are partly contradictory. As a result, the performance and consumption, particularly of gasoline engines with direct injection, will improve even further.

### 11.6.1 Technology description

The air cycling valve (ACV) represents a procedure that has been around for a long time [44, 45, 46]. This procedure is at a predevelopment stage and makes it possible to control the charge exchange process freely and flexibly. A valve that switches rapidly between the accumulator and the intake valve is used in each intake pipe of the intake manifold as shown in **figure 11.49**. The charge exchange process with the conventional, mechanical valve-train becomes more flexible due to an air cycling valve that can be freely controlled electromagnetically. Thereby, the characteristics of the electromechanical valve-train systems, with regard to control timings and thus the thermodynamic potential, can be achieved to a large extent without the need for completely re-designing the cylinder head mechanism and the combustion chamber geometry. A large number of publications, patent applications and research works about air cycling valves have been published recently. These sources of information reflect the potential assessments of this technology from various perspectives e.g. [44, 48, 49, 50, 51, 52, 55, 57].

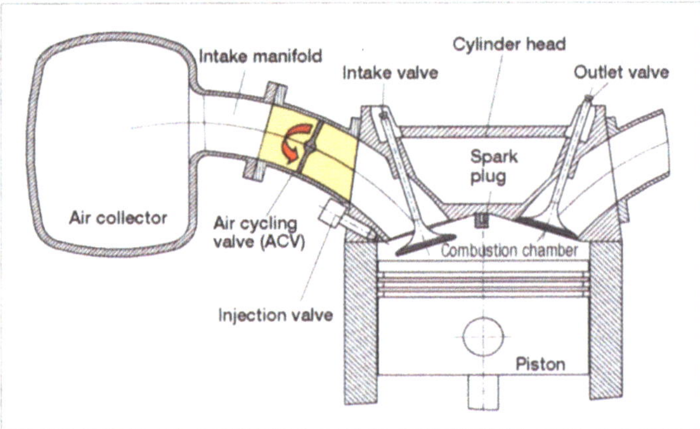


**Figure 11.49:** Air cycling valve in a gasoline engine with direct injection

The use of air cycling valve creates the following possibilities for controlling the charge exchange work in gasoline engines with direct injection:

- **Throttle-free load control:** reduction of the intake air mass without throttle losses by metering with the air cycling valve; analogous to the reduction in the valve lift period with mechanically variable valve-trains.
- **Intake-side cylinder shut-off:** flexible intake side cylinder shut-off for reducing the charge exchange work by means of load point displacement; alternating switch prevents cooling of individual cylinders.
- **Pulse supercharging:** increased engine performance in lower vehicle speed range by inducing pressure waves in the intake manifold and including it in the combustion chamber.
- **Temperature control:** warming and cooling of the intake air by compressing and expanding it in the combustion chamber and in the intermediate volumes between the air cycling valve and the intake valve by piston movement.

## 11.6.2 Design principle and boundary conditions

The air cycling valve is an additional valve in front of the cylinder head and can be freely controlled electromagnetically. Together with the intake and exhaust valve of the engine, the opening and closing action of the valve defines the charge exchange process. A variety of valve concepts (swivel damper, flat-seat valve, and rotary disc valve) can be used due to the basic spatial separation of the air cycling valve from the combustion sequences in the combustion chamber. The concept is an add-on solution for all existing basic engines; **figure 11.49**. In this way, the engine manufacturer's production processes are maintained for the entire engine mechanism. This aids the further diffusion of the technology. On the other hand, the supplier can then be responsible for the operational and quality assurance of the air cycling valves to be adapted.

The technical realization of the air cycling valve for future mass production presents a major challenge. The required switching time, the sealing requirement, the installation space available and the necessary flow cross-section must be taken into consideration in the design; a drive for flexible and precise switching in the engine cycle must also be available.

Due to their highly dynamic functioning, the air cycling valves must have very quick switching-time for closing and opening the intake cross-section. The typical switching time for realizing dethrottled engine operation in the range close to the idle speed range and for pulse charging is approximately 2.5 to 3 ms. The cross-section and internal geometry of the air cycling valve must be designed in such a way that there are no limitations during the nominal power of the engine. A critical feature is to maintain minimal valve leakage – in a range less than 1 kg/h at a pressure difference of $\Delta p$ = 600 mbar – by using the closed sealing element that should remain more or less constant throughout the service life of the air cycling valves.

The air cycling valves are used for dethrottling when the engine is operated in part load and for pulse supercharging in full load operation, and therefore operate almost continuously. This means that they have very high endurance requirements. The valves must be designed for approximately 200 to 300 million switching operations in a temperature range of $T_{ambient}$ = –40 to +130 °C. Power can be supplied from the 12 V vehicle electrical system. Depending on the vehicle speed, the pressure in the intake manifold and the switching strategy, the typical power consumption of each air cycling valve (i.e. per cylinder) is approximately 10 to 60 W. Air cycling valves require an electronic control with "soft-touch" functions, also used in the electromechanical valve actuation, to ensure a soft dead stop of the magnets in the drive; especially for acoustical reasons. The specifications for an air cycling valve used as a rotary valve are given in **figure 11.50**. These high requirements for

**Figure 11.50:**
Air cycling valve by MAHLE
Principle: solenoid-controlled spring-mass resonator with pivoting angle of 45°
Switching time: 2.7 ms,
Leakage: < 1 m³/h at $\Delta p$ = 600 mbar,
Port cross-section: 1300 mm²,
Self-adaptive current control without sensors
Supply from vehicle electrical system: 12 V

the fast and frequent switching of the valves with lesser power loss necessitate extensive development of new drives for this application [47, 48, 49, 53].

### 11.6.3 Thermodynamic potential

Some of the methods that can be implemented with the air cycling valve are explained below in detail. The basic theory is described and substantiated with some test results.

**Dethrottling for fuel economy in gasoline engine with direct injection and homogenous operation**

The air cycling valve represents the control system for metering the required air mass when dethrottling gasoline engines. The air cycling valve takes over the function of the throttle valve, and meters the fresh air. Control strategies in sync with advance closing of the intake; **figure 11.51**, and retarded closing of the intake can be implemented using the air cycling valve along with a conventional valve-train. A combination of advanced and retarded air intake is also possible in two strokes of the air cycling valve in the engine cycle.

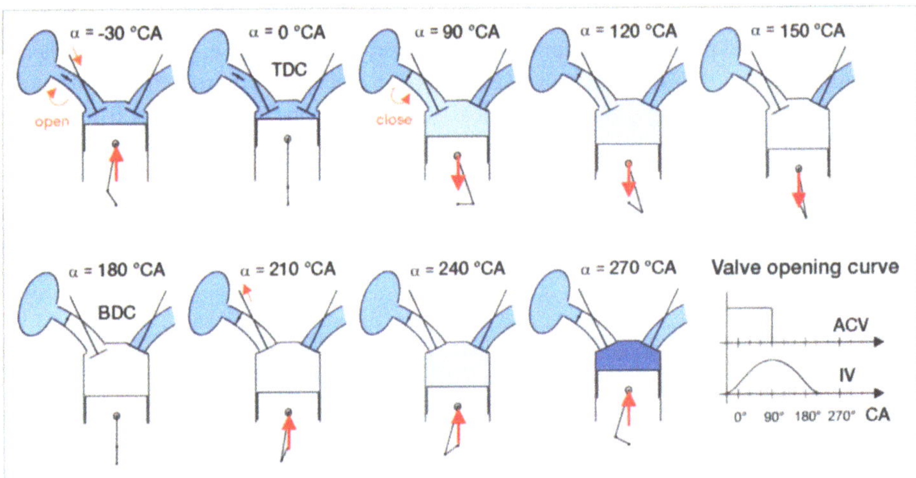


**Figure 11.51:** Dethrottled engine operation with ACV analogous to advance closing of intake

The minimum opening period (hence a very short switching time) of the air cycling valve plays only a secondary role for the dethrottling, because the range, in which the air cycling valve and the intake valve open at the same time can be specified by positioning the opening window of the air cycling valve appropriately. The fresh air is taken in only during this period. However, a short switching time (< 3 ms) offers a larger scope for selecting the position of the ACV opening window relative to the intake phase. Important aspects for dethrottling are the valve sealing and the dead volume between the intake valve and the air cycling valve. These play a significant role with regard to controlling, especially at low loads and consequently small air masses: a minimum quantity of air that leaks into the combustion chamber is already available in the form of leakage air and the air in the dead volume. If a smaller amount of air is required, this cannot be adjusted by the air cycling valve. Throttling is once again required in such a case in order to allow less air into the dead volume and thus the combustion chamber.

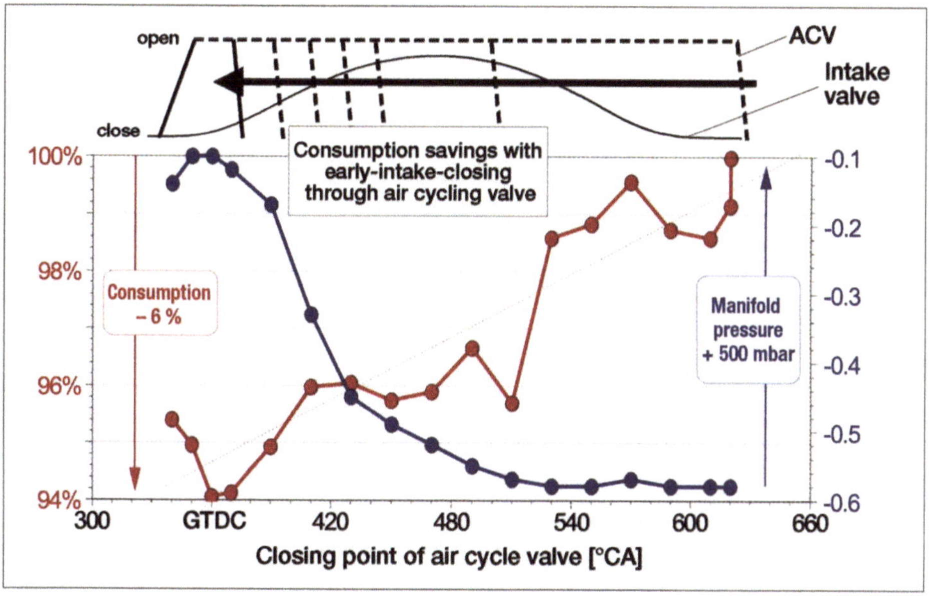


**Figure 11.52:** Reduction in consumption by dethrottling with ACV in a four-cylinder 2 l gasoline engine with direct injection (homogenous) at $n = 2000\ \text{min}^{-1}$ and $p_{me} = 2.0$ bar

**Figure 11.52** shows the test results for dethrottling with the air cycling valve. Conventionally, the load point is maintained by closing the throttle valve if the air cycling valve is opened for as long as the intake valve. If the opening duration of the air cycling valve reduces during a constant intake valve lift, the throttle valve can be opened more and the pressure in the intake manifold increases almost up to the ambient pressure. In an opened throttle valve, the charge exchange work is reduced due to the reducing throttle losses. The minimized loss work reduces the fuel consumption by about 6% for the four-cylinder gasoline engine with direct injection, **figure 11.52**, as compared to the basic engine in the throttle operation.

The dethrottling process has greater potential to save fuel for engines with large displacement, as the engine is often run in part load in the actual driving mode. Measurements with the air cycling valve for dethrottled load control in an eight-cylinder gasoline engine with direct injection were carried out to demonstrate the full potential; the results are shown in **figure 11.53** and **figure 11.54**.

**Figure 11.53** shows a load section for the engine speed of 2000 min$^{-1}$. In this case, the consumption of the test engine that is dethrottled using the air cycling valve is plotted against that of the basic engine. Additional measures to increase the charge movement in the combustion chamber were implemented for operation with the air cycling valve. The measurements show that the reduction in consumption is between 3% and 9%, depending on the load. The charge exchange work was reduced by approximately 50% in the operat-

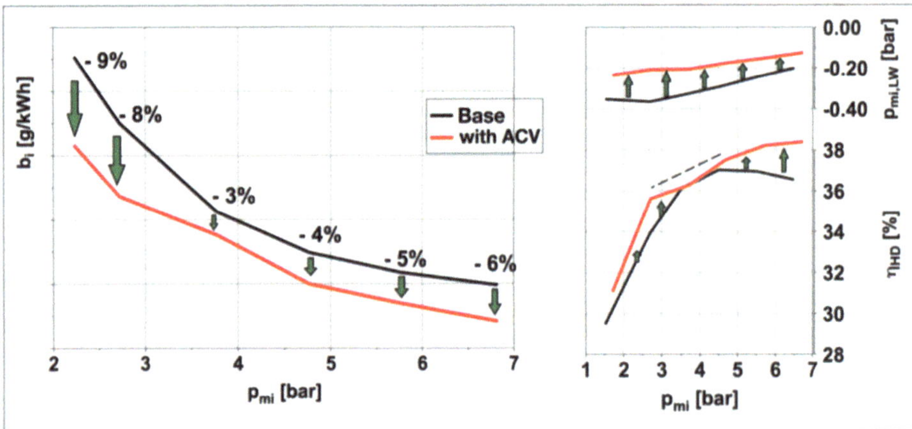


**Figure 11.53:** Load section measurements of fuel consumption with dethrottling by the air cycling valve in a V8 gasoline engine with direct injection at $n = 2000$ min$^{-1}$

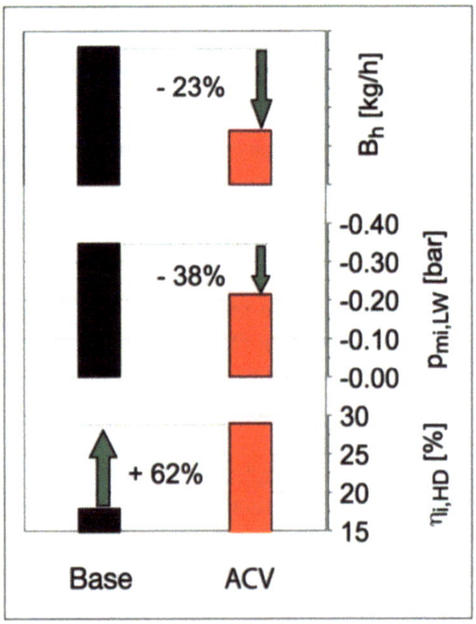


**Figure 11.54:** Fuel savings by dethrottling with the air cycling valve in combination with intake-side cylinder shut-off (4/8) in a V8 gasoline engine with direct injection in idle mode at $n = 700$ min$^{-1}$

ing points tested. Likewise, the air cycling valve operation increases the efficiency at high pressure. It was improved by approximately 0.5 to 1.5%. With this data, fuel savings of more than 5% are expected in the NEDC [New European Driving Cycle]. The possibility of exploiting further potential by process control optimized for operating the air cycling valve and with favorable combustion chamber geometry seems realistic.

It is also important to note here that the represented values were measured at stationary operating points. A transfer to transient driving operation is not fully possible for many other technologies. However, a large amount of fuel can also be saved in transient operation by dethrottling with the air cycling valve operation. Derivative action in the calibration of the engine ECU is not required due to the fast switching-time of the air cycling valve, and the ambient pressure is already present in the intake manifold due to the dethrottling. Therefore, the time-consuming process of charging the intake manifold and the accumulator is not required.

At the idle-speed of $n = 700$ min$^{-1}$ in the eight-cylinder gasoline engine with direct injection, **figure 11.54**, a consumption reduction of 23% as against the basic engine is achieved. For this, four air cycling valves were permanently closed to implement an intake-side cylinder shut-off, while the four active cylinders with the air cycling valves were dethrottled. Here, minimization in the consumption results in a higher load due to reduction in the charge exchange work on the one hand, and a drastic increase in the high pressure efficiency of the active cylinders resulting from the load level displacement on the other hand.

#### The "Miller cycle" in relation to turbocharging

The Miller cycle in a gasoline engine is a procedure in which the intake air temperature in the combustion chamber of supercharged engines is specifically reduced in sync with the early closing of the intake [58]. The temperature decrease can be used to reduce the knocking tendency of the supercharged engine. The air cycling valve blocks the air mass required for combustion in the combustion chamber while the piston is still in the intake phase. The reduced opening period of the intake cross-section must be compensated for by the corresponding increase in the charge air pressure in order to supply the required air mass to the combustion chamber before the air cycling valve closes. The air that has been precooled by the heat exchanger is further cooled by means of expansion due to the piston movement and blocked in the combustion chamber by the intake valve. A higher compression ratio can be realized in the supercharger with the resulting higher compression temperatures, and this can be re-used to increase the engine torque and power without further thermal load on the combustion chamber.

#### Pulse supercharging to increase torque at low engine speeds

Gasoline engines with direct injection are used as supercharged engines and as naturally aspirated engines. In both the models, the air cycling valve can also be used for pulse supercharging. This improves torque response, especially in the lower engine speed range. Increasing the torque at low engine speeds without increasing the displacement is an important objective in the new engine concepts. It is necessary for ensuring good accelerating behavior by vehicles with continuously increasing total weight as well as its desired elasticity. Furthermore, high torque at low engine speeds is the basic prerequisite for future downsizing concepts. Economical vehicles with sufficient driving performances can facilitate the implementation of new consumption strategies.

Using the air cycling valve, pressure waves are created in the intake tract and are used to realize pulse supercharging via specific control of flow and vibration processes. This way, an ideal intake runner charging can be realized for each engine speed. Thus the engine always receives maximum charging in full load. The air cycling valve is closed at the beginning of the intake phase. After the exhaust valve is closed, the upward motion of the piston creates a strong negative pressure in the combustion chamber and in the area between the air cycling valve and the intake valve. Fresh gas flows at high speed into the cylinder as the air cycling valve opens. In the intake manifold, a negative pressure wave is released that flows in the direction of the accumulator and is reflected at the open intake manifold end. It flows back to the combustion chamber in form of a compression wave, where it is blocked by the closing air cycling valve or the closing engine intake valve.

High pressure is not only formed in the combustion chamber now but also in the prechamber between the air cycling valve and the intake valve. Until the beginning of the next charge exchange process it is then available for purging the residual gas from the cylinder via the exhaust valve.

The effect of pressure wave generation for supercharging in the intake system can be utilized in two different ways at low engine speeds, which reduces the problems associated with the feasibility of using increased volumetric efficiency in torque (see further below). Pulse supercharging with the air cycling valve can be used in gasoline and diesel engines and can especially help in significantly improving the dynamics of vehicles with exhaust-gas turbocharging, as shown in [50] in detail.

Test results, given in **figure 11.55**, of an eight-cylinder gasoline engine with direct injection show the potential of pulse supercharging with the air cycling valve to increase the torque in the lower engine speed range. With this method, it was possible to considerably increase the torque as compared to the basic value without air cycling valve operation. An increase of 23% was also achieved at $n$ = 1000 rpm $min^{-1}$. The mentioned values were achieved with the same fixed, short intake manifold lengths. This torque can be erratically built up from one cycle to another as the charging process of the intake system retards the torque build-up.

However, one problem with pulse supercharging is that the increased volumetric efficiency does not often lead to an equal increase in torque. Due to its technique (compression without intercooling), pulse supercharging causes an increase in the temperature of the fresh air in the combustion chamber which, in turn, increases the knocking tendency. As a result, the engine with pulse supercharging cannot be operated with optimum charging at very low engine speeds. The dual opening of the air cycling valve in the intake phase achieves corresponding

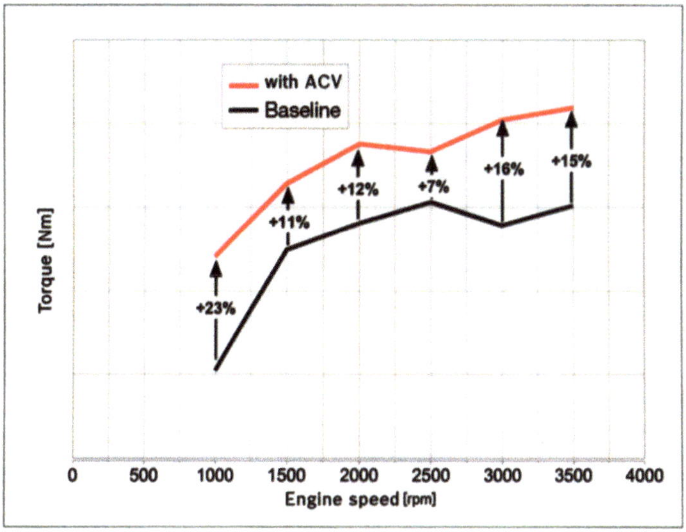


**Figure 11.55:** Torque increase by pulse supercharging with the air cycling valve in an eight-cylinder gasoline engine with direct injection and fixed, short intake manifold length

charge effects. In the intake phase only low pressure differences must be built-up via the air cycling valve during each switching process.

### Thermal loading

The air cycling valve opens very late in the intake cycle, once, for a short duration, close to the bottom dead center of the piston. While it is open, the already discussed high flow rates occur in the combustion chamber, and fresh air warms up. While this warming of the air is an unwanted side effect of pulse supercharging, which increases the knocking tendency, it is especially required and used in the case of thermal loading.

**Figure 11.56** shows the temperature plotted against the crankshaft angle during a thermal loading process in comparison to the conventional intake process with a completely open air cycling valve. Initially in thermal loading operation, with closed air cycling valve and opened intake valve, the temperature decreases drastically, due to the expansion. After the air cycling valve opens and the intake valve has closed completely, the air temperature increases during the inflow in a short period of time by an average of $\Delta T_{ES} = 23$ K. In the process shown, the air mass increases by 3% as compared to the intake process with open flaps. This results in a high air mass due to the considerably higher air temperature in the combustion chamber. Due to the thermal loading, the temperature at the end of the compression is approximately $\Delta T_{VE} = 40$ K higher; this assists the cold start and warm-up process.

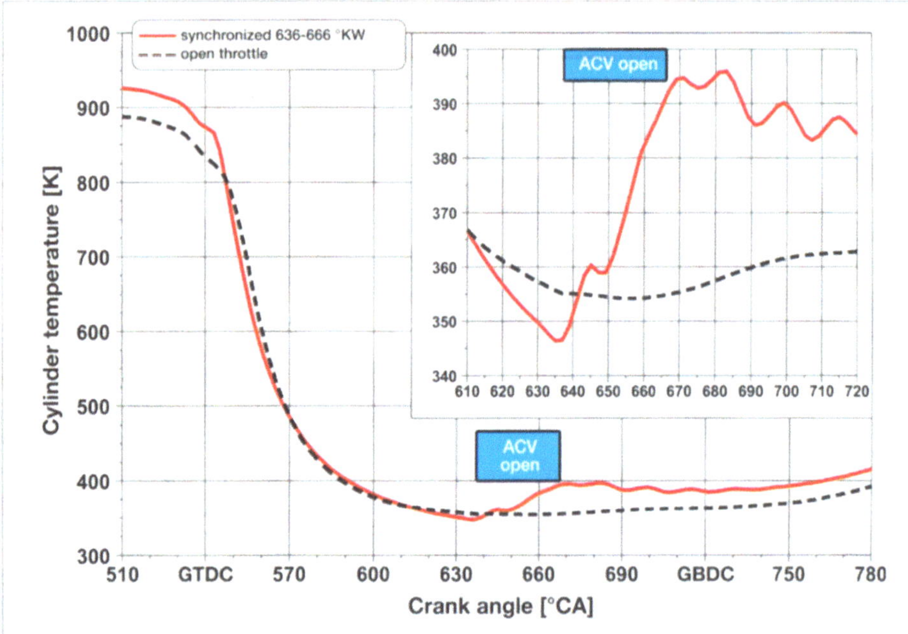


**Figure 11.56:** Thermal loading in a four-cylinder gasoline engine with direct injection (2.0 l. 4 V) at $n = 1000\ \text{min}^{-1}$; calculation of the average gas temperature by charge exchange work analysis

**Summarized assessment**

The air cycling valve can have various effects on the charge exchange process of a gasoline engine as compared to a conventional valve-train. This makes it possible to combine the air cycling valve with different current and future development concepts. Two possible technology combinations are shown in **figure 11.57**.

The initial approach is the use of the air cycling valve for dethrottling, mostly for large-displacement gasoline engines. This eliminates much of the charge exchange work, leading to increased fuel savings in part load and the current test cycles. With homogenous operation, a complete system for worldwide use in gasoline engines with direct injection can be realized. Using the air cycling valve for dethrottling is very viable economically as compared to the competing dethrottling technologies, as shown in **figure 11.58**.

Moreover, nothing but already proven and cheap three-way catalytic converter technology is now required for exhaust-gas aftertreatment. Cold start assistance and pulse supercharging during acceleration can also be used.

The second method shown in **figure 11.57** is the effective power assistance of downsizing strategies for gasoline engines. Dethrottling plays a minor role in this case, as the engine is generally operated by lesser displacements at higher load levels. A large dimensioned

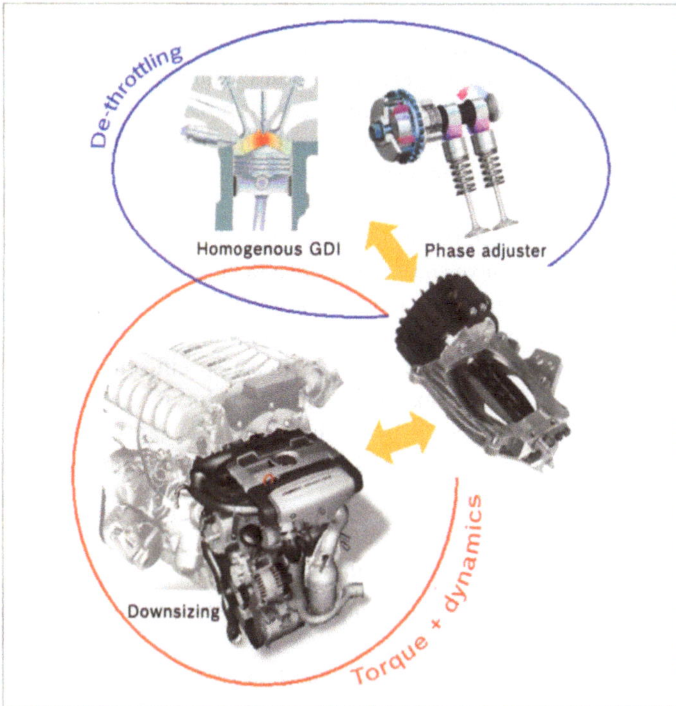


**Figure 11.57:** Engine development concepts, which can be effectively supported by the air cycling valve

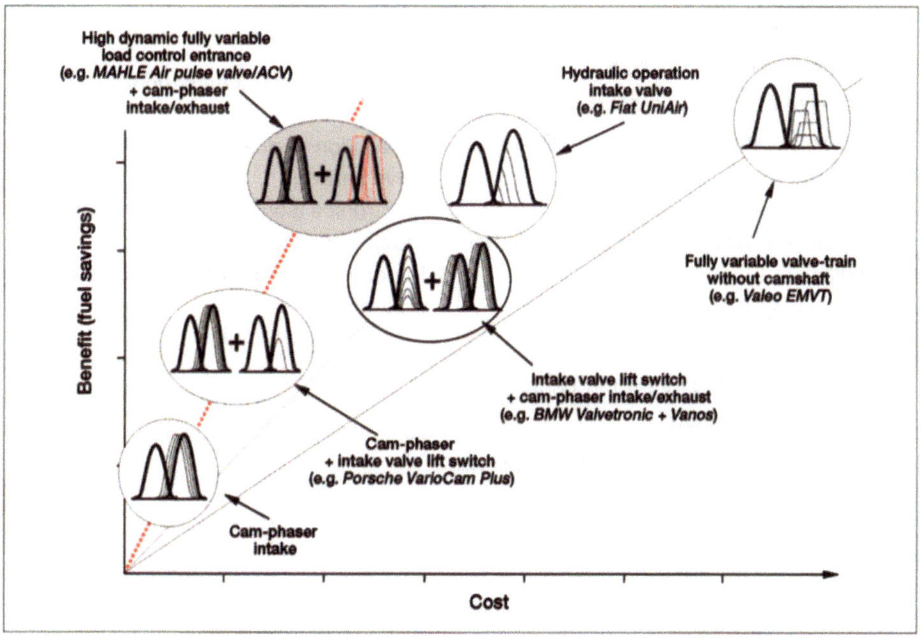


**Figure 11.58:** Classification of the air cycling valve technologies (*air cycling valve*) for dethrottling of gasoline engines based on the diagram in [59]

turbocharger, which has the known disadvantages of low torque in the lower engine speed range and during transient engine operation, is used in a better way to ensure nominal power. Here, the required low end torque can be provided by the air cycling valve. Moreover, much faster torque build-up than in throttle valve guided systems is possible by immediately (already in the next cycle) adapting the air cycling valve switching periods to the sudden load variation requirement. The thermal loading options an also be used in this case.

The only significant disadvantage is the significant amount of development required in the mechanics and control of the air cycling valve and to calibrate it to the engine. On the other hand, the extensive variability resulting from the freely controllable switching periods in the engine cycle can be used for all the procedures and applications that use the air cycling valve; **figure 11.59**. The possibility of a cycle-specific and cylinder-specific circuit should be highlighted here. Tolerance dependent on unequal load distribution to the individual cylinders can be adjusted later by the ECU. Similarly, if the boundary conditions change, the mode of operation can be changed from one cycle to another, e.g. by switching the early closing and late opening of the air cycling valve.

| **Advantages** | **Disadvantages** |
|---|---|
| Fully variable (cycle selective and cylinder selective switching)<br>Air-side control of load<br>Switching from e.g. early intake valve closing (EIVC) to late intake valve opening (LIVO) from one working cycle to the next<br>No influence on power output<br>Balancing of function tolerances<br>Dry-running operation without breakdown (c.f. EMVT)<br>Known processes for component manufacture in valve-train area, despite full variability<br>Function diagnosis while operational<br>Good cost/benefit ratio<br>Diversity (dethrottling, pulse turbocharging, cold start support, etc.) | High development costs in the following areas:<br>♦ Adjustment of piston geometry and injection nozzle characteristics to air pulse valve operation<br>♦ Charge movement measures<br>♦ Acoustics<br>♦ Technology for manufacturing plug and sealing<br>♦ Prevention of dual stop armature/ valve sealing<br>♦ Electronics (sensor-free control with soft touch function<br>♦ Function development for engine control unit |

**Figure 11.59:** Functionality and economy aspects

# 11.7 Canister purge

The boiling characteristic of conventional gasoline fuel mean that a fraction of the fuel is present in vapor form at thermodynamic equilibrium at the temperatures prevailing during vehicle operation. As presented in detail in [60] and [61] this could result in hydrocarbons with low boiling points reaching the atmosphere via the tank ventilation, particularly at high ambient temperatures and after vehicle shut-off. In the past, evaporative losses contributed significantly to air pollution, particularly in overcrowded urban areas. Limits to evaporative emissions were therefore fixed in early 1970 in all important markets based on the first legal regulations in the USA, compliance with which is verified by the SHED test (*Sealed Housing of Evaporative Determination*) for vehicle certification.

To fulfill the legal requirements for evaporative emissions, the gaseous hydrocarbons escaping from the tank through the tank ventilation are collected in the activated carbon of an activated carbon filter (ACF). **Figure 11.60** shows the structure of a canister purge system of a vehicle for the US market. The tank is connected with the ACF via an operating vent line. This line is used to direct the gases loaded with hydrocarbons emanating from the tank to the ACF at idle and during vehicle operation. The gaseous hydrocarbons are largely adsorbed by the activated carbon of the ACF and the air escapes into the atmosphere through an opening in the ACF. The function of the roll-over valve located near the tank is to prevent liquid fuel from escaping via the vent line if the vehicle has an excessive slant or rolls over. When ambient pressure rises, (e.g. during descent) and when the fuel tank is being emptied (during vehicle operation), the tank is ventilated by the ACF and the vent line due to the reversed flow direction. The filter must be purged with ambient air on a regular basis in order to prevent the ACF becoming overloaded with hydrocarbons, which leads to a breakthrough of fuel components. To do this, some of the intake airflow is drawn in by the filter when the engine is running, due to the pressure difference between the intake manifold (behind the throttle valve) and atmosphere. As the pressure differences between the atmosphere and the intake manifold vary in several areas during the vehicle operation, the amount of purge air is metered out, depending on the load and rotational speed of the engine, by an electric canister purge valve driven by the engine management system. The amount of ACF purge air is usually limited to approximately 1% of the amount of intake air drawn in by the engine. The metered addition of the ACF purge air to the intake air directly influences the mixture formation in the engine. Since, in addition, the operating life of the canister purge valve is approximately equal to the service life of the vehicle, demands with respect to the long term stability of the metering accuracy of the canister purge valve are high. As a good compromise between cost and functionality, electromagnetic stroke valves, pulse-width modulated with frequencies between 10 and 15 Hz and with variable pulse-pause ratio, are prevalent in the market as canister purge valves. Proportional valves have the basic advantage of continuous delivery flow through the valve when compared to stroke valves. Proportional valves have not found favor because of the high technical effort necessary to ensure stable long term compensation for interfering factors in the characteristic curve (temperatures, vehicle system voltages, friction, source and core setting in the valve seat elastomer).

There is a legal requirement for vehicles intended for the US market which states that both the hydrocarbons escaping from the tank during idle periods and during vehicle operation and the gases released from the vehicle during refueling must be contained. In these ORVR concepts (*Onboard Refueling Vapor Recovery*), a refueling vent line with a considerably larger diameter than an operating vent is provided e.g. going from the tank filler neck to the ACF. The volume of the ACF in the ORVR concepts increases by a factor of around four because of the comparatively large amount of vapor that has to be stored for a short time. The amount of purge air through the ACF also increases accordingly, where the purge air has to be added to the intake air without negative effect on mixture formation, the emission of toxic exhaust gases and the operational behavior of the engine.

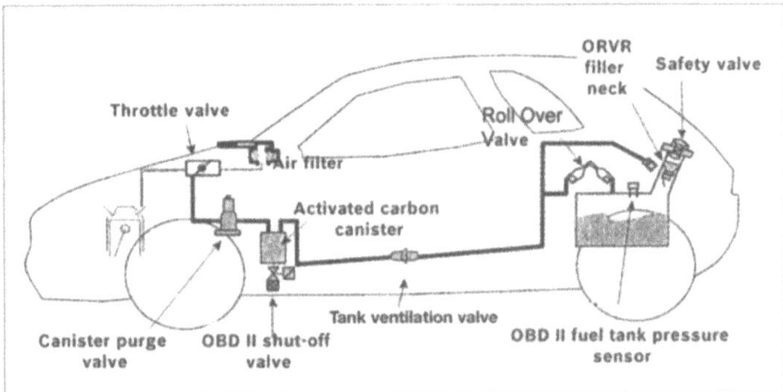


**Figure 11.60:** Diagram of a tank ventilation system for the US market

Specific demands are made of characteristic curves of canister purge valves in the case of engines with gasoline direct injection, which operate at part load with stratified charge. As these engines have to be operated at low intake manifold vacuums in large ranges of the map as shown in detail in [61], both the canister purge valve and the lines should have large cross-sections so that the ACF can be purged sufficiently during vehicle operation. On the other hand, exact metering of small amounts of purge air has to be possible in homogeneous operation with high intake manifold vacuums. A solution for this is to have a valve with pneumatic precontrol. In this valve, the intake manifold vacuum acts on a diaphragm, which prethrottles the volumetric flow during high vacuum at the exhaust nozzle of the valve by means of a valve needle. The metering of the gas flow is carried out in the same way as in conventional stroke valves by the pulsed solenoid armature. A large dynamic range can be achieved by connecting the solenoid armature and choker valve in series. **Figure 11.61** shows a diagram of this type of valve.

As compared to naturally aspirated engines, the degree of complexity of the canister purge in turbocharged gasoline engines with direct injection also increases in the same way as in the crankcase ventilation systems covered in section 11.5. As shown in **figure 11.62**, a check valve is placed in the vent line between the canister purge valve and the area of the intake system that is charged by the charging pressure in order to prevent a flow from the intake manifold across the canister purge valve and the ACF into the atmosphere. With the objective of being able to regenerate the ACF in the charging operation, this vent line is

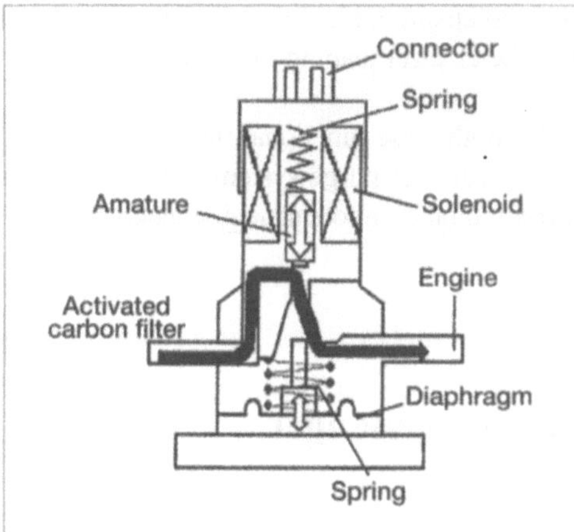


**Figure 11.61:**
Operational chart of a pneumatic precontrolled duty cycle solenoid valve

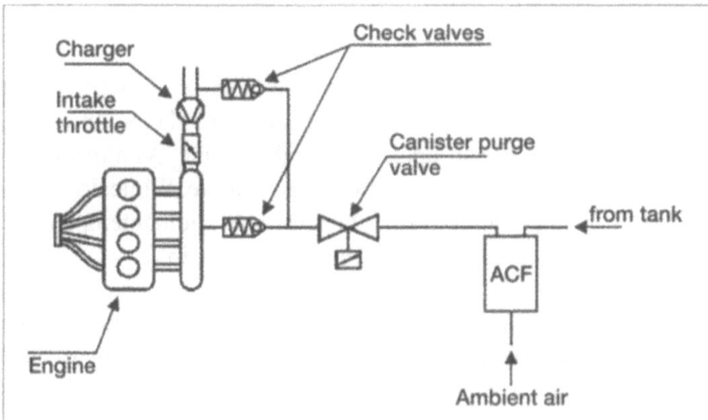


**Figure 11.62:**
Operational chart of a canister purge system for charged gasoline engines

bifurcated at the back of canister purge valve and the second line is conducted in the flow direction in front of the charger. Another check valve placed in this line prevents a bypass flow around the throttle valve in case of largely closed throttle valve during intake operation.

Since small amounts of purge air have to be metered out exactly at the canister purge valve during the natural aspirated operation of the engine at high pressure differences, and on the other hand, large amounts of purge air are desirable for the ACF during charging operation at small pressure differences between the atmospheric pressure and the pressure in the induction pipe of the charger, an alternative ACF purging concept is available for charged gasoline engines in which canister purge valves (controlled independently to each other by the engine control) are positioned in both vent lines leading to the intake system in front of and behind the charger. In this case, if the check valves are removed then the canister purge valve, which is effective during intake operation, can be designed for ensuring small volumetric flows at high pressure differences, whereas the canister purge valve, which is effective during charging operation, can be optimized to ensure large volumetric flows at small pressure differences in the vent lines in front of the charger leading to the intake system.

It should be noted that the overall system costs for both these functionalities can be reduced by functionally integrating the components of the canister purge system in the crankcase ventilation system and, if required, removing individual components e.g. check valves.

## 11.8 Auxiliary heating systems for passenger compartments

### 11.8.1 Introduction

The increase in the efficiency of engine performance in the case of gasoline engines with direct injection implies that less dissipated heat will enter the coolant circulation. In winter, this implies an extension of the warm-up period upon starting the engine. As a result, the passenger compartment is heated up comfortably only after a longer drive. While driving in part load range around the city or in stop-and-go traffic, the desirable coolant temperature of a minimum of 85°C is probably not achieved at any time during the trip.

Meanwhile, this heating power deficit – and along with it the losses that resulted from the defogging system and heating up of the passenger compartment – has been known since the beginning of 1990 in the case of diesel engines with direct injection. The solution lies in auxiliary heating systems, which combine demands for comfort and safety with environmental protection achieved through a concomitant reduction in fuel consumption.

### 11.8.2 Auxiliary heating systems

The following list names only those systems that are engaged while the engine is running. Nor does the list claim to be complete; it is restricted to series applications with significant volume. Systems that compensate the heating power deficit through modifications in the combustion internal engine are not considered here, neither are the heat accumulator and latent heat accumulator.

- Electric auxiliary TC heater, air-side
- Electric coolant heater
- Auxiliary fuel heater
- Visco heater
- Auxiliary exhaust-gas heater.

The auxiliary PTC heater is discussed in detail because it is becoming increasingly popular. Approximately two thirds of all the auxiliary heaters in consumption-optimized vehicles (diesel and gasoline engines with direct injection) today are auxiliary PTC (*Positive Temperature Coefficient*) heaters. Different types of auxiliary heaters are described below.

### 11.8.3 Electric auxiliary PTC heater, air-side

**Integration**

In most of the applications, the PTC auxiliary heater, as part of the air conditioning system, is placed parallel to the water-cooled heat exchanger in the direction of airflow as the last of the components for the air stream to pass through. Heated air is tapped for defrosting, cockpit air vent and leg room immediately after passing the auxiliary PTC heater; **figure 11.63**.

In addition to integrating them in blower cabinets, there is an increased demand for improvement of comfort by placing heat sources locally within the vehicle interior. In these "decentralized heating concepts", the PTC auxiliary heater is inserted in an air duct to provide additional heat, e.g. in the rear leg space.

**Assembly**

PTC ceramic modules are located between aluminum lamellae. Typically, PTC ceramics modules are rectangular, parallel-plane modules with laminar metallization and their numbers define the performance. The electrical connection is realized by metals sheets in

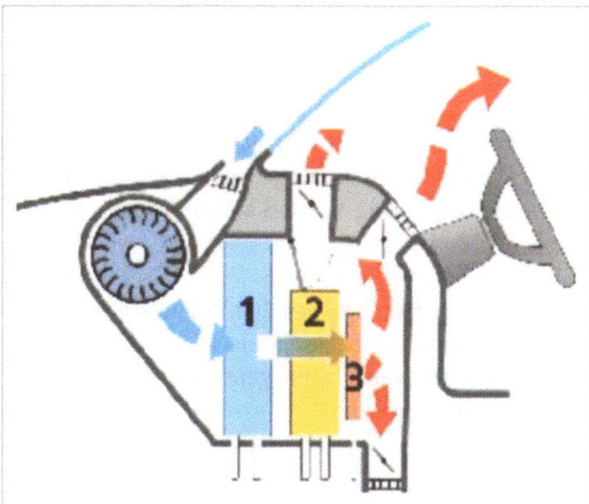


**Figure 11.63:**
Schematic diagram [Source: catem]
1 Evaporator,
2 Water cooled heat exchanger
3 PTC auxiliary heater

a plastic housing, one end of which functions as an electrical interface. Stacking several lamellae with the corresponding number of PTC ceramic modules makes it possible to assemble different heat levels separately. The package stacked layer-wise is finally pressed together by a spring element and held together by a plastic case; **figure 11.64**. The electrical contact of PTC heating resistors connected in parallel is guaranteed over the service life by the continuous compression with spring pressure. The spring pressure also causes very good heat conductivity between the PTC ceramic modules, which heat up when power is supplied and provide heat to the airflow via the aluminum lamellae.

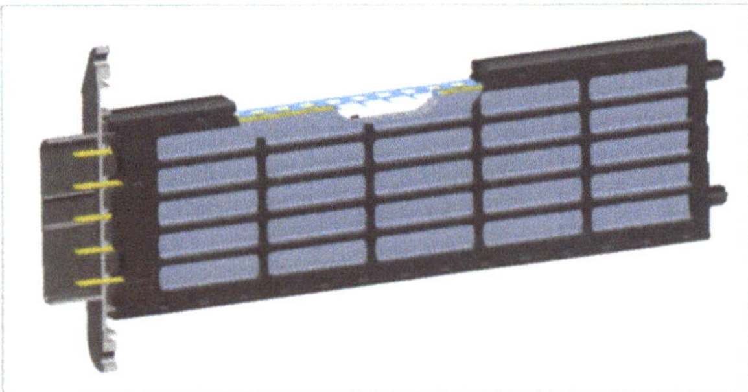

**Figure 11.64:**
Laminar PTC auxiliary heater
[Source: catem]

### Operating mode of PTC characteristic

The electrical resistance of the PTC modules is temperature-dependent; **figure 11.65**. During start-up in cold conditions the electrical resistance is low and the resulting power consumption high. The electrical resistance of the module increases with increasing temperature (by means of prewarmed air because of the increase in the coolant temperature at the water-cooled heat exchanger) and the auxiliary heater is self-regulated automatically. The result is lower current consumption. The characteristic resistance curve ensures suitable dynamics for the heating power requirement and prevents excess temperatures in the auxiliary heater or surrounding components.

### Power outputs

The definition of power in 12 V vehicle electric systems extends from 400 W heating elements to a maximum of 2 kW. The generator is the limiting element. For high vehicle system voltages, far higher electrical power outputs (42 V to 3.5 kW; 500 V to 6 kW) can be implemented.

Decentralized heating elements have a power output, between 200 W and 400 W and are inserted in pairs per vehicle. A considerable side effect makes the electrical auxiliary heater more effective: the additional load on the engine created by the generator reduces the cold start phase and carries increased coolant temperature with it.

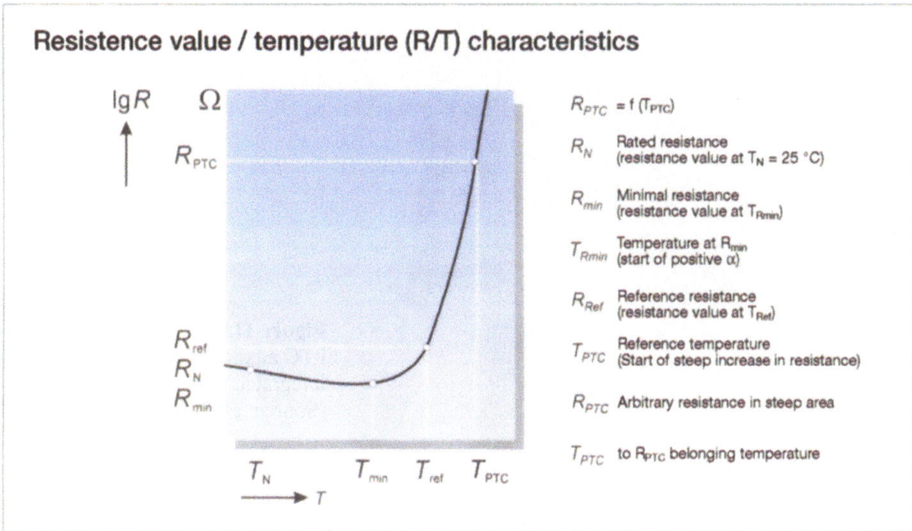


**Figure 11.65:** Resistance/temperature profile curve [Source: catem]

### Control concepts

The laminar PTC auxiliary heater is switched on or off by means of relay based on need and existing starting conditions. The concept is to distribute the heating power with the help of two relays into the heat level 1 = 1/3; heat level 2 = 2/3 and both together as heat level 3.

The integration of power control in the PTC auxiliary heater makes it possible to efficiently utilize the available energy and manage energy by taking into consideration the load levels in different driving situations. Such heating elements communicate with the bus system present in the vehicle e.g. LIN Bus or CAN Bus and have additional functions as required e.g. diagnostic capability and overload protection. The heating power is regulated by the DF signal (DF = dynamo field, exciter field) provided in the generator and regulates the maximum possible heating power up to approximately 100% generator capacity.

This continuously variable load control is realized by a periodic switching on and off (PWM, pulse width modulation) of the heating circuit. The electrical loads on the vehicle electric system are minimized by means of suitably transferred control of individual heating circuits of the PTC auxiliary heater, and simultaneous improvement of the EMC behavior (EMC = electromagnetic compatibility). The power dissipation of the transistors is introduced into the airflow by an efficient cooling concept; **figure 11.66**. Another advantage of the PTC auxiliary heater with integrated load control is that it distributes heat across the entire heat exchanger surface. This implies better temperature homogeneity in the air flowing out of the auxiliary heater.

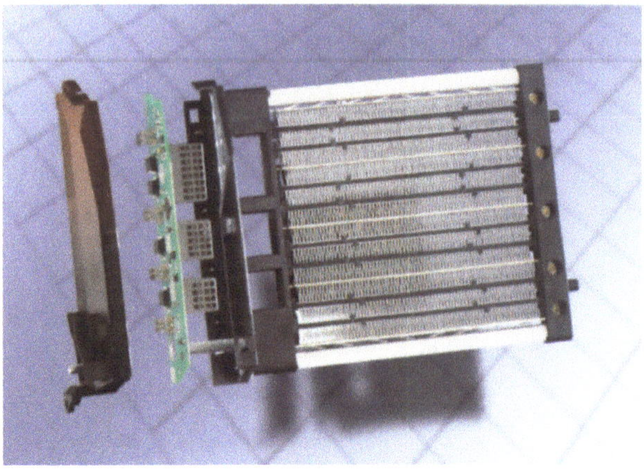

**Figure 11.66:**
PTC auxiliary heater with integrated power control [Source: catem]

The characteristics of the PTC auxiliary heater with integrated power control are as follows:

- High power density with less weight
- Spontaneous heating of the air stream
- Intrinsic safety through automatic control and limited maximum surface temperature
- Flexible, long-lasting concept
- Good integration capability in vehicle heating systems
- Cost-effective system
- Availability of different control possibilities
- Additional warming of the engine
- Appropriate generator power is required.

## 11.8.4 Electric cooling water heater

### Integration

To integrate the heating power in the coolant, the heater is either integrated as a flow heater with hose connections, **figure 11.67**, or as flange connection at the engine in the coolant circuit. The compact structure and simple integration also make it possible to incorporate the heater in existing vehicles at various places in the engine compartment.

### Assembly

The individual heating cores, whose number can be changed as per the power desired, consist of a coiled heating wire. This is thermally coupled to the aluminum casing and makes an electrical contact at the lower end, while the other end leads to an electric control module.

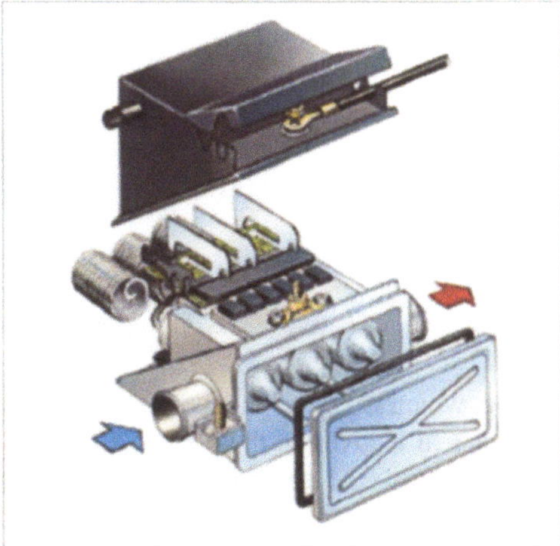

**Figure 11.67:**
Electric coolant heater [Source: catem]

The control module is electrically insulated and fixed to the aluminum casing. A plastic cover ensures the necessary protection of the electric and/or electronic components and connects the control electronics with the vehicle via the integrated signal connector. The coolant heater is connected with the vehicle mass via the aluminum casing. A bolt mounted at the power electronic provides the positive terminal.

### Operating mode

The power electronics control the individual heating cores linearly depending on the electric power available by taking into consideration the DF signal in the vehicle. The heat emerging from the heater coil is dissipated to the coolant flowing via the aluminum casing. The dissipated power of the control electronics is transferred to the casing in a similar manner by an optimized internal structure in order to be able to operate the auxiliary heater for as long as possible and avoid premature thermal regulation. Due to the entry of heating power into the coolant and the losses associated with it over the surface of the engine and cables, the effect is delayed during cold start. In order to use the auxiliary heating power of cooling water as effectively as possible, integration into the heat flow is advisable.

### Power outputs

The electric coolant heater is subject to the same limitations in power output as the electric PTC auxiliary heater. The normal power outputs are between 1.5 kW and 2.0 kW.

The characteristics of the electric coolant heater are as follows:

- Compact design
- Simple integration
- Robust, weight-optimized concept
- Electronic control

- Additional warming of the engine
- Delayed operating mode through heating up of the coolant
- Corresponding generator power is necessary.

### 11.8.5 Auxiliary fuel heater

#### Integration

The auxiliary fuel heater enables auxiliary heating, which is engine independent, through the use of primary energy fuel. The heat energy gained is given to the vehicle's heat exchanger and distributed in the cabin by existing air channels. The engine is warmed by the residual heat in the coolant.

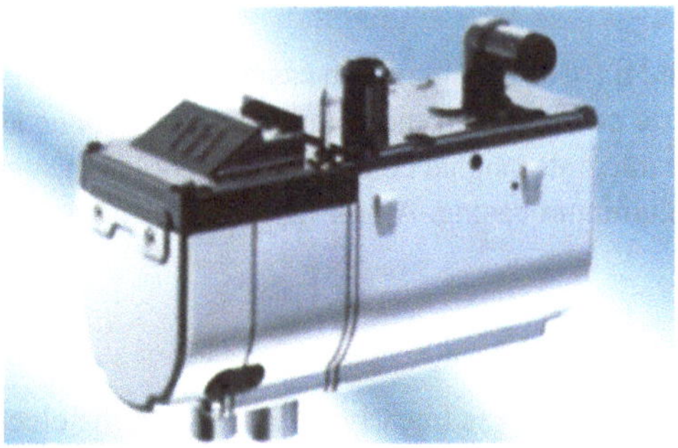

**Figure 11.68:**
Fuel auxiliary heater [Source: Eberspaecher]

#### Assembly

The central component is the combustion chamber, which is fitted with a glow plug and surrounded by a coolant. Other components include a combustion air blower and sensors such as flame and temperature sensors. The casing contains the ECU and displays the connections for fuel supply and combustion air. Integration into the coolant circuit is carried out via hose connections. The exhaust gas is conveyed through a muffler; **figure 11.68**.

#### Operating mode

Combustion air from the surroundings and fuel from the vehicle tank are mixed in the combustion chamber and ignited. The heat exchanger transfers the heat energy to the vehicle-specific coolant system with a very high efficiency of > 80%. The fuel auxiliary heater is the only system described here that can be upgraded to an independent car heating system with a rebuilding kit.

**Power outputs**

Very large heating power outputs can be realized with the independent system. The power outputs in passenger cars are usually from 3 to 5 kW.

The characteristics of the auxiliary fuel heater are as follows:

- Engine independent, no substantial loading of vehicle electric system
- Large heating power outputs can be achieved
- High efficiency
- Requires complex infrastructure for connection to fuel supply, coolant circuit, control and power supply and the emission of exhaust gas
- Additional warming of the engine
- Delayed operating mode due to heating of the coolant.

## 11.8.6 Visco heater

**Integration**

The visco heater requires propulsion by the engine using a belt and it is integrated in the coolant circuit of the vehicle.

**Assembly**

The mechanical system contains a casing with hose connections, filled with viscous oil and a belt pulley which transfers rotational motion to the movable plates; **figure 11.69**.

**Figure 11.69:**
Visco heater [Source: Ventech]

### Operating mode

At engine start, the movable plate in the casing is set in rotational motion by propulsion of the belt pulley and mixes the oil. The shear forces arising from it release heat energy, which is dissipated into the circulating coolant.

### Power outputs

The power output largely depends on the speed of rotation. Approximately 1 kW of heating power output is realized in idling position. The rated power output of 3 kW is achieved from a rotational speed of 1500 $min^{-1}$. Another heating effect arises due to the additional engine load.

The characteristics of the visco heater are as follows:

- Simple heating system
- Additional warming of the engine
- Suitable for vehicles with low generator output
- Delayed operating mode due to heating up of the coolant
- Heating power output dependent on speed of rotation.

## 11.8.7 Auxiliary exhaust-gas heater

### Integration

The auxiliary exhaust-gas heater is integrated in the exhaust branch and connected to the coolant circuit.

### Assembly

The system consists of a heat transmitter which is embedded in one of the two tracts. The second tract forms the gas bypass. A controllable exhaust-gas flap is placed before this division of exhaust gas flow into two; **figure 11.70**.

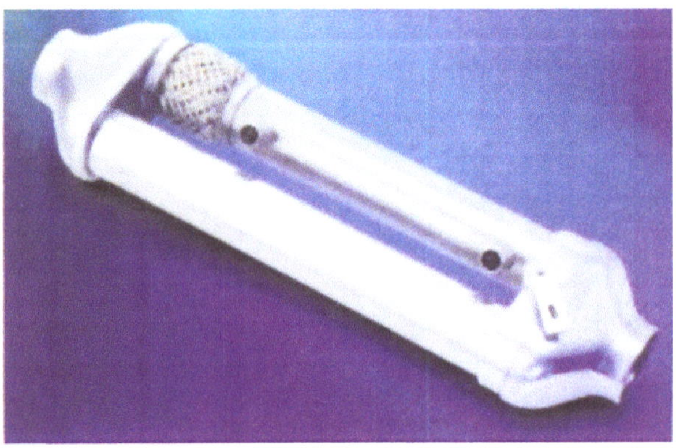

**Figure 11.70:**
Exhaust-gas auxiliary gas heater (Source: ArvinMeritor]

**Operating mode**

The auxiliary exhaust-gas heater uses the heat energy contained in the exhaust gas. This amounts to one third of the heating value of the fuel used. The exhaust gas temperature is transferred through the heat exchanger to the coolant. To regulate the heat input, the exhaust-gas flap situated in front of the heat transmitter and bypass continuously conducts the exhaust gas mass flow across the respective tracts. Under strongly fluctuating operating conditions, such as very high or low exhaust gas temperature, the bypass protectively prevents evaporation of the coolant in the heat transmitter and stops condensate formation in the exhaust-gas cable.

**Power outputs**

The delivered power output in the coolant fluctuates strongly, due to its dependence on engine speed and engine load and does not always ensure that auxiliary heating is available. In gasoline engines, the heating power output can go from 500 W in the idling position up to a maximum output of 5 kW.

The characteristics of the auxiliary exhaust-gas heater are as follows:
- Delayed operating mode due to heating of the coolant
- Heavy dependence of the heating power output on engine speed and engine load
- Complex system
- Installation position places extremely high demands on materials and design.

## 11.8.8 Summary and future prospects

The task of compensating for the heating power deficit has to be solved for all car classes and propulsion systems. Auxiliary heating systems as described above offer an effective contribution here so as to meet customer demands for comfort. The main criteria for selecting auxiliary heating systems are cost, application effort, response characteristic and heating power output as required.

Auxiliary fuel heaters offer the highest heating power outputs in modern vehicles and are the only heaters that can be upgraded to independent vehicle heaters. The auxiliary exhaust-gas heater is less popular than the visco heater because of the high cost of integrating it in the vehicle. In addition, the response characteristic is delayed in the systems that use a coolant as a transmitting medium.

Taking into consideration that approximately 50% of all car trips take place within a range of 6 km and a duration of 15 minutes, the PTC auxiliary heater is the fastest means to remove windscreen fogging and supply the heat needed to warm up the vehicle cabin.

Combined with an efficient generator, electric auxiliary heating systems are basically in a position to compensate for the heating power deficit of fuel-saving vehicles. Here the PTC auxiliary heater has many advantages to offer in the cost/benefit analysis, and this accounts for its popularity amongst auxiliary heating systems. The range, from a simple PTC auxiliary heater built up in stages to the bus-capable PTC auxiliary heater with integrated power control, covers every claim to comfort.

Without any limitation of the available electric energy in the present-day 12 V vehicle electric system, still larger heating power outputs with corresponding comfort gain would be feasible. Here the hybridization of vehicles offers a chance to create additional values. Vehicle electric systems with much higher voltage ranges, one would be in a position to install electric auxiliary heaters with to 6 kW heating power output.

## Literature

[1] *Wilstermann, H.*: Wechselspannungszuendung mit integrierter Ionenstrommessung als Sensor fuer die Verbrennungs- und Motorregelung. Dissertation, Karlsruhe, 1999

[2] *Maly, R.*: Die Zukunft der Funkenzuendung. In: MTZ 59 (1998), Nr. 8

[3] Bosch: Zuendung im Ottomotor. Fachwissen Kfz-Technik. Motorsteuerung fuer Ottomotoren. 2002

[4] *Zimmermann, D.; Kerek, Z.; Wirth, M.; Gansert, K. P.; Grzeszik, R.; Storch, A.; Josefsson, G.; Sandquist, H.*: Entflammung geschichter Gemische im strahlgefuehrten Benzin-DI-Motor: Optimierung mit Hilfe von Hochgeschwindigkeits-Visualisierung. 6. Internationales Symposium fuer Verbrennungsdiagnostik. Baden-Baden, 2004

[5] *Xander, B.*: Grundlegende Untersuchungen an einem Ottomotor mit Direkteinspritzung und strahlgefuehrtem Brennverfahren. Dissertation, Universitaet Karlsruhe (TH). Logos-Verlag, 2006

[6] *Dreizler, A.*: Physikalisch-chemische Grundlagen der Zuendprozesse. Vorlesungsskript, Institut fuer technische Verbrennung der Universi-taet Stuttgart, WS 97/98

[7] *Maly, R.*: Fuel Economy in road vehicles powered by spark ignition engines. Chapter spark ignition: its physics and effect on the internal combustion engine, S. 91–148, Plenum Press, 1984

[8] *Maly, R.; Saggaue, B.; Spaude, H. W.; Vogel, M.; Wagner, E.; Greiner, E.; Horch, E. J.*: Die Auswirkungen einer verbesserten elektri-schen Entflammung auf die Verbrennung im Ottomotor. In: Automobil Industrie (1978), Heft 3, S. 37–41

[9] *Albrecht, H.; Maly, R.; Saggaue, B.; Wagner, E.*: Neue Erkenntnisse ueber elektrische Zuendfunken und ihre Eignung zur Entflammung brennbarer Gemische, Teil 1. In: Automobil Industrie (1977), Heft 4, S. 45–50

[10] *Herden, W.; Maly, R.; Saggaue, B.; Wagner, E.*: Neue Erkenntnisse ueber elektrische Zuendfunken und ihre Eignung zur Entflammung brennbarer Gemische, Teil 2. In: Automobil Industrie (1978), Heft 2, S. 15–22

[11] *Pischinger, F.*: Motorische Verbrennung. Abschlußbericht des SFB 224, 2001

[12] *Bertling, H.; Schwarz, H.*: Anforderungen an Zuendsysteme zur Entflammung magerer Gemische. In ATZ 80 (1978), Nr. 4 (Teil 1) und Nr. 5 (Teil 2)

[13] *Gansert, K. P.; Storch, A.*: Ignition and inflammation at spray guided combustion systems. 1st Ignition Symposium, 2004

[14] *Kaiser, T.; Hoffmann, A.*: Einfluss der Zuendkerzen auf das Entflammungsverhalten in modernen Motoren. In: MTZ 61 (2000), Nr. 10

[15] *Miyashita, N.; Matsubara, Y.; Iwata, K.; Ishikawa, M.*: Spark Plugs for Gasoline Direct Injection Engines. SAE 2001-01-1200, 2001
[16] *Hori, T.; Shibata, M.; Okabe, S.; Hashizume, K.*: Super ignition spark plug with fine center & ground electrodes. SAE 2003-01-0404, 2003
[17] *Daniels, F. C.; Scilzo, B. M.*: The effects of electrode design on mixture ignitability. SAE Technical Paper 960606, 1996
[18] *Pischinger, S.; Heywood, J. B.*: Einfluss der Zuendkerze auf zyklische Verbrennungsschwankungen im Ottomotor. In: MTZ 52 (1991), Nr. 2
[19] *Trachte, D.; Kaiser, T.*: Trends in spark plug engineering. 1st Ignition Symposium, 2004
[20] *Latsch, R.*: The swirl-chamber spark plug: a means of faster, more uniform energy conversion in the spark-ignition engine. SAE 840455, 1984
[21] *Pischinger, S.; Geiger, J.; Neff, W.; Boewing, R.; Thiemann, J.; Koß, H. J.*: Einfluss von Zuendung und Zylinderinnenstroemung auf die ottomotorische Verbrennung bei hoher Ladungsverduennung. In: MTZ 63 (2002), Nr. 5
[22] *Vormstein, H.; Stellwagen, K.*: TCG 2020 V20 neuer Hochleistungs-Gasmotor fuer Containereinsaetze. 3. Dessauer Gasmotoren-Konferenz, 2003
[23] *Behrens, R.; Henn, R.; Stellwagen, K.*: TBG 632 – Ein neuer Großgasmotor mit hohem Wirkungsgrad. In. MTZ 62 (2001), Nr. 2
[24] Multitorch GmbH – Industrial Spark Plugs, Sinsheim, www.multitorch.de
[25] *Geringer, B.; Klawatsch, D.; Graf, J.; Lenz, H. P.; Schuöcker, D.; Liedl, G.; Piock, W. F.; Jetzinger, M.; Kapus, P.*: Laserzündung – Ein neuer Weg für den Ottomotor. In: MTZ 65 (2004), Nr. 3
[26] *Geringer, B.; Klawatsch, D.; Graf, J.; Lenz, H. P.; Schuöcker, D.; Liedl, G.; Piock, W. F.; Jetzinger, M.; Kapus, P.*: Laserinduzierte Zündung an einem Otto-DI-Brennverfahren der zweiten Generation. 24. Internationales Wiener Motorensymposium, 2003
[27] Miniaturlaser soll Zündkerze künftig ersetzen, www.all4engineers.com
[28] Firma Iplas www.cyrannus.com
[29] *McIntyre, D. L.; Thompson, G. J.; Smith, J. E.*: The coaxial cavity resonator as a RF IC engine ignition source. SAE 2001-01-0987, 2001
[30] *Zima, S.*: Motorkolben. Verlag Vieweg, 2005
[31] *van Basshuysen, R.; Schäfer, F.* (Hrsg.): Handbuch Verbrennungsmotor, 4. Aufl. Verlag Vieweg, 2007
[32] *van Basshuysen, R.; Schäfer, F.* (Hrsg.): Lexikon Motorentechnik, 2. Aufl. Verlag Vieweg, 2006
[33] *Geyer, A.; Bosch, H.; Schellhase, T.; Ruch, R.*: Beanspruchung und Lösungsansätze im tribologischen System Ventil-Ventilsitzring-Ventilführung. VDI Fachtagung „Ventiltrieb und Zylinderkopf“, Sept. 2004
[34] *N.N.*: Fuel additives and the environment. technical committee of petroleum additive manufacturers in Europe ATC, January 2004
[35] *v. Känel, A.; Abele, M.; Grahle, P.*: A new concept for steel-composite lightweight valves. SAE 2000-01-0906, 2000
[36] *Abele, M.; Lechner, M.*: Das MAHLE Leichtbauventil. 23. Internationales Wiener Motorensymposium, 2002
[37] *Wild, S.*: Thermodynamik von Luftansaugsystemen. Handbuch Verbrennungsmotor, 3. Auflage. Vieweg-Verlag, 2005, Kap. 7.20, S. 275–278
[38] *Weber, O.; Wild, S.*: Leistung plus Drehmoment – Optimierte Sauganlage mit vollvariablen Resonanzrohren. Wiener Motorensymposium, 2001
[39] *Weber, O.; Paffrath, H.; Cedzich, W.; Beutnagel, J.*: Thermodynamische und akustische Auslegung von Ansaugsystemen für Fahrzeugmo-toren unter Berücksichtigung fertigungstechnischer Belange. Wiener Motorensymposium, 1998
[40] *Jessberger, T.*: Funktionsintegration bei Ansaugsystemen aufgeladener Motoren. Haus der Technik-Fachbuch: Ansaugsysteme aufgelade-ner Motoren. Expert Verlag, 2006, S. 1–8

[41] *Meinig, U.; Pietschner, S.; May, T.*: Kurbelgehäuseentlüftungssysteme aktueller und zukünftiger Fahrzeugmotoren, Teil 1 und 2. In: MTZ 65 (2004), Nr. 10, S. 768–777 und Nr. 11, S. 910–915

[42] *van Basshuysen, R.; Schäfer, F.* (Hrsg.): Handbuch Verbrennungsmotoren, 3. Auflage. Wiesbaden: Vieweg Verlag, 2005

[43] *Krause, W.*: Ölabscheidung in der Kurbelgehäuseentlüftung. Dissertation Universität Kaiserslautern, 1995

[44] *Schatz, O.*: Lufttaktventil. Patent DE 4308931, 1993

[45] *Schatz, O.*: Gesteuertes Rückschlagventil. Patent DE 3737828, 1987

[46] *Elsäßer, A.; Schilling, W.; Schmidt, J.; Brodesser, K.; Schatz, O.*: Impulsaufladung und Laststeuerung von Hubkolbenmotoren durch ein Lufttaktventil. In: MTZ 62 (2001), Nr. 12

[47] *Schneider, E.; Scholten, L.; Wallrafen, W.; Zentgraf, M.*: Ein integriertes Saugmodul zur Anwendung der Impulsaufladung in der Großse-rie. 12. Aachener Kolloquium Fahrzeug- und Motorentechnik, 2003

[48] *Findeisen, H.; Linhart, J.; Wild, S.*: Development of an actuator for a fast moving flap for impulse charging. SAE-Paper 2003-01-0402, 2003

[49] *Elsäßer, A.; Schilling, W.; Schmidt, J.; Kallenbach, E.; Beyer, F.; Baumbach, J.*: Entwurf schnellwirkender magnetischer Aktoren für Lufttaktventile. Tagung Innovative Klein- und Mikroantriebstechnik, 6. EGT-/GMM-Fachtagung, Darmstadt, 2004

[50] *Koch, A.; Kreuter, P.; Wensing, M.; Metzger, O.; Bey, R.; Nüsser, J.*: Impulsaufladung: Umsetzbarkeit unter realen Fahrbedingungen und in Großserie am Beispiel eines Ottomotors mit Abgasturboaufladung. In: Variable Ventilsteuerung II – Ein Verfahren zur Reduzierung von Kraftstoffverbrauch und Emissionen. Haus der Technik, Fachbuch Band 32. Expert Verlag, 2004

[51] *Sarikoc, F.; Kettner, M.; Velji, A.; Spicher, U.; Krause, A.; Elsäßer, A.*: Potential of reducing the NOx emissions in a spray guided DI gasoline engine by stratified exhaust gas recirculation (EGR). SAE-Paper 2006-01-1261, 2006

[52] *Takahashi, H.; Harada, T.; Yamaki, T.; Oikawa, T.*: Study on impulse charger for enhancement of volumetric efficiency of SI engine. SAE-Paper 2006-01-0191, 2006

[53] *Dingelstadt, R.; Elsäßer, A.; Schilling, W.; Schmidt, J.; Kallenbach, E.; Beyer, F.; Baumbach, J.; Otto, R.; Kucera, U.*: Modellbasierte Optimierung von Magnetantrieben für Lufttaktventile. Tagung Mechatronik 2005, Wiesloch, 2005

[54] *Werler, A.*: Steuerventil zur regelbaren Brennluftzufuhr. Patent DE 10204019980A1, 2005

[55] *Schmolla, W.; Schreiner, M.*: Ventil und Ventiltrieb. Patent DE102004007793A1, 2005

[56] *Leiber, H.; Leiber, T.*: Hubkolbenmotor mit einem in einem Saugrohr angeordneten Lufttaktventil. Patent DE1031782A1, 2004

[57] *Lulley, M.; Schnüpke, H.; Schröer, D.*: Vorrichtung zur Steuerung der Fluidströmung in einem Kanal. Patent DE10303985A1, 2004

[58] *Schatz, O.*: Verbrennungsmotor der Kolbenbauart und Verfahren in seinem Betrieb. Patent DE 102 40 913 A1, 2004

[59] *Golovatai-Schmidt, E.; Schwarzenthal, D.; Geiger, U.; Haas, M.; Scheidt, M.*: Technologies for variable valve trains; a contribution to modern S.I. engines. CONAT20041005, International Automotive Congress, Brasov, Romania, 2004

[60] *Meinig, U. et al.*: Emissionen aus Kraftstoffsystemen von Pkws. Renningen: Expert Verlag, 2002

[61] *Meinig, U.; Heinemann, J.*: Neue Anforderungen an Tankentlüftungssysteme. In: ATZ 101 (1999), Heft März, S.166–173

[62] Bosch: Ottomotorenmanagement. Wiesbaden: Vieweg Verlag, 1998

[63] *Beetz, K.; Bohlender, F.*: Elektrische Zuheizsysteme. Reihe: Wärmemanagement des Kraftfahrzeugs, hrsg. von N. Deußen. Expert Verlag, 1998

[64] *Kampf, H.; Taxis-Reischl, B.; Weinbrenner, M.*: Zuheizsysteme für Fahrzeuge mit kraftstoffsparenden Motoren. Reihe: Wärmemanage-ment des Kraftfahrzeugs, hrsg. von N. Deußen. Expert Verlag, 2000

# 12 Gasoline Direct Injection in Electric Hybrid Systems

## 12.1 Introduction

Since the diesel engine with direct injection gained a large market share in Europe because of its fuel economy, there have been numerous technical developments in the area of the gasoline engine in order to reduce fuel consumption. Gasoline engines are especially suitable for hybridization. Downsizing and direct injection have shown considerable potential for fuel savings as well as reduction of friction losses, optimization of the ancillaries, and the application of energy management and new combustion processes. **Figure 12.1** illustrates the paths of consumption reduction by hybridization. Two points directly affect the powertrain: the generation of energy at load points with better efficiency, and downsizing with load point shifting. The positive effect of the hybrid components more than compensates for the disadvantage of additional weight.

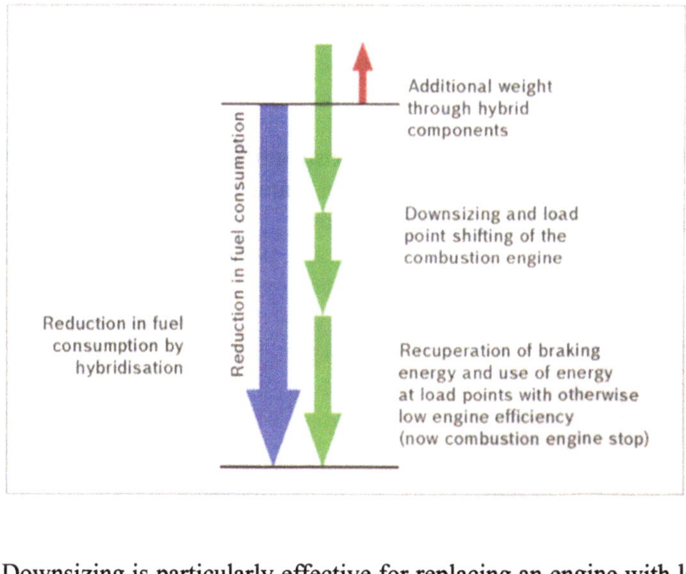


**Figure 12.1:** Consumption reduction path, mid-size passenger car [1]

Downsizing is particularly effective for replacing an engine with large displacement with one that has small displacement but high power and good torque. This applies particularly to gasoline engines with direct injection, since the fuel consumption is also very low. In order to achieve the necessary driving performance, appropriate torque is also required at low speeds. The 1.4 l gasoline engine with double charging, recently brought into series production, is an example with small displacement and superior torque as mentioned in [2]. A turbocharger/mechanical supercharger combination is used to achieve high torque at low

The parallel hybrid enables simultaneous drive by both powertrain units. The serial hybrid always has an electromotive drive, which makes it necessary to generate "on-board" current. This can be achieved by a diverse range of units, e.g. by gasoline or diesel engines, gas turbines or fuel cells. The advantage is that the power generating unit can operate at an optimal efficiency and produce optimal emissions.

The mixed hybrid is a combination of the serial and parallel hybrid, whereby the power distribution hybrid drive represents an additional, if not extensive, variant of all the hybrid systems. The conventional transmission (manual or automatic) is replaced by a single-stage or multi-stage planetary transmission with direct coupling of at least two E-machines. This extremely compact unit constitutes a continuously variable transmission (ECVT), with which all the hybrid functions such as start/stop, recuperation and partial or complete electrical driving can be executed. The advantages and disadvantages of various systems are compared in **figure 12.5**.

| **Serial hybrid** | **Parallel hybrid** | **Starter/generator (mild hybrid)** |
|---|---|---|
| *Purely electrical axle drive* | *Combined combustion engine-electrical axle drive* | *E-machine tightly coupled with combustion engine* |
| + Starter/generator function<br>+ Purely electrical drive<br>+ Recuperation with full power output<br>+ Current supply at 0 km/h<br>+ Optimum energy management<br>+ Comfortable engine start also at $v > 0$ km/h<br>- Great weight (three times installed nominal output)<br>- Boost function only electrically by overload<br>- High costs<br>- Great weight<br>- Expensive packaging | + Starter/generator function<br>+ Boost function<br>+ Recuperation<br>+ Purely electrical drive<br>+ Optimum energy management possible<br>+ Immediate start of combustion engine at $v > 0$ km/h<br>+ Relatively small pulse modulation of E-engines<br>+ Current supply also at 0 km/h<br>- Costs<br>- Great weight<br>- Complicated structure<br>- Longer powertrain | + Starter/generator function<br>+ Boost function (ISG)<br>+ Simple structure<br>+ Low costs<br>+ 42 V rated power output<br>+ Also with belt drive (RSG)<br>- Combustion engine cannot be decoupled<br>- Limited recuperation (drag torque of the engine)<br>- No pure E-operation<br>- Low fuel savings |

**Figure 12.5:** Technical assessment of hybrid drive according to [5]

Besides the technical evaluation, market requirements play an important role.

The global reasons for the application of alternatively driven vehicles are analyzed in [5] from a marketing perspective. To begin with, the political and legal framework is very different:

### Europe

- Reduction in fleet consumption by 2008: 140 g $CO_2$/km (voluntary commitment)
- Consumption target by 2010: 120 g $CO_2$/km by $CO_2$ assessment

## USA

- Compliance with ZEV mandate (*Zero Emission Vehicles*) from 2010: Mercedes Benz requires 2% ZEVs (FC: Fuel cell, EV: Electrical Vehicles) and 2% AT-PZEVs (*Advanced Technology Partial Zero Emission Vehicles*, (Hybrid)) or 4% ZEVs (law)
- Tightening of the CAFE limit values (*Corporate Average Fuel Economy*) to 30 mpg (approximately 8 l/100 km) by 2010 (draft bill with high probability)
- Regulations for reducing $CO_2$ emissions of passenger cars and light commercial vehicles in California from 2009 (draft bill)

## Japan

- Class dependent $CO_2$ tax from 2010 (law): [JAMA/KAMA announcement (*Japan/Korea Automobile Manufacture Association*): Prefulfillment of limit values from 2005]
- Reduction in consumption to 140 g $CO_2$/km from 2009 (JAMA/KAMA voluntary commitment)
- Governmental requirement of alternative drives through tax benefits (Hybrid: up to 73% less vehicle acquisition tax, up to 75% less annual vehicle tax) and direct incentives

## China

- As there is already a huge demand for energy in China, alternative fuel (renewable) and economic vehicle concepts (hybrid) are important there.

However, traffic behavior also forms a basis for different technological demands:

| | | |
|---|---|---|
| **Europe** | • Balanced mix of freeway traffic (high speeds) and city traffic<br>• Dynamic operational profile | Hybrids<br>Gasoline hybrid (possible)<br>Diesel hybrid (possible)<br>Diesel (preference) |
| **USA** | • Significant proportion of freeway traffic<br>• Characteristically uniform traffic flow in the moderate speed range | Gasoline hybrid or diesel<br>Diesel engines would be ideal but restrictive legislation on exhaust-gas emissions and insufficient fuel supply |
| **Japan** | • City traffic with high proportion of stop/go<br>• High speed instances rare<br>• No diesel fuel for passenger cars<br>• Short routes on average | Hybrids!<br>Gasoline hybrids are logical consequence; also supported by enthusiasm for technology |
| **China** | • High proportion of city traffic<br>• Few sources of oil owned by China<br>• Almost no high speeds over long routes | Alternative fuel<br>Hybrids |

Forecasts of the market share of hybrid vehicles vary widely. An optimistic study carried out in the USA [6] shows that in the year 2022, approximately 50% of the vehicles launched in the US market will be hybrid vehicles. Even so, this would be 6 million per year. By contrast, predictions of the number of vehicles in Japan and Europe are modest. Regardless of these predictions, it is evident that hybrid technology is advancing. Toyota and Honda are market leaders today; both have already reached the second generation. The Toyota Prius, RX 400h and other models are full hybrids, while the Honda Civic is a mild hybrid. As shown in **figure 12.6**, Toyota in the USA sold almost 300,000 hybrid vehicles in 2005. Meanwhile the Chevrolet Silverado also introduced mild hybrid technology [7] with ISAD (integrated starter, alternator and damper). Other companies will follow.

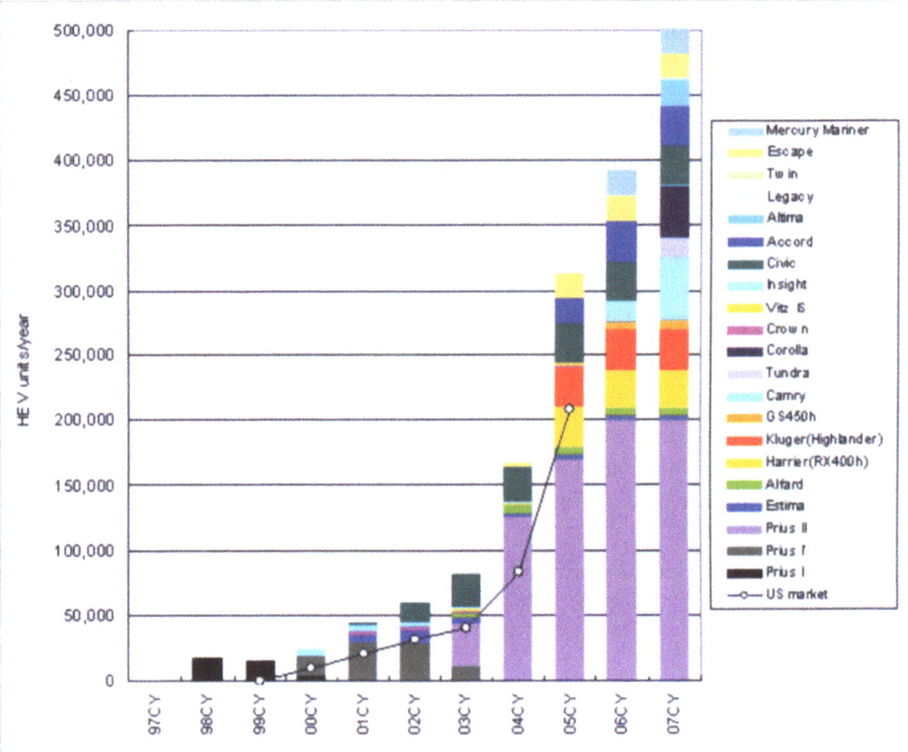


**Figure 12.6:** Hybrid market in the USA (as at July 2005)

The following list shows other hybrid vehicles available or scheduled for mass production:

| | |
|---|---|
| Micro | Citroen C2 Start/Stop (2005) |
| | Citroen C3 Start/Stop (2005) |
| | |
| Mild | Chevrolet Silverado (2004) |
| | Citroen C4 Hybrid Hdi (2006) |
| | GMC Sierra (2004) |
| | Honda Civic IMA (2004) |
| | Honda Civic IMA (2006) |
| | Honda Insight IMA (1999) |
| | Peugeot 307 Hybrid Hdi (2006) |
| | Audi Q7 Mild (or Full) Hybrid (2008) |

| | |
|---|---|
| Full | Lexus RX 400h (2005) |
| | Lexus RX 250h (2007) |
| | Lexus GS 600h (2007) |
| | Toyota Prius (1999) |
| | Toyota Prius (2004) |
| | Toyota Prius (2006) |
| | Ford Escape Hybrid (2005) |
| | Ford Mercury Mariner (2006) |
| | Chevrolet Tahoe (2007) |
| | CMC Yukon (2008) |

The Toyota hybrid system THS II has succeeded THS I (Prius), which went into series production in December 1997 [8]. The sudden success of hybridization can be attributed to a change in strategy by Toyota. The Prius II performs even better than the Prius I. More than 120,000 THS I vehicles have been sold worldwide. The system structure of Prius II [9] is given in **figure 12.7**.

**Figure 12.8** compares the characteristics of the THS II and the THS I.

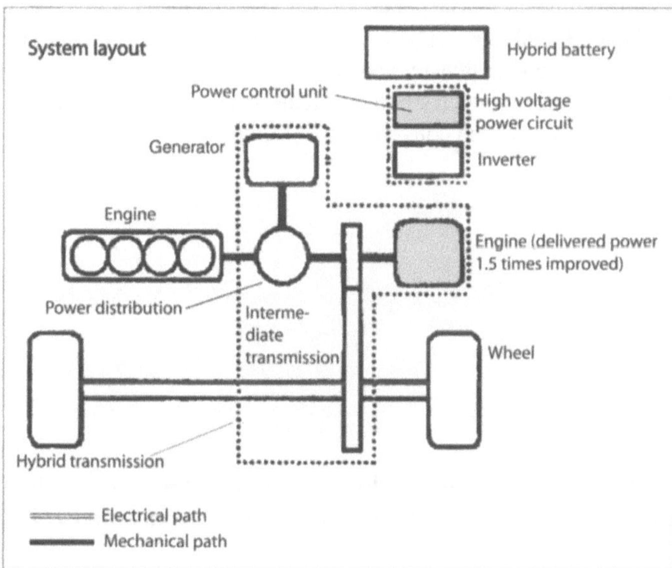


**Figure 12.7:** System layout of Toyota Prius II [9]

| | | THS I | THSII |
|---|---|---|---|
| Engine | Type | 1.5 l gasoline engine (high expansion rate) | 1.5 l gasoline engine (high expansion rate) |
| | Maximum power (kW) | 57/5.000 | 53/4.500 |
| | Maximum torque (Nm) | 115/4.200 | 115/4.200 |
| E-motor | Type | AC-Synchro | AC-Synchro |
| | Maximum power (kW) | 50/1.200-1.540 | 33/1.040-5.600 |
| | Maximum torque (Nm) | 400/0-1.000 | 350/0-400 |
| System | Maximum power (kW) / vehicle speed (km/h) | 82/85 or higher | 74/120 or higher |
| | Maximum power (kW) at 85 km/h | 82 | 65 |
| | Maximum torque (Nm) / vehicle speed (km/h) | 478/22 or lower | 421/11 or lower |
| | Maximum torque at 22 km/h | 478 | 378 |
| Battery | Type | Nickel - metal hydride | Nickel - metal hydride |
| | Maximum power (kW) | 21 | 21 |

**Figure 12.8:** Comparison of characteristics of THS II and THS I

Improvements in the gasoline engine, which will be the appropriate partner for hybridization in the future, as well the increase in the power and torque of electrical drives with higher voltage from THS I to THS II, should be particularly emphasized. The voltage was increased from 274 to 500 V. The AC synchronous permanent engine shows optimum values in terms of the weight and volume used. The generator was also enhanced, e.g. increase in rotational speed from 6,500 to 10,000 $min^{-1}$. The same applies to the enhancement of the nickel metal hydride battery, whose power density was increased from 880 to 1250 W per kg. The fuel consumed during the driving cycles in the official driving modes is specified in the table, for which different data has been converted to l/100 km.

Consumption in official test mode:

| Test mode | THS II Prius 2 | | THS Prius 1 | |
|---|---|---|---|---|
| 10–15 M | 35.5 (km/l) | 2.82 l/100 km | 31 (km/l) | 3.23 l/100 km |
| US-Comb | 65.8 (mpg) | 3.57 l/100 km | 57.6 (mpg) | 4.08 l/100 km |
| EU | 104 (g $CO_2$/km) | 4.33 l/100 km | 120 (g $CO_2$/km) | 5.00 l/100 km |

However, fuel consumption when the vehicle is actually being driven is very different from the formal test. A comparison carried out by the German magazine *auto motor sport* [10] resulted in the following values:

Fuel consumption in practice in standard driving cycle (*auto motor sport*):

| Vehicle type | l/100 km | g $CO_2$/km |
|---|---|---|
| Audi A3 1.6 FSI | 6.6 | 158 |
| Honda 1.3 IMA | 5.4 | 130 |
| Toyota Prius II | 5.5 | 133 |
| Volkswagen 1.9 l TDI | 4.9 | 131 |

It should be noted that the gasoline engine with direct injection (FSI/fuel stratified injection) has the highest fuel consumption while the diesel engine with direct injection has the lowest. Hybrid vehicles, compared to gasoline and diesel vehicles, are midway in terms of fuel consumption; their $CO_2$ emissions, on the other hand, are equivalent to that of diesel vehicles.

As both the systems have considerable development potential, the end result of the "race" cannot be predicted. At the same time, the low acceleration capability of the Honda as compared to other vehicles has to be taken into consideration. As opposed to the Toyota Prius II, the Honda IMA concept works as a mild hybrid [11], with a 15 kW electric motor placed between the engine and the transmission. The direct current motor works with 144 V, and a nickel metal hydride battery with 158 V and 5.5 Ah is used as an energy accumulator. **Figure 12.9** shows the general design of the system.

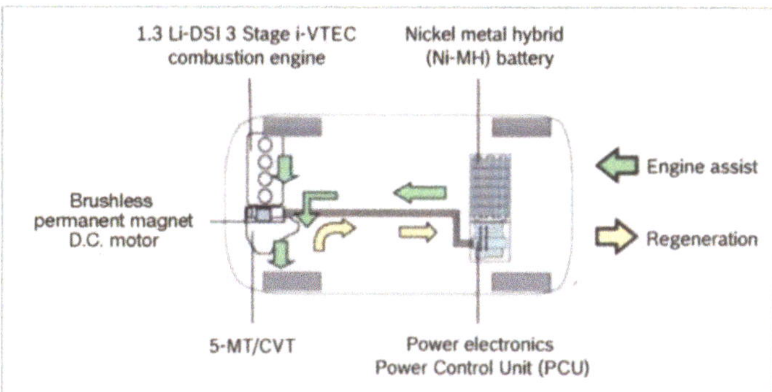


**Figure 12.9:** The Honda IMA concept [11]

Toyota has combined hybrid technology with electrical rear-wheel drive in the Lexus GS 450h [12]. The GS 450h is equipped with 250 kW (340 hp) power output (a combination of direct and intake manifold injection) and accelerates from zero to 100 km/h in six seconds. Its average fuel consumption is specified at 7.9 l of premium gasoline per 100 km. Apart from the combustion chamber and the power distribution transmission assembly – known for its use in Toyota Prius – with electric motor and generator, the Lexus RX 400h has an additional E-motor at the rear axle in order to generate the all-wheel drive. In comparison to the Toyota Prius II, the RX 400h also has a higher system voltage apart from the second driving motor. Additionally, the power of the front electric motor was doubled by enhancing the engine speed range. The power of the nickel metal hydride battery was also almost doubled; **figure 12.10** [12].

**Figure 12.10:** Cross-section of the hybrid transmission of the Lexus RX 400h [12]

Lexus introduced the LS 600h in 2007. The V8 gasoline engine (a combination of direct and port injection) produces 320 kW (435 hp) in addition to the power of the electric motor. This means that the LS 600h should have the power output of a 12-cylinder gasoline engine. The power outputs in Lexus vehicles are achieved partly by the working method of the electric motors. The Lexus electric motors work at twice the engine speed of those in the Prius, i.e. at 13,000 rpm. The operating voltage in the current Prius is more than 500 V; in the Lexus it is 650 V.

Another critical factor for hybrid drives is the battery. The following overview shows the different battery variants [13]:

| Battery type | Spec. energy Wh/kg | Spec. power W/kg | Energy density Wh/l | Power density W/l | Costs EUR/KWh |
|---|---|---|---|---|---|
| Lead acid | 32 | 430 | 68 | 910 | 100 - 150 |
| Nickel cadmium | 35 | 700 | 100 | 2000 | 225 - 350 |
| Nickel metal hydride | 40 | 1200 | 100 | 3000 | 225 - 300 |
| Lithium ion | 70 | 2000 | 150 | 4200 | ? |

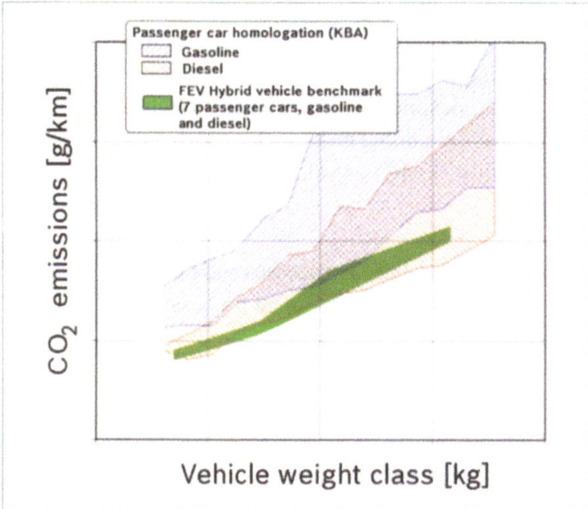


**Figure 12.11:**
$CO_2$ emissions in the NEDC cycle [1]

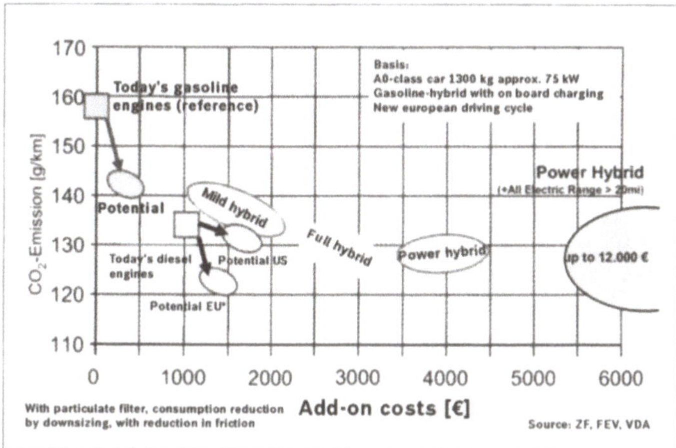


**Figure 12.12:**
$CO_2$ emissions as a function of hybrid drives and costs [12]

The FEV [1] compared the $CO_2$ emissions in the NEDC. **Figure 12.11** shows the results for seven hybrid vehicles.

The estimated reduction in fuel consumption and additional system costs are given in **figure 12.12**. These are considered less important if features that have more customer value, such as automatic parking, can be realized through hybridization.

Due to the debates on the issue of $CO_2$ and also due to the technical possibilities as described above, all the vehicle manufacturers are intensively working on introducing hybridization in new vehicles.

As is already evident from the detailed descriptions of hybridization, there is almost no area where it has not been implemented. **Figure 12.13** shows the individual assemblies and functions influenced by hybridization. Gasoline engines with direct injection which provide more torque than engines with larger displacement, despite the downsizing brought about by the electrical drive, are particularly suitable. Cross-functional tasks such as noise vibration harshness (NVH) and electromagnetic compatibility (EMC) are also among the functions influenced by hybridization.

Due to the complexity of hybrid systems, the model-based processes and both individual and complete system analyses are an important part of the development process [16, 17].

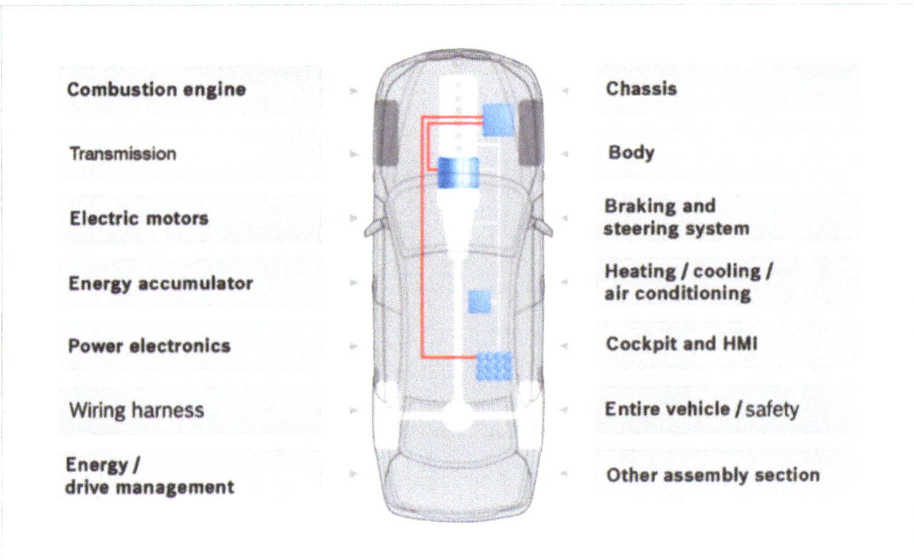


**Figure 12.13:** Vehicle areas influenced by hybridization [15]

## Literature

[1] FEV-Motorentechnik Broschuere. Aachen, May 2006

[2] *Krebs, R.*: Twincharger und Downsizing – Neue Wege zu einer nachhaltigen $CO_2$-Minderung. VDA, Technischer Kongress. Frankfurt, 2006

[3] *Koehle, S.*: Entwicklungsziele, Fahrzeugbeschreibung und erste Messergebnisse des VW Bora mit Hybridantrieb und Energiemanagement. Braunschweig: GZVB, 2004;
*Antony, P. et al.*: Handbuch der Kraftfahrzeugtechnik. Wiesbaden: Vieweg, 2005
*Noreikat, K. E. et al.*: Hybride Fahrzeugantriebe. Die Evolution zum Mehrwerthybrid. VDI-Bericht 1565, Duesseldorf

[4] *Borgmann, K. et al.*: Effiziente Dynamik als Loesung des Zielkonflikts zwischen Kundenwunsch und Gesetzesanforderung. In: Technischer Kongress VDA (Tagungsband). Frankfurt, 2006

[5] *Miska, J.*: Hybridfahrzeuge – Produktauspraegungen als Antwort auf Kundenanforderungen. Braunschweig: GZVB, 2004
[6] National Research Council / National Academy of Sciences, USA: The hydrogen economy: opportunities, costs, barriers, and R&D needs. 2004
[7] *Kozlowski, F. et al.*: Der ISAD-Starter-Generator. Eine Innovation im Antriebsstrang zur Reduktion des Treibstoffverbrauchs geht in Serie. Braunschweig: GZVB, 2004
[8] *Harada, I.*: Entwicklung eines neuen Toyota Hybrid Fahrzeuges. In: Motor und Umwelt, AVL, Graz, 7.+8. September 2000 (Tagungsband)
[9] *Muta, K. et al.*: Development of a new generation hybrid system THS II – drastic improvement of power performance and fuel economy. SAE, SP 1833, 2004
[10] *Hack, G. et al.*: Die Liter-Leistung. www.auto-motor-sport.de;
Honda: Pressematerial, 2004
[11] *Brachmann, Th.*: Development of powertrain for the Civic IMA. A gasoline-electric hybrid vehicle. Offenbach: Honda RD-Center, 2004;
*Iigima, T.*: Development of Hybrid Systems for 2006 Compact Sedan. SAE International SP_2008, Warren, 2006
[12] *Adachi, M. et al.*: Development of a new hybrid transmission for RWD CAR. SAE International SP_2008. Warren, 2006
[13] *Koehler, U.*: Batterien fuer Elektro- und Hybridfahrzeuge in Hybrid-, Batterie und Brennstoffzellen-Elektrofahrzeuge. Expert Verlag, 2005
[14] *Rabe, M.*: Synthetische Kraftstoffe fuer nachhaltige Antriebe von morgen. In: Technischer Kongress, VDA (Tagungsband). Frankfurt, 2006
[15] *Nietschke, W.*: Hybrid-Hype oder technologische Chance. MTZ-Konferenz Motor 2006. Wiesbaden: Vieweg Verlag, 2006
[16] *Kraft, D. et al.*: Modellgestuetzte Entwicklung von Komponenten für elektrische Hybridfahrzeuge. Robert Bosch GmbH
[17] *Kraft, D. et al.*: Hybridkonzepte aus Sicht der Robert Bosch GmbH. Tag des Hybrids, 2005

# 13 Fuels and Lubricants

## 13.1 Gasoline fuels

### 13.1.1 Introduction

A large number of different products are manufactured with a crude oil base. Processing of crude oil (purifying, distilling, post processing and refining) yields two fuel types, amongst other by-products: gasoline and diesel. They are used in different types of engines based on their suitability. Their high energy density, liquid form, manageability and good value for money make them suitable for supplying energy to automobiles. They have survived for more than 100 years and their technical signatures are constantly being adapted, which enables engine development.

### 13.1.2 General principles

**Crude oil as a raw material for fuels and lubricants**

Crude oil consists of hydrocarbons, the predominant ones being alkanes (paraffin) and cycloalkanes. Crude oil is a mixture of countless hydrocarbons, each characterized by a certain molecular structure, i.e. by the arrangement of carbon and hydrogen atoms in different "atomic structures". They have intrinsic characteristics that differ from each other, e.g. a certain boiling point that can lie between –163°C and +600°C or more, a certain density or other physical characteristics such as anti-knocking quality, ignition quality, vapor pressure, viscosity and low temperature characteristics, which are very important for their use in combustion engines. Most types of crude oil contain oxygen, sulfur and nitrogen compounds in addition to the hydrocarbons.

**General information about the processing procedure / refining technology**

In the refineries that manufacture fuels, lubricants and other by-products, the crude oil components are separated based on the boiling point. They are altered both chemically and physically by treating with special chemical processes. The desired end products are generated by combining various processes. Distilling, absorbing, filtering and extracting are the commonly known separation processes. Polymerization, isomerization, hydrocracking, alkylation, reformation and ultimately conversion (cracking) come under transformation processes; **figure 13.1**. The chemical treatment processes help to remove the unwanted components from the crude oil or its fractions or to convert them to neutral compounds. Refining as aftertreatment for the purpose of "purification" is also considered a part of this process.

Crude oil is broken down into various heavy components by fractioned distillation and condensation in crude oil distillation equipment and the primary products are delivered for further processing. Distillation is the first processing step in every refinery. Distillation is a purely physical process, as the hydrocarbon molecules are only separated according to their sizes, but not changed chemically. The crude oil is heated in the tube furnace at more than 100°C for this purpose and transferred to the atmospheric distillation columns. Here, fractions of the crude oil are brought to a gaseous state. That which remains in a liquid state gathers at the bottom of the distillation columns.

Individual components condense continuously and the respective products are drained off laterally from the individual levels of the distillation columns and are directed into the storage tank after cooling down. As well as the gaseous components, light gasoline fuels, naphtha, petroleum, light- and medium-heavy gas oil are obtained.

What is known as "atmospheric residue" accumulated at the bottom of the atmospheric column serves as raw material for the vacuum section of the distillation equipment. Here, the crude oil components with higher boiling points are distilled further, with respect to molecules, in vacuum. The residue is reheated in a second oil heater and the vapor is transferred to a vertical distillation column, which is in a vacuum state. Fractioned condensation takes place in a method similar to atmospheric distillation, in the various levels of the columns.

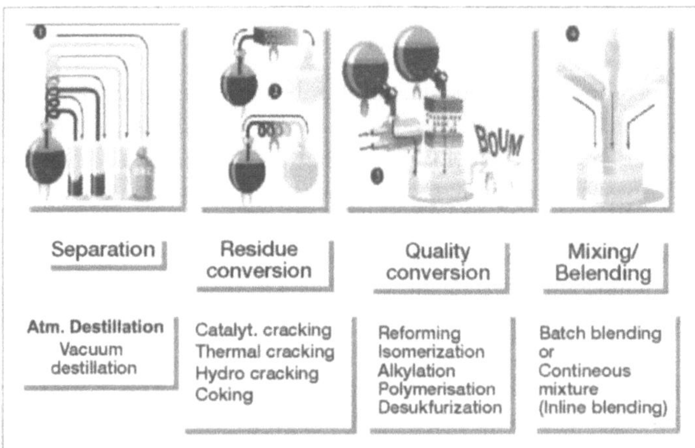


**Figure 13.1:** Main features of refinery techniques

Cracking, reforming and polymerization are the further essential steps of the refining process. The breaking of the large, long-chain (high boiling point) molecules into smaller molecules is known as cracking. A short-chain alkane or alkenes are formed here by cracking the long-chain alkane. The raw material is heated in a furnace at above 400°C in a thermo-catalytic cracking process. In a reactor, the raw material comes into contact with the catalyst, which is made up of silicon pearls or spheres and is at a temperature of approximately 550°C. In this way, the products that are put in the reactor are cracked into fuel gas, liquefied petroleum gas, gasoline fuel, diesel, fuel oil and unwanted residual carbon. The carbon is deposited on the surface of the catalyst, which needs to be regenerated again later, as it be could deactivated by carbon deposits.

Most of the reforming processes that take place in the presence of a catalyst are, in principle, catalytic cracking or hydrocracking processes with subsequent molecular restructuring (conversion), which can take place in different ways. Hydrogen is mostly split off (dehydrogenating process) and entire molecular groups are relocated.

Polymerization is a process, with the help of which the gaseous hydrocarbon compounds present in large amounts are bound together to form larger molecules. This process is also usually carried out in the presence of catalysts.

Equipment in most modern refineries is very complex. The refineries have multiple conversion plants where a large proportion of the heavy product flows are converted into the lighter products that are now in great demand. In terms of quantity, fuels are the most important of all the products or product groups that are manufactured in the refineries. **Figure 13.2** gives an overview of the products of a refinery in relation to the amount of raw material (crude oil, additives etc.) used.

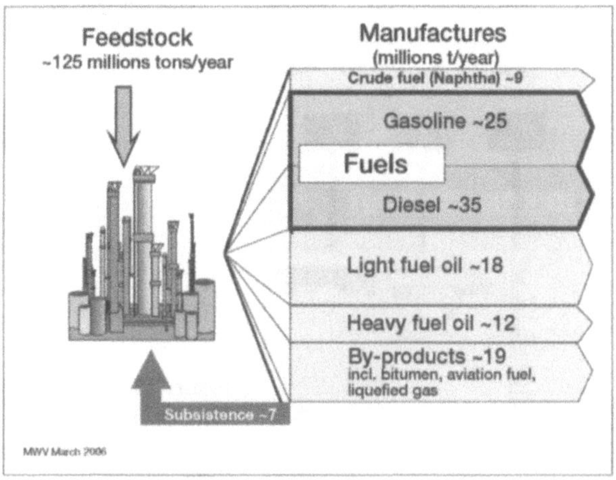


**Figure 13.2:**
Products of German refineries in 2005

## 13.1.3 Gasoline fuel components and gasoline fuel mixture

Many diverse components are produced in the refinery processes described above; they consist of various hydrocarbon compounds and are mixed into gasoline fuels. Availability, prices and usability characteristics are vital parameters for determining the composition of the gasoline fuel mixture. The demands on the gasoline fuel mixture have drastically increased due to its application (more complex engine technology such as direct injection and improvements in the emission and comfort characteristics.)

The gasoline fuel mixture is liquid at room temperature, with a boiling point between 30 and 200°C and consisting of approximately 200 hydrocarbons. In addition to the hydrocarbons, aromatics, paraffin/naphthenes and olefins, it may also contain compounds with oxygen (alcohol and ether) such as ethanol, tertiary butanol and methyl tertiary butyl ether MTBE. Nowadays, the hydrocarbons are obtained almost exclusively from crude oil.

The combination of the individual components affects the chemical-physical characteristic values of the fuel that in turn affect the usability characteristics.

The combination of the individual components affects the chemical-physical characteristic values of the fuel, which in turn affect the usability characteristics. After the basic fuel has been prepared inclusive of the typical refinery additives such as antioxidants etc. – this already corresponds to the minimum requirements of the fuel consumption standard EN 228 – special chemically active agents (additives) are added to improve its usability characteristics. A modern branded fuel becomes "marketable" only after these company-specific additive packages have been mixed in (this generally happens during the filling of the tanker); **figure 13.3**. The effectiveness of the additives is an essential distinguishing feature of the fuel quality of various providers, even in the case of branded manufacturers.

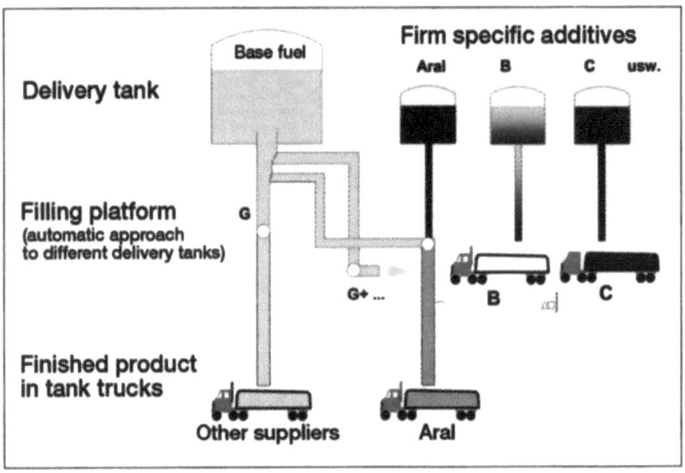


**Figure 13.3:** Marketability of fuels by performance additives

Gasoline fuel is currently offered in the qualities/types superplus, super, standard and premium products, such as e.g. "ultimate 100"; all of which are lead-free for environmental reasons. Apart from a few exceptions, leaded fuels have disappeared from the European markets. An essential distinguishing feature of these types is the octane number, a measure of the anti-knocking quality of gasoline fuel. Put simply, the products from various manufacturing processes are combined differently during mixing for the preparation of the various types of fuels.

### 13.1.4 The practical significance of gasoline fuel requirements

The composition and characteristics of the extremely complex mixture of the product flows from the refinery, oxygenated components and additives, described in the previous section, must naturally be adapted for practical use. This imposes a variety of different requirements for problem-free use in vehicle engines, for storage (sometimes for long periods) and for distribution.

Certain specifications that provide information about the quality of a fuel must be met in order to satisfy these requirements. The minimum gasoline fuel quality requirements are regulated by DIN EN 228, March 2004 edition. The essential specifications for gasoline fuel are summarized and their significance for practical vehicle operation explained briefly in **figure 13.4**.

| Value | Normal (NUK) | Super (SUK) | SuperPlus (SUP) | Influence on vehicle operation |
|---|---|---|---|---|
| Anti-knock properties (Octane number) | min. 91.0 ROZ | min. 95.0 ROZ | min. 98.0 ROZ | Knocking for low and medium engine speed |
| | min. 82.5 MOZ | min. 85.0 MOZ | min. 88.0 MOZ | Knocking for high engine speed and high load |
| Density at 15 °C from<br>to | 720 kg/m³<br>775 kg/m³ | 720 kg/m³<br>775 kg/m³ | 720 kg/m³<br>775 kg/m³ | Fuel consumption, emissions |
| Lead content | max. 5 mg/l | max. 5 mg/l | max. 5 mg/l | Deposits, catalytic converter |
| Vapor pressure (DVPE)<br>Summer<br>Winter / Transition period | 45.0-60.0 kPa<br>60.0-90.0 kPa | 45.0-60.0 kPa<br>60.0-90.0 kPa | 45.0-60.0 kPa<br>60.0-90.0 kPa | Cold start, hot start, Evaporative emissions |
| Distillation range<br>Transit to 70 °C (=E70)<br>Summer<br>Winter / Transition period | 20.0-48.0 % (V/V)<br>22.0-50.0 % (V/V) | 20.0-48.0 % (V/V)<br>22.0-50.0 % (V/V) | 20.0-48.0 % (V/V)<br>22.0-50.0 % (V/V) | Cold start, hot start, Driving behavior with hot and cold engine |
| Distillation range<br>Transit to 100 °C<br>Summer<br>Winter / Transition period | 46.0-71.0 % (V/V)<br>46.0-71.0 % (V/V) | 46.0-71.0 % (V/V)<br>46.0-71.0 % (V/V) | 46.0-71.0 % (V/V)<br>46.0-71.0 % (V/V) | |
| End of boiling | max. 210 °C | max. 210 °C | max. 210 °C | Residue formation, exhaust gases, erosion |
| Volatility indicator<br>VLI = 10 VP + 7 E70<br>Transition period | max. 1150 | max. 1150 | max. 1150 | Start and driving behavior, with hot engine |
| Evaporation residue | max. 5 mg/100ml | max. 5 mg/100ml | max. 5 mg/100ml | Residue formation |
| Sulfur content | max. 50/10 mg/kg | max. 50/10 mg/kg | max. 50/10 mg/kg | Corrosion, catalytic converter |
| Corrosion impact on copper | max. 1 | max. 1 | max. 1 | Corrosion |
| Aromatic content | max. 35.0 % (V/V) | max. 35.0 % (V/V) | max. 35.0 % (V/V) | Exhaust emission |
| Benzene content | max. 1.00 % (V/V) | max. 1.00 % (V/V) | max. 1.00 % (V/V) | Exhaust emission |
| Total oxygen content | max. 2.7 % (m/m) | max. 2.7 % (m/m) | max. 2.7 % (m/m) | Driving behavior, consumption, emissions |

**Figure 13.4:** Specifications of gasoline fuel (Extract from DIN EN 228, March 2004 edition)

The quality and prices of the fuels in Germany are regulated by the 10th Federal Emission Control Ordinance (BImSchV) and the adherence to the legal specifications is defined in the related Fuel Quality Regulation (KQV). The Fuel Quality Regulation (KQV) includes what are known as rejection thresholds for each characteristic along with all the requirements from the product requirement standard DIN EN 228. The rejection thresholds account for the lack of precision of each of the test processes that are used to check a fuel characteristic. If, for example, the environmental department officials determine, while checking the fuel quality, that a threshold has been exceeded or not met, then this can be penalized only if the deviation is larger than the higher tolerance allowed by the rejection threshold. These minimum requirements of the DIN EN 228 standards have been progressively tightened over the years, making the reduction of the sulfur content more significant.

**Sulfur – a natural but unwanted component of crude oil**

On the one hand, sulfur content has an influence on the emission of sulfur dioxide, trioxide and particles and on the other, high sulfur content has negative effects on the lifespan and the efficiency level of exhaust-gas catalytic converters. Thus, sulfur-free fuel (maximum sulfur content of 10 ppm) represents an essential prerequisite for introducing emission-free gasoline engines with direct injection. Frequent regeneration of the $NO_x$ storage catalytic converter is necessary for gasoline fuel with a higher sulfur content. As a result, the potential for minimizing fuel consumption in gasoline engines with direct injection cannot be fully realized. Sometimes, sulfur components even contaminate the active surfaces of $NO_x$ catalytic converters irreversibly.

Therefore, the SuperPlus type with a maximum sulfur content of 50 ppm was spontaneously offered nationwide by the petroleum industries in Germany from January 1, 2000 which went beyond the requirements of the legislation (a maximum sulfur content of 50 ppm has been valid across the EU since January 1, 2005). The German BP GmbH was even in a position offer sulfur-free (maximum sulfur content of 10 ppm) SuperPlus gasoline for the first time in the Munich area in April 2000. The spontaneous reduction of a maximum sulfur content of 10 ppm (sulfur-free) in SuperPlus fuel was extended nationwide by March 2001.

Further measures to achieve a faster decrease in the sulfur content had already been put in place by the German government by means of tax incentives in 1999. Thus, an additional tax of approximately 0.015 €/l was charged for all fuels with a sulfur content of more than 50 ppm (low sulfur) from November 2001 and all fuels with a sulfur content of more than 10 ppm from January 2003 respectively. This meant that practically all the gasoline and diesel in Germany have been offered with a sulfur content of a maximum of 10 ppm i.e. "sulfur-free" since the beginning of 2003; **figure 13.5**.

The introduction of sulfur-free gasoline was seen as a prerequisite by the vehicle manufacturers for introducing gasoline engines with direct injection and for the highly effective exhaust-gas aftertreatment systems that will be needed in the future to meet the requirements set by legislation on exhaust-gas emissions. Less sulfur content, i.e. sulfur-free fuel, lengthens the lifespan of the catalytic converter, even for conventional gasoline engines.

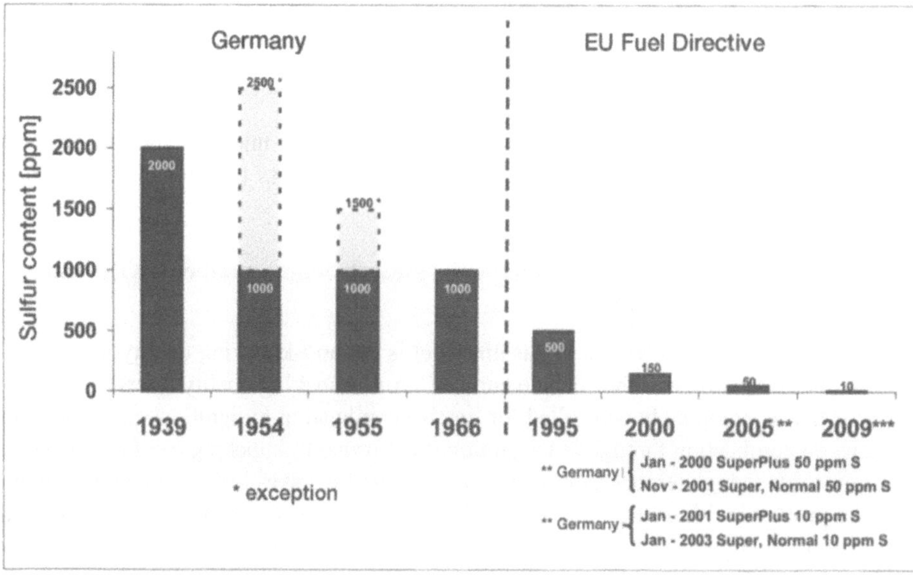


**Figure 13.5:** Development of the sulfur content in gasoline

### Volatility – a basic prerequisite

The tendency of gasoline fuel to vaporize, i.e. its volatility, is the essential prerequisite for its use in the gasoline engines and at the same time an important quality feature. Volatility is characterized by two fuel characteristics: boiling curve and vapor pressure.

As gasoline is a mixture of various hydrocarbons, it does not have a definite boiling point but a boiling range that lies between about 30 and 200°C. The boiling curve describes the part of the fuel that has vaporized at increasing temperature and together with vapor pressure is important for the start and drive behavior well as for uniform "smooth" gasoline engine operation. The vaporized volume up to 70°C is very important. This component should be significant enough to make the engine easy to start, but not so significant as to cause a fuel vapor lock. The fuel component vaporized up to 100°C, in particular, determines the operational reliability and acceleration response of a warm engine in addition to the warm-up phase. The component vaporized up to 180°C is important for cold start behavior and the component from approximately 180°C must be kept to a minimum in order to prevent dilution of the lubricating oil. In this way, the engine-specific boiling curve requirements leave almost no room for optimization with regard to emission behavior.

The vapor pressure describes the inclination of fuel to vaporize. Along with the fuel component vaporized up to 70°C, it affects the behavior of the engine during cold start and once it has warmed up. Based on the weather, gasoline fuels must adapt to the seasons, and seasons in different regions. This way, there is lower vapor pressure in summer than in winter. The lower vapor pressure is necessary in summer in order to prevent a fuel vapor lock and the resulting disturbances in engine operation. At the same time, a lower vapor pressure is preferable for preventing possible vapor emissions from the fuel system (i.e. from the environmental point of view).

Certain requirements can arise in gasoline engines with direct injection and especially in the case of systems with stratified charge, because there is less time for mixture formation.

**Anti-knocking quality and octane numbers: research octane number (RON) and motor octane number (MON)**

The most common characteristic of gasoline fuel is the anti-knocking quality and the associated measure known as the "octane number". Anti-knocking quality refers to gasoline behavior that is completely controlled for precise combustion by ignition sparks and not unregulated combustion through auto-ignition. Fuel prone to knocking can be inclined to auto-ignite due to high temperatures and stronger pressure waves in the combustion chamber. The problem with unregulated combustion is that it causes thermal and mechanical overload of the engine. The auto-ignition generates a shock wave with extreme pressures and high temperatures, due to which the engine parts can start vibrating. This generates the typical "knocking" or "ringing" noise. The consequences of this range from a noticeable loss of power to serious engine damage, e.g. possibly a burnt piston. Fuels not prone to knocking do not show such spontaneous combustions. These gasoline fuels can be identified by their high octane number.

The measure for the octane number has been randomly selected. It lies between 0 and 100. This is compared with a mixture of anti-knocking isooctane (octane number 100) and knocking n-heptane (octane number 0) in order to obtain an exact value for a given fuel. They can be differentiated as motor octane number (MON) and research octane number (RON). The motor octane number (MON) assesses the anti-knocking quality of the fuel precisely due to stricter conditions. The RON always turns out to be higher than the MON as the high pre-warming of the fuel mixture reduces the value in the engine system: the more sensitive a fuel is to thermal load, the lower the MON.

The basic rule is that the octane number should be as high as possible for engine operation, as it determines the maximum possible compression of the engine, and most importantly, the power achievable and the fuel consumption. A reduced octane number is needed in gasoline engines with direct injection due to internal cooling of the system by the injected gasoline. The engine compression can therefore be increased to improve the efficiency levels.

**Detergents and friction modifiers – performance spectrum of the additives**

Along with a careful selection of fuel components, chemical additives, known as "performance additives", guarantee smooth combustion, which results in low emissions. In practice, admixtures from 0.1 to 0.3% are normally used. In the end, it is the combination of well-chosen high quality basic fuels and high powered additives that creates the high tech fuels.

The development of new additives that are constantly adapted to current engines is an essential function of the research and development departments in the petroleum industry. A powerful additive must show a broad spectrum effect in this process, i.e. it should be effective in different fuel types as well as in different kinds of vehicles under all operating conditions, and must not give rise to harmful side effects.

A variety of different chemical compounds are used as additives. They are mixed as an additive package in the basic fuel while filling the tanker. These are basically purifying and cleansing additives (detergents), anti-corrosion additives, and friction modifiers.

The engine concepts of the vehicle manufacturers changed by the beginning of the 1970's due to the introduction of exhaust gas thresholds. Because the combustion was always precisely controlled, it became necessary to avoid the deposit of unwanted residue in all the problem zones – from the carburetor and injection valves to the inlet valves and pistons. This was done using detergent additives. A complete cleansing effect in the intake system should be ensured even under stringent conditions, more so in view of the advancements in engines. An additional requirement of the detergent additive is to reduce the deposits already formed (e.g. by using a non-additive fuel). Long-term tests with various engines show that this is certainly possible to varying degrees. The effectiveness of the additive is not only a question of unwanted residue but also of the resulting engine disturbances. The greater cleanliness in the inlet valves, combustion chamber and injectors (particularly the injectors of gasoline engines with direct injection) achieved due to additives is reflected in lower instances of repairs, an improvement in exhaust-gas emissions and a reduction in fuel consumption.

Engine manufacturers conceiving gasoline engines with direct injection are faced with special challenges due to crankcase ventilation and increased exhaust-gas recirculation rates. The detergent additives present in the fuel no longer come in contact with the inlet valves during direct injection of the fuel and as a result, can neither prevent deposits forming on the inlet valves nor remove the deposits that have already formed.

Since the fuel inevitably comes into contact with oxygen and moisture during transportation, storage and usage, this can cause corrosion in the pipelines and containers. The corroding products can cause damage and disturbance in the distribution chain and during operation. This is where anti-corrosion additives (antioxidants) are used. These provide a reliable and long lasting protection against the oxidation of metal surfaces with a very thin protective layer.

Certain fuel suppliers even use additives for reducing friction or as wear protection. They can prevent premature wear while operating with extremely low sulfur fuels in the case of certain fuel pumps, especially those for engines with direct injection. In addition, the reduction in frictional resistance enables a decrease in fuel consumption, which is especially evident in engines with large displacement and low speeds.

### 13.1.5 Effects of the fuel on exhaust emissions

Contaminants in the exhaust gas cannot be avoided completely as the fuel combustion in the engine is always incomplete. The composition and characteristics of the fuel affect the exhaust emissions. Effective additives, which keep the engine and the fuel-carrying components clean, also have a positive effect on the exhaust emissions.

The European commission developed the "Auto Oil I Program" (AOPI) in cooperation with the petroleum and automobile industry between 1992 and 1996. In the process, new thresholds for vehicle emissions and new regulations for fuel quality with EU validity were defined from 2000 onwards. The new fuel regulations were sanctioned in DIN EN 228 for gasoline fuels and show considerable potential for reducing emissions. The essential changes in the legislation for gasoline fuels from 2000 onwards addressed the different thresholds, which are represented in detail in **figure 13.6**.

| | DIN EN 228 bis 31.12.99 | DIN EN 228 ab 01.01.00 | DIN EN 228 ab 01.01.05 |
|---|---|---|---|
| Benzene (SUP max – vol % | 5(1*) | 1 | 1 |
| Sulfur max – ppm | 500 | 150 | 50 |
| Aromatics max – vol % | --- | 42 | 35 |
| Alkene (SU / BI) max – vol % | --- | 18 / 21 | 18 |
| Vapor pressure (So) max – kPa | 70 | 60 | 60 |
| E 100 °C min – vol % | 40 | 46 | 46 |
| E 150 °C min – vol % | --- | 75 | 75 |

* optional

**Figure 13.6:** Essential changes in the legislation for gasoline fuel (DIN EN 228)

The thresholds were made more stringent from 2005 onwards. Thus, further decrease in the maximum sulfur content and maximum aromatic content are predetermined today and have already been implemented.

A peculiar problem is that changes to the fuel often have a negative effect on the various contaminants. Thus, highly volatile gasoline, in particular, has a positive effect on the formation of contaminants. This, in turn, has a negative effect on the evaporative emissions. These contradictory effects on the different contaminants, which are inseparable, are applicable for all the possible measures that aim to reduce the contaminants in the exhaust gas by modifying the fuel. In addition, demands from production and application technology further restrict the room to maneuver. In spite of these constraints, which make fuel improvement extremely challenging, the new quality fuels have been remarkably successful time and again.

Thus, Aral succeeded in providing fuels by mid 2004 by introducing its new premium products "ultimate 100" and "ultimate diesel" that satisfy all the requirements of the Worldwide Fuel Charter, Category 4 for highly developed and futuristic vehicles. In this charter, the worldwide automobile industry defines its quality standards for fuels – an important prerequisite for developing more efficient and more environmentally friendly engines. As a result, the course is set for a new generation of fuels that enable lower fuel consumption with more power and lower emissions.

### 13.1.6 Gasoline fuels with bio-components: bio-ethanol and ethyl tertiary butyl ether (ETBE)

The term "biogen", derived from the Greek/Neo-Latin languages means as much as "originated by the actions of living beings, formed out of dead organisms". Thus, biogenic material has a natural origin. In Directive 2003/30/EC, the EU requires its member countries to ensure that a minimum of 2% by the end of 2005 and 5.75% by December 31, 2010 of the fuels (intended for use in transportation) come from renewable sources and are hence predominantly of biogenic origin. At present, various activities are taking place in the EU member countries to introduce fuels with bio components.

There are numerous ways of using or altering biomass. It can be used in a direct energy application (heat/current) by combustion. A CO and H2 enriched gas mixture can also be obtained by a non-stoichiometric, thermal conversion of the biomass with air (gasification). This gas mixture can be used either for an energy application or for a material application. For material application, the product of vaporization, which is also described as synthesis gas, is converted to methanol in chemical processes or to hydrocarbons by Fischer-Tropsch synthesis. Alternatively, the biomass can also be converted to ethanol by fermenting alcohol which can then be converted to ethyl tertiary butyl ether (ETBE) by further chemical conversions (etherification).

**Fuels with bio components: bio-ethanol**

The process of conversion of different sugar compounds to alcohol using yeast is an age-old one and is described as fermentation. Nowadays, almost all vegetable products such as corn, maize, potatoes and sugar beet, as well as ligneous material (lignocelluloses) can be considered as raw material. A major disadvantage of all processes is that yeast perishes in approximately 14% ethanol content (as in the case of wine fermentation) and the alcohol formed until then must be separated by distillation. Based on the raw material and production methods, the environmental protection advantage of ethanol varies as compared to gasoline fuel. A saving in $CO_2$ formation is not achieved in case of ethanol production from wheat by using lignite as processing energy. The $CO_2$ emissions can be reduced up to 90% if ethanol is produced from sugarcane instead, and the residual biomass (bagasse) is used as processing energy.

A possible use of ethanol is in the "pure form" as E85 (85% ethanol and 15% gasoline). E85 is used in special vehicles, known as *flexible fuel vehicles*, which have found a measure of popularity in the USA and in Sweden so far. The use of bio-ethanol as a fuel component (admixtures up to 5% as E5 or 10% as E10) is associated with technical hurdles, particularly in relation to behavior with water and what is known as the vapor pressure anomaly. This comes about because alcohol and water mix completely. The behavior of ethanol with water changes if it is mixed with gasoline fuel. A separation in the upper hydrocarbon-alcohol phase and a lower alcohol-water phase takes place if the water quantity exceeds the threshold, which depends on the alcohol content and the temperature and decreases with decreasing temperature. This lower phase is not suitable for engine operation. A similar problem that is not easily overcome is presented by the vapor pressure behavior of ethanol. If ethanol is added to the gasoline fuel, it results in a higher vapor pressure than would be expected for an ideal mixture of two liquids. The vapor pressure of the mixture can be more than that of the gasoline fuel used even if the vapor pressure of ethanol lies below that of the gasoline fuel. This also affects the quality of the fuels – both fuels that contain ethanol and low-ethanol fuels – set at a maximum vapor pressure value and mixed within the distribution chain. The etherification by means of the stock flow available in a refinery is preferred as an alternative that produces the high quality fuel component ETBE. At present, it is possible to add up to a maximum of 5 Vol % of ethanol content to the gasoline fuels adhering to the maximum oxygen content specified and as per the standards for gasoline fuels (EN228) in Europe. An enhancement of up to a maximum of 10 Vol % (E10) is being worked upon. However, there are constraints from the automobile industries to increasing ethanol content, as significant problems can arise, including problems with the aluminum components and corrosion at the fuel pumps of gasoline engines with direct injection.

**Fuels with bio components: ethyl tertiary butyl ether (ETBE)**

ETBE is produced from bio-ethanol by etherification. ETBE has fuel characteristics similar to the MTBE (methyl tertiary butyl ether) that has been successful as a fuel component (octane number booster) for many years. ETBE possesses very high octane numbers (RON and MON) but without the prominent vapor pressure anomalies of ethanol. The emissions of gasoline fuel containing ETBE are also comparable to or even better than those of fuel containing MTBE. Therefore, German refineries started using ETBE as a blending component for gasoline engines in March 2004.

# 13.2 Lubricants for gasoline engines

The development of modern engine concepts with higher power density with reduced fuel consumption and emission values, as for example in a gasoline engine with direct injection, places high demands on lubricants. Lubricants are therefore structural elements without which combustion engines and the transmission would not have functioned reliably. Modern high-performance lubricants are able to address even the highest demands, especially against the backdrop of high thermal loads of the units and increasingly longer oil change intervals.

## 13.2.1 Lubricant types

Lubricants for motor vehicles are subdivided into the following groups based on their use:

- Engine oils for four-stroke engines
- Engine oils for two-stroke engines
- Transmission oils for manual transmission
- Transmission oils for automatic transmission
- Transmission oils for axle drive
- Hydraulic oils
- Grease
- Universal oils for agricultural machines.

The group of engine oils for four-stroke engines is used to lubricate passenger car gasoline engines with direct injection. The requirements and performance characteristics of such engine oils are described below.

## 13.2.2 Functions of engine oils

The engine oil has to fulfill various functions and must be unrestricted in its efficiency under different operating conditions, i.e. during cold start as well as in the case of high thermal load. The engine oil must fulfill the following main functions:

- Lubrication of components that slide against each other for reducing wear and friction.
- Cooling of the engine during heat dissipation
- Transfer of power of the piston/connecting rod to the crankshaft
- Sealing between pistons, piston rings and cylinder liner
- Corrosion protection prevents rust formation by neutralizing acidic combustion products.

### 13.2.3 Engine oil requirements

The main emphasis during the development of modern engine concepts is the reduction of fuel consumption and emission, simultaneously with high power density and vehicle dynamics as well as significantly extended oil change intervals. Here, superior engine oil performance is of vital importance to ensure continuous optimum and safe operation. Extensive demands are made on the engine oil quality due to this.

- Anti-wear protection: reduction of wear in the engine components such as valve-train, pistons, piston rings, cylinder tracks, bearings, gear wheels as far as possible
- Engine cleanness: prevention of deposits, especially in the piston area, as well as avoiding sludge formation in the oil circuit (oil pumps, oil sumps, valve cover, cylinder head)
- Ageing stability: avoiding the formation of oil ageing products at high oil temperatures
- Elastomeric compatibility: neutrality against elastomeric materials
- Stable viscosity state: optimum viscosity-temperature relation at all operational temperatures or operational conditions with good shear stability and low evaporation loss.

In addition to that, environmental aspects such as the potential for reduction of fuel and oil consumption and catalytic converter compatibility are also important.

### 13.2.4 Engine oil composition

Modern engine oils are made up of one or more basic oil types and different additives depending on the engine power class. Multi-grade oils, which comprise the following components, are now used predominantly, **figure 13.7**.

| Components | Approximate proportion in [%] |
|---|---|
| Basic oil | 78 |
| VI conditioner | 10 |
| Detergents | 3 |
| Dispersers | 5 |
| Anti-wear protection | 1 |
| Miscellaneous | 3 |

**Figure 13.7:**
Components of a typical four-stroke engine oil [6]

Careful selection and harmonization of the basic oil components and additives is of vital importance and yields high-performance engine oil.

### 13.2.5 Viscosity

Viscosity is the characteristic of a liquid to offer resistance to flow. Light oils (low viscosity) show a low flow resistance (smooth-running effect), while the viscous oils (high viscosity) have a higher flow resistance for equally good load-carrying capacities of the lubricating film.

Viscosity is a temperature-dependent variable, i.e. the viscosity reduces with increasing temperature and increases at low temperatures. The viscosity-temperature behavior is described by the viscosity index (VI) parameter. The fewer the changes in viscosity at different temperatures, the higher the viscosity index of the engine oil; **figure 13.8**. The natural viscosity index of oil can be increased by adding a VI conditioner; **figure 13.9**.

Lubricants whose viscosity changes only under the influence of temperature and pressure and not at different shear velocities are called Newtonian liquids. The single-grade engine oils that do not contain any VI conditioner belong to this class. In multi-grade engine oils, which contain VI conditioners, change in viscosity depends not only on temperature and pressure, but also on shear velocity, i.e. the viscosity reduces with increasing shear velocity. These oils are known as the non-Newtonian liquids; **figure 13.10**.

The shear velocity is defined as the oil speed at the moving part divided by the thickness of the lubricating film; the unit is [$s^{-1}$]; **figure 13.11**.

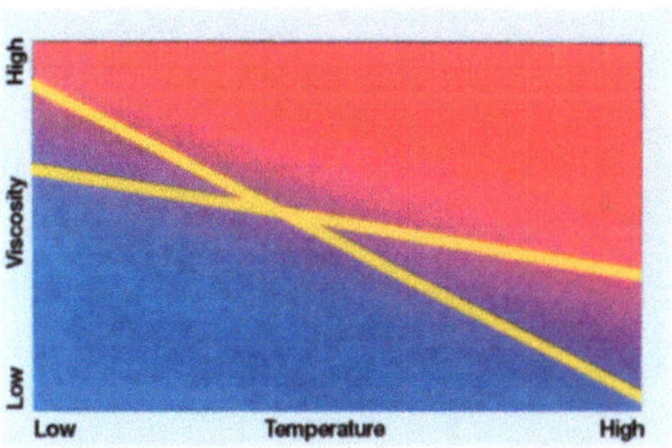


**Figure 13.8:** Temperature-viscosity behavior [7]

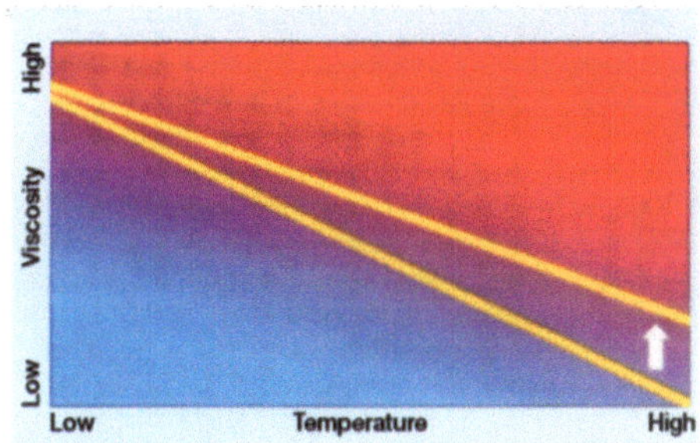


**Figure 13.9:** Change in the viscosity caused by VI conditioner [7]

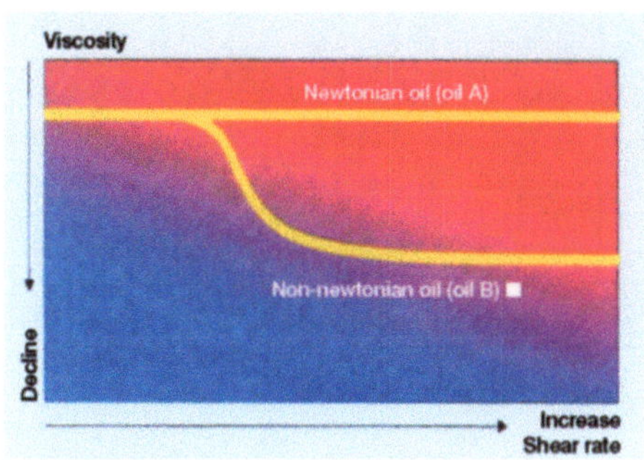


**Figure 13.10:** Viscosity and shear velocity [7]

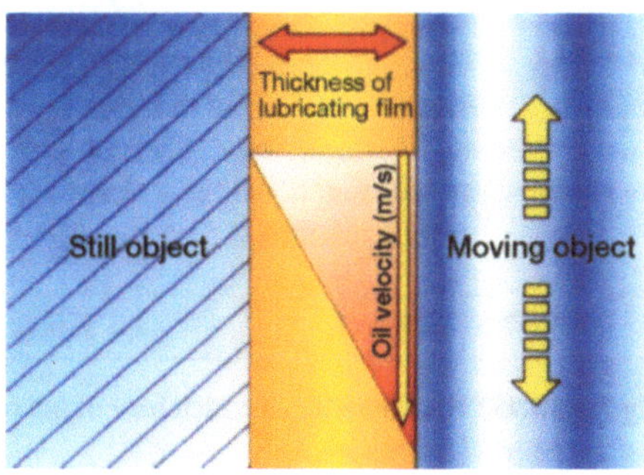


**Figure 13.11:** Shear velocity [7]

A relatively high shear velocity of approximately $10^{-5}$ to $10^{-6}$ $s^{-1}$ occurs in the crankshaft main bearing and in the pistons and cylinder walls at equally high temperature levels. Here, the HTHS viscosity (*High Temperature High Shear*) is introduced in order to achieve realistic viscosity information. For this, the viscosity is determined at an oil temperature of 150°C and a shear velocity of $10^{-6}$ $s^{-1}$ in a rotation viscometer; the unit is [mPa · s].

Ultimately, the lubrication of the engine oil at high temperatures and high shear velocity in the engine is ensured by fixing the thresholds for the HTHS viscosity. The decrease in the threshold of the HTHS viscosity leads to further potential for reduction in fuel consumption. However, the prerequisites for this must be provided by design changes in the engine.

### Viscosity classes for engine oils

Viscosity is the best-known characteristic value of engine oils that differentiates single-grade and multi-grade engine oils. Today, multi-grade oils are the most frequently used type of oil, as they fulfill the viscosity requirements in the respective low temperature range as well as those at 100°C. Thus, multi-grade engine oils can be used all year round.

The classification of engine oils is described with the SAE (*Society of Automotive Engineers*) classes; **figure 13.12**. Multi-grade engine oils in a low viscosity class, such as SAE 0W-30 or SAE 0W-40, are known as smooth-running oils and reduce the fuel consumption as the friction decreases.

| SAE viscosity grade | Low temperature viscosity | | Kinematic viscosity at 100°C | | HTHS 150°C |
|---|---|---|---|---|---|
| | Cold cranking viscosity (mPa · s) at temp. (°C) max | Cold pumping viscosity (mPa · s) at temp. (°C) max | (mm²/s) min | (mm²/s) max | (mPa · s) min |
| 0W | 6200 at −35 | 60000 at −40 | 3.8 | – | – |
| 5W | 6600 at −30 | 60000 at −35 | 3.8 | – | – |
| 10W | 7000 at −25 | 60000 at −30 | 4.1 | – | – |
| 15W | 7000 at −20 | 60000 at −25 | 5.6 | – | – |
| 20W | 9500 at −15 | 60000 at −20 | 5.6 | – | – |
| 25W | 13000 at −10 | 60000 at −15 | 9.3 | – | – |
| 20 | – | | 5.6 | < 9.3 | 2.6 |
| 30 | – | | 9.3 | < 12.5 | 2.9 |
| 40 | – | | 12.5 | < 16.3 | 2.9 1) |
| 40 | – | | 12.5 | < 16.3 | 3.7 2) |
| 50 | – | | 16.3 | < 21.9 | 3.7 |
| 60 | – | | 21.9 | < 26.1 | 3.7 |

1) applies to 0W-40, 5W-40, 10W-40 2) applies to 15W-40, 20W-40, 25W-40

**Figure 13.12:** SAE viscosity grades for engine oils - SAE J 300

### 13.2.6 Basic oils

Nowadays, mineral oil raffinates, hydrocrack oils and synthetic hydrocarbons are mainly used in the four-stroke engine oils. These basic oil components are used individually as well as in different mixtures and the engine oils formed thereby are known colloquially as mineral oil-based, partly synthetic or completely synthetic oils.

The basic oils selected determine important characteristics such as viscosity, evaporation loss and the responsiveness of additives. Accordingly, the basic oils are subdivided into five groups as per ATIEL (*Association Technique de l'Industrie Europeenne des Lubrifiants*); **figure 13.13**.

| Group | Composition | Sulfur content | Viscosity index |
|---|---|---|---|
| I | < 90% (m/m) saturated hydrocarbons | > 0.03% (m/m) | ≥ 80 < 120 |
| II | ≥ 90% (m/m) saturated hydrocarbons | ≤ 0.03% (m/m) | ≥ 80 < 120 |
| III | ≥ 90% (m/m) saturated hydrocarbons | ≤ 0.03% (m/m) | ≥ 120 |
| IV | Polyalphaolefin (PAO) | | |
| V | All others not included in the I, II, III and IV groups (e.g. esters) | | |

**Figure 13.13:** Classification of the basic liquids as per ATIEL [8]

#### Mineral oil raffinates

This group of basic oils is still used frequently in engine oils as they are relatively cost-effective to manufacture. They are manufactured from crude oil by atmospheric distillation, vacuum distillation, solvent refining, deparaffinization and hydrofinishing. Here, the raffinates occur with different viscosities, and the adjoining distillation sections preferred for use in engine oils are selected and mixed to set the viscosity. These raffinates show a normal ageing resistance, relatively high evaporation losses, especially the less viscous oils, average low temperature behavior and a viscosity index of 90 to 100.

#### Hydrocrack oils

Inputs for the hydrocrack oils are vacuum gas oil or long-chain normal paraffins from the deparaffinization of the raffinates. The long-chain molecular structures of these products are then broken down (cracked) into smaller lubricating material molecules in special cracking plants and the unsaturated broken bonds are saturated with hydrogen (hydrated). The resulting products undergo vacuum distillation (different viscosity sections) and deparaffinization. The hydrocrack oils have a much higher viscosity index – from 120 and 150 – than the mineral oil raffinates and behave better at low temperatures.

**Synthetic hydrocarbons**

The manufacturing process for synthetic hydrocarbons requires hydrocarbons from crude oil (as in the case of mineral oil raffinates and hydrocrack oils). Petroleum is used as input, and first undergoes the cracking process, in which short-chain olefins are produced. These products can be processed into different synthetic oils with appropriate processes.

- Synthetic hydrocarbons PAO (Poly-α-Olefin) and PIB (Poly-Iso-Butene)
- Synthetic esters
- Polyglycoles.

The most common basic liquid for engine oils is PAOs, although some synthetic esters are also used. Due to their custom-made molecular structure, synthetic hydrocarbons have a very high natural viscosity index, very good low temperature behavior, lower evaporation loss and good ageing resistance.

The different manufacturing processes of the basic oils for use in vehicles are summarized below in a very simplified form; **figure 13.14**.

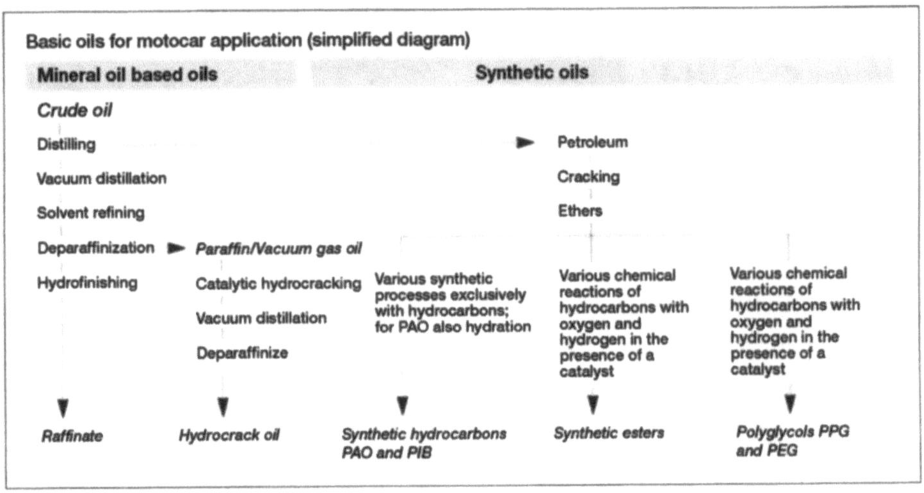


**Figure 13.14:** Basic oils for vehicle applications [7]

## 13.2.7 Additives

The basic oil alone cannot fulfill the varied and high engine oil demands, so additives, which improve the efficiency of the finished product, are mixed with the engine oil.

Additives are roughly differentiated into surface-active substances, interface-active substances and those which improve the basic oil characteristics. The surface-active and interface-active additives possess an oleophillic part and a polar part. The oleophillic part is responsible for the solubility in the basic oil and the polar part contains the respective active components that get deposited on the surfaces. Organic (ash-less) or organo-metallic (ash-forming) components are used; **figure 13.15**.

The basic oil characteristics can also be improved by the additive components, where these components have a positive effect on the temperature-viscosity relation, low temperature behavior and elastomeric resistance. In addition, the additive components provide sufficient anti-ageing protection to the engine oil. This is very important in the case of extended oil change intervals and very high thermal load.

The use of foam inhibitors reduces foam generation in the engine oil as it accelerates oil ageing and changes viscosity and compressibility. This has a negative effect on the efficiency of the engine oil (hydraulic control processes, pump capacity/oil pumps); **figure 13.16**.

| Additive type | Typical component group | Function |
|---|---|---|
| Detergent | Sulfonate, carbamate | Prevention of deposits containing varnish and carbon, especially in the piston ring area (keep clean/clean up/ peptization)<br>Neutralization of acidic combustion residue |
| Disperser | Phenolate, polyisobutenyl-succinic acid derivatives, phosphates | Prevention of sludge formation and deposits by solubility and peptization |
| Extreme pressure/ anti-wear | Zinc dithiophosphate, phosphorus compounds and sulfur compounds | Anti-wear prevention in case of extreme load and high temperature level by forming a protective layer on the surface of the sliding counterpart |
| Friction modifier | Synthetic ester, phosphonate | Reduction of frictional coefficients, especially by mixed friction of the sliding counterparts by forming a protective layer |
| Corrosion/ rust protection | Sulfonate, grease acid amides | Protection against corrosive attack by oxygen and water (acids) by forming a barrier on the metal surfaces |
| Metal deactivators | Benzotriazole, thiadiazole | Prevents the occurrence of catalytically effective metal ions in the engine oil by forming a neutralizing protective layer on the surface |

**Figure 13.15:** Surface/interface-active additives

| ACEA | Status | Functional range | Oil type |
|---|---|---|---|
| A1 | Current | Passenger car gasoline engines | Smooth-running engine oils |
| A2 | Withdrawn in 2004 | | Standard engine oils |
| A3 | Current | | Premium engine oils |
| A5 | Current | | Premium smooth-running engine oils |
| B1 | Current | Passenger car and light commercial diesel engines | Smooth-running engine oils |
| B2 | Withdrawn in 2004 | | Standard engine oils |
| B3 | Current | | Premium engine oils |
| B4 | Current | | Premium engine oils |
| B5 | Current | | Premium smooth-running engine oils |
| C1 | Current | | Low SAPS engine oils |
| C2 | Current | | Low SAPS engine oils |
| C3 | Current | | Low SAPS engine oils |
| E1 | Expired in 2001 | Heavy commercial diesel engines | Standard engine oils |
| E2 | Current | | Standard engine oils |
| E3 | Withdrawn in 2004 | | Premium engine oils |
| E4 | Current | | Premium engine oils |
| E5 | Withdrawn in 2004 | | Standard/premium engine oils |
| E6 | Current | | Premium engine oils |
| E7 | Current | | Premium |

**Figure 13.17:** Overall idea of the ACEA specifications

| ACEA | Requirements |
|---|---|
| A1/B1 – 04 | Multi-grade engine oil XW-20, XW-30 and XW-40 for gasoline and lightweight diesel engines. Suitable for the use of frictionless, low viscosity oils with a HTHS viscosity between 2.6 and 3.5 mPa · s at 150°C. These oils are not suitable for all engine types. Use only with appropriate manufacturer approval. |
| A3/B3 – 04 | Shear stable, stay-in grade engine oils for gasoline and lightweight diesel engines with high performance and with / without extended oil change intervals if they are specified by the manufacturer. HTHS viscosity > 3.5 mPa · s at 150°C. Low viscosity all-season oils for complicated operating conditions as per manufacturing specifications. |
| A3/B4 – 04 | Shear stable, stay-in grade engine oils for gasoline and DI diesel engines with high performance. Also suitable for use as described in B3. HTHS viscosity > 3.5 mPa · s at 150°C. |
| A5/B5 – 04 | Shear stable, stay-in grade engine oils for extended oil change intervals in gasoline and lightweight diesel engines with high performance, which are suitable to use the friction-less and low viscosity oils with a HTHS viscosity of 2.9 to 3.5 mPa · s at 150°C. These oils are not suitable for all engine types. To be used only with manufacturer approval. |

**Figure 13.18:** ACEA specifications for gasoline engines and diesel engines (Date 2004) [9]

## API classifications

The API classifications differentiate between the categories S (*Service Fill*) for gasoline engines and C (*Commercial*) for diesel engines. The following **figure 13.19** gives an overall idea of the individual categories.

| Gasoline engines (S = Service Fill) | | | Diesel engines (C = Commercial) | | |
|---|---|---|---|---|---|
| Category | Status | Year of initiation | Category | Status | Year of initiation |
| SM | valid | 2004 | CI-4 PLUS | valid | 2005 |
| SL | valid | 2001 | CI-4 | valid | 2002 |
| SJ | valid | 1997 | CH-4 | valid | 1998 |
| SH | | 1994 | CG-4 | valid | 1995 |
| SG | | 1989 | CF-4 | valid | 1990 |
| SF | | 1980 | CF-2 | valid | 1994 |
| SE | | 1972 | CF | valid | 1994 |
| SD | obsolete | 1968 | CE | | 1985 |
| SC | | 1964 | CD-II | | 1985 |
| SB | | 1930 | CD | obsolete | 1955 |
| SA | | 1925 | CC | | 1961 |
| | | | CB | | 1994 |
| | | | CA | | The 40s |

**Figure 13.19:** Complete overview of the API classifications

The API classifications that are valid currently for gasoline engines are predominantly based upon the test sequences with American engine technology and driving profiles; **figure 13.20**. Since the introduction of the SH category, the respective test cycles must be reported and registered as per the CMA Code of Practice (*Chemical Manufacturers Association*).

| API | Requirements |
|---|---|
| SJ | Corresponds to the API SH with additional requirements for high temperature deposits and evaporation loss. Strict read-across rules for exchangeability of different basic liquids. |
| SL | Corresponds to the API SJ with intensified high temperature resistance, evaporation loss and anti-wear protection requirements. |
| SM | Corresponds to the API SL with intensified anti-wear and anti-corrosion protection and aging behavior requirements. Restriction of phosphorus and sulfur content in the engine oil. |

**Figure 13.20:** Current API classifications for gasoline engines

| Manu-facturer | Specification | Functional range/requirements |
|---|---|---|
| BMW | Long-life-01 | Specific gasoline engines from B2001<br>with flexible oil change intervals<br>HTHS > 3.5 mPa · s<br>ACEA A3 + in-house tests with increased anti-wear requirements, gas content in the oil, sludge, increase in viscosity, wear |
| | Long-life-01 FE | Specific gasoline engines from B2001<br>with flexible oil change intervals<br>HTHS > 3.5 mPa · s<br>ACEA A3 + in-house tests with increased anti-wear requirements, gas content in the oil, sludge, increase in viscosity, wear, specific minimum fuel consumption test |
| | Long-life-04 | Specific gasoline engines from B2004<br>with flexible oil change intervals<br>HTHS > 3.5 mPa · s<br>ACEA A3 + in-house tests with increased anti-wear requirements, gas content in the oil, sludge, increase in viscosity, wear |
| MB | Sheet 229.1 | Gasoline engines<br>HTHS > 3.5 mPa · s<br>ACEA A2/A3 + specific sludge requirements |
| | Sheet 229.3 | Gasoline engines – flexible oil change intervals 15,000 to 20,000 km depending on engine oil quality<br>HTHS > 3.5 mPa · s<br>ACEA A3 + specific sludge requirements, minimum fuel consumption, seal compatibility |
| | Sheet 229.5 | Gasoline engines – flexible oil change intervals up to 30,000 km<br>HTHS > 3.5 mPa · s<br>ACEA A3 + specific requirements in the M111E sludge test, minimum fuel consumption, seal compatibility |
| Opel | GM-LL-A-025 | Gasoline engines with flexible oil change intervals<br>Up to 30,000 km<br>HTHS > 3.5 mPa · s<br>ACEA A3 + in-house tests with intensified oxidation stability, wear, air refining state, minimum fuel consumption requirements |
| VW<br>AUDI<br>SEAT<br>SKODA | Norm 502 00 | Gasoline engines with 15,000 km oil change intervals<br>HTHS > 3.5 mPa · s<br>Specific tests with increased requirements for wear, sludge and seal compatibility |
| | Norm 503 00 | Gasoline engines with extended maintenance interval (WIV)<br>Smooth-running oils, HTHS 2.9 to 3.4 mPa · s<br>Specific tests with increased requirements for: sludge, oxidation resistance, fuel economy, wear, seal compatibility |
| | Norm 503 01 | Gasoline engines with extended maintenance interval (WIV)<br>Smooth-running oils, HTHS > 3.5 mPa · s<br>Specific tests with increased requirements: sludge, oxidation resistance, fuel economy, wear, seal compatibility |
| | Norm 504 00 | Gasoline engines with extended maintenance interval (WIV)<br>MID SAPS engine oil, HTHS min. 3.5<br>Specific test requirements: sludge, oxidation resistance, fuel economy, wear, seal compatibility |

**Figure 13.21:** Manufacturer specifications gasoline fuel engine oils

The API classifications are no longer significant in Europe as they cannot address the specific requirements of the European vehicle technologies, driving profiles and operational conditions in their entirety; but they are sometimes mentioned by some European vehicle manufacturers in their fuel and lubricant specifications.

### 13.2.9 Fuel and lubricant specifications of passenger car manufacturers

The fuel and lubricant specifications of the independent vehicle manufacturers and their requirements are specifically adjusted to current engine technologies and applications. Here, the thermal load of the units, extended oil change intervals, minimum fuel consumption values and the effect of the engine oil on the exhaust-gas aftertreatment system play an increasingly important role. The requirements mentioned here generally more than meet the ACEA specifications and the API classifications. **Figure 13.21** gives a general overview of some vehicle manufacturers' current specifications for engine oil for use in gasoline engines with and without direct injection.

The engine specifications must also be adapted in tandem with advances in engine and lubricant technology. Therefore, the possibility that separate performance classes for gasoline engines with direct injection will be introduced in the future cannot be ruled out.

Here, catalytic converter resistance and deposit formation in the intake tract, and especially the inlet valve cleanliness, should be taken into consideration as the fuel cleaning additives are not effective in the inlet valve area of gasoline engines with direct injection. This requires careful harmonization of the basic oils and the additives, or additive dosing of the engine oil. In addition, more anti-wear protection requirements must be taken into consideration, including requirements for local oil dilution, friction reduction, ability to remove contamination, neutralizing the $NO_x$ intake and longer oil change intervals.

### 13.2.10 Future prospects

The gasoline engine with direct injection has fuel-savings potential that can be further improved by using high quality, efficient smooth-running engine oils. Here, engine oil as structural element also has a vital contribution to make in order to ultimately offer customers environmental-friendly, high-performance engine technology.

## Literature

[1] Deutsche BP AG: Das Buch vom Erdöl, 1989
[2] Mineralöl-Wirtschaftsverband: MWV-Jahresbericht/Mineralöl-Zahlen 2005. Hamburg: MWV, 2006
[3] ARAL AG: Ottokraftstoffe – Anforderungen, Qualitäten, Perspektiven. Bochum: ARAL AG, 2000
[4] ACEA: Worldwide fuel charter 2002. Brussels: ACEA, 2002
[5] *van Basshuysen; Schäfer*: Handbuch Verbrennungsmotor, 4. Auflage. Wiesbaden: Vieweg Verlag, 2007
[6] Castrol: Schmierstoffwissen, Grundwissen Motorenöl. www.bp.com
[7] Aral: Schmierstoffe, Grundlagen. Fachreihe Forschung und Technik, 1997
[8] Aral: Archiv
[9] ACEA: ACEA European oil sequences. www.acea.be 2004

# 14 Engine Concepts

## 14.1 Series concepts

### 14.1.1 History

From experience with aircraft engine design, where gasoline direct injection has been used since 1920, the companies Gutbrod and Goliath developed a two-stroke engine with direct injection in collaboration with Bosch, which was used in the series production vehicles Gutbrod Superior and Goliath GP 700. Both vehicles used a two-stroke engine having two cylinders with 600 ccm and 26 horsepower (hp) developed since 1949, under the guidance of Hans Scherenberg. The engines were equipped with the Bosch modified diesel injection system. This helped the vehicles achieve very good mileage and favorable fuel consumption, which was approximately 30% better than that of carburetor variants.

Against the background of the positive results achieved with gasoline direct injection in two-stroke engines, Daimler Benz began development of gasoline direct injection for a 3.0 l passenger car engine at the beginning of 1952. So many positive results were available even by the end of that year that vehicle tests of this engine concept could be started. After a successful series development process, the Daimler-Benz 300 SL sports car equipped with 3.0 l fuel injection engine (W198) was manufactured in series from the summer of 1954; **figure 14.1**. **Figure 14.2** shows a cross-section of the six-cylinder inline engine of the 300 SL [1].

**Figure 14.1:**
Mercedes-Benz 300 SL

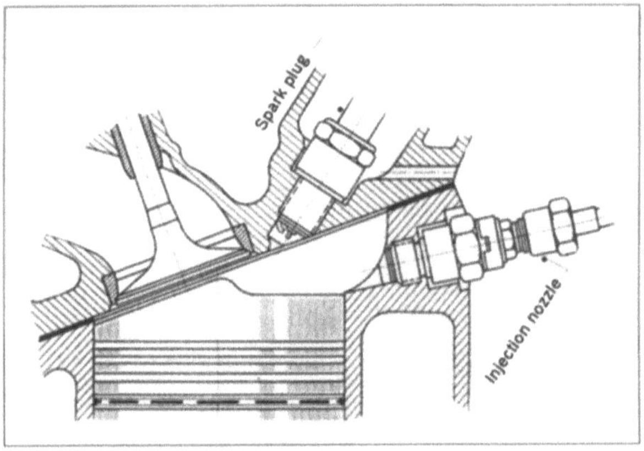


**Figure 14.2:**
Mercedes 300 SL; combustion chamber of the production engine [1]

Specifications for W198:

| | |
|---|---|
| **Design** | Six-cylinder inline engine with direct injection |
| **Displacement** | 2996 ccm |
| **Bore x stroke** | 85 mm x 88 mm |
| **Power** | 215 hp at 5800 rpm |
| **Torque** | 280 Nm at 4600 rpm |
| **Compression ratio** | 8.55:1 |
| **Mixture formation** | Direct injection, mechanically controlled, six-plunger piston injection pump |
| **Injection pressure** | 45 bar |
| **Valves** | Overhead, 2 per cylinder, overhead camshaft, operation through duplex roller chain |
| **Lubrication** | Dry sump lubrication |

Bosch supplied the throttle valve housing used for mixture formation, the six-plunger injection pump equipped with a leakage oil lock in the inline design, and the injection nozzles without a leakage line. In terms of basic structure, they are identical to or differ only slightly from the components used in the two-stroke engines mentioned above. However, the control unit of injection pumps was provided with a barometric cell for adjustment of the air fuel mixture at high altitudes (mountain passes). The injection pressure of the injection nozzle is 45 bar. In spite of considerable reduction in fuel consumption by means of direct injection, the maximum power of the engine could be raised up to approximately 10%. The points to be highlighted with regard to this engine concept are high elasticity at low engine speeds and a direct, instantaneous response of the engine to load changes.

The development activities in gasoline direct injection were virtually completely disrupted in the 1960's and 1970's due to relatively low fuel prices and the development of injection technology in the direction of manifold injection in particular.

### 14.1.2 Stratified charge concepts: wall-guided and air-guided combustion processes

Initially gasoline direct injection was of interest to several vehicle manufacturers because of the development of common rail injection systems and modern, efficient engine controls, which offered a range of ways to influence and adapt the injection. One of the first representatives of the modern generation was the GDI engine (*Gasoline Direct Injection*) with gasoline direct injection introduced by Mitsubishi in the IAA (International Motor Show) in 1995; **figure 14.3**. It was launched in Japan in 1996 and in Europe in 1997. The exhaust gas was purified by a selective, continuously working reduction catalytic converter with iridium coating and a three-way catalytic converter. The $NO_x$ conversion rate was only approximately 60%.

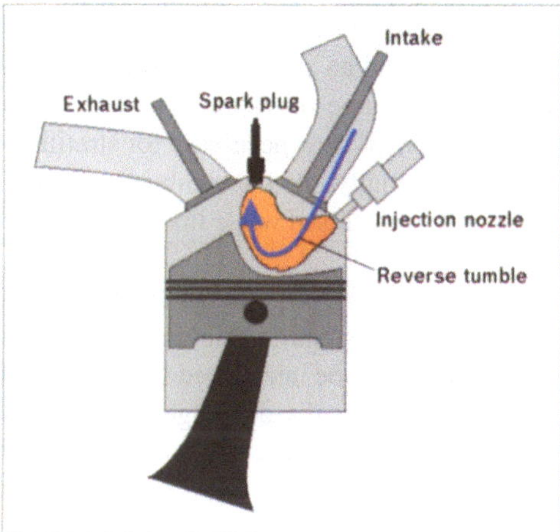


**Figure 14.3:**
Principle of Mitsubishi GDI

Specifications:

| | |
|---|---|
| **Displacement** | 1.8 liter |
| **Power** | 90 kW/122 hp at 5500 rpm |
| **Number of cylinders** | 4 (16 valves) 5 gear manual transmission |
| **Injection system** | Gasoline direct injection (GDI) |
| **Maximum speed** | 200 km/h |
| **0 to 100 km/h** | 10.4 s |
| **Consumption** | approximately 8.5 to 9 liters at 100 km |

Unlike the Mercedes 300 SL, in which direct injection was given priority in order to increase power output, stratification of charges for reducing fuel consumption is focused on direct injection concepts. The Mitsubishi engine uses the wall-guided process, in which the fuel is injected in the spark plug by a special piston recess assisted by a reverse tumble;

**figure 14.3**. A stratification of cylinder charge can thus be achieved in the lower load and speed range so that an ignitable air/fuel mixture is available around the spark plug, whereas only air is present in the cylinder periphery. The charge exchange can be dethrottled and the charge exchange losses can be reduced by using a global lean mixture. Thus, fuel consumption savings of up to 10% are possible in practice.

A selective, continuously working reduction catalytic converter with iridium coating is used for exhaust-gas purification. This catalytic converter, along with a conventional three-way catalytic converter, is integrated into a tandem catalytic converter. The catalytic converter is mounted on the underside because it has low temperature resistance. However, a $NO_x$ conversion rate of only 60% is achieved with the help of this catalytic converter principle.

In 1996, Toyota introduced an engine with stratified charge, named D4, into the Japanese market. Though the wall-guided process is used, as at Mitsubishi, at Toyota, a swirl flow was used to assist mixture formation instead. The piston had an asymmetric, plate-shaped recess into which the injector injected during stratified charge operation. The latter was placed below the discharge valves as at Mitsubishi. The swirl flow necessary for stratification was created with the help of a swirl port leading tangetially into the combustion chamber. The second intake port, also called the charge port, was closed while this took place.

Two primary catalytic converters near the engine as well as a $NO_x$ storage catalytic converter mounted on the underside were used to purify the exhaust gas, achieving a $NO_x$ conversion rate of more than 90%. However, a rich mixture had to be introduced frequently for the regeneration of the storage catalytic converter giving rise to disadvantages with regard to fuel consumption. The engine was not permitted in Europe and the USA, mainly due to insufficient exhaust-gas purification.

A third manufacturer, Nissan, introduced an engine with direct injection to the market in 1998. The engine had a V6 engine with stratified charge as in the wall-guided process. Like Toyota, Nissan also used swirl support for charge stratification. The somewhat flat, unsymmetrical piston recess was supposed to reduce the variations in airflow. A *Casting Net Injector*, which was mounted below the inlet valves, produced an asymmetric fuel spray in order to create a uniform fuel film at the piston head that was able to vaporize lightly. A three-way catalytic converter on the underside was used for exhaust-gas purification. The Japanese $NO_x$ exhaust emissions standards were met only in the showroom condition. The engine was not introduced to Europe and the USA because of its emissions levels.

In 2000, PSA Peugeot Citroën (Peugeot Société Anonyme) introduced Peugeot 406 and Citroen C5 with the 2.0 HPI (*High Pressure Direct Injection*) engine with direct injection to the market. The engine was essentially based on the Mitsubishi GDI engine and also used a wall-guided process supported by reverse tumble. With a displacement of 1997 ccm, the engine gave power output of 103 kW and was, according to PSA, was 10% more economical than the manifold variant but produced higher torque and lower emissions. The injection pressure of the fuel high pressure injection lay between 30 and 100 bar. As opposed to Mitsubishi, however, PSA used a combination of a three-way catalytic converter and a $NO_x$ storage catalyst, in addition to two small precatalytic converters. The storage catalyst had to be regenerated for three seconds every 60 seconds in rich operation.

Also in 2000, another engine with wall-guided charge stratification process arrived on the market: the VW FSI (*Fuel Stratified Injection*) in Lupo; **figure 14.4**. It was a further development of the MPI engine EA111 with a displacement of 1.4 liters and power of 74 kW. The power could be increased up to 77 kW with direct injection. An FSI process involved a wall-guided process with air support. The piston was provided with a fuel recess on the intake side and a flow recess on the exhaust side which diverted the air in the desired direction [2].

The intake ports were designed as charge ports, but they could be selectively closed in the lower port cross-section with a flow valve so that a strong tumble flow was developed which assisted mixture formation. The high pressure injection valve with a spray angle of 70° and a spray inclination angle of 20° was positioned below the intake ports. The spark plug was placed at the center as in the basic engine.

**Figure 14.4:** Lupo FSI [2]

The compression ratio could be raised from 10.5 to 12 in the basic engine by internal cooling of the fuel. The fuel consumption benefit when the compression was increased was approximately 2.3%. In the stratified operating mode, the engine operated almost in a dethrottled manner ($p_{intake} > 900$ mbar) and achieved an air/fuel mixture ratio between 1.8 and 3. The fuel consumption benefits in the lean stratified operating mode depended on the operating point and amounted to 10% in medium part load and achieved up to 44% when idling, according to VW.

The intensive charge movement in the FSI engine also led to a very good EGR (exhaust-gas recirculation) compatibility. Thus, EGR rates of up to 25% could be used in homogenous operation without a noticeable increase in hydrocarbon emissions. This led to additional fuel savings.

Although the Euro 3 legislation was in force at the time of introduction of Lupo FSI, the engine had already achieved the stringent Euro 4 values. For this purpose, a primary catalytic converter with three-way coating placed near the engine combined with a $NO_x$ storage catalytic converter placed on the underside was used. VW was the first vehicle manufacturer in the world to use a $NO_x$ sensor placed after the $NO_x$ catalytic converter in order to achieve optimum conversion of nitrogen oxide.

Specifications VW FSI:

| | |
|---|---|
| **Displacement** | 1390 ccm |
| **Number of cylinders** | 4 |
| **Compression ratio** | 12:1 |
| **Power** | 77 kW at 6200 rpm |
| **Torque** | 130 Nm at 4250 rpm |
| **Exhaust-gas recirculation** | Electric EGR valve with central feed |
| **Camshaft adjustment** | Intake camshaft with 40° adjustment range |
| **Engine control** | Bosch MED7 |
| **Ignition** | Separate spark coil |
| **In-cylinder flow** | Switchable tumble flap system |

In the period that followed, FSI engines with additional cylinder capacities of 1.6 l, 2.0 l, and 3.2 l were offered in other VW models (Golf, Passat). However, the present day FSI engines run only in the homogenous mode and can operate completely without the stratified operating mode, which is also the case with all supercharged FSI engines.

In 2002, the subsidiary company Audi also introduced an engine with direct injection named FSI into the market. However, this engine had to undergo a complete rework of the four-valve aluminum cylinder head and the piston geometry when compared to the VW version, since it involved a purely air-guided process; **figure 14.5**. A strong tumble flow was generated using a continuously adjustable charge movement flap and diverted by the piston recess in such a way that the injected fuel was transported to the spark plug by the airflow.

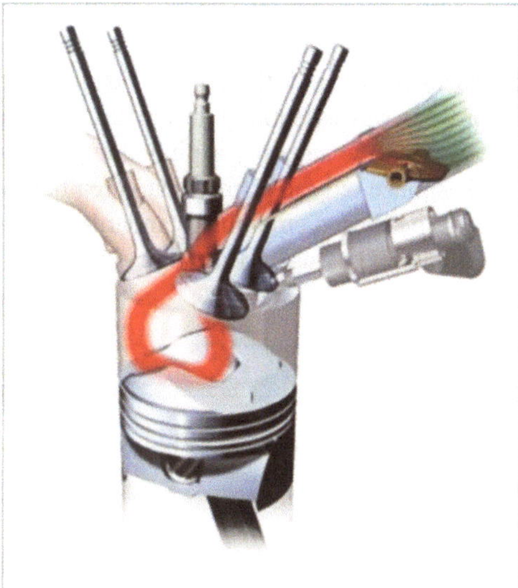

**Figure 14.5:**
Air-guided FSI concept – Audi [3]

Specifications Audi FSI:

| | |
|---|---|
| Design | Four-cylinder inline engine with direct injection |
| Displacement | 1984 ccm |
| Bore x stroke | 82.5 mm x 92.8 mm |
| Valves per cylinder | 4 |
| Camshafts | 2 |
| Rated power | 110 kW at 6000 rpm |
| Maximum torque | 200 Nm at 3250 to 4250 rpm |
| Emissions level | Euro 4 |
| $CO_2$ emissions | 170 g/km |

The fuel was transported to the high pressure valves using a demand-controlled Bosch high pressure pump (type HDP2). The fuel was injected into the combustion chamber at a flat angle of 22.5°, thus avoiding contact with the piston head.

The Audi engine possessed two catalytic converters for purifying the exhaust gas. Besides the three-way catalytic converter near the engine, a $NO_x$ storage catalytic converter was located on the underside which had a $NO_x$ sensor at the outlet of catalytic converter just as in the VW engine. The storage catalytic converter was controlled by map and temperature.

Alfa Romeo made a debut in the same year with the type 156 with a 2 l four-cylinder direct injection engine called JTS (*Jet Thrust Stoichiometric*). The injectors are located below the injection ports, which are designed purely as charge ports and do not allow any influence of charge movement. Like most others so far, the process in the Alfa Romeo is a wall-guided process. The piston has a small recess moved towards the exhaust side in order to direct the fuel towards the spark plug. The process was essentially designed to obtain high torque and power output values and outperformed the naturally aspirated engine by approximately 10%. The charge stratification takes place only in a small map area below 2000 rpm and an effective mean pressure of 4 bar. The Euro 4 emissions standards can be adhered to without special $NO_x$ exhaust-gas aftertreatment by using very high exhaust-gas recirculation rates.

Specifications Alfa Romeo JTS:

| | |
|---|---|
| **Design** | Inline |
| **Number of cylinders** | 4 |
| **Displacement** | 1970 ccm |
| **Power** | 121 kW (165 hp) at 6400 rpm |
| **Maximum torque** | 206 Nm at 3250 rpm |
| **Bore x stroke** | 83.0 mm x 91.0 mm |
| **Compression ratio** | 11.3:1 |
| **Valves/cylinders** | 4 |

With the M271 DE, DaimlerChrysler brought an engine with "first generation" direct injection onto the market in the autumn of 2002; **figure 14.6**. The engine was available as 200 CGI (*Charged Gasoline Injection*) in the C class models. It is a 1.8 l, four-cylinder engine

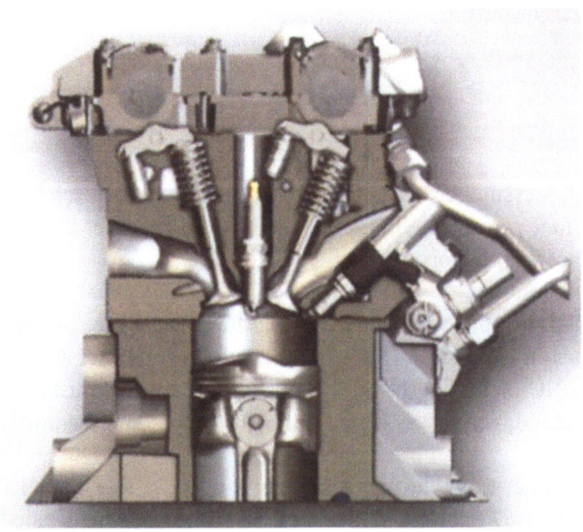

**Figure 14.6:**
M271 DE with wall-guided direct injection [4]

with two adjustable camshafts, Lanchester balancing shaft, mechanical supercharging and wall-guided direct injection that DaimlerChrysler calls direct injection of the first generation [4]. Direct injection of the second generation refers to spray-guided direct injection.

Specifications DaimlerChrysler 200 CGI:

| | |
|---|---|
| **Design** | Inline |
| **Number of cylinders** | 4 |
| **Displacement** | 1796 ccm |
| **Power** | 125 kW (170 hp) at 5300 rpm |
| **Bore x stroke** | 82.0 mm x 85.0 mm |
| **Compression ratio** | 11:1 |
| **Valves/cylinders** | 4 |
| **Maximum torque** | 250 Nm at 3000 rpm |
| **Injection pressure:** | maximum 120 bar |

The four-valve cylinder head has two tumble ports, one of which can be closed with a tumble flap for generating swirl. The spark plug is located at the center while the injection nozzle, inclined at 42°, is located below the intake valve. The piston has an eccentric recess at the intake side as in Toyota's process. A firewall catalytic converter near the engine, a connected cooling section and a $NO_x$ storage catalytic converter with $NO_x$ sensor separate from the engine are arranged together for exhaust-gas aftertreatment.

Ford presented its 1.8 liter SCi (*Smart Charge Injection*) engine with wall-guided direct injection at the 24$^{th}$ International Vienna Motor Symposium in 2003. Series production followed in the same year. Since then the engine has been available in Mondeo vehicles.

Specifications Ford 1.8 SCi:

| | |
|---|---|
| **Design** | Inline |
| **Number of cylinders** | 4 |
| **Assembly position** | Diagonal to the front engine |
| **Displacement** | 1798 ccm |
| **Power** | 96 kW (130 hp) at 6000 rpm |
| **Bore x stroke** | 83.0 mm x 83.1 mm |
| **Compression ratio** | 11.3:1 |
| **Valves/cylinders** | 4 |
| **Maximum torque** | 175 Nm at 4250 rpm |
| **Fuel** | Super (95 ROZ) |

The power unit is based on the four-cylinder Duratec line and is made entirely of aluminum. The piston has an asymmetric recess. A swirl flow can be created by continuously adjustable charge exchange flaps in the secondary ports. This assists the combustion and transport of fuel to the spark plug in the stratified operating mode. The stratified mode extends up to 50% of the load range and 60% of the speed range of the entire engine map. The maximum injection pressure of 120 bar is created with a single-cylinder high pressure injection pump (HDP2) with flow regulation provided by Bosch. Particularly in the stratified operating mode, the high energy ignition system is tuned with coil-on plugs and surface gap spark plugs with quadruple electrodes. The engine achieves the Euro 4 emissions standard with two three-way catalytic converters and an additional $NO_x$ storage catalytic converter.

### 14.1.3 Homogeneous mixture concepts

With the experience of the initial wall-guided and air-guided combustion processes, which showed that the theoretical fuel consumption benefits of direct injection could not be realized in the expected magnitude during actual driving operation and that exhaust-gas aftertreatment was problematic ($DeNO_x$ cat), several vehicle manufacturers proceeded to introduce concepts with homogeneous direct injection and stoichiometric mixture. They already make use of parts of the advantages of injecting fuel directly into the combustion chamber, but cannot use gasoline direct injection's full potential to reduce fuel consumption. The homogeneous and stoichiometric direct injection has the advantage of allowing the use of a conventional three-way catalytic converter.

With the 2.0 l 6V IDE (*Injection Direct Essence*, internal name F5R, 103 kW), Renault was the first European manufacturer to introduce an engine with direct injection to the market in 2000; **figure 14.7**. Indeed, it was Renault who initiated a purely homogeneous direct injection concept as opposed to the Japanese competition [5]. Here the plugs and injection nozzles were arranged very close to each other just as in the spray-guided process.

Specifications Renault IDE:

| | |
|---|---|
| **Displacement** | 1998 ccm |
| **Power** | 103 kW/ 140 hp at 5500 rpm |
| **Number of cylinders** | 4 (16 valves) |
| **Injection system** | Gasoline direct injection, 40 to 100 bar |
| **Torque** | 195 Nm |
| **Compression ratio** | 11.5:1 |

**Figure 14.7:**
Cross-section of Renault 2.0 16V IDE [5]

The piston had a symmetric recess. The load was controlled only quantitatively. The Renault engine was advantageous only because of the increase in compression ratio made possible through internal cooling of the fuel inside the combustion chamber. The exhaust gas was purified with the help of a conventional three-way catalytic converter. In addition, an exhaust-gas recirculation system was installed. The engine achieved Euro 3 emissions standards with this.

In May 2003, Opel introduced a 2.2 l engine with direct injection in its Signum model. The term DIRECT ECOTEC in the engine name refers to a homogeneous concept with lateral nozzles, central spark plug and flat piston. At Opel, an engine with a charge stratification with piston recess and swirl flow was also developed. However, a comparative study showed that the high system costs and the high price of the SuperPlus fuel required for the stratified variants through which benefits in fuel consumption are achieved during customer use, could not be compensated when compared with homogeneous stoichiometric variants.

Specifications Opel DIRECT ECOTEC:

| | |
|---|---|
| **Number of cylinders** | 4 inline |
| **Displacement** | 2198 ccm |
| **Valves/cylinders** | 4 |
| **Bore x stroke** | 86.0 mm x 94.6 mm |
| **Compression ratio** | 12.0:1 |
| **Maximum power** | 114 kW (155 hp) at 5600 rpm |
| **Maximum torque** | 220 Nm at 3800 rpm |
| **Fuel** | Super (ROZ 95) |
| **Emissions standard** | Euro 4 |

In January 2003, BMW also commenced series production of gasoline engine with direct injection. 12-cylinder engines with homogeneous direct injection are available in the models 760i and 760Li. The fuel is injected with a single-hole injector which is mounted below the intake ports. The piston has a light recess at the intake side which assists the airflow and thus mixture formation. The intake ports were designed to give optimum power output. By using the Valvetronic variable valve timing system, the fuel consumption can be lowered by about 10% as compared to conventional four-valve engines with 3% higher power output and 5% higher torque.

Specifications BMW 760:

| | |
|---|---|
| **Number of cylinders** | V12 |
| **Valves per cylinder** | 4 |
| **Displacement** | 5972 ccm |
| **Bore x stroke** | 80 mm x 89 mm |
| **Power** | 327 kW (445 hp) at 6000 rpm |
| **Maximum torque** | 600 Nm at 3950 rpm |

VW and Audi's FSI engines were introduced in section 14.1.2. At present, they are widely operated in the homogeneous mode. Only in Audi A3 are the 1.6 and 2.0 l engines still operated in the stratified mode. The piston has a different recess geometry in engines with the homogeneous concept. The recess is symmetric and designed at the center of the piston; **figure 14.8**.

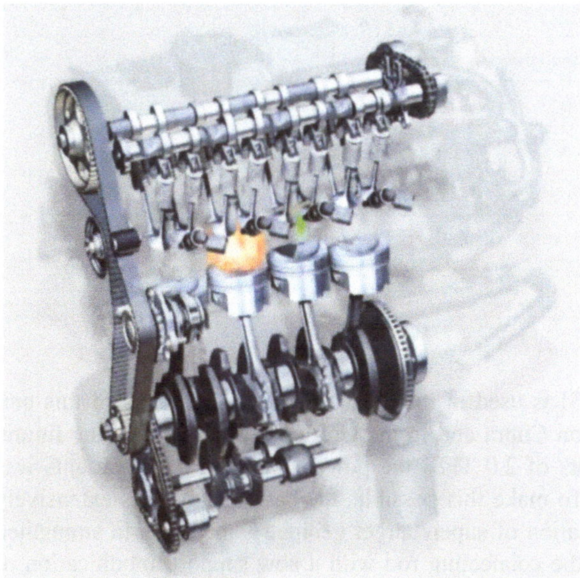

**Figure 14.8:**
Audi 2.0 FSI
Homogeneous concept

Toyota subsidiary Lexus is the only manufacturer to combine gasoline direct injection with intake manifold injection in GS 450h and the forthcoming LS 460. Either the direct injection system or both systems together are used, based on power requirement. The advantages of both the processes can be combined in this way. A 12-hole nozzle is used in the intake manifold, which injects the fuel at 4 bar. Double slit injectors are used for direct injection. The injection pressure here may go up to 130 bar. The intake ports and the specially molded piston head create a swirl, which has to swirl the air/fuel mixture at an optimum level based on system activity. The manifold injection ensures optimum mixture formation at low load demand. At higher load, however, the direct injection has to ensure optimum engine output. The air density increases with internal cooling, while the knocking tendency decreases, which leads to an increase in the compression ratio and to improvement in the efficiency in particular. The engine operates in homogeneous mode except in cold start conditions. At cold start, a stratified charge is produced with an additional direct injection in the compression cycle in order to increase the exhaust gas temperature and reduce the light-off time of the catalytic converter.

### 14.1.4 Supercharging concepts

All engines with supercharging and direct injection in the market so far have concepts with homogeneous mixture formation, e.g. the Turbo FSI engines by VW and Audi. In the case of VW, the 2.0 TFSI engine is available in Golf, while for Audi it is available in A4. The inline four-cylinder engine has a turbocharger with charge air intercooling. The four valves for each cylinder are controlled by two overhead camshafts [7].

Specifications VW; Audi TFSI:

| | |
|---|---|
| **Engine** | Four-cylinder inline engine |
| **Displacement** | 1984 ccm |
| **Bore x stroke** | 82.5 mm x 92.8 mm |
| **Maximum power** | 147 kW (200 hp) at 5100 rpm |
| **Maximum torque** | 280 Nm at 1800 to 5000 rpm |
| **Compression ratio** | 10.3:1 |
| **Emissions standard** | Euro 4 |

A high power variant of 2.0 TFSI is used in the new Audi S3. Derivatives of this unit will also be used in the Seat Leon Cupra and in the Golf GTI Edition 30 in the future. In comparison with basic variants of 2.0 TFSI the power output of these variants was increased from 147 to 195 kW. To make this possible, the basic engine was extensively reworked. This included optimization of supercharger geometry, measures to strengthen the crankcase, strengthening of the connecting rod with a new support, modification of pistons and a new high-temperature-proof cylinder head made of aluminum alloy. The combustion chamber design was optimized, to prevent problems such as pre-ignition and extreme knocking. Peak pressures of up to 110 bar and medium pressures up to 22 bar are still encountered. **Figure 14.9** shows a photo of one of these engines.

**Figure 14.9:**
Audi 2.0 TFSI high power variant in S3 with direct injection and exhaust-gas turbocharging [6]

Specifications Audi TFSI:

| | |
|---|---|
| **Engine** | Four-cylinder inline engine |
| **Displacement** | 1984 ccm |
| **Bore x stroke** | 82.5 mm x 92.8 mm |
| **Maximum power** | 195 kW (265 hp) at 6000 rpm |
| **Maximum torque** | 350 Nm at 2500 to 5000 rpm |
| **Compression ratio** | 9.8:1 |
| **Emissions standard** | Euro 4 |

VW has introduced a 1.4 l engine with combined turbocharging and compressor supercharging on the market using the abbreviation TSI. This is the first time that these two supercharging technologies have been used in combination in an engine with direct injection. The target was to achieve low fuel consumption and high torque at the same time, **figure 14.10**.

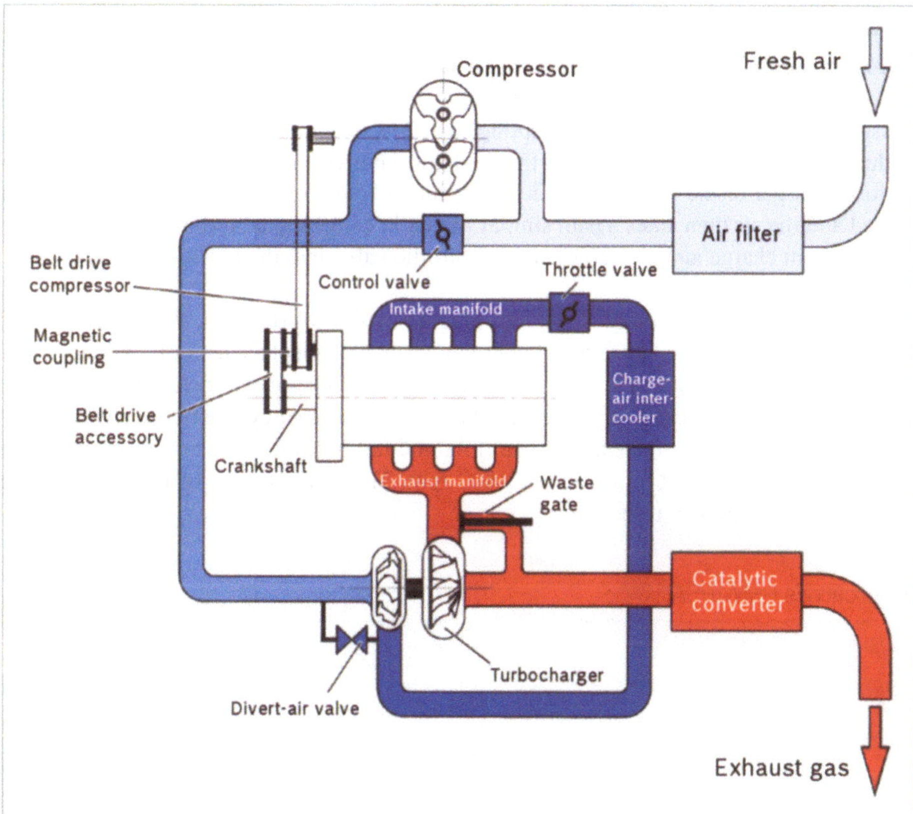


**Figure 14.10:** VW 1.4 l TSI engine, schematic diagram of the charge flow [7]

A low specific fuel consumption and thus higher efficiency is achieved by displacement reduction and the related lower frictional losses. Since engines with exhaust-gas turbochargers normally have a pronounced starting problem, known as turbo lag, a mechanical supercharger that can provide sufficient air to the engine during low engine speeds was installed in addition to the conventional exhaust-gas turbocharger.

Specifications VW TSI:

| | |
|---|---|
| **Engine design** | Four-cylinder inline engine |
| **Displacement** | 1390 ccm |
| **Bore x stroke** | 76.5 mm x 75.6 mm |
| **Maximum power** | 125 kW (170 hp) at 6000 rpm |
| **Maximum torque** | 240 Nm at 1500 to 4750 rpm |
| **Compression ratio** | 10.0:1 |
| **Emissions standard** | Euro 4 |

A compressor, driven mechanically with a belt, helps to increase torque at low engine speeds. In the twin charger, the turbocharger with waste gate control is used in addition at high engine speeds. The compressor and turbocharger are connected in series. The compressor is operated by a magnetic coupling, which is integrated with the water pump in a module. A control valve ensures that the fresh air required for the operating point reaches the turbocharger or the compressor. In a pure turbocharger operation, the control valve is opened and the air then takes a path similar to that in conventional turbocharged engines – over the front charge air intercooler and the throttle valve into the intake manifold.

BMW developed a new family of engines together with PSA (Citroën, Peugeot). The first three variants will be used as four-cylinder engines with a displacement of 1.6 liters in the Mini. The Mini Cooper S will have an engine with twin scroll turbocharging and direct injection. In the twin scroll turbocharger, the ports of the two cylinders are always separated from each other in the exhaust manifold and turbocharger. The arrangement causes the charging effect to set in already at 1400 rpm. The fast response of the supercharger prevents the well-known turbo lag.

Specifications BMW/PSA:

| | |
|---|---|
| **Engine design** | Four-cylinder inline engine |
| **Displacement** | 1600 ccm |
| **Maximum power** | 128 kW (175 hp) at 5500 rpm |
| **Maximum torque** | 240 Nm at 1400 to 4000 rpm |
| **Compression ratio** | 10.5:1 |

BMW has the first six-cylinder inline engine with twin turbocharging in the market [8, 9] since September 2006. The engine with an aluminum crankcase is available in the top model of the third series, i.e. 335i Coupe. According to BMW, spray-guided gasoline direct injection with homogeneous mixture formation is available for the first time with the "high-precision injection", which was developed together with Siemens, and leads to a noticeable reduction in fuel consumption. The piezo-controlled injector is placed centrally between the valves; **figure 14.11**. The fuel can thus be injected into the combustion chamber in a conical and uniform manner. Due to the use of two superchargers, the engine reacts more spontaneously to the load cycle changes desired by the driver.

Specifications BMW 335i:

| | |
|---|---|
| **Engine design** | Six-cylinder inline engine |
| **Displacement** | 2979 ccm |
| **Bore x stroke** | 84 mm x 89.6 mm |
| **Maximum power** | 225 kW (306 hp) at 5800 rpm |
| **Maximum torque** | 400 Nm at 1300 to 5000 rpm |
| **Compression ratio** | 10.2:1 |

**Figure 14.11:**
Cross-section of BMW engine, central position of injector and spark plug

In their new edition of the GT model (from 2007), Opel have used a variant of the previously mentioned 2.2 l ECOTEC DIRECT engine with a smaller displacement. The new power unit is provided with a displacement of 2.0 liters, variable valve control, a charge air intercooler and a double flow turbocharger that promises fast response at low engine speeds.

Specifications (temporary values) Opel:

| | |
|---|---|
| **Number of cylinders** | 4 inline |
| **Displacement** | 1998 ccm |
| **Valve system** | 16V |
| **Bore x stroke** | 86.0 mm x 86.0 mm |
| **Compression ratio** | 9.2:1 |
| **Maximum power** | 194 kW (264 hp) at 5300 rpm |
| **Maximum torque** | 353 Nm at 2000 to 5200 rpm |

## 14.1.5 Stratified charge concepts: spray-guided combustion processes

When compared to the earlier engines with stratified charge having wall-guided and air-guided combustion processes, which were unable to realize the expected fuel consumption benefits in practice, the introduction of the next, so-called second generation of engines with gasoline direct injection with spray-guided combustion processes promises even greater potential for saving on fuel consumption.

DaimlerChrysler was the first manufacturer to introduce an engine of the second generation of gasoline direct injection with spray-guided direct injection and charge stratification into the market in 2006. The M272 DE 35 engine is available in the CLS 350 CGI vehicle [10]; **figure 14.12**. It is a 3.5 l V6 engine with a piezo injector placed at the center in the four-valve cylinder head. The piezo injector has a nozzle which opens outwards, known as an

**Figure 14.12:**
DaimlerChrysler M272 DE [11]

A-nozzle and developed together with Bosch. The spark plug was shifted from its central position towards the exhaust valve and slightly inclined so that its electrodes are positioned at the boundary of the nozzle jet. The fuel is introduced to the combustion chamber at up to 200 bar by three port injectors. A volume-controlled three-piston high pressure pump is used to generate pressure. When compared to wall-guided and air-guided combustion processes, larger fuel savings are achieved by the increased charge stratification range in spray-guided direct injection than by its first-generation predecessors.

A double flow, external exhaust-gas recirculation system is used to reduce exhaust-gas emissions; **figure 14.13**. Two three-way catalytic converters near the engine ensure that the light-off temperature is quickly attained in cold start conditions in the entire double flow exhaust system. The $NO_x$ storage catalysts which come next are mounted on the underside to protect against high temperatures. They have an active temperature window from 250 to 500°C. For precise control of the $NO_x$ catalytic converters, there is a temperature sensor upstream and a $NO_x$ sensor downstream of each $NO_x$ catalytic converter. The latter also ensures adherence to the $NO_x$ limit values over the entire service life of the vehicle.

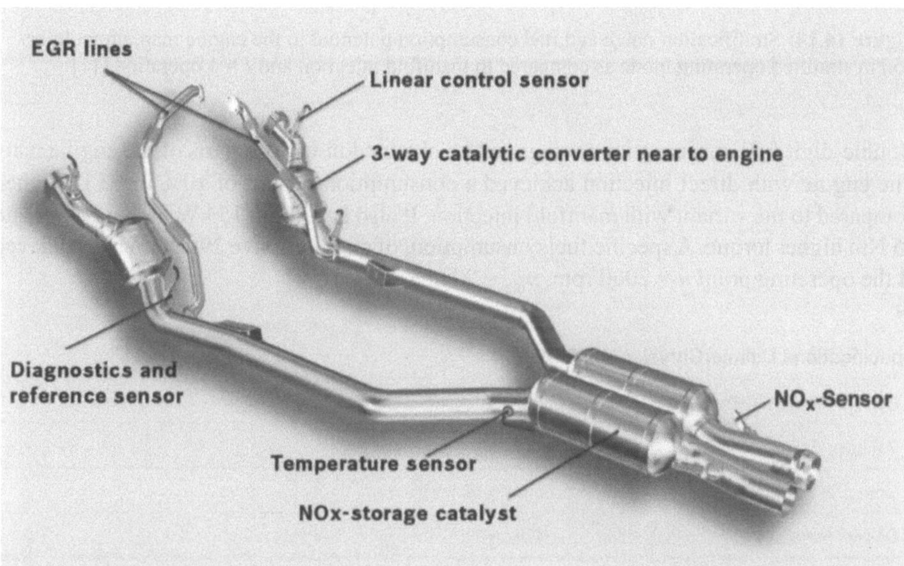


**Figure 14.13:** Exhaust system of CLS 350 CGI [11]

The spray-guided combustion process has been very successful in expanding the map range in which charge stratification is possible, when compared to the combustion process of the first generation. Speeds up to greater than 120 km/h can be easily achieved in the stratified mode in the vehicle CLS 350 CGI [10]. The fuel consumption potential of the engine as compared to the variant with manifold injection is shown in **figure 14.14**.

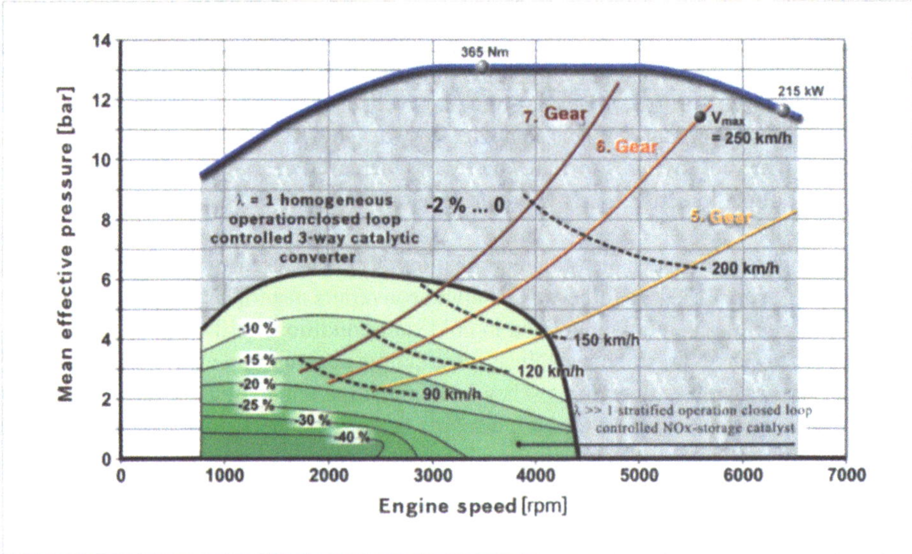


**Figure 14.14:** Stratification range and fuel consumption potential in the engine map, direct injection in stratified operating mode as compared to manifold injection and $\lambda = 1$ operation [11]

Double-digit fuel consumption savings can be obtained in various parts of the engine map. The engine with direct injection achieved a consumption benefit of 10% in NEDC when compared to the variant with manifold injection. It also had about 15 kW higher power and 15 Nm higher torque. A specific fuel consumption, of an impressive 290 g/kWh is achieved at the operating point $n = 2000$ rpm, $p_{me} = 2$ bar.

**Specifications DaimlerChrysler M272 DE[10]:**

| | |
|---|---|
| Cylinder arrangement/number | V6 |
| V angle | 90° |
| Bore x stroke | 86 mm x 92.9 mm |
| Displacement | 3498 ccm |
| Valves per cylinder | 4 |
| Compression ratio | 12.2:1 |
| Power | 215 kW (292 hp) at 6400 rpm |
| Torque | 365 Nm at 3000 to 5100 rpm |
| Spec. consumption at best point | 240 g/kWh |
| Consumption at 2000 rpm, 2 bar | 290 g/kWh (manifold injection: 360 g/kWh) |

BMW has planned to introduce new engines with spray-guided direct injection into the market in the first half of 2007, at first in the new third series Cabrio. In addition to the engine already described with a six-cylinder inline engine, twin turbocharging and gasoline direct injection *High Precision Injection* (HPI), the two air intake variants of the engine with stratified operating mode will also be introduced. These engines are all equipped with injection systems identical to the homogeneously operating turbocharged engine. At the same time a two 2 l four-cylinder engine also with gasoline direct injection and stratified operating mode is planned. The engines are distinguished by a central injector position and a spark plug which is placed in the immediate vicinity of the engine. An outwards opening piezo injector with an oil-damped thermal compensator is used as injector [8]. The known advantages of this technology are: a low coking tendency and very low switching times of 200 µs of the injector, which enable full and partial strokes in addition to multiple injections. A three-cylinder axial piston pump with integrated fuel supply control valve for controlling pressure and flow increases the fuel pressure up to a maximum of 200 bar. The fuel reaches the injectors via small individual lines over a common rail with a pressure sensor. The important specifications of new engines with direct injection from BMW are summarized in **figure 14.15**.

**Figure 14.16** shows the stratification range of the engine with gasoline direct injection and spray-guided combustion process. The engine can operate at optimal fuel consumption with a lean mixture in a large load and speed range. The best point is defined at an effective specific fuel consumption of approximately 240 g/kWh, which corresponds to a high quality engine with manifold injection.

| | Stratified | | | Homogeneous |
|---|---|---|---|---|
| Type | 320i | 325i | 330i | 335i (Twin turbo) |
| Engine design/number of cylinders | Inline/4 | Inline/6 | | |
| Displacement [ccm] | 1995 | 2996 | | 2979 |
| Bore x stroke | 90.0 x 84.0 | 88.0 x 85.0 | | 89.6 x 84.0 |
| Power [kW (hp)]<br>at [rpm] | 125 (170)<br>6700 | 160 (218)<br>6100 | 200 (272)<br>6700 | 225 (306)<br>5800 |
| Maximum torque [Nm]<br>at [rpm] | 210<br>4250 | 270<br>2400–4200 | 320<br>2750–3000 | 400<br>1300–5800 |
| Compression ratio :1 | 12.0 | | | 10.2 |
| ECE consumption [l/100 km] total | 6.6 | 7.9 | 8.1 | 9.9 |

**Figure 14.15:** Specifications [12, 13]

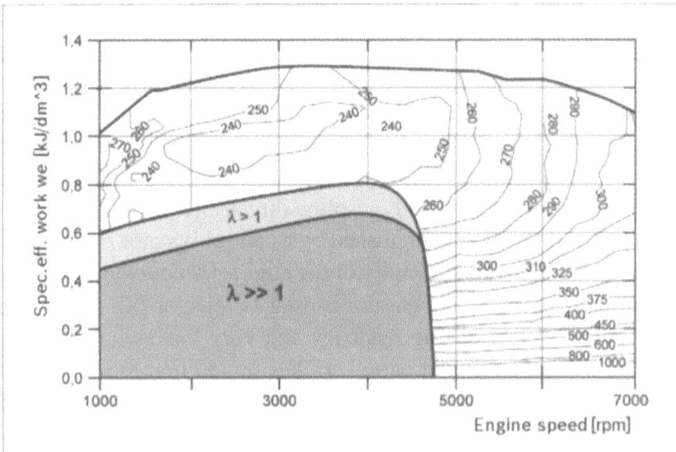


**Figure 14.16:** Stratification range of BMW combustion process in the engine map [14]

The high fuel consumption advantage of the engine with spray-guided combustion process can be seen in the reference point at $n = 2000$ rpm and $p_{me} = 2$ bar also in the FEV scattering range; **figure 14.17**. When compared to the VALVETRONIC engine with a variable valve timing system, which is reasonably well positioned in the scattering range, fuel consumption can be reduced by approximately 13 % to 295 g/kWh. This emphasizes the high potential of gasoline direct injection with spray-guided combustion processes for reduction in fuel consumption and increase in efficiency at part load. Fuel consumptions equal to those of diesel engines are achieved in this region.

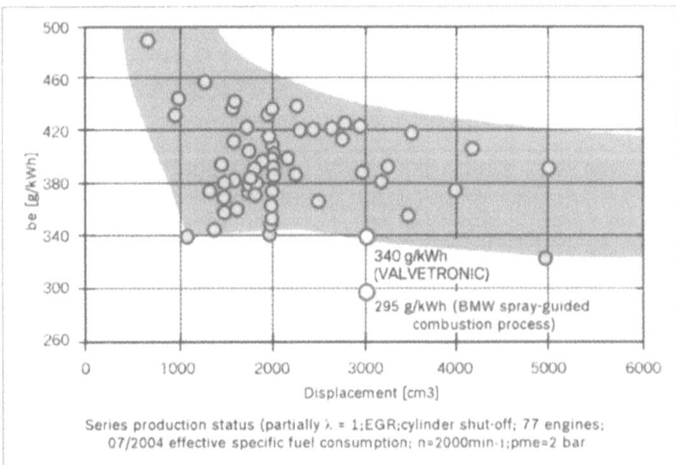


**Figure 14.17:** Fuel consumption of the engine with spray-guided combustion process in FEV scattering range for gasoline engines [14]

# 14.2 Motor sports concepts and impetus for series development

## 14.2.1 Introduction

Audi developed a racing gasoline engine with direct injection and stratified charge with which the superiority of the direct injection was to be proved in motor sport. This engine was used both in the Le Mans 24-Hour Race and in the American Le Mans Series AMLS, which is based on the same regulations. The regulations for sports prototypes in Le Mans permit both turbocharged and naturally aspirated engines whose capacities are approximated by means of air mass limiter and displacement-dependent charge pressure limitation in accordance with their technical complexity. This ensures that different technical concepts have equal opportunities.

Racing engines are developed according to their very specific application. The main areas of development, however, almost always apply, irrespective of the application area; **figure 14.18**. While the first three variables mentioned are largely determined by the regulations, the others are to some extent the result of the geometrical dimensions or depend on the efficiency. Positive differentiation can therefore be carried out only according to highest possible efficiency which in turn basically consists of mechanical and thermal efficiency.

| | |
|---|---|
| Size | As small as possible |
| Weight | As low as possible |
| Endurance | Race distance + 10 m |
| Center of gravity | As deep as possible |
| Power | As much as possible |
| Torque | As much as possible |
| Efficiency | As high as possible |
| Consumption | As low as possible |
| Drivability | As easy to control as possible |

**Figure 14.18:** General development points for a racing engine

The improvement of thermal efficiency was the motivation for the development of the FSI technology to be used in motor racing. The aim was and is the better usage of fuel energy by direct injection than is possible with conventional mixture preparation and combustion processes.

## 14.2.2 Gasoline direction injection in motor racing

Gasoline direct injection was already in use more than 100 years ago and was developed by different companies for both sports engines and road vehicle engines in the 1950's. However, these efforts never got beyond the experimental stage or required exotic fuels.

Audi undertook the task of developing an injection technology that could prove to the world its serviceability for its series in the most challenging sports car endurance race. The extremely high full load operation environment of more than 72% and the race distance of more than 5600 km, which corresponds to a complete Formula 1 season, are typical for Le Mans. **Figure 14.19** shows the operating range of the map for the Le Mans 24-Hour Race.

A careful analysis of these boundary conditions led to the choice of a homogeneous mixture formation process. For this a tumble must be generated by means of carefully adjusted intake ports, i.e. a directed air movement in the cylinder which achieves a mixing of air and fuel as homogeneously as possible in connection with a corresponding injection valve. **Figure 14.20** shows the mixture distribution in the combustion chamber for the Le Mans 24-Hour Race. The effects of the process, which has to function with standard fuel in the Le Mans 24-Hour Race, will be described in more detail below.

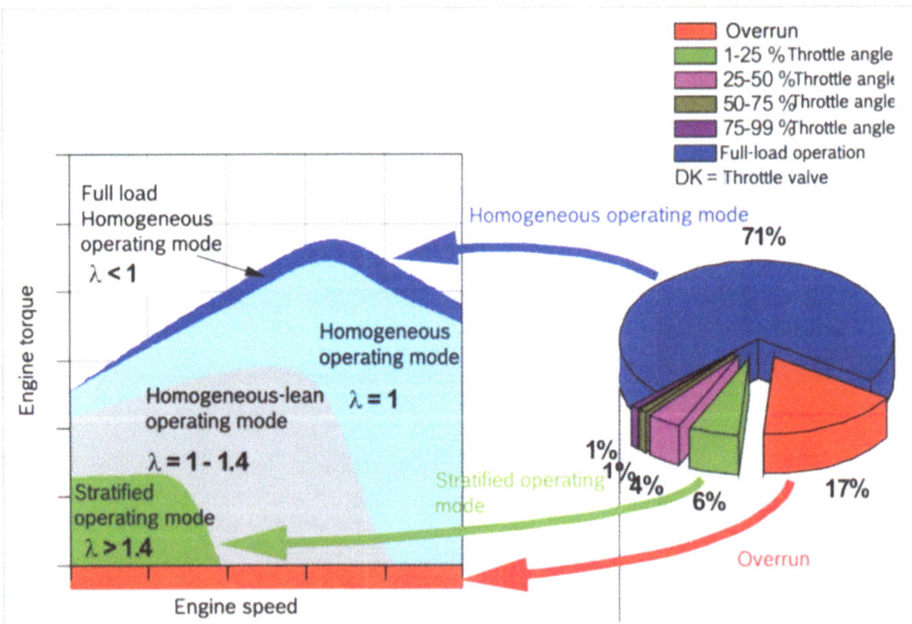


**Figure 14.19:** Operating ranges for gasoline engine with direct injection and load percentages during a Le Mans round

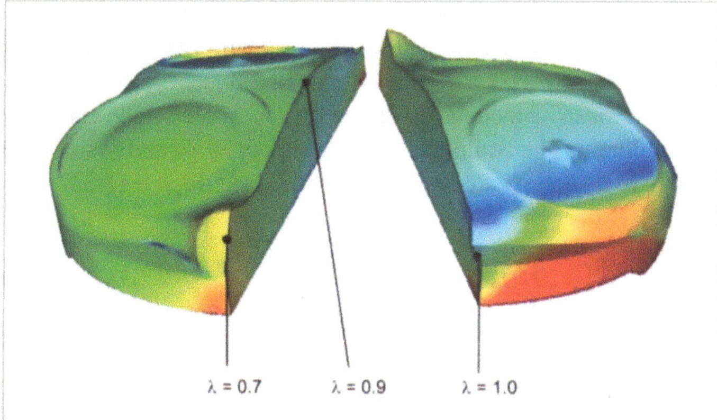


**Figure 14.20:**
Mixture distribution in the combustion chamber

## 14.2.3 V8 bi-turbo with direct injection for the Le Mans 24-Hour Race

The vehicle designed for the 24-Hour Race, the "R 8", has become the most successful sports car for this task due to its sophisticated technology. The decision to develop a turbocharged engine for this application was based on the requirement to achieve highest possible efficiency through low engine speeds. Furthermore, a turbocharged engine offers an excellent performance to weight ratio. The very successful 3.6 l V8 bi-turbo engine developed for these requirements thus provided a good basis for the further development of direct injection; **figure 14.21**.

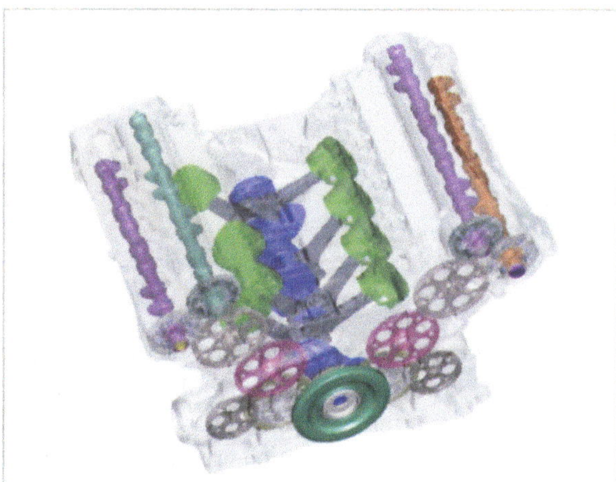

**Figure 14.21:**
X-ray image of the V8 engine

This engine, incidentally the first pure racing engine made by Audi since the days of the legendary Silver Arrows of the 1930's, was designed as a 90° V8 with load bearing structure. The crankshaft drive consists of a 180° crankshaft with H connecting rods made of high-strength steel. The four overhead camshafts are driven by gear wheels. The crankcase and dry sump are made of aluminum. The complete engine weighs 175 kg and is smoothly integrated into the vehicle due to its compact design; **figure 14.22** and **figure 14.23**.

**Figure 14.22:**
The engine in the vehicle

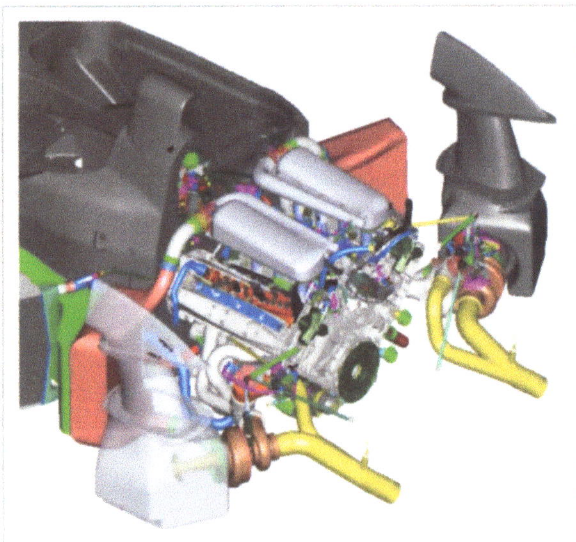

**Figure 14.23:**
The engine in the virtual front end

The cylinder head has four valves per cylinder, which are actuated by the two overhead camshafts through barrel tappets. The plug is located centrally in the roof-shaped combustion chamber; **figure 14.24**. The piston is flat except for the recesses for the intake and exhaust valves. The compression ratio is 12.2:1. Since direct injection was already taken into account during the design of the cylinder head, the adjustment at this point was relatively easy; **figure 14.25.**

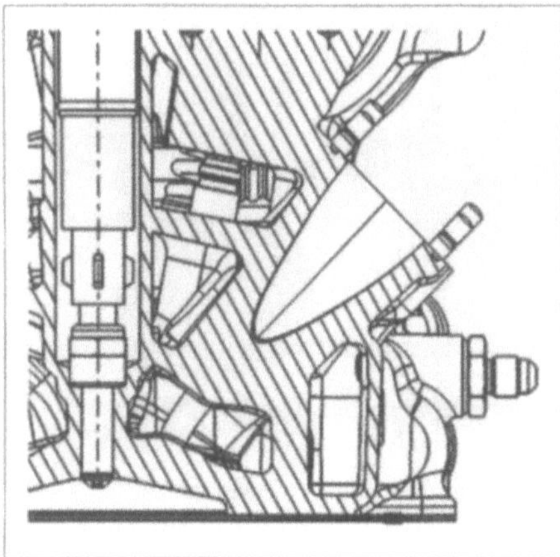

**Figure 14.24:**
MPI cylinder head

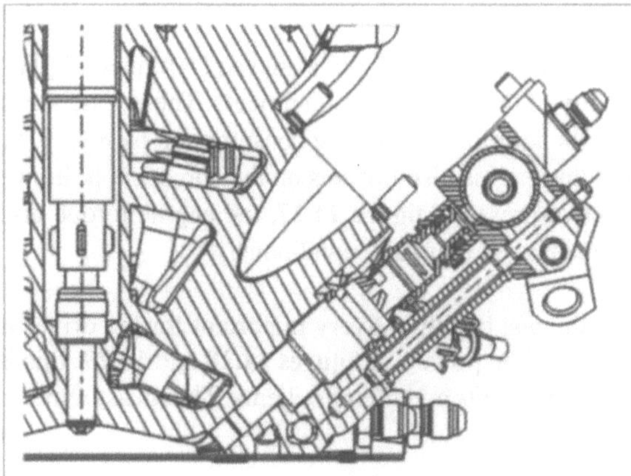

**Figure 14.25:**
FSI cylinder head

Together with the port geometry already mentioned, the injection nozzle plays the second key role in the direct injection process. It supplies the required fuel as finely atomized as possible within 2 to 6 ms. A multi-hole nozzle is the solution for this task. For this purpose, tests with different nozzles and spray configurations were required; **figure 14.26**.

For a good mixture preparation of fuel and air, controlled air movement (tumble) is required in the combustion chamber for the direct injection process of the V8 engine. This tumble is generated by the geometry of the intake ports and forms, in addition to the nozzle geometry, the heart of the process. The objective of the port development, which was enhanced by tests and 3D CFD calculation, is to create a special tumble flow over the entire engine speed range with losses as small as possible.

All components involved were optimally adjusted to each other both on the test bench and through simulation calculation in order to achieve a very homogeneous mixture.

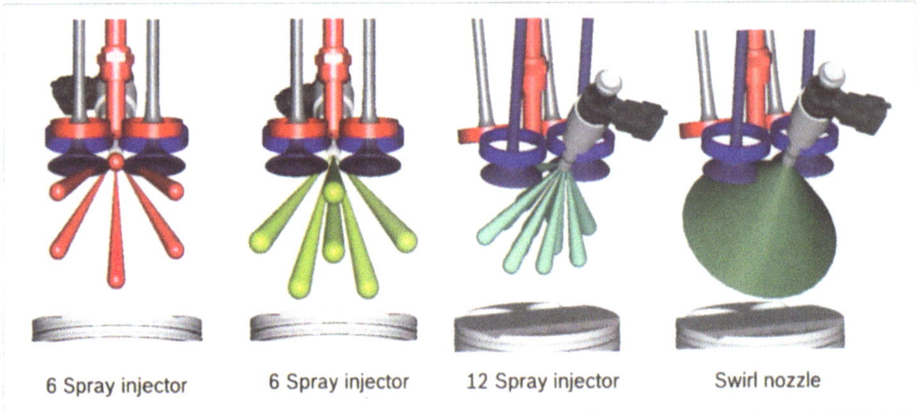


**Figure 14.26:** Nozzle variants

By July 2006, this engine had won 48 of 81 endurance races and achieved 50 runner-up placings, and proved the performance capability of the direct injection process. The engine speed range used goes from 3000 to 7500 rpm. In addition to reducing consumption by 8 to 10%, the implementation of direct injection in this engine also impacts on the combustion. In line with the increased indicated mean effective pressure, **figure 14.27**, the engine torque also increased by up to 8% under otherwise unchanged boundary conditions.

A benefit that is difficult to measure on the test bench but very important for the driver is the decisively better controllability of the engine power. In **figures 14.28** and **14.29**, this is shown in the different accelerator pedal movements and their gradients. The measurements

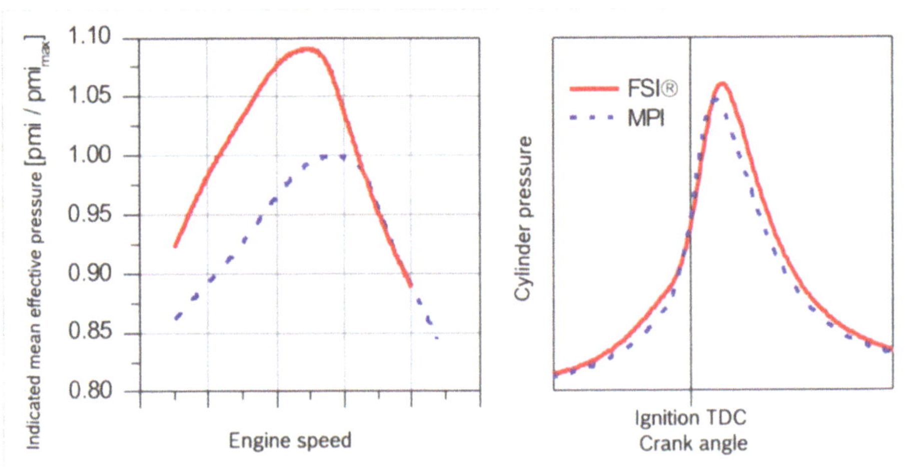


**Figure 14.27:** Comparison of combustion for MPI and FSI® (pressure characteristic)

were taken simultaneously in the same curve with two racing cars which were driven by an MPI engine or an engine with direct injection.

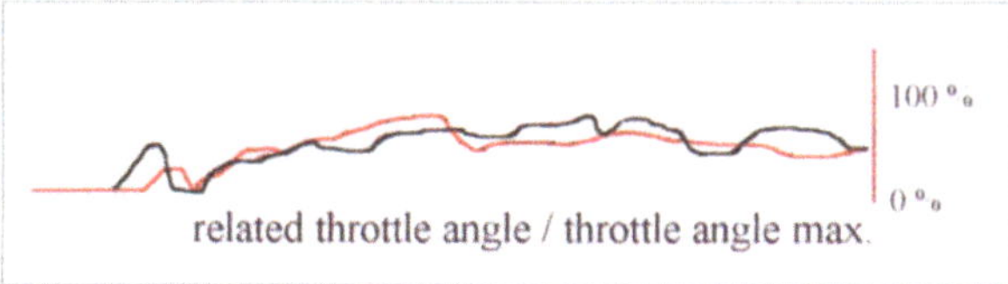


**Figure 14.28:**
Related throttle angle/maximum throttle angle

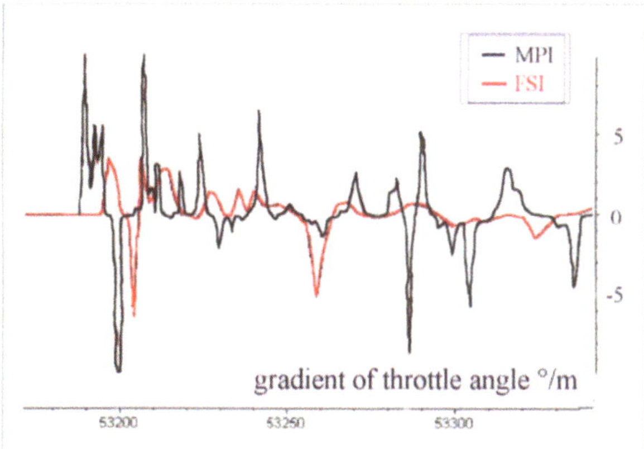


**Figure 14.29:**
Gradient of throttle angle [°/m]

## 14.2.4 Use in a 2 l production engine with direct injection and turbocharging

Most of the general development points for racing engines mentioned in the beginning can be applied to production engines almost without restriction; **figure 14.30**. Significant differences can be found in the desired service life in different climatic boundary conditions and especially in the fact that these engines are not operated by race drivers. Nevertheless the 2.0 l production engine with direct injection and turbocharging, **figure 14.31**, and the 3.6 l bi-turbo engine for Le Mans have much in common. Both work with homogeneous mixture formation which offers driving fun and sporty performance with little consumption. The injection pressures are in the range of up to 120 bar for both engines. The compression in the production engine is a little lower than that of the V8 engine; however, with a ratio of 10.5:1, it is also high. The resulting efficiency is therefore correspondingly favorable.

| | |
|---|---|
| Size | As small as possible |
| Weight | As low as possible |
| Power | As much as possible |
| Torque | As much as possible |
| Efficiency | As high as possible |
| Consumption | As low as possible |
| Drivability | As easy to control as possible |

**Figure 14.30:** General development points for a racing engine/production engine

**Figure 14.31:**
2.0 l production engine with direct injection and turbocharging

The TFSI engine is equipped with a 4 V cylinder head. The two overhead adjustable camshafts are driven by a chain. The valves are actuated by roller-type cam followers. The spark plug is in the center of the combustion chamber, the injection nozzle in almost the same position as in the V8 3.6 l engine; **figure 14.32**. The air movement is generated through tumble ports. At low engine speeds close to low idle, this is supported by adjustable tumble flaps.

Since this engine was developed as a successor to the successful 1.8 l 5-valve MPI engine, the improvements brought by direct injection can be shown directly by adjusting the different displacements; **figure 14.33**. In addition to the improvement of the torque by approximately 10% as compared to the 1.8 l MPI engine scaled up to 2 l, it is particularly remarkable that the torque build-up takes place 300 to 500 rpm earlier. This spontaneity can be also seen in the charge pressure gradient.

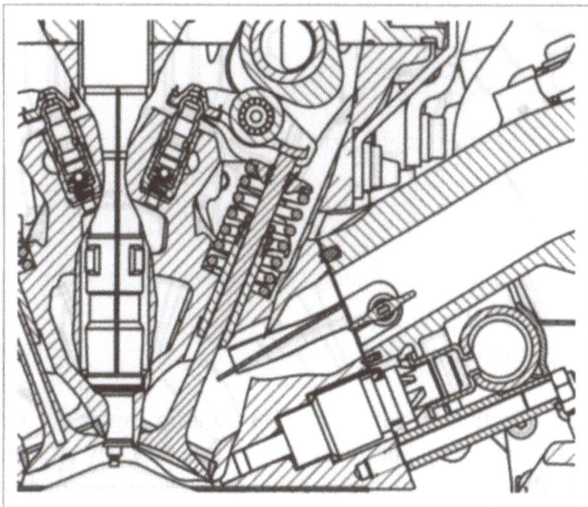

**Figure 14.32:**
Sectional view of the cylinder head on injection nozzle level (2.0 l production engine with direct injection and turbocharging)

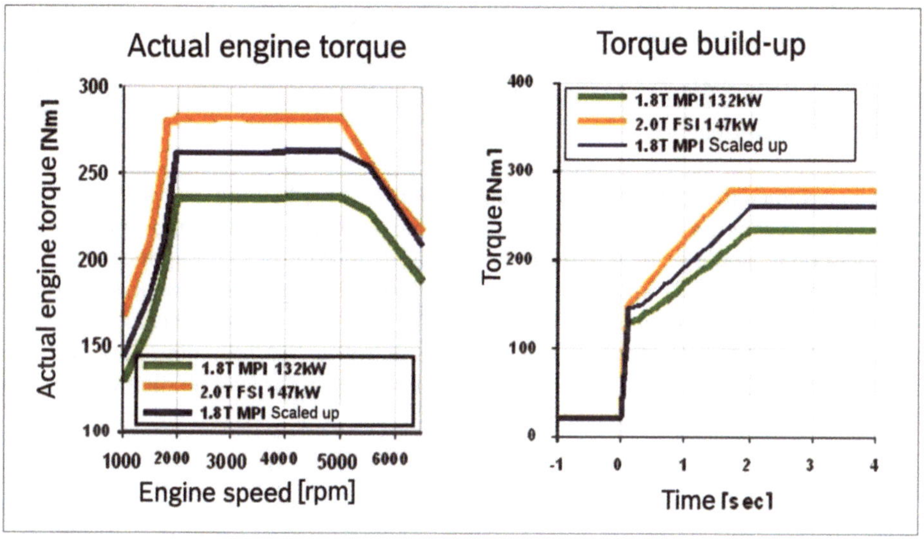


**Figure 14.33:** Full load comparison 1.8 T MPI engine/2.0 l engine with direct injection and turbocharging

## 14.2.5 Comparison of engines with direct injection for motor racing and for series production

Given that the principles of the two engines are very similar, it comes as no surprise that the improvements are within a comparable range as compared to the respective MPI predecessor engine with conventional manifold injection and mixture preparation; **figure 14.34.**

| | R8 | 2.0 l turbo |
|---|---|---|
| Fuel consumption | -8% to -10% | -10% to -12% |
| Power and torque | +9% | +10% |
| Drivability and vehicle response | + | + |

**Figure 14.34:** Results of the development

### 14.2.6 Summary and future prospects

The successful introduction of direct injection into the 3.6 l V8 bi-turbo engine for Le Mans is a milestone in sports engine development. For the first time since the early tests on passenger car engines in the 1950's, it has been possible to demonstrate a mixture preparation process with standard unleaded fuel and at engine speeds common in day-to-day traffic that meets even extreme requirements. The advantages with regard to consumption reduction with simultaneous increase in torque and power were proven, as were the clear improvements in vehicle response and controllability. Having been proven in motor racing, this technology was used first in the 3.2 l V6 engine and is now also being successfully applied in the 2.0 l engine.

It can be assumed that further improvements will be achieved during the further development of this technology. There is still considerable potential for the combustion process and injection technology, be it in the reduction of the nozzles or in the application of the piezo technology already known from the 3.0 l 1V6 TDI engine. In connection with an injection pressure much higher than the 100 to 120 bar currently used, it can enhance the mixture preparation even further. Thus quality control is also becoming possible. The related dethrottling of the intake area and a great capability for lean running will lead to a further major reduction in consumption.

## 14.3 Two-stroke gasoline engines with direct injection

### 14.3.1 Introduction

For gasoline engines, the two-stroke principle was used very early on, even before being patented by Nikolaus August Otto. The main features of today's light vehicle two-stroke engine were presented by Julius Soehnlein when he applied for a patent in 1891: in this engine, for the first time, the intake process, the scavenging process and the exhaust process were controlled solely by a working piston.

Nowadays two-stroke engines with internal mixture formation with effective efficiency of greater than 50% are the most efficient thermal engine; however, they work on the basis of the diesel process. For these engines used in large ships – slow-speed engines with single-cylinder displacements, sometimes of more than 1.5 $m^3$, and cylinder power of more than 3000 kW at engine speeds of approximately 100 rpm, the two-stroke principle with direct injection is the standard process. For gasoline engines, the desire to avoid the significant scavenging losses – this is possible with direct injection even for piston-port controlled engines, i.e. light, compact engines in line with the virtues of the simple two-stroke engine – was obvious and was aptly described [19] as far back as 1950: "The economic gain to be expected which means fuel savings of about 25% makes the effort for an operational internal mixture formation in two-stroke engines especially attractive".

## 14.3.2 History

In line with this statement, which is still valid today, extensive work on two-stroke gasoline engines with direct injection had already begun in the time between the two world wars [20]. The extraordinary potential of the two-stroke engine led to the very first series application of direct injection in the vehicle in 1952. However, the concept of the two-stroke gasoline engine with direct injection, shown in **figure 14.35**, enjoyed only moderate success with a small number of units manufactured in spite of the indicated consumption reduction of up to 30%.

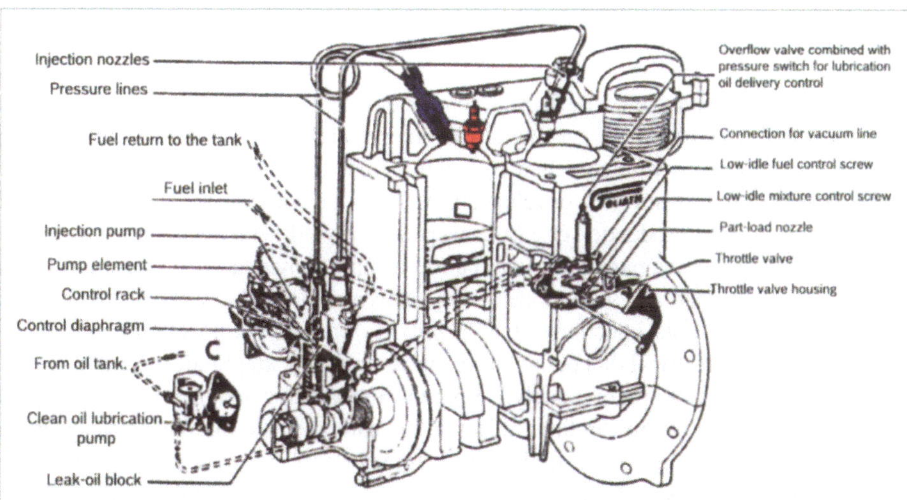


**Figure 14.35:** Goliath/Gutbrod superior two-cylinder two-stroke engine for passenger car power-train ($V_h$ = 0.9 $dm^3$; 28 kW / 4000 rpm)

The subsequent oil crisis and in particular the introduction of general emissions regulations triggered new attempts to realize the two-stroke engine with internal mixture formation. The French company Motobecane put a lot of effort into developing a two-wheel drive. The engine control was carried out in an interesting optoelectronic way, since the option of fully electronic map-based control did not exist at that time. The application started in small-scale series production in 1971 with a single-cylinder motorcycle engine with a displacement of 125 ccm; in 1978, the three-cylinder motorcycle engine with a displacement of 350 ccm (28 kW/7200 rpm, 36 Nm/6500 rpm) as shown in **figure 14.36** followed.

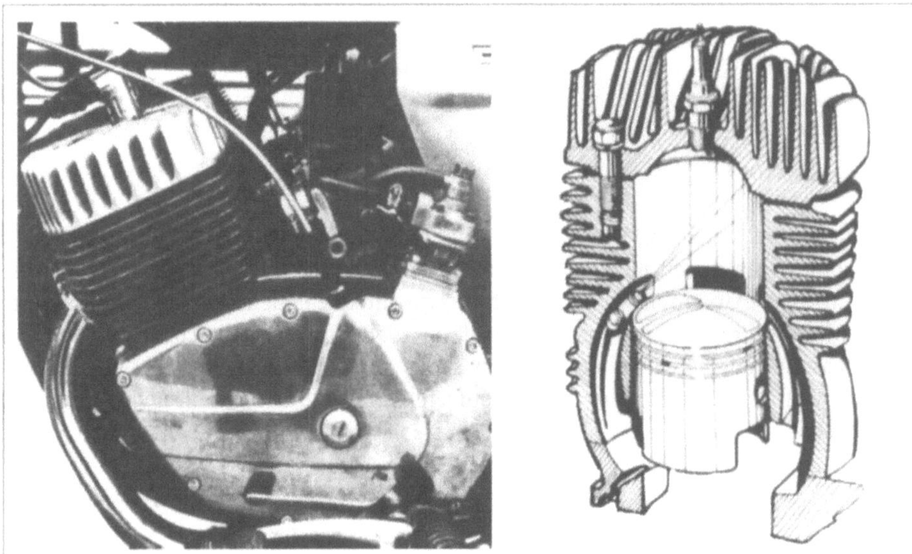

**Figure 14.36:** Motobecane 350 ccm (left), sectional view of single-cylinder (right) [21]

Approaches which can use cost-effective components already available in large-scale series production from passenger car four-stroke engines and also use electronic engine control units were pursued with so called semi-direct injection concepts; the 50 ccm engine by PUCH [22, 40], **figure 14.37**, developed in cooperation with the TU Graz in 1986 is an example of a close-to-production engine concept . It can work with injection in the transfer ports and, due to the comparatively long period of time provided for mixture formation, under low thermal load and with moderate injection pressure, which enables the use of a standard intake manifold injection valve. In order to provide the injection pressure efficiently without an electrical fuel pump, the crankcase pressure amplitude is used to provide the required pressure of approximately 2.5 bar via a reed valve and pressure transmission [22]. A disadvantage of this "indirect injection" is naturally that the fuel injection in the cylinder is limited by the control timing of the transfer port, which does not enable complete prevention of the scavenging losses. In spite of this limitation, an important improvement was achieved with this as opposed to the concepts with external mixture formation; **figure 14.38.**

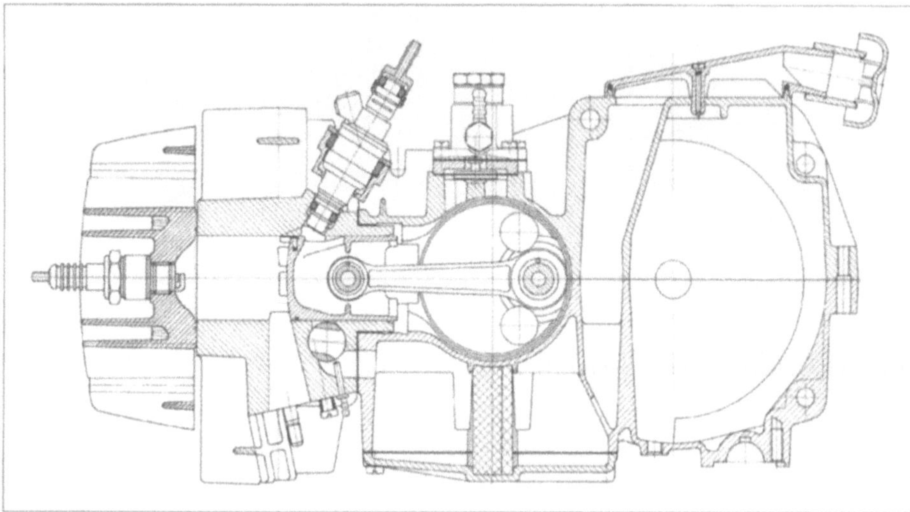

**Figure 14.37:** Prototype engine PUCH Maxi 50 ccm [22]

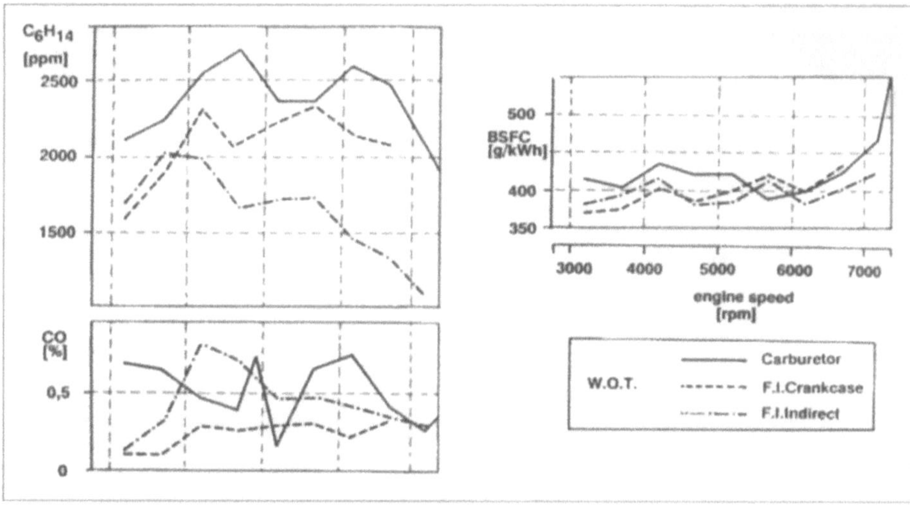


**Figure 14.38:** Full load comparison of PUCH prototype engine vs. carbureted engine and crank-case injection with regard to emissions and spec. consumption [22]

At the end of the 1980's, several development units undertook basic work on direct injection for two-stroke engines [23, 24, 25], on the basis of which the chances of the two-stroke engine as a passenger car prime mover were evaluated and possible solutions were identified [26, 27]. **Figures 14.39** and **14.40** show the geometric assemblies examined in [24] and the function results achieved with them.

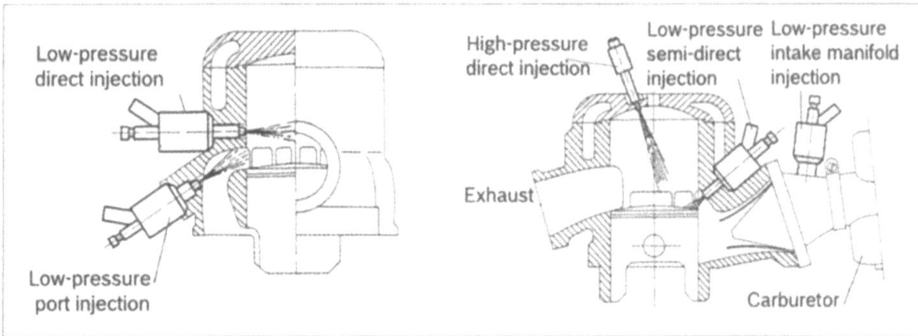


**Figure 14.39:** Injector arrangements [24]

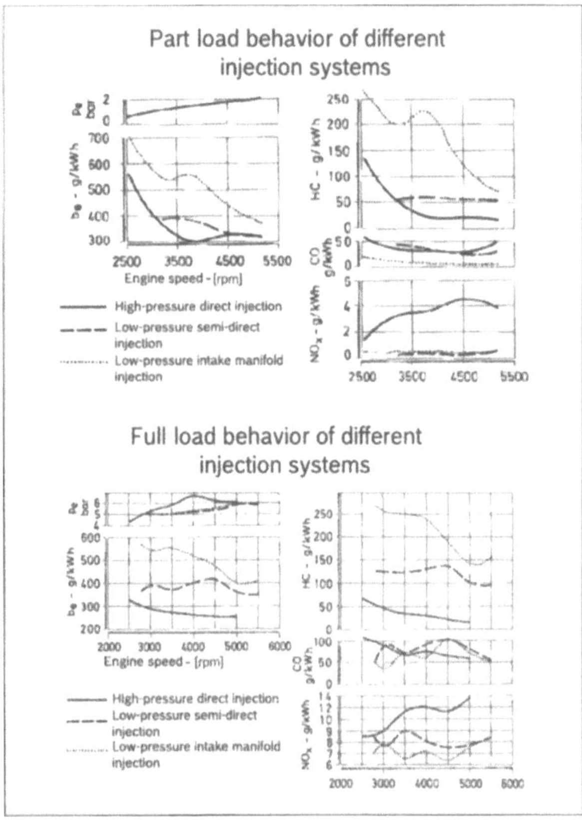


**Figure 14.40:**
Part load behavior and full load behavior of different injection systems according to figure 14.39 [24]

The very short period of time available for the injection and mixture formation process in the two-stroke engine (from the mixture formation phase of less than 60° crankshaft angle between exhaust closing time and ignition point at high engine speeds results a time in the range of ≈ 1 ms!), which is due to the engine concept, makes extraordinarily high demands of the fuel injection system. The injection systems used during the first applications on

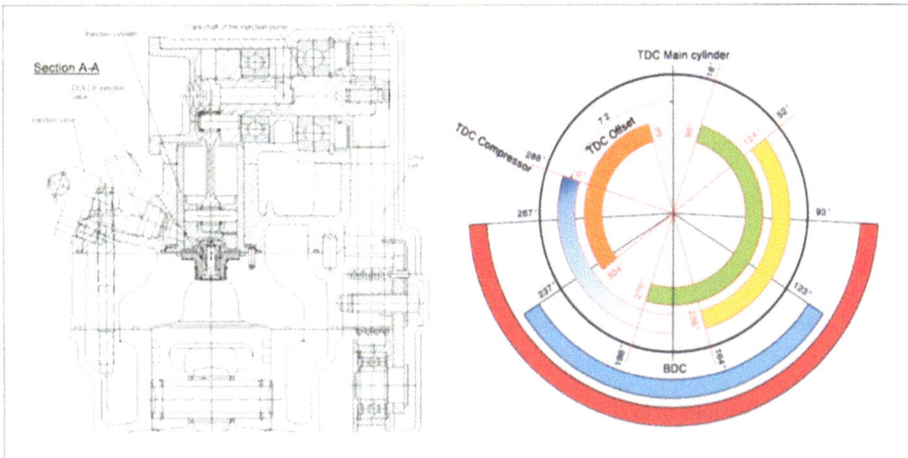


**Figure 14.41:** Sectional view of the cylinder head of the 125 ccm two-stroke engine (DOLCE) with direct injection based on the FAST principle [31]

the basis of diesel systems formed an obstacle for series implementation due to their costs and complexity, in particular. To deal with these disadvantages of conventional high pressure injection, different concepts for high pressure generation [28] and for supporting the dispersion through air supply outside the combustion chamber [26, 29, 30] were tested and presented.

An interesting approach in this context was examined and further developed in the framework of the EU project DOLCE (collaboration of Piaggio, Institut Francais du Petrole, Magneti Marelli, ROSI and TU Graz) [31]. This approach includes the insertion of fuel in a compressor cylinder via a low pressure injection valve. The air/fuel mixture generated there is injected in the main combustion chamber by a mechanically driven injection piston via a pressure-actuated injection valve at the desired time; **figure 14.41**. The time sequence of such a working cycle is also shown in **figure 14.41**.

**Figure 14.42** shows the emissions achieved with a DOLCE prototype vehicle with catalytic converter and compares it with series-production vehicles with a conventional two-stroke engine, a four-stroke engine and a prototype engine with secondary air system and catalytic converter. [32, 33, 34, 35], among others, give an overview of the numerous development directions and trends

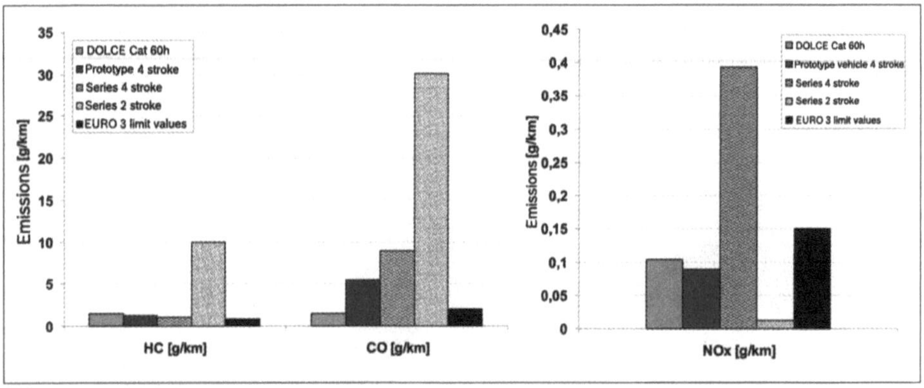


**Figure 14.42:** Emissions in the driving cycle ECE R40 for the DOLCE vehicle and different reference vehicles [31]

## 14.3.3 Series concepts

Good prerequisites for series application in two-stroke engines can be found in air assisted systems which enable a good mixture formation at low injection pressures even for extremely short time frames for the fuel injection. In 1995, such a system was used in an outboard engine for the first time. Further intensive development with regard to its use in a vehicle was pursued by several manufacturers, **figure 14.43** [34], and led to fleet tests; however, series application in the passenger car has not yet taken place, for various reasons, and is not foreseeable either.

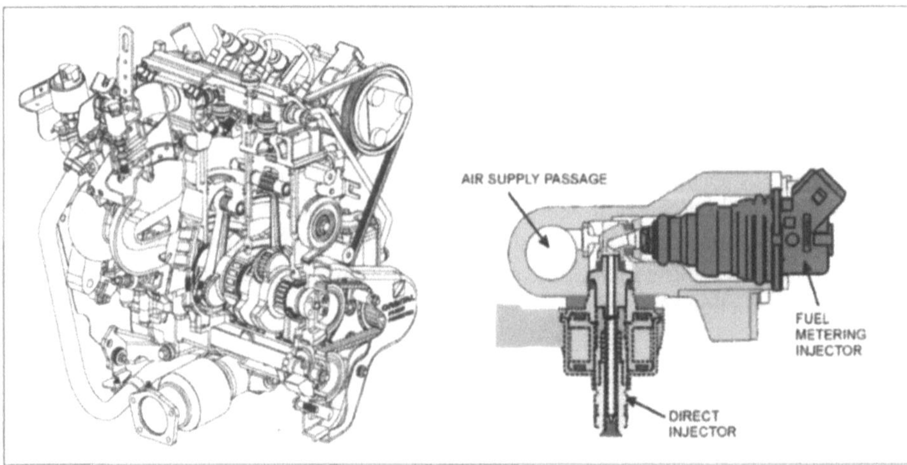


**Figure 14.43:** Orbital three-cylinder passenger car engine [34] $V_h = 0.8$ dm$^3$, $P_{eff}$=58 kW/4500 rpm and respective orbital injection unit [36]

A fundamentally different high pressure injection concept was introduced in series in a V6 outboard engine under the trade name Johnson about the same time (1995). Its electronically controlled Ficht fuel injection system EFI [28. 37], **figure 14.44**, uses the kinetic energy of a moving mass which, converted into pressure energy due to sudden delay, is applied to the fuel, which is subsequently injected.

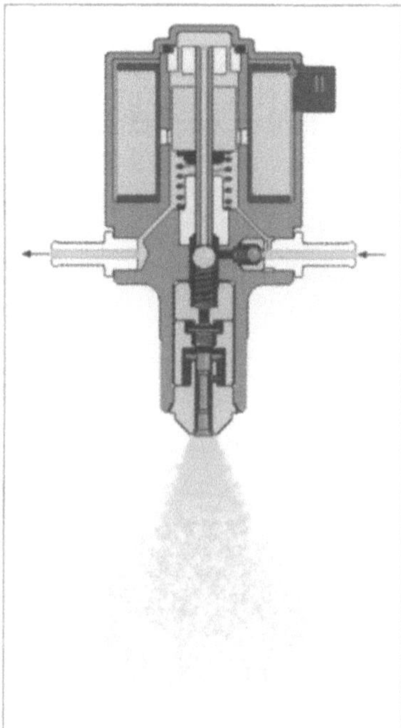

**Figure 14.44:**
Ficht fuel injector [37]

A pressure wave injection based on the reversal of the process, i.e. the deceleration of a liquid column, and therefore corresponding to the principle of a hydraulic hammer has also been or is in the process of being developed for the application in gasoline and diesel engines [32].

For small displacement two-wheel drives, the cost aspect is of crucial importance. The first series concept for internal mixture formation implemented in the two-wheeler, the Piaggio 50 ccm FAST Engine [30, 38], used a carburetor and a mechanical injection similar to that described in **figure 14.41** in 1997.

Today, several two-stroke gasoline engines for two-wheelers are equipped with electronically controlled direct-injection systems, which is remarkable in view of the major extra expenses and the additional costs for the engine of about 80 to 100% for small engines. **Figure 14.45** shows an implemented example of a small displacement two-wheeler engine by Aprilia; the scheme of the injection system is depicted in **figure 14.46**.

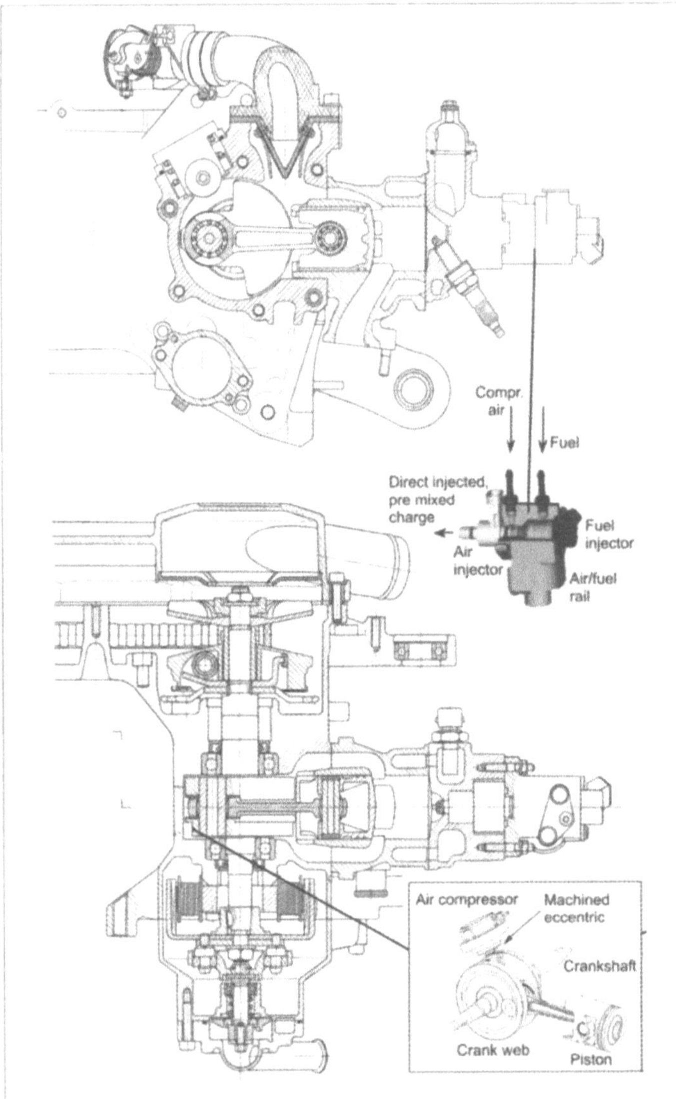


**Figure 14.45:** Sectional view of the Aprilia DiTech 50 ccm two-stroke engine

A high pressure direct injection corresponding to the systems of four-stroke passenger car engines is also used in two-stroke outboard engines, **figure 14.47**.

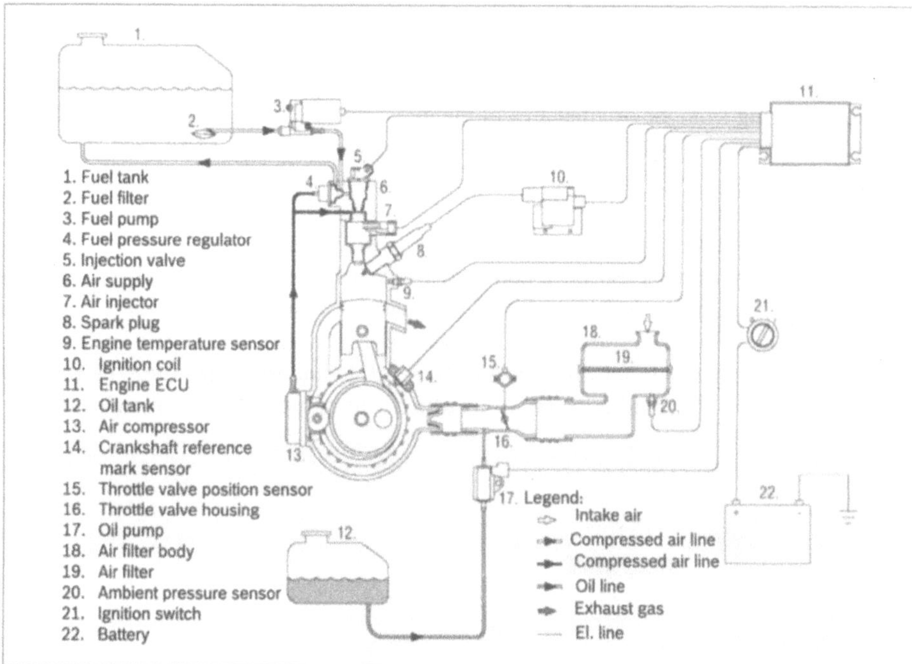


**Figure 14.46:** Scheme of the aSDI two-stroke injection [33]

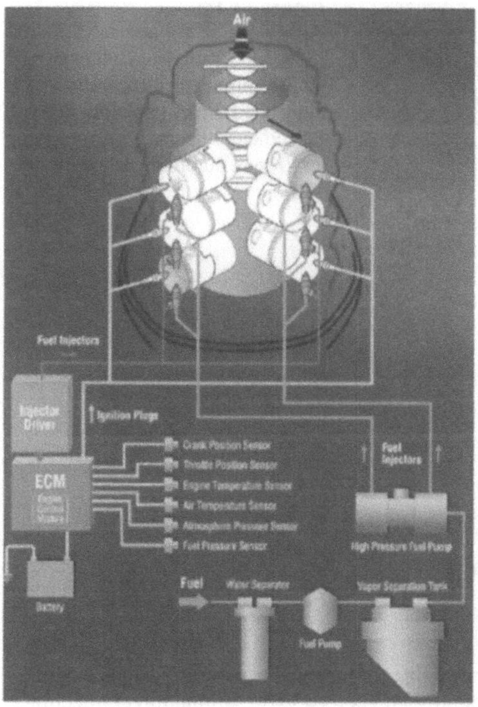


**Figure 14.47:**
Yamaha V6 two-stroke engine with high pressure direct injection HPDI [19]

### 14.3.4 Applications and future prospects

Two-stroke engines with internal mixture formation are currently used for different applications ranging from the original area of the two-stroke drive, **figure 14.48**, to outboard engines [40], **figure 14.47**, jet-ski drives [41], **figures 14.49** and **14.50**, and drives for sledges.

The variety of concepts and injection systems for two-stroke gasoline engines is just as diverse as the applications. So far a convergence to a few or standardized technologies cannot be observed nor is it foreseeable in the future. The high pressure direct injection technology that has become established in passenger car engines meanwhile has provided suitable series components for high pressure generation and fuel injection. The research project initiated by Bombardier Rotax [44], which will describe both challenges and op-

**Figure 14.48:**
Aprilia SR 50 R Factory (MY 2006), single-cylinder two-stroke engine, electronically controlled direct injection, displacement 49 ccm, power 4 kW at 7750 rpm, automatic transmission

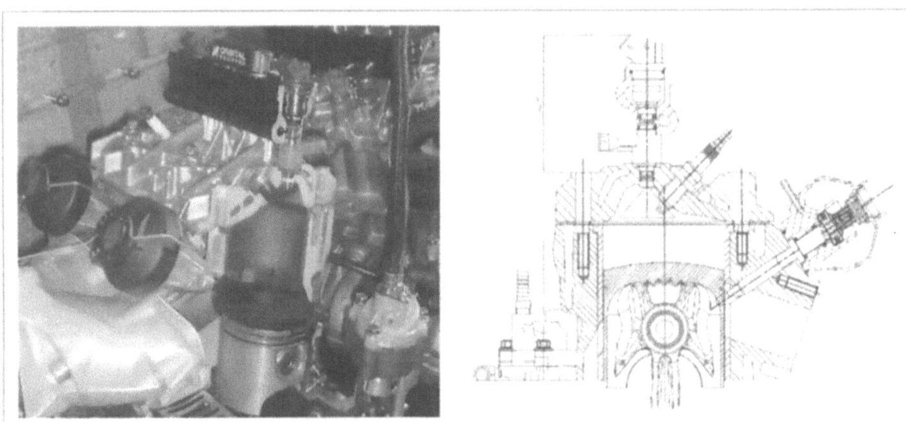

**Figure 14.49**: Rotax double-cylinder two-stroke engine for personal watercraft, displacement 951 ccm, 93 kW, fresh water cooled, intake is controlled via reed valve, orbital direct injection

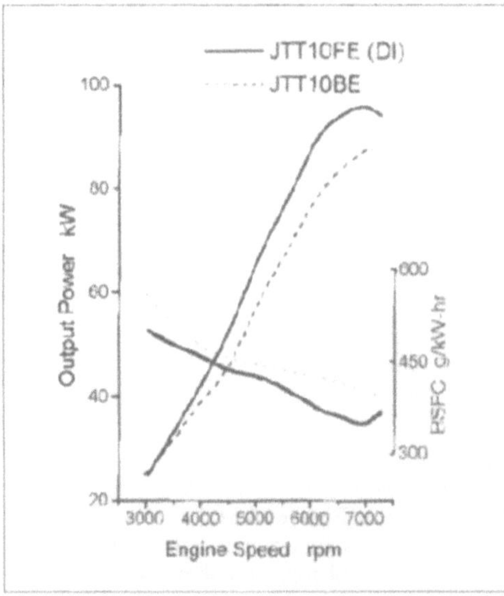


**Figure 14.50:**
Kawasaki three-cylinder two-stroke engine for personal watercraft. Direct injection, displacement 1071 ccm, B x H = 80 x 71 mm, $P_{max}$ = 96 kW, $M_{max}$ = 138 Nm [41]

portunities for high pressure direct injection to further reduce emissions even at highest engine speeds and specific power, is one example.

The tests were carried out with an adapted automotive solenoid valve system with a fuel pressure of 130 bar; the conical swirl nozzle used had an angle of 60° and a static flow of 19.6 g/s. As test engine, a single-cylinder unit with a displacement of 500 ccm and a compression ratio $e$ =11.6 was used, which reaches a mean effective pressure of about 9 bar at the rated speed of 8000 rpm.

**Figure 14.51** shows the control times used for this engine and the injection phases possible for internal mixture formation in a circle diagram. They can be seen for the two-stroke or four-stroke engine for an early (homogeneous) and a late (stratified) injection strategy [43]. For the arrangement of the injection nozzle and the spark plug [34], the geometrical shape of the combustion chamber and the design of the nozzle, charge movement and mixture formation were examined and designed using 3D CFD simulation; the sectional view is shown in **figure 14.52**.

**Figure 14.53** shows the calculation grid with reed valve chamber (left), transfer ports, crankcase including the volume below the piston, combustion chamber (two variants) and exhaust ports. The results in **figure 14.54** show, for a geometrical design, the charge composition at ignition point for the injection strategies "early (homogeneous)" and "late (stratified)" injection. The much more homogenous charge for the advanced injection as compared to the late injection can be clearly seen. In the exhaust area (right), however, the advantage of the retarded injection becomes apparent. Due to the extremely late injection start, no air/fuel mixture reaches the exhaust port. Thus the scavenging losses typical for

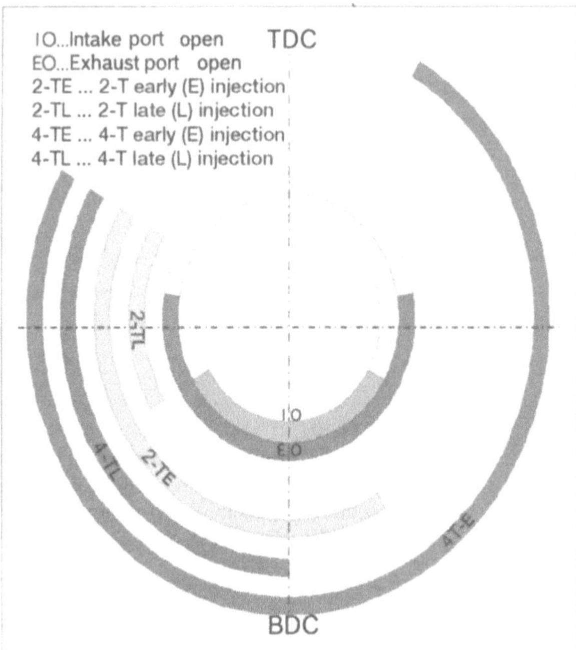


**Figure 14.51:**
Circle diagram of the injection phases for high pressure direct injection for two-stroke engines (2-T) and four-stroke engines (4-T) for early and late injection strategy [43]

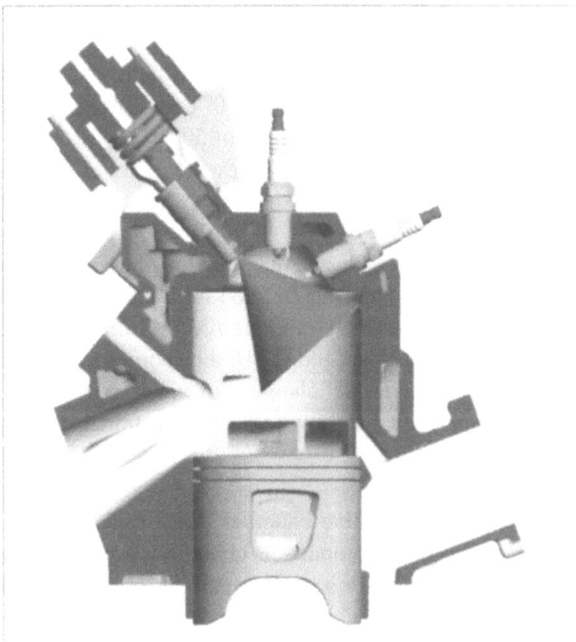

**Figure 14.52:**
Sectional view of the cylinder with different spark plug positions [44]

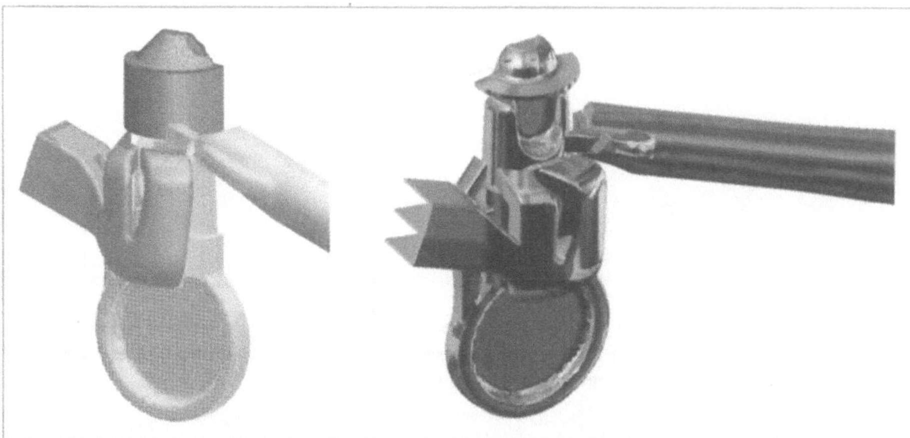

**Figure 14.53:** Calculation grid of the 3D CFD simulation of the complete engine [44]

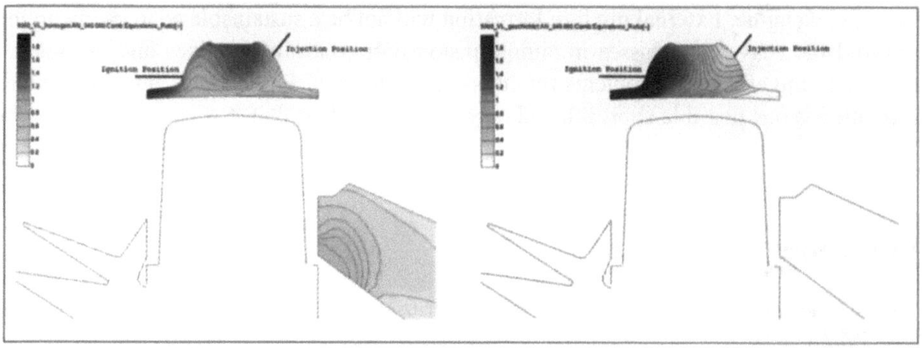


**Figure 14.54:** Charge composition ("equivalence ratio", $1/\lambda$) at ignition point for early injection, left (injection start = 160° CA), and for late injection, right (injection start = 250° CA) [45]

the two-stroke engine can be almost entirely avoided when the late injection strategy is used.

The functional results from the experimental tests, **figure 14.55**, show the advantages of high pressure direct injection as compared to other mixture formation systems. In comparison to conventional mixture preparation (carburetor ≥ 100%), high pressure direct injection achieves significant advantages both for consumption and hydrocarbon emissions. The hydrocarbon emissions can be reduced by approximately 95% by means of late injection. The higher hydrocarbon emissions that result, which are a disadvantage of the early injection, can be more than compensated for by using an oxidation catalytic converter. With early injection, even a small advantage can be achieved, as well as the prerequisite for rated speeds well over 7000 rpm in the mean effective pressure.

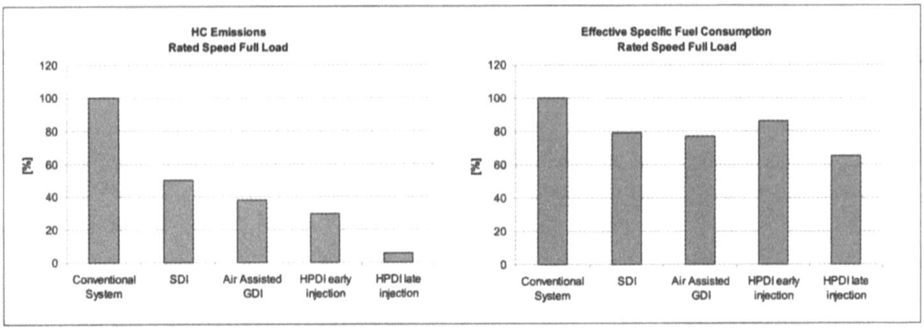


**Figure 14.55:** Comparison of the hydrocarbon emissions and the consumption of different mixture formation systems at rated speed and nominal torque [44]

Finally it can be stated that a convincing solution of the problem of emissions, with regard to both function and costs, is the decisive point for the long term survival of the gasoline two-stroke engine. External mixture formation will not be a sustainable concept due to the unavoidable scavenging losses in simple piston-port controlled engines and the related emissions and efficiency problems for emissions-controlled applications. Internal mixture formation is one possible approach and may even be the best one.

## Literature

[1] *Scherenberg, H.*: Der Erfolg der Benzin-Einspritzung bei Daimler-Benz. In: MTZ 22 (1961), Nr. 7

[2] *Winterkorn, M.; Bohne, P.; Spiegel, L.; Soehlke, G.*: Der Lupo FSI von Volkswagen. In: ATZ 102 (2000), Nr. 10

[3] *Buechling, J.*: Der neue 2,0-l-FSI-Motor von Audi. In: MTZ 63 (2002), Nr. 10

[4] *Lueckert, P. et al.*: Weiterentwicklung der Benzin-Direkteinspritzung bei Mercedes Benz. 13. Aachener Kolloquium Fahrzeug- und Motorentechnik 2004, Sonderdruck

[5] *Goppelt, G.*: Der neue Ottomotor mit Direkteinspritzung von Renault. In: MTZ 60 (1999)

[6] *Dornhoefer, R. et al.*: Der neue R4 2,0l 4V TFSI-Motor im Audi S3. Aufladetechnische Konferenz. Dresden, 2006

[7] *Krebs, R. et al.*: Neuer Ottomotor mit Direkteinspritzung und Doppelaufladung von Volkswagen. In: MTZ 66 (2005), Nr. 11

[8] *Welter, A.; Unger, H.; Hoyer, U.; Bruener, Th.; Kiefer, W.*: Der neue aufgeladene BMW Reihenmsechszylinder Ottomotor. 15. Aachener Kolloquium Fahrzeug- und Motorentechnik. Aachen, 2006

[9] *Luttermann, Ch.; Missy, St.; Schwarz, Ch.; Klauer, N.*: High Precision Injection in Verbindung mit Aufladung am Beispiel des neuen BMW Twin Turbo Ottomotors. 15. Aachener Kolloquium Fahrzeug- und Motorentechnik. Aachen, 2006

[10] *Lueckert, P.; Waltner, A.; Rau, E.; Vent, G.; Schaupp, U.*: Der neue V6-Ottomotor mit Direkteinspritzung von Mercedes-Benz. In: MTZ 67 (2006), Nr. 11

[11] *Waltner, A.; Lückert, P.; Schaupp, U.; Rau, E.; Kemmler, R.; Weller, R.*: Die Zukunftstechnologie des Ottomotors: strahlgefuehrte Direkteinspritzung mit Piezo-Injektor. 27. Internationales Wiener Motorensymposium, Fortschrittberichte VDI, Reihe 12, Nr. 622. Duesseldorf: VDI Verlag, 2006

[12] Auto Motor und Sport (2006), Heft 23, S. 16 f.
[13] Internet: www.bmw.com
[14] *Klueting, M.*: Ottomotorische Konzepte zur Steigerung der „Effizienten Dynamik". Tagung: „Entwicklungstendenzen bei Ottomotoren"
[15] *Kotauschek, W.; Diel, H.; Baretzky, U.; Ullrich, W.*: Der Audi V8 FSI Bi-turbo fuer das 24 h Rennen in Le Mans. 24. Wiener Motoren Symposium, 2003
[16] *Baretzky, U.; Andor, T.; Diel, H.; Ullrich, W.*: The direct injection system of the 2001 Audi turbo V8 Le Mans engines. SAE Motorsports Engineering Conference, Indianapolis 2002
[17] *Baretzky, U.; Pfeffer, T.; Kuntz, P.; Ullrich, W.*: Optimization of combustion of an Audi racing engine utilizing simulation calculation and indication. Internationales Symposium fuer Verbrennungsdiagnostik. Baden-Baden, 2004
[18] *Wurms, R.*: Der neue Audi 2.0l TFSI. 13.Aachener Kolloquium Fahrzeug und Motorentechnik, 2004
[19] *List, H.*: Der Ladungswechsel der Verbrennungskraftmaschine. Zweiter Teil: Der Zweitakt. Wien: Springer-Verlag, 1950
[20] *Schmidt, U.; Froede, W.*: Zweitaktmotoren mit Benzineinspritzung. In: Deutsche Kraftfahrforschung, Zwischenbericht Nr. 68. 1938
[21] Homepage: www.zweitakt.de/modelle/strasse/motobec350
[22] *Laimboeck, F. J.*: The potential of small loop-scavenged spark ignition single cylinder two-stroke engines. Habilitation, TU Graz, 1990
[23] *Nuti, M.*: Direct fuel injection for two-stroke S.I. engines – exhaust emissions abatement. 2. Grazer Zweiradtagung (Tagungsband), 1986
[24] *Plohberger, D.; Mikulic, L.; Landfahrer, K.*: Der Zweitaktmotor mit innerer Gemischbildung – Entwicklungsstand und Zukunftsaussichten. 9. Internationales Wiener Motorensymposium, 1988
[25] *Plohberger, D.; Mikulic, L.; Landfahrer, K.*: Development of a fuel injected two-stroke gasoline engine. SAE Technical Paper No. 880170, 1988
[26] *Schlunke, K.*: Der Orbital Verbrennungsprozess des Zweitakt-Motors. 10. Internationales Wiener Motorensymposium. VDI Verlag, 1989
[27] *Plohberger, D., Mikulic, L.*: Der Zweitaktottomotor als PKW-Antriebskonzept – Anforderungen und Loesungsansaetze. 10. Internationales Wiener Motorensymposium. VDI Verlag, 1989
[28] *Heimberg, W.*: Ficht pressure surge injection system. SAE Technical Paper No. 931502, 1993
[29] *Cobb, W.*: Compression Wave Injection: A mixture injection method for two stroke engines based on unsteady gas dynamics. SAE Technical Paper No. 2001-01-1817/4237
[30] *Nuti, M.; Pardini, R.*: Twenty years of Piaggio direct injection research to mass produced solution for small 2T SI engines. SAE Technical Paper No. 980760, 1998
[31] *Kirchberger, R.*: Abgasoptimierung durch Einsatz von luftunterstuetzter Direkteinspritzung beim Zweitakt-Kleinmotor. Dissertation, TU Graz, 2003
[32] *Stan, C.*: Direkteinspritzsysteme fuer Ottomotoren – Konzepte und Entwicklungstrends. Essen: Expert-Verlag, 1999
[33] *Archer, M.; Bell, G.*: Advanced electronic fuel injection systems – an emissions solution for both 2- and 4-stroke small vehicle engines. Technical Paper No. 2001-01-0010, 2001
[34] *Meinig, U.*: Standortbestimmung des Zweitaktmotors als Pkw-Antrieb – Teil 3: Zweitaktmotor. In: MTZ 62 (2001), Nr. 10
[35] *Dell'Orto, P.; Colombo, P.; Chiavazzo, E.; Manganiello, G.; Marchetti, M.; Migliaccio, M.; Grassia, P.*: The Dell'Orto Direct Injection System for Small Engines
[36] *Houston, R.; Cathcart, G.*: Combustion and emissions characteristics of Orbital's combustion process applied to multi-cylinder automotive direct injected 4-stroke engines, Orbital Engine Company. SAE Technical Paper No. 980153, 1998
[37] Webpage Provenion: http://provenion.de/ffi.htm
[38] *Nuti, M., et al.*: FAST injection system: Piaggion solution for ULEV two stroke engines. SAE Technical Paper No. 970362, 1997
[39] http://www.yamaha-motor.com/outboard/products/subcathome/14/home.aspx
[40] *Sogawa, M.; Kato, M.*: Development of the high pressure direct injection (HPDI) – system for

two-stroke outboard Motor. SAE Technical Paper No. 2001-01-1786/4209, 2001

[41] *Fukami, Y.; Watabe, S.; Jinja, Y.; Matsuda, Y.; Takahashi, K.; Noda, T.*: Superiority of newly developed direct-injection engine over conventional engine for personal watercraft. SAE Technical Paper No. 2001-01-1787/4210, 2001

[42] www.Rotax.at

[43] *Schmidt, S.*: Auslegung, thermodynamische Analyse und Entwicklung von Zweitakt-Brennverfahren mit Hochdruck-Direkteinspritzung. Dissertation, TU Graz, 2005

[44] *Schmidt, S.; Eichlseder, H.; Kirchberger, R.; Nimmervoll, P.; Ohrnberger, G.; Wagner, M.*: GDI with high-performance 2-stroke application: concepts, experiences and potential for the future. SAE Technical Paper No. 2004-32-043, 2004

[45] *Schmidt, S.; Winkler, F.; Schoegl, O.; Pontoppidan, M.*: Development of a Combustion Process for a High Performance 2-Stroke Engine with High Pressure Direct Injection. SAE Technical Paper No. 2004-01-2942, 2004

[46] *Bartsch, C.*: Ein Jahrhundert Motorradtechnik. Duesseldorf: VDI Verlag, 1987

# 15 Gasoline Direct Injection in the USA, Europe and Japan

## 15.1 Introduction

Gasoline direct injection, an issue of secondary importance in automobile engineering until the mid-1990's, has established itself as something with which all major automobile manufacturers have been dealing since 2000. Due to different general conditions as well as different development and original equipment manufacturer (OEM) strategies, the markets in Europe, the USA and Japan are not developing simultaneously, nor are they producing identical solutions. In future, gasoline direct injection will play an even more significant role in markets with strict requirements with regard to emissions, power and fuel consumption and will help gasoline engines to improve their competitiveness against rival propulsion-related concepts.

## 15.2 The markets: USA, Europe and Japan

The approaches for implementation of GDI in automobile engineering outlined in chapter 2 History of Gasoline Direct Injection did not become established in the 1950's. It was not until 1996, with the introduction of the first Mitsubishi model with gasoline direct injection that efforts suddenly intensified. In Japan, but especially in the Western European automobile and systems supply industry, everyone was surprised by the aggressively advertised 25% reduction in fuel consumption. The first car series developed by European manufacturers under great time pressure failed to meet expectations, as did Japanese attempts. The initial euphoria was quickly followed by disillusionment. The promised reduction of 25% in fuel consumption in the consumption cycle could not be met even in theoretical derivations. Instead, there was often no relevant improvement at all in the consumption values when actually driving the vehicle. The additional benefit of a slight increase in torque and power as compared to port fuel injection was not proportionate to the high system costs. Besides high development efforts and cost-intensive applications of the concept of lean combustion, the system costs resulted from high component costs including expensive exhaust-gas aftertreatment ($NO_x$ catalysts), which did not offer any economies of scale during the ramp-up phase of production. Systems with wall-guided combustion concept, especially, were not noticeably superior to good port fuel injection systems even though they were much more expensive to manufacture. Several other system approaches were also unattractive with regard to cost benefit aspects. This led to a great sense of uncertainty among suppliers about the future of gasoline direct injection. Some newly introduced engines and several engine projects were cancelled, postponed, stopped in the early stages or did not reach production numbers as quickly as predicted. The majority of passenger car manufacturers limited their activities to preliminary development and small projects, explored the technical possibilities and observed the market. Press coverage was neutral and even negative in the first six years, as the product performance did not meet expectations.

Several parallel developments finally brought about a change and new faith in the potential of GDI. One determining factor was the pressure to reduce fuel consumption in the wake of rising crude oil prices and legislative initiatives to reduce $CO_2$ emissions. Another was the lowering of emissions limits, which especially affected diesel engines. Although they had better $CO_2$ values, they also produced more particle and nitrogen oxide emissions than gasoline engines and therefore required more complex exhaust-gas treatment, which made them even more expensive.

What was decisive, however, were the new findings about additional feasible advantages, which were discovered through technology. In addition to the clear reduction in pollutant emissions as compared to the first generation, fuel savings were realistically estimated at 15% with correct system design. There are alternative technical approaches to achieve this. One is displacement downsizing by means of turbocharging, which offers a high potential for synergy in gasoline engines with direct injection. The alternative, or additional, approach is the spray-guided combustion concept, which can prove to be attractive for engines with six or more cylinders. Even the first generation of the aforementioned turbocharged engines with direct injection, developed by Audi as power versions and widely introduced in the Volkswagen group from 2003 onwards, led to very positive press reactions. The models introduced did not focus primarily on fuel savings through reduction in displacement but on high engine torque over a wide engine speed range. As a result, public perception was successfully changed and become positive, and a competitive alternative to the diesel engine, which had managed to increase its market share from 15 to 50% in Europe between 1989 and 2005, appeared possible. Direct injection and turbocharging therefore paved the way to success for the gasoline engine, just as they had done years earlier for the diesel engine.

In Japan, market development was less erratic. Although Toyota, Nissan and Honda followed with engine projects, the number of units produced and the product range offered remained very limited with the exception of Toyota. With a 90% share of the Japanese market for vehicles with gasoline direct injection, Toyota also changed from a lean burn combustion concept to stoichiometric combustion in 2003. Mitsubishi, the pioneer on the market, had backed out by this point in time and no longer had a model on offer. Another major Japanese OEM, Mazda, introduced a gasoline engine with direct injection and turbocharging in the European, USA and Japanese markets in 2005, as a power variant in small volumes.

Finally, the US market was opened up. BMW was first, with its 12-cylinder engine as homogenous mode concept. Toyota and then Audi followed with considerable volumes. GM was the first of the major US American OEMs to act, with a supercharged high power engine in 2006 and subsequently with a 3.6 l V6 engine. Higher power requirements with low fuel consumption and good emissions, while also attempting to avoid expensive secondary air injection, drove this development. A decisive factor for success in the USA will be whether downsizing, with a cylinder reduction from eight to six cylinders and turbocharging, will be accepted by the end customers and will therefore be able to compete with the expensive hybrid and diesel engine concepts. If low sulfur fuels are made available and system costs are reduced, spray-guided lean burn combustion concepts will also have a good chance for success in US engines with big power output per liter in the medium term. **Figure 15.1** shows the first market launch classified by manufacturers and combustion concepts.

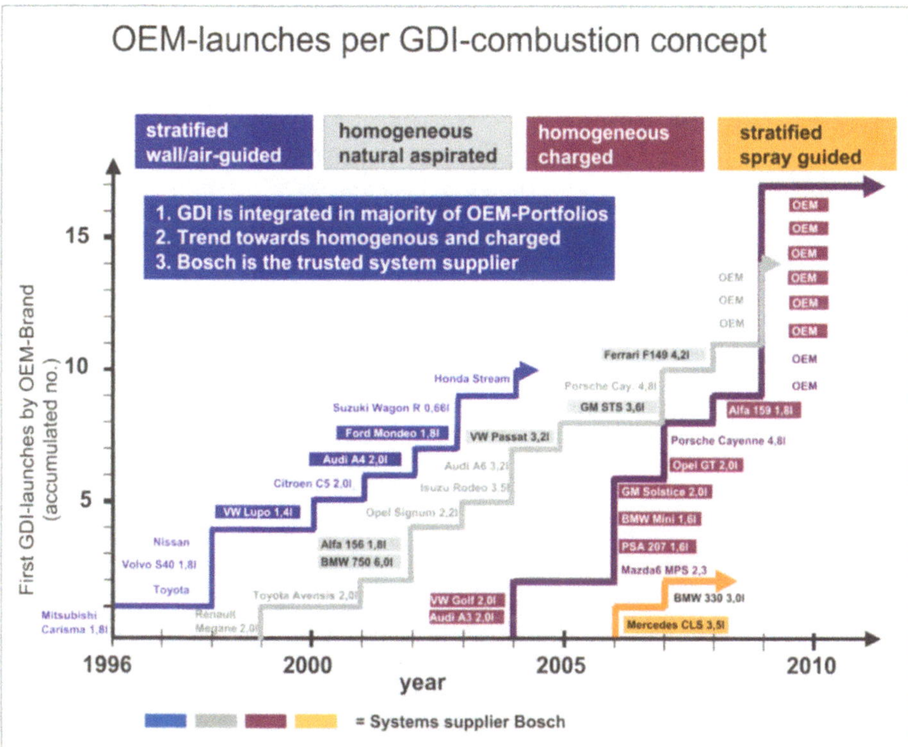


**Figure 15.1:** OEM market launch by combustion concepts

## 15.3 Future prospects

In the markets of Europe, Japan and the USA, gasoline direct injection will continue to establish itself and will be part of the portfolio of all well-known OEMs. The drivers for this development, besides $CO_2$ reduction, are good torque development and power increase, especially in the context of synergetic turbocharging. The latter has already paved the way for the breakthrough of the diesel engine in Western Europe. Since the emissions resulting from stoichiometric combustion, with the use of known three-way catalytic converters, meet even the strictest emissions standards worldwide, there is no significant barrier to its distribution. Increasing cost effectiveness in catalytic converter technology for denoxation will also open up potentials for spray-guided lean combustion processes, especially in engines with more than four cylinders.

While gasoline direct injection is limited in its suitability to function as an attribute for a unique profile and brand image improvement, it will have a significant role to play in keeping the gasoline engines competitive and meeting the increased worldwide requirements regarding consumption, emissions and power at realistic market prices. **Figure 15.2** shows the market development.

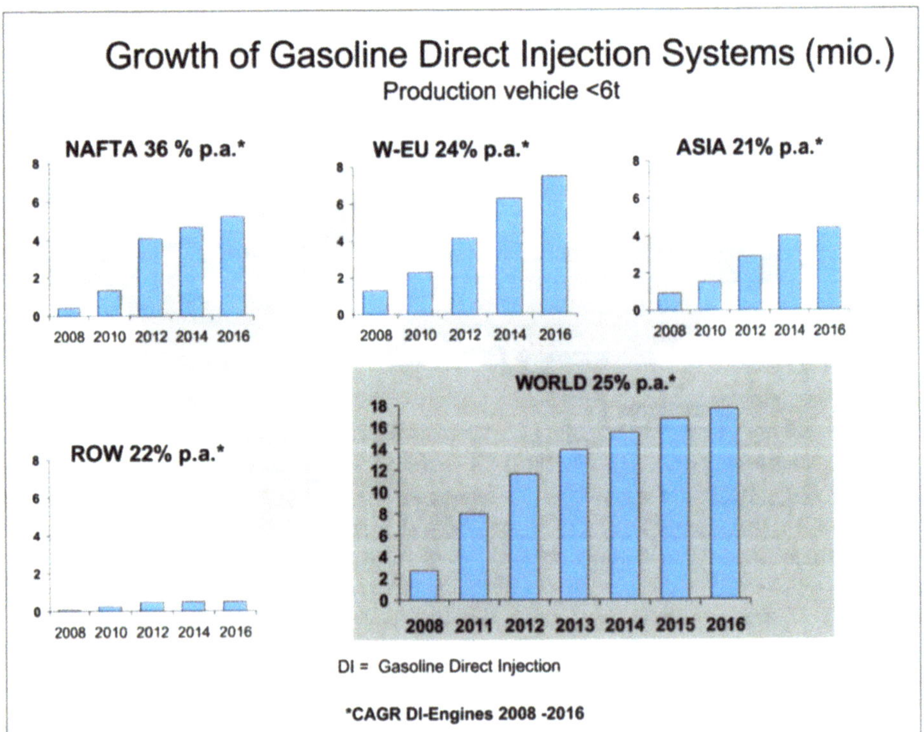


**Figure 15.2:** Market development of gasoline direct injection

# 16 Future Prospects

Reducing fuel consumption and minimizing pollutant emissions is already the focus of engine development for automotive vehicles and this will continue into the future. Both diesel and gasoline engines with direct injection still have enormous potential in terms of reducing fuel consumption, despite the present high technical standard. Current statements from experts [1] confirm that consistent further development of the gasoline engine and implementation of its technical potential have made it capable of achieving fuel savings that can compete with those of current and future diesel engines. With the clear trend towards direct injection in gasoline engines in combination with supercharging, the gasoline engine not only promises less fuel consumption and low pollutant emissions, but also its share of driving pleasure.

The spray-guided mixture formation and combustion process is particularly suitable for operation in the low and medium load and engine speed ranges, in which stratified charge is useful and will be used more frequently in the future. In view of the present state of development and knowledge, extensive research and development work will be required before the full potential of this process with regard to increased efficiency and minimum pollutant emissions can be exploited. In the next few years, the aim for this type of injection must be to achieve reliable mixture formation so that an ideal stratification between fresh mixture and ambient air and/or ambient exhaust gas is attained through exhaust-gas stratification under all boundary conditions for which a stratified charge mode is desirable. DaimlerChrysler and BMW brought the first engines with spray-guided direct injection onto the market [2, 3]; other manufacturers will certainly follow this trend over the next few years.

The injection system is of particular importance for optimal engine operation with spray-guided direct injection. Today's state-of the-art injection pressures of up to 200 bar were already achieved by DaimlerChrysler and BMW in 2006 and 2007 when they launched the first two engines featuring these pressures. Research into significantly higher injection pressures in the range above 300 bar – possibly up to 500 bar and more – is already being carried out [4] and shows some further potential for reducing fuel consumption and achieving minimum pollutant emissions. The need and time frame for the development of injection systems that can cope with these pressures over the long term will be determined mainly by whether the higher pump driving power that is required can be compensated for and the potential converted to an effective benefit.

So far, research and development relating to direct injection in gasoline engines is still only in the early stages and is not sufficiently advanced for spray-guided direct injection with lean operation and stratified charge to be used worldwide. For this reason, direct injection in gasoline engines will be launched by almost all automobile manufacturers – now and for the next few years – with homogeneous mixture formation and in many cases in

few years, given current levels of expertise and experience in its development. In [6], it is noted that the fuel cell will not be used on a larger scale as a drive unit in automobiles for the next 25 to 30 years. All automobile manufacturers share this opinion. Some of them, including GM and Honda, think that the fuel cell will be ready for series production from 2015, whereas others, like Volkswagen and Nissan, are of the opinion that it will not be sufficiently developed to be suitable for series application in vehicles before 2030 [6, 7]; **figure 16.3**. The very high cost of the fuel cell drive is a major problem. Moreover, the required fuels, e.g. hydrogen, are problematic with regard to both their production and the creation of the necessary infrastructure, as they are cost-intensive and inefficient. This applies to hydrogen generated from both fossil fuels and renewable energy sources.

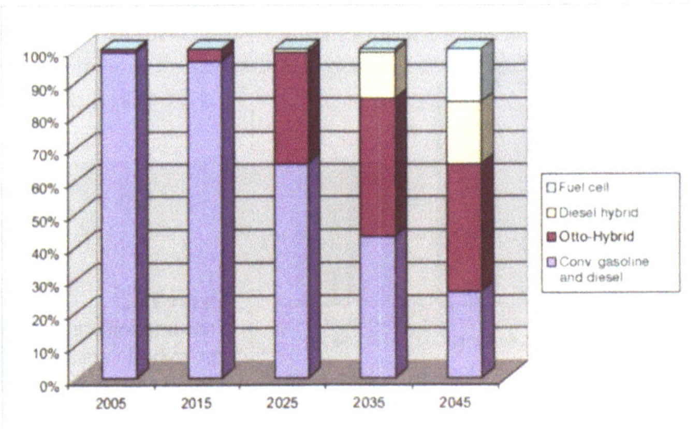


**Figure 16.3:** Development of drive concepts for passenger cars [7]

**Figure 16.3** shows that the hybrid drive as a combination of the gasoline engine and the electric engine will be used extensively from 2015 onwards. It is envisaged that the hybrid drive as a combination of diesel and electric engines will be used from 2025. The first fuel cell drives are expected to be used in larger quantities in passenger cars only from 2035 to 2045, because there is almost no hydrogen available from renewable sources at this stage, and it is too expensive and inefficient to generate hydrogen from fossil fuels.

If the gasoline engine with direct injection is developed further in the coming years as consistently as the diesel engine has been over the last 20 years, it will enable process efficiencies and fuel consumption in the range of today's most economical diesel engines while also complying with the strictest emission standards. Even now, after 130 years of its existence, the gasoline engine will remain the dominating drive unit for passenger cars for a long time yet, due to its untapped potential. If the trend towards hybrid drives as a combination of combustion and electric engines proves to have been accurately predicted in [7], the gasoline engine with direct injection in combination with an electric engine with little power, as compared to the combustion engine without hybrid elements, can be seen as the best solution for future passenger car drives in terms of energy efficiency and exhaust emissions – as long as the cost is reasonable and still acceptable. The gasoline engine would have spray-guided direct injection and should be used only for propulsion while driving. The electric engine should be used both as a generator for converting the braking energy generated during vehicle deceleration into electrical energy and for drive when starting, for starting processes and for running in areas with reduced traffic. The electric engine should also take on a booster function during acceleration processes, if required and, in combination with the battery, should ensure that energy is supplied to the vehicle's electrical system. This relieves the combustion engine of tasks for which it is currently used in addition to drive but for which it is not really suited. If the future combustion engine is used as a gasoline or diesel engine with direct injection primarily to drive the vehicle, while the other needs in the vehicle relating to energy supply are met through hybridization, pollutant emissions at almost zero level are possible with fuel consumption of up to 50% less than the current level. Driving enjoyment will still be guaranteed.

## Literature


[1] *List, H.*: Eroeffnungsvortrag. 18. Tagung Motor und Umwelt. Graz, 2006

[2] *Waltner, A.*: Die Zukunftstechnologie des Ottomotors: strahlgefuehrte Direkteinspritzung mit Piezo-Injektor. 26. Wiener Motorensymposium, 2006

[3] *Klueting, M.; Melcher, T.*: Effiziente Dynamik – ein integraler Ansatz fuer die BMW Antriebe der Zukunft. 14. Aachener Kolloquium Fahrzeug- und Motorentechnik, 2005

[4] *Spicher, U.; Kubach, H.; Haensche, J. P.*: Die strahlgefuehrte Direkteinspritzung als Zukunftskonzept fuer Ottomotoren. MTZ-Konferenz Motor 2006, Stuttgart, 2006

[5] *Mikulic, L.*: Die Zukunft des Automobils: Herausforderungen fuer die Mercedes Car Group. 13. Aachener Kolloquium, 2004

[6] *Rabe, M.*: Keine Begrenzung nach oben und unten. In: Automobil Industrie (2006), Nr. 9

[7] Aussage und Bericht von Nissan im Februar 2006

# Subject Index

## F



## G



## H

## I

**Z**

MIX
Papier aus verantwortungsvollen Quellen
Paper from responsible sources
FSC® C105338

If you have any concerns about our products,
you can contact us on
**ProductSafety@springernature.com**

In case Publisher is established outside the EU,
the EU authorized representative is:
**Springer Nature Customer Service Center GmbH**
**Europaplatz 3, 69115 Heidelberg, Germany**

Printed by Libri Plureos GmbH
in Hamburg, Germany